VOLUME SIX HUNDRED AND EIGHT

METHODS IN ENZYMOLOGY

Enzymes in Synthetic Biology

METHODS IN ENZYMOLOGY

Editors-in-Chief

ANNA MARIE PYLE
Departments of Molecular, Cellular and Developmental Biology and Department of Chemistry
Investigator, Howard Hughes Medical Institute
Yale University

DAVID W. CHRISTIANSON
Roy and Diana Vagelos Laboratories
Department of Chemistry
University of Pennsylvania
Philadelphia, PA

Founding Editors

SIDNEY P. COLOWICK and NATHAN O. KAPLAN

VOLUME SIX HUNDRED AND EIGHT

METHODS IN ENZYMOLOGY

Enzymes in Synthetic Biology

Edited by

NIGEL SCRUTTON

Manchester Institute of Biotechnology,
The University of Manchester

ELSEVIER

ACADEMIC PRESS

An imprint of Elsevier

Academic Press is an imprint of Elsevier
50 Hampshire Street, 5th Floor, Cambridge, MA 02139, United States
525 B Street, Suite 1650, San Diego, CA 92101, United States
The Boulevard, Langford Lane, Kidlington, Oxford OX5 1GB, United Kingdom
125 London Wall, London, EC2Y 5AS, United Kingdom

First edition 2018

ISBN: 978-0-12-815148-8
ISSN: 0076-6879

For information on all Academic Press publications
visit our website at https://www.elsevier.com/books-and-journals

Publisher: Zoe Kruze
Acquisition Editor: Fiona Pattison
Editorial Project Manager: Shellie Bryant
Production Project Manager: Divya Krishna Kumar
Cover Designer: Mark Rogers

Typeset by SPi Global, India

CONTENTS

Section III
Enzyme Families: Parts and Platforms for Chemical Diversity

CONTRIBUTORS

Rudolf K. Allemann
School of Chemistry, Cardiff University, Cardiff, United Kingdom

Jamie Auxillos
Manchester Institute of Biotechnology (MIB), School of Chemistry, The University of Manchester, Manchester, United Kingdom

Arren Bar-Even
Max Planck Institute of Molecular Plant Physiology, Potsdam-Golm, Germany

Guangkai Bian
Key Laboratory of Combinatorial Biosynthesis and Drug Discovery, Ministry of Education and School of Pharmaceutical Sciences, Wuhan University, Wuhan, PR China

Agata Brzezniak
Manchester Institute of Biotechnology, School of Chemistry, University of Manchester, Manchester, United Kingdom

Yizhi Cai
Manchester Institute of Biotechnology (MIB), School of Chemistry, The University of Manchester, Manchester, United Kingdom

Kate Campbell
Department of Biology and Biological Engineering; Novo Nordisk Foundation Center for Biosustainability, Chalmers University of Technology, Gothenburg, Sweden

Pablo Carbonell
Manchester Centre for Synthetic Biology of Fine and Speciality Chemicals (SYNBIOCHEM), Manchester Institute of Biotechnology, The University of Manchester, Manchester, United Kingdom

Yun Chen
Department of Biology and Biological Engineering; Novo Nordisk Foundation Center for Biosustainability, Chalmers University of Technology, Gothenburg, Sweden

Sebastian C. Cosgrove
Manchester Institute of Biotechnology, School of Chemistry, University of Manchester, Manchester, United Kingdom

Andrew Currin
School of Chemistry, Faculty of Science and Engineering, Manchester Synthetic Biology Research Centre for Fine and Speciality Chemicals (SYNBIOCHEM), Manchester Institute of Biotechnology, University of Manchester, Manchester, United Kingdom

Junbiao Dai
Centre for Synthetic Genomics, Institute of Synthetic Biology, Shenzhen Institutes of Advanced Technology, Chinese Academy of Sciences, Shenzhen, China

Thomas Duigou
Micalis Institute, INRA, AgroParisTech, Université Paris-Saclay, Jouy-en-Josas, France

Mark S. Dunstan
Manchester Centre for Synthetic Biology of Fine and Speciality Chemicals (SYNBIOCHEM), Manchester Institute of Biotechnology; School of Chemistry, The University of Manchester, Manchester, United Kingdom

Jean-Loup Faulon
Manchester Centre for Synthetic Biology of Fine and Speciality Chemicals (SYNBIOCHEM), Manchester Institute of Biotechnology; School of Chemistry, The University of Manchester, Manchester, United Kingdom; Micalis Institute, INRA, AgroParisTech, Université Paris-Saclay, Jouy-en-Josas, France

Marco W. Fraaije
Groningen Biomolecular Sciences and Biotechnology Institute, University of Groningen, Groningen, The Netherlands

Scott P. France
Manchester Institute of Biotechnology, School of Chemistry, University of Manchester, Manchester, United Kingdom

Max J.L.J. Fürst
Groningen Biomolecular Sciences and Biotechnology Institute, University of Groningen, Groningen, The Netherlands

Eva Garcia-Ruiz
Manchester Institute of Biotechnology (MIB), School of Chemistry, The University of Manchester, Manchester, United Kingdom

Hazel M. Girvan
Manchester Institute of Biotechnology, University of Manchester, Manchester, United Kingdom

Lucy Green
School of Chemistry, Faculty of Science and Engineering, Manchester Synthetic Biology Research Centre for Fine and Speciality Chemicals (SYNBIOCHEM), Manchester Institute of Biotechnology, University of Manchester, Manchester, United Kingdom

Chen Guo-Qiang
MOE Lab of Bioinformatics, School of Life Sciences; Center for Synthetic and Systems Biology; Tsinghua-Peking Center for Life Sciences, Tsinghua University, Beijing, China; Manchester Institute of Biotechnology, Centre for Synthetic Biology, The University of Manchester, Manchester, United Kingdom

Rachel S. Heath
Manchester Institute of Biotechnology, School of Chemistry, University of Manchester, Manchester, United Kingdom

Florence Huynh
School of Chemistry, Cardiff University, Cardiff, United Kingdom

Laura N. Jeffreys
Manchester Institute of Biotechnology, University of Manchester, Manchester, United Kingdom

Adrian J. Jervis
Manchester Centre for Synthetic Biology of Fine and Speciality Chemicals (SYNBIOCHEM), Manchester Institute of Biotechnology; School of Chemistry, The University of Manchester, Manchester, United Kingdom

Michael C. Jewett
Department of Chemical and Biological Engineering; Chemistry of Life Processes Institute; Center for Synthetic Biology, Northwestern University, Evanston; Robert H. Lurie Comprehensive Cancer Center; Simpson Querrey Institute, Northwestern University, Chicago, IL, United States

Yin Jin
MOE Lab of Bioinformatics, School of Life Sciences; Center for Synthetic and Systems Biology; Tsinghua-Peking Center for Life Sciences, Tsinghua University, Beijing, China

Ashty S. Karim
Department of Chemical and Biological Engineering; Chemistry of Life Processes Institute; Center for Synthetic Biology, Northwestern University, Evanston, IL, United States

Leonard Katz
Joint BioEnergy Institute, Emeryville; QB3 Institute, University of California, Berkeley, CA, United States

Jay D. Keasling
Biological Systems and Engineering Division, Lawrence Berkeley National Laboratory; QB3 Institute; Department of Bioengineering; Department of Chemical and Biomolecular Engineering, University of California, Berkeley; Joint BioEnergy Institute, Emeryville, CA, United States; Novo Nordisk Foundation Center for Biosustainability, Technical University of Denmark, Hørsholm, Denmark; Center for Synthetic Biochemistry, Shenzhen Institutes for Advanced Technologies, Shenzhen, China

Douglas B. Kell
School of Chemistry, Faculty of Science and Engineering, Manchester Synthetic Biology Research Centre for Fine and Speciality Chemicals (SYNBIOCHEM), Manchester Institute of Biotechnology, University of Manchester, Manchester, United Kingdom

Mathilde Koch
Micalis Institute, INRA, AgroParisTech, Université Paris-Saclay, Jouy-en-Josas, France

Tianyi Li
Centre for Synthetic Genomics, Institute of Synthetic Biology, Shenzhen Institutes of Advanced Technology, Chinese Academy of Sciences, Shenzhen, China

Jiazhang Lian
Department of Chemical and Biomolecular Engineering, University of Illinois at Urbana-Champaign, Urbana, IL, United States; Key Laboratory of Biomass Chemical Engineering of Ministry of Education, College of Chemical and Biological Engineering, Zhejiang University, Hangzhou, China

Steffen N. Lindner
Max Planck Institute of Molecular Plant Physiology, Potsdam-Golm, Germany

Quanli Liu
Department of Biology and Biological Engineering; Novo Nordisk Foundation Center for Biosustainability, Chalmers University of Technology, Gothenburg, Sweden

Tiangang Liu
Key Laboratory of Combinatorial Biosynthesis and Drug Discovery, Ministry of Education and School of Pharmaceutical Sciences, Wuhan University; Hubei Engineering Laboratory for Synthetic Microbiology, Wuhan Institute of Biotechnology, Wuhan, PR China

Nikola Lončar
Groningen Biomolecular Sciences and Biotechnology Institute, University of Groningen, Groningen, The Netherlands

Tian Ma
Key Laboratory of Combinatorial Biosynthesis and Drug Discovery, Ministry of Education and School of Pharmaceutical Sciences, Wuhan University, Wuhan, PR China

Juan Mangas-Sanchez
Manchester Institute of Biotechnology, School of Chemistry, University of Manchester, Manchester, United Kingdom

Caterina Martin
Groningen Biomolecular Sciences and Biotechnology Institute, University of Groningen, Groningen, The Netherlands

Kirsty J. McLean
Manchester Institute of Biotechnology, University of Manchester, Manchester, United Kingdom

David J. Miller
School of Chemistry, Cardiff University, Cardiff, United Kingdom

Sarah L. Montgomery
Manchester Institute of Biotechnology, School of Chemistry, University of Manchester, Manchester, United Kingdom

Andrew W. Munro
Manchester Institute of Biotechnology, University of Manchester, Manchester, United Kingdom

Jens Nielsen
Department of Biology and Biological Engineering; Novo Nordisk Foundation Center for Biosustainability, Chalmers University of Technology, Gothenburg, Sweden; Novo Nordisk Foundation Center for Biosustainability, Technical University of Denmark, Kongens Lyngby, Denmark

Bo Pang
Joint BioEnergy Institute, Emeryville, CA, United States

Jeremy I. Ramsden
Manchester Institute of Biotechnology, School of Chemistry, University of Manchester, Manchester, United Kingdom

Christopher J. Robinson
Manchester Centre for Synthetic Biology of Fine and Speciality Chemicals (SYNBIOCHEM), Manchester Institute of Biotechnology; School of Chemistry, The University of Manchester, Manchester, United Kingdom

Joanna C. Sadler
School of Chemistry, Faculty of Science and Engineering, Manchester Synthetic Biology Research Centre for Fine and Speciality Chemicals (SYNBIOCHEM), Manchester Institute of Biotechnology, University of Manchester, Manchester, United Kingdom

Carl Schultz
Department of Chemical and Biomolecular Engineering; Carl R. Woese Institute for Genomic Biology, University of Illinois at Urbana-Champaign, Urbana, IL, United States

Nigel S. Scrutton
Manchester Centre for Synthetic Biology of Fine and Speciality Chemicals (SYNBIOCHEM), Manchester Institute of Biotechnology; School of Chemistry, The University of Manchester, Manchester, United Kingdom

Neil Swainston
School of Chemistry, Faculty of Science and Engineering, Manchester Synthetic Biology Research Centre for Fine and Speciality Chemicals (SYNBIOCHEM), Manchester Institute of Biotechnology, University of Manchester, Manchester, United Kingdom

Eriko Takano
Manchester Centre for Synthetic Biology of Fine and Speciality Chemicals (SYNBIOCHEM), Manchester Institute of Biotechnology; School of Chemistry, The University of Manchester, Manchester, United Kingdom

James Titchmarsh
Hamilton Sales & Service UK Limited, Birmingham, United Kingdom

Nicholas J. Turner
Manchester Institute of Biotechnology, School of Chemistry, University of Manchester, Manchester, United Kingdom

Sebastian Wenk
Max Planck Institute of Molecular Plant Physiology, Potsdam-Golm, Germany

Chen Xiangbin
MOE Lab of Bioinformatics, School of Life Sciences; Center for Synthetic and Systems Biology; Tsinghua-Peking Center for Life Sciences, Tsinghua University, Beijing, China

Jiang Xiao-Ran
MOE Lab of Bioinformatics, School of Life Sciences; Center for Synthetic and Systems Biology; Tsinghua-Peking Center for Life Sciences, Tsinghua University, Beijing, China

Oren Yishai
Max Planck Institute of Molecular Plant Physiology, Potsdam-Golm, Germany

Tao Yu
Department of Biology and Biological Engineering; Novo Nordisk Foundation Center for Biosustainability, Chalmers University of Technology, Gothenburg, Sweden

Satoshi Yuzawa
Biological Systems and Engineering Division, Lawrence Berkeley National Laboratory, Berkeley; Joint BioEnergy Institute, Emeryville, CA, United States

Amin Zargar
Biological Systems and Engineering Division, Lawrence Berkeley National Laboratory; QB3 Institute, University of California, Berkeley, CA, United States

Huimin Zhao
Department of Chemical and Biomolecular Engineering; Carl R. Woese Institute for Genomic Biology, University of Illinois at Urbana-Champaign, Urbana, IL, United States

PREFACE

The emergence of synthetic biology as an interdisciplinary science and its potential impact on the bioeconomy have presented major challenges to the field of enzymology. The industrialization of biology is at the core of the emerging bioeconomy, and the rapid discovery, early implementation, and optimization of enzyme catalysts in synthetic biology programs are central to future biomanufacturing strategies. These challenges include the development of novel biocatalytic processes and tools for strain engineering that can be used to generate natural products, chemicals, materials, fuels, and other high-value products in an economically affordable and sustainable way. Rapid discovery, engineering, implementation, and exploitation of enzymes are at the core of Design–Build–Test–Learn (DBTL) platforms required for the engineering of biology. This volume provides examples of cutting-edge methods across these domains with a focus on the development of synthetic biology platforms for the production of natural products, chemicals, and materials.

This volume features contributions from leading practitioners in the field that cut across five basic themes. In the *Enzyme Discovery* domain, in silico methods are presented to assist in the combinatorial design of new metabolic pathways to target products based on retrosynthesis. These computational methods offer insight into enzyme selection, cluster discovery, and pathway design. This is followed by two contributions on *Enzyme Engineering Tools and Platforms* focused on synthetic biology design and synthesis of new biocatalysts and the use of cell-free synthetic biology methods for enzyme and pathway prototyping. The section on *Enzyme Families: Parts and Platforms for Chemical Diversity* provides a comprehensive set of methods for the generation of chemical diversity using important classes of biocatalyst (terpene cyclases/synthases; amine oxidases; flavoproteins; and P450 biocatalysts). These examples are selected because of their widespread use in synthetic biology programs. Insights into genome editing and genome assembly are then provided in the section on *Genome Engineering for Enzyme Pathways* followed by specific examples of pathway engineering in the section on *Bioengineered Strains With Deigned Enzyme Pathways.*

The contributions were selected to illustrate the different methodological approaches being used in synthetic biology programs across DBTL platforms and to discuss enzymes as novel "Parts" and as enabling "Tools."

The intention was for this selection to be illustrative of activity in the synthetic biology field and to be a useful resource to students, instructors, and researchers in the field. I wish to express my thanks to all authors who willingly contributed to this volume. Editing the volume was both informative and enjoyable.

NIGEL SCRUTTON
Manchester Institute of Biotechnology, The University of Manchester

SECTION I

Enzyme Discovery

CHAPTER ONE

Enzyme Discovery: Enzyme Selection and Pathway Design

Pablo Carbonell*, Mathilde Koch†, Thomas Duigou†, Jean-Loup Faulon*,†,‡,1

*Manchester Centre for Synthetic Biology of Fine and Speciality Chemicals (SYNBIOCHEM), Manchester Institute of Biotechnology, The University of Manchester, Manchester, United Kingdom
†Micalis Institute, INRA, AgroParisTech, Université Paris-Saclay, Jouy-en-Josas, France
‡School of Chemistry, The University of Manchester, Manchester, United Kingdom
[1]Corresponding author: e-mail address: jean-loup.faulon@jouy.inra.fr

Contents

Abstract

In this protocol, we describe in silico design methods that can assist in the engineering of production pathways that are based on enzymatic transformations. The described protocols are the basis for automated processes to be integrated into an iterative Design–Build–Test–Learn cycle in synthetic biology for chemical production. Selecting the right enzyme sequence for a desired biocatalytic activity from the extensive catalogue of sequences available in databases is challenging and can dramatically influence the success of bioproducing chemical compounds. A method for enzyme selection is presented that helps identifying candidate enzyme sequences through a scoring approach that considers not only sequence homology but also reaction similarity. Selecting a viable biochemical pathway for compound production requires screening large sets of reactions in a process involving combinatorial complexity. A method for pathway design using retrosynthesis is presented. The protocol allows the discovery

Methods in Enzymology, Volume 608
ISSN 0076-6879
https://doi.org/10.1016/bs.mie.2018.04.005

of alternative chemical pathways leading to the final product by using reaction rules of selectable degree of specificity. The protocols can be reversed through clustering discovery and product identification processes. The integration of these protocols into a general pipeline provides a toolbox for enhanced automated synthetic biology design and metabolic engineering.

1. INTRODUCTION

Industrial biotechnology is facing two main challenges to develop more sustainable alternatives to petroleum-based chemistry: (1) its high R&D costs and (2) the limited range of compounds currently available for bioproduction. Computational strategies could help in addressing both these issues by the increasing number of tools integrated into a unified pipeline to develop engineered organisms for the production of high-value compounds (Carbonell, Currin, Jervis, et al., 2016). Enzymes are the essential building blocks enabling the chemical biotransformations leading to formation of the desired compounds. Several tools exist to select candidate sequences for the enzymes at each step of a given pathway including antiSMASH for biosynthetic gene clusters (Weber et al., 2015), as well as tools based on reaction homologies like EC-Blast (Rahman, Cuesta, Furnham, Holliday, & Thornton, 2014) and *E*-zyme (Yamanishi, Hattori, Kotera, Goto, & Kanehisa, 2009) or based on machine learning (Li et al., 2018; Mellor, Grigoras, Carbonell, & Faulon, 2016).

Alternative metabolic pathways are first in silico designed and assessed before being built and tested (Hadadi & Hatzimanikatis, 2015; Lee & Kim, 2015; Medema, van Raaphorst, Takano, & Breitling, 2012; Moriya et al., 2010). While some computationally driven strategies make combinations of known metabolic reactions albeit not necessarily in the same chassis (Rodrigo, Carrera, Prather, & Jaramillo, 2008), others allow to design pathways that use novel reactions not stored in metabolic databases through the use of promiscuity hypotheses (Campodonico, Andrews, Asenjo, Palsson, & Feist, 2014; Carbonell, Parutto, Baudier, Junot, & Faulon, 2014; Carbonell, Planson, Fichera, & Faulon, 2011; Delépine, Duigou, Carbonell, & Faulon, 2018; Hadadi, Hafner, Soh, & Hatzimanikatis, 2016; Henry, Broadbelt, & Hatzimanikatis, 2010; Liu et al., 2014; Yim et al., 2011).

In this protocol, we describe in silico design methods associated with the engineering of pathways for chemical production through enzymatic transformations with the goal of automating processes that can be integrated into iterative Design–Build–Test–Learn cycles like those present in synthetic

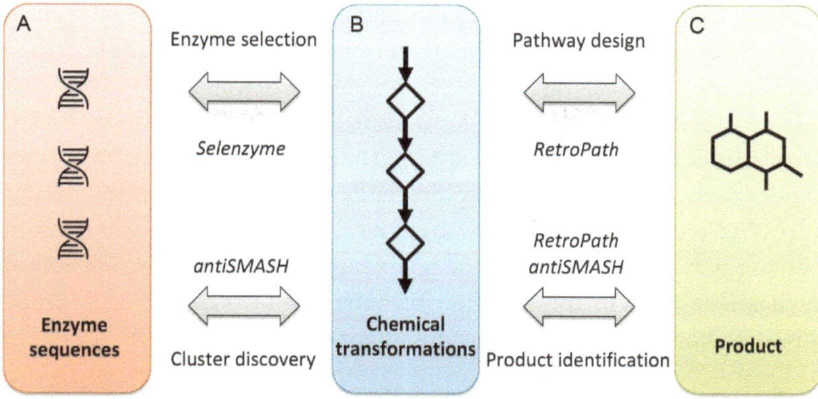

Fig. 1 Enzyme selection, cluster discovery, and pathway design. (A) Enzyme selection identifies candidate enzyme sequences. (B) Pathway design allows discovering alternative chemical transformations leading to the product. (C) The discovery process is reversed when the final product is unknown, cluster discovery allows annotating chemical transformation for a given gene cluster, and product identification predicts chemical products of the pathway.

biology projects. As depicted in Fig. 1, we consider several scenarios where computational methods can provide solutions to enzyme-related design problems often found in the context of bioproduction of chemicals: enzyme selection, cluster discovery, and pathway design.

2. ENZYME SELECTION

2.1 Introduction to Enzyme Selection

This protocol considers the scenario where a production pathway and its associated product are known. In that case, we use computational methods to replace individual enzymes by more efficient and/or specific alternatives. Choosing appropriate enzyme sequences to efficiently catalyze the set of reactions that are contained in production pathways is a relevant design problem whose solutions will depend on the specifications, for instance whether the system will work on cell or cell-free environments or the process conditions. Selecting the right starting enzyme sequences for each case can dramatically influence the success of the overall project. Low-performance enzymes can create bottlenecks in the pathways and lead to poor titers, yields, and productivity. When bottlenecks are present, enzyme engineering provides strategies like directed evolution for adaptation of the enzyme sequence toward the desired activity. However, applying directed evolution might not be always feasible and, even when it is the case, choosing the right initial parent enzyme is an important requirement to increase success rates.

Here, we describe a method for automated sequence selection with the ability to mine reactions that are expressed in generalized rules. Our approach provides a systematic way for addressing either natural or nonnatural chemical diversity of biosynthetic pathways by limiting ourselves not only to natural biochemical reactions but also to de novo engineered pathways. The method can therefore assist in the selection of heterologous candidate enzymes to express for both natural and nonnatural target reactions.

2.2 Protocol Description (Selenzyme)

Enzyme sequence selection is often performed based on sequence homology, i.e., by mining protein databases like UniProt through BLAST searches. This approach is useful in order to identify families and classes of enzymes that potentially share similar functions. Multiple sequence alignments of homologous can help us to identify alternative parent enzymes by looking at conserved regions and hot spots for directed evolution through (Carbonell & Trosset, 2015). However, this approach has still some room for improvement, as described in this protocol by looking closer at the chemical transformations that are achieved by these homologous sequences. Notably, looking for enzymes with similar annotated chemical reactions helps us to identify alternative enzymes not found by performing a search that is based solely on sequence alignments.

2.2.1 Preparation Steps

2.2.1.1 Pathway Representation and Use of Generalized Transformations

The goal of this method is to have a proper methodology that can be applied to systematically search for enzyme sequences and for any given pathway such as the ones identified by protocols described in this chapter. Most importantly, the approach should be able to work with reactions for which no known sequences are available in enzyme databases. The method should be able to provide a list of suggested enzyme sequences as parent templates for enzyme discovery of sought activities.

The first step is to encode each of the reactions in the pathway in a meaningful way that can be processed to compare chemical similarity.

1. Reactions can conveniently be represented using the SMILES/SMARTS notation ("Daylight Theory Manual, Version 4.9. http://www.daylight.com/ (accessed September 1, 2017)," 2017), as it provides an intuitive notation and can be either sketched by hand or through a chemical editor.

2. A reaction SMILES is formed by combining together the SMILES symbols for reactants, agents (optional) such as catalysts, solvents, etc., and products (a detailed example is shown in Fig. 4). Whereas SMILES representation for chemicals is commonly found in online reaction databases like KEGG, MetaCyc, etc., reactions represented in SMILES format are found less frequently. For instance, the reaction database Rhea (http://rhea-db.org) allows downloading any reaction in MDL RXN format rather than SMILES.
3. Molecular format converters like OpenBabel (O'Boyle et al., 2011) or molecular toolkits like RDKit (Landrum, 2017) can read and interconvert between reaction formats.
4. Moreover, a useful feature of the SMARTS representation is that it allows for generalized transformations, generally called reaction rules. Reaction rules are a powerful tool in this context because they allow us to define a class of reactions and therefore to consider alternative enzyme candidates. In this chapter, the power of using reaction rules for retrosynthesis is shown, and Fig. 4 provides examples of SMARTS encoding for reactions.

2.2.2 Computing Reaction Similarity

Several approaches are possible in order to define a metrics that measures similarity between reactions. Notably, Reaction Decoder Tool (RDT) (Rahman et al., 2016) (available at https://github.com/asad/ReactionDecoder) provides one type of reaction comparison based on bond changes, reaction centers, or substructures. This tool requires atom–atom mapping (AAM) in order to be able to identify reaction centers and bond changes (see Section 3.2.1.2 about atom–atom mapping in this chapter). Here, we take a different approach to describe a simplified algorithm that does not require AAM and is based on computing similarities between reactants.

2.2.2.1 Description of the General Workflow

In order to compute similarity between two chemical species, we employ a fingerprint approach as described in the following workflow and summarized in Fig. 2.

1. Fingerprints are binary vectors where each chemical feature is represented by one or more components of the vector as bits (Maggiora & Shanmugasundaram, 2011). The features spanned by the vector are generally related to topological elements of the chemical structure like fragments, pharmacophores, etc. Since the number of components that exist in the chemical space can become very large, fingerprint vectors

A

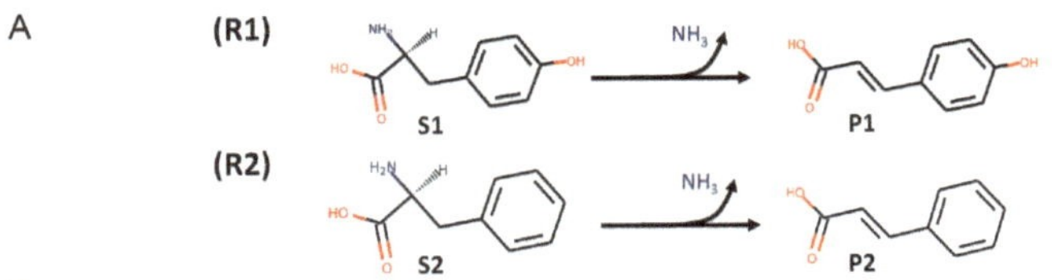

B

S1=Tyrosine	1111111111111111111101111111111111111111111111111111111111111111
P1=Coumarate	1111111111111111111111111111110111111011111111111111111111110111
S2=Phenylalanine	1111111111111111111101111111111111111111110111111111111111111111
P2=Cinnamate	1111101111101111110111111110100111111011111111111111111111110110
S=S1 AND S2	1111111111111111111111111111111111111111110111111111111111111111
P=P1 AND P2	1111101111101111111111111110101111111111111111111111111111111110
L1=S1 AND NOT P1	0000000000000000000000000000001000000100000000000000000000001000
L2=S2 AND NOT P2	0000010000010000001000000001011000000100000000000000000000001001
L=L1 AND L2	0000000000000000000000000000001000000100000000000000000000000000
R1=P1 AND NOT S1	0000000000000000000010000000000000000000000000000000000000000000
R2=P2 AND NOT S2	0000000000000000000010000000000000000000001000000000000000000000
R=P1 AND P2	0000000000000000000010000000000000000000000000000000000000000000

Fig. 2 Computing reaction similarity. (A) Similarity between tyrosine ammonia lyase EC 4.3.1.23 (*top*) and phenylalanine ammonia lyase EC 4.3.1.24 (*bottom*) is compared by using fingerprints. (B) For clarity, short fingerprints of 64 bits were computed using RDKit limited for the two substrates (tyrosine and phenylalanine) and the two products (coumarate and cinnamate). A first approach consists on computing similarity between left (S) and right (P) reactants by counting the number of common bits in the fingerprint. A second approach consists on computing left (L1, L2) and right (R1, R2) fingerprints for the reaction centers (bits not in common between left and right reactants) and then to count number of common bits in the left (L) and right (R) reaction centers.

are usually limited in size through a compression technique known as "hashing." Hashed fingerprints can be computed by several chemoinformatics toolboxes like RDKit for Python or through workflow systems like KNIME (Fillbrunn et al., 2017).

2. To compute similarity between chemicals, we compare their fingerprints. For that purpose, there are several metrics available (Maggiora & Shanmugasundaram, 2011), the Tanimoto similarity being one of the most widely used. For binary fingerprints, Tanimoto similarity can be computed as the ratio of common bits divided by the total number of bits in the molecules, making this calculation very efficient.
3. In order to compute similarity between reactions, we can use two types of strategies based on reactant or reactions fingerprints.

Reactant fingerprints

- For each left reactant, we compute pairwise similarities to each right reactant and keep the closest one.
- For each right reactant, we compute pairwise similarities to each left reactant and keep the closest one.

Reaction fingerprint

- Define an overall reaction fingerprint as the difference between the sum of fingerprints of the products and the sum of reactants.
- Alternatively, it can be computed by using OR and XOR logical operations instead of computing the sum and difference, respectively. However, this last approach has the drawback that it does not make a difference between chemical features that are created or transformed through the chemical reaction.

4. Finally, reaction similarity is computed by comparing the resulting fingerprints.
   - For reactant fingerprints, average both left and right closest similarities in order to obtain an overall similarity between the two reactions.
   - For reaction fingerprints, compute similarity between the overall fingerprints.
   - An alternative solution for reaction fingerprints is to keep a table with two fingerprint vectors, one consisting on the substrate features that are transformed and another one for the product features and measuring similarity in a similar approach as for reactant fingerprints (see Fig. 2).

2.2.2.2 Reaction Directionality

In the previous description, we assumed a preferred direction for the reactions. When this is not the case, we may compute similarity in both directions in order to keep the one that is the closest one. In principle, databases like MetaCyc, KEGG, or Brenda (Altman, Travers, Kothari, Caspi, & Karp, 2013; Chang et al., 2014) contain information about reaction's preferred direction, reversibility and, increasingly, data about free Gibbs energy. When this information is not available, several strategies are possible in order to assign a preferred direction to the reaction.

1. Estimation of reaction thermodynamics by using group contribution methods (Jankowski, Henry, Broadbelt, & Hatzimanikatis, 2008). Each chemical group in the reaction contributes to the free Gibbs according to some tabulated experimental data. The balance between left and right groups provides roughly an estimate of free energy and eventually allows

determining the reaction direction. This is the strategy used by some databases like MetaCyc (Caspi et al., 2012) and in some organism models (Feist et al., 2007).

2. Knowledge-based reaction direction inference. If a set of reactions with a high level of confidence knowledge of their direction are available, we can use such information in order to infer the most plausible direction for another given reaction. The algorithm is as follows:
   - **A.** Start with initial set of reactions with curated direction information.
   - **B.** Collect information about reactants and cofactors for each reaction in the reference set.
   - **C.** Extract a list of currency metabolites (common cofactors) from a reference database.
   - **D.** FOR each nonannotated reaction:
     - **D1.** Identify left and right cofactors and main reactants based on information from A, B, C.
     - **D2.** Search for top similar annotated reactions.
     - **D3.** Extract frequency of left/right cofactor pairs.
     - **D4.** Set direction with highest frequency in top similar reactions based on left/right cofactor pairs.

2.2.3 Screening, Ranking, and Selection

2.2.3.1 Database Screening

In order to identify candidate enzyme sequences for a given chemical transformation, screen the target reaction against the reactions in an annotated metabolic database like Metacyc, MetaNetx (Moretti et al., 2016), or Biochem4j (Swainston et al., 2017).

1. The calculation should be performed in a computationally efficient way. A similarity algorithm like the one described in the previous section based on fingerprints is therefore required.
2. Reactions in the database are classified in descending order of similarity according to the fingerprint algorithm. For each reaction, add annotated sequences to the candidate list until either the maximum number of allowed sequences is reached or some similarity minimum threshold has been exceeded.

2.2.3.2 Properties' Calculation

The list of candidate sequences obtained in the previous step might contain redundancies as well as overrepresentation for some reaction classes. Therefore, the next step is to analyze the set in order to characterize each protein according to key properties and to ultimately provide some ranking leading to candidate selection.

1. Based on the information contained in the database, reactions can be classified according to their reaction EC class. This information can help in order to perform a quick initial preselection.
2. Redundancy in sequences can be easily identified through CD-HIT (Li & Godzik, 2006), which provides a fast way of clustering large sets of sequences.
3. A multiple sequence alignment of the sequences is then applied to get a more detailed comparative analysis of the sequences with the advantage that it can help to identify conserved regions in the protein, using T-Coffee (Taly et al., 2011) or ClustalW.
4. Additional position-based and global properties can be computed using packages like EMBOSS (Rice, Longden, & Bleasby, 2000), amino acid indexes (Kawashima & Kanehisa, 2000), or online tools like PredictProtein (https://www.predictprotein.org/).
5. Other properties: PSSM (position-specific scoring matrices) using PSI-BLAST, functional domains using Pfam collections and HMMER (http://pfam.xfam.org/).

2.2.3.3 Ranking of Sequence Candidates

Once the list of sequence candidates with calculated properties has been generated, the list needs to be prioritized. The criteria for ranking might vary depending on the application. For enzyme sequences that are going to be expressed in heterologous host, a ranking can be formed based on the following criteria:

1. Protein properties: percentage of secondary structure (helices, sheets, turns), molecular weight, isoelectric point, percentage of polar amino acids, etc.
2. Functional properties: target reaction similarity, UniProt protein evidence, sequence conservation in the alignment, etc.
3. Host-specific properties: sequence taxonomic distance to host organism, solubility (http://scratch.proteomics.ics.uci.edu/), toxicity (http://exploration.weizmann.ac.il/pandatox).

By combining together these properties through weighted sums, we can obtain a score for the sequence. Weights in the score can be heuristically fitted by downloading the proteome sequences of the host organism and calculating typical values obtained for each of these parameters. A more complex scoring function can be also developed by employing machine learning using host-specific values for computed protein properties as a training set.

2.2.4 Selenzyme: Online Enzyme Selection Tool

Selenzyme is a free online tool for enzyme selection (http://selenzyme.synbiochem.co.uk) that integrates the aforementioned features to efficiently search for enzyme sequences starting from some target reaction (Carbonell et al., 2018).

1. User's queries consist of a target reaction expressed in SMILES or RXN format or cross-referenced to an external database identifier or EC classification.
2. The software outputs a table of candidate sequences that can be ranked based on different criteria. In addition, the user can manually add or remove additional sequences to the table. A multiple sequence alignment of the output table can be visualized through MSAViewer (Yachdav et al., 2016).

An example of output table from Selenzyme is shown in Table 1. The input query was the reaction SMARTS shown in Fig. 4. Several properties were computed for the sequences that are annotated for the closest reactions to the target. A score is calculated for each sequence based on decreasing order by reaction similarity, taxonomic distance, sequence conservation, and protein evidence.

2.2.4.1 Integration With RetroPath2.0 and Other Workflows

Selenzyme has been integrated into the SYNBIOCHEM-automated Design–Build–Test–Learn pipeline for fine chemical production (Carbonell, Currin, Dunstan, et al., 2016). The output of the pathway discovery tool RetroPath2.0, described in this chapter, can point to Selenzyme queries allowing enzyme sequence discovery for each step in the pathway.

Selenzyme provides a RESTful service allowing query submission through web-based applications. A KNIME node (Fillbrunn et al., 2017) making use of the RESTful service is available at http://www.myexperiment.org/packs/734. The reaction query can be generated using the chemoinformatics workflows in KNIME and the resulting tables containing sequences can be easily processed downstream.

2.2.5 Other Applications of the Protocol

1. The described protocol has been focused on enzymes in production pathways. However, the application of the protocol can be extended

Table 1 Output Table From Selenzyme Based on Reaction SMARTS for EC 1.1.1.37

Score	Seq. ID	Organism Source	Tax. Distance	Protein Evidence	Consv. Score	% Helices	% Sheets	% Turns	% Coils	Mol. Weight	Isoelec. Point	Polar (%)
127.7	P61891	*Escherichia coli* O157:H7	3	3	58	39.5	33.8	7.4	24.7	32,337.3	5.4603	39.423
98.7	P61895	*Raoultella terrigena*	6	3	59	35.9	38.3	8.6	24.9	23,055.38	5.043	40.889
97.7	Q83Q04	*Shigella flexneri*	6	3	58	39.2	33.8	7.8	24.7	32,349.36	6.5453	39.423
77.7	Q59838	*Salmonella muenchen*	8	3	58	41.2	31.8	8.6	24.3	29,502.96	5.7944	40.636
69.7	P61897	*Pectobacterium carotovorum* subsp. *carotovorum*	9	3	60	28.3	40.6	13.8	29	16,148.7	9.297	42.208
65.7	C5BF98	*Edwardsiella ictaluri* (strain 93–146)	9	3	56	38.2	35.1	6.8	25.3	32,348.45	4.9809	36.859
57.7	A1JIV0	*Yersinia enterocolitica* serotype O:8/biotype 1B (strain NCTC 13174/8081)	10	3	58	35.6	34.6	9.5	25.8	32,581.62	5.815	40.193
56.7	P37226	*Photobacterium profundum*	10	3	57	41.6	34.5	7.1	22.3	32,293.2	4.7141	37.5
46.7	B5FGF5	*Vibrio fischeri* (strain MJ11)	11	3	57	43.1	29.2	8.8	24.4	32,252.09	4.5494	37.942
46.7	A8FRU0	*Shewanella sediminis* (strain HAW-EB3)	11	3	57	47.1	31.2	4.4	22.7	32,010.79	4.5659	37.299
43.7	Q5WU94	*Legionella pneumophila* (strain Lens)	11	3	54	36.6	30.6	14.3	23.6	36,063.95	5.7759	45.758
39.7	A4G5Z9	*Herminiimonas arsenicoxydans*	12	3	60	47.6	27.5	9.9	20.1	35,364.65	7.0784	41.641

Continued

Table 1 Output Table From Selenzyme Based on Reaction SMARTS for EC 1.1.1.37—cont'd

Score	Seq. ID	Organism Source	Tax. Distance	Protein Evidence	Consv. Score	% Helices	% Sheets	% Turns	% Coils	Mol. Weight	Isoelec. Point	Polar (%)
38.7	Q5H496	*Xanthomonas oryzae* pv. *oryzae* (strain KACC10331/KXO85)	12	3	59	42.3	29.5	9.6	23.7	34,925.04	5.4464	39.939
34.7	B0BQN0	*Actinobacillus pleuropneumoniae* serotype 3 (strain JL03)	12	3	55	51.5	29.2	6.6	17.9	33,404.75	5.4643	38.17
31.7	Q14IT0	*Francisella tularensis* subsp. *tularensis* (strain FSC 198)	12	3	52	45.9	33.7	11.6	14.2	34,090.77	7.3844	40.752
29.7	Q8XXW5	*Ralstonia solanacearum* (strain GMI1000)	13	3	60	47.3	30.4	6.7	20.8	35,422.8	6.6924	39.818
29.7	Q393V1	*Burkholderia lata* (strain ATCC 17760/LMG 22485/NCIMB 9086/R18194/383)	13	3	60	54.5	26.9	9	14.7	35,131.36	5.7875	40.549
28.7	A1VRQ1	*Polaromonas naphthalenivorans* (strain CJ2)	13	3	59	51	22.1	8.3	23.7	34,845.99	5.4269	38.415
24.7	Q2GCH6	*Neorickettsia sennetsu* (strain Miyayama)	13	3	55	51.2	30.4	5.7	18.1	33,851.77	8.3718	39.683

to other applications such as in the development of enzyme-mediated biosensors (Delépine, Libis, Carbonell, & Faulon, 2016), in the design of transporters, etc.

2. The protocol described here provides a first step when selecting for some target reaction. The resulting list of candidate enzymes as well as the resulting multiple sequence alignment (Fig. 3) can be used as a starting point in order to carry out structure-based enzyme analysis of design.

2.2.6 Improvements of the Protocol

Several improvements are possible on the described protocol.

1. Estimating enzyme efficiency based on kinetic parameters when they are available. Database like BRENDA provides kinetic values obtained by enzymatic assays, which can provide a first estimate of enzyme efficiency.
2. Using machine-learning techniques to predict enzyme efficiency using kinetic parameters in databases as training set (Mellor et al., 2016).
3. Balancing multienzyme pathways: enzyme selection for multistep pathways should consider overall pathway performance, i.e., searching some compatibility in the reactions (same source, kinetics, etc.). This approach can also be used to guide pathway tuning through transcriptional and translational parts either to refactor a natural gene cluster or to balance a de novo pathway from multiple heterologous sources.

3. PATHWAY DESIGN

3.1 Introduction to Retrosynthesis

This protocol considers the scenario where a desirable product is known, but no natural pathway leading to it is known. In such a case, we use retrosynthesis to discover possible pathways. Such algorithm generates networks linking the target compounds that one desires to bioproduce (the source) and the metabolites of the chassis strain (the sink) by applying chemical rules. These networks are then processed to extract biologically relevant information. For example, pathways can be enumerated (Carbonell, Fichera, Pandit, & Faulon, 2012) and ranked based on several criteria including enzyme availability and performance, product and intermediate compound toxicities (Planson, Carbonell, Paillard, Pollet, & Faulon, 2012), or the theoretical yield of the desired compound (Campodonico et al., 2014; Carbonell, Parutto, Herisson, Pandit, & Faulon, 2014; Cho, Yun, Park, Lee, & Park, 2010; Liu et al., 2014; Paricharak, Klenka, Augustin, Patel, & Bender, 2013). Users of retrosynthesis-based solutions often face a challenge common

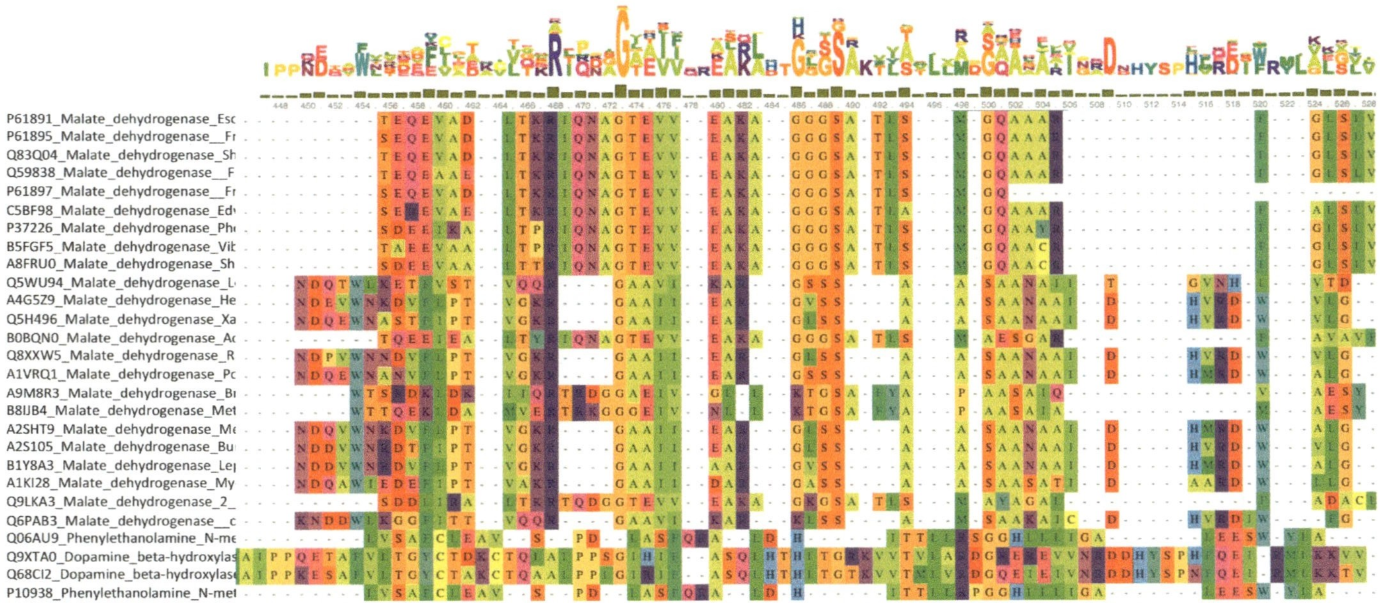

Fig. 3 Detail of a multiple sequence alignment displayed by Selenzyme for the example given in Table 1 for EC 1.1.1.37.

to most of these tools: the algorithms and the underlying data are often not fully documented and released. In most cases, authors provide fine-tuned webservers (Campodonico et al., 2014; Carbonell, Parutto, Baudier, et al., 2014; Carbonell, Parutto, Herisson, et al., 2014; Hadadi et al., 2016; Jeffryes et al., 2015) filled with pregenerated data that focuses on some exemplar cases. On the contrary, Delépine et al. (2018) provide the open-source workflow RetroPath2.0 based on the KNIME analytics platform (Fillbrunn et al., 2017) that uses community nodes and is therefore fully modular and can easily be tuned to a user's need.

We describe first a protocol based on the RetroPath2.0 workflow and the necessary steps to generate and use retrosynthesis for bioproduction. The tool is available at myExperiment.org (https://www.myexperiment.org/workflows/4987.html) along with a set of reaction rules and some classic metabolic engineering examples to test RetroPath2.0 features.

3.2 Protocol Description (RetroPath2.0)

3.2.1 Preparation Steps

3.2.1.1 Sink and Source Definition

The first step to generate a retrosynthesis map is to encode all compounds of interest in a format that will allow processing by the retrosynthesis algorithm. Source compounds are the compounds the workflow starts iterating upon (compounds one desires to produce) and the sink compounds are the compounds where the algorithm stops (either the metabolome of the chassis organism or compounds easily supplemented in the media).

1. Gather compounds from a whole-cell metabolic model of the chassis organism of interest. It must include structural data for compounds.
2. Filter out compounds with incomplete structure. They often stand to define a class of compounds and cannot be processed further.
3. Select the sink compounds: either the whole set of compounds from the chassis organism or a subset selected based on expert knowledge. One can, for example, remove compounds belonging to blocked pathways by performing a flux-balance analysis. Sink compounds can also include molecules easily supplemented in the media.
4. Choose source compounds that one desires to bioproduce and collect associated structural information in order for the algorithm to process them.

3.2.1.2 Reaction Rules

The second step is to encode (bio)chemical reactions that will be used to perform the retrosynthesis. RetroPath2.0 uses reaction SMARTS to encode reactions. It is a SMIRK-like reaction rule ("Daylight Theory Manual, Version 4.9. http://www.daylight.com/ (accessed September 1, 2017)," 2017) format defined by RDKit (Landrum, 2017).

1. Select a database containing reaction information. Such database should at least provide the structure of each compound involved in a reaction.
2. Remove all reactions that do not modify the structure of a compound (transport reactions for example) and reactions involving incomplete structures (class of compounds, R-groups, etc.). Remove stereochemistry.
3. Perform an AAM to identify the reaction center (i.e., the part of the molecule that changes during the reaction). This AAM is also necessary later on to compute the reaction SMARTS.
4. Before building the monocomponent rules, one should consider the reactions in the reverse direction, i.e., consider the natural products as substrates and natural substrates as products. This is only needed for a retrosynthetic usage. Decompose multiple substrate reactions into components. There are as many components as there are substrates and each component gives the transformation between one substrate and the products. Each product must contain at least one atom from the substrate according to the AAM. This strategy enforces that only one substrate can differ at a time from the substrates of the reference reaction when applying the rule. Cosubstrates and coproducts that are currency cofactors (such as water, CO_2, ATP, NADP, etc.) can be ignored from the rules under the assumptions that they are available in the cell and that there is no gain for retrosynthesis analysis in modeling promiscuity on these compounds.
5. Compute reaction rules as reaction SMARTS for each component. Do it for varying diameters around the reaction center (2–16 in RetroPath2.0) by removing from the reaction components all atoms that were not in the spheres around the reaction center atoms (Fig. 4).

3.2.2 Building a Retrosynthesis Network

Once the user has performed the preparation steps, the rules, sink, and source are provided as inputs to RetroPath2.0 that builds the retrosynthesis network. The following section describes the steps followed by the algorithm. It will help highlighting key tunable parameters for the advanced user: scoring of the enzymes, number of compounds kept for the next iteration, maximum number of steps, as well as the role of the reaction diameter.

A

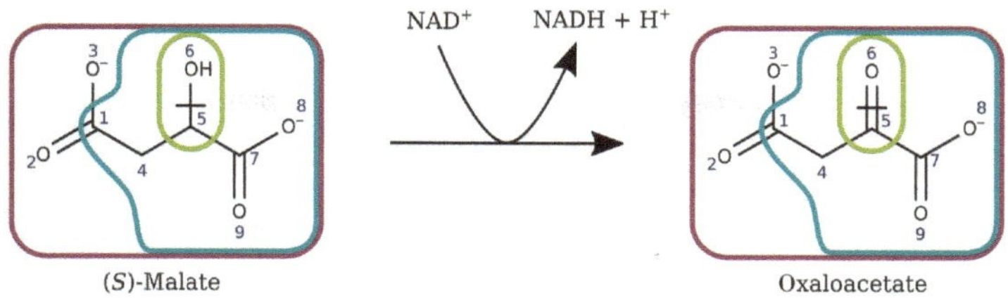

B

$D=\infty$
```
[#8:6]-[#6:5](-[#6:4]-[#6:1](-[#8-:3])=[O:2])-[#6:7](-[#8-:8])=[O:9]
>>[#8-:3]-[#6:1](=[O:2])-[#6:4]-[#6:5](=[O:6])-[#6:7](-[#8-:8])=[O:9]
```

$D=4$
```
[#6:1]-[#6:4]-[#6:5](-[#8:6])-[#6:7](-[#8-:8])=[O:9]
>>[#6:1]-[#6:4]-[#6:5](=[O:6])-[#6:7](-[#8-:8])=[O:9]
```

$D=0$
```
[#6:5]-[#8:6]>>[#6:5]=[O:6]
```

Fig. 4 RetroPath2.0 rules and corresponding SMARTS for reaction 1.1.1.37 at various diameters. (A) Reaction 1.1.1.37 with atom mapping. The level of promiscuity is modeled by the diameter (∞ *purple*, 4 *blue*, or 0 *green*), i.e., the number of bonds to consider around the reaction center (atoms 5 and 6), defining the atoms kept in the rules. (B) The SMARTS corresponding to the reaction rules at the various diameters. For readability, cofactors are not considered.

3.2.2.1 General Workflow Presentation

The RetroPath2.0 workflow essentially follows the algorithm described below, proceeding in an iterative manner. For the sake of clarity, we present here the algorithm for one initial source compound, but it can process multiple compounds in parallel.

1. Verify the source compound is not already present in the sink.
2. Start iteration by applying the rules to each compound of the source set. The workflow starts by applying rules at the biggest diameter allowed by the user, typically $d=16$. If no product is formed, the diameter is decreased to the available diameter value just below, allowing matching for shorter fragments and therefore more promiscuity. This step is repeated until some products are formed or until all possible diameters have been exhausted. For each compound, the products are computed using the RDKit KNIME nodes (Landrum, 2017). Products are standardized and duplicates are merged. All substrate–product pairs are added to the growing network along with the reaction rules linking them.
3. (Optional) Products can be scored according to a rule prioritization scheme. The one used in RetroPath2.0 will be briefly described. Products are ranked according to the score of the rule used to produce them at the current iteration.
4. In the next iteration, the set of products becomes the new source set. However, before iterating, the workflow removes from the new source

set all compounds that belong to the sink (as these are already solutions and there is no need to iterate) and the workflow adds the product set to the sink in order to avoid applying reactions on the same products during subsequent iterations.

5. Apply steps 2–4 either a predefined number of iterations or until the source set is empty.
6. The final produced graph is composed of the list of links between substrates and products annotated with their corresponding reaction rule. Products belonging to the initial sink set are annotated as such.

3.2.2.2 Remark: Rule Scoring by Enzyme Sequence Consistency

As mentioned in the optional step 3 of the algorithm, products can be scored according to a rule prioritization scheme. In RetroPath2.0, this is done by scoring the confidence one can have that a rule is indeed linked to a trustworthy enzymatic sequence. More specialized scoring strategies are possible depending on the selected enzyme sequence. A description of the Selenzyme protocol can be found in this chapter.

3.2.3 Pathway Enumeration Between Two Pools of Compounds

3.2.3.1 Computing the Scope

The aim is to compute the metabolic scope connecting the source compounds to the sink compounds, i.e., the set of compounds and reactions that are involved in at least one pathway.

1. Forward search: starting from source compounds, find all reachable compounds. A compound is considered as reachable whenever it is in the initial sink or if it is producible using a fireable reaction.
2. Backward search: starting from the sink compounds, add to the scope all reactions that can be fired, as long as the reaction substrates are all reachable.
3. (Optional) Simplify the scope by only keeping the shortest paths from the compounds of interest to the sink.
4. Visualize and explore the scope thanks to ScopeViewer, a web application provided in Delépine et al. (2018).

3.2.3.2 Enumerating Pathways

For each source compound, proceed to the following steps to find pathways producing this source compound.

1. Build the stoichiometric matrix. The stoichiometric matrix describes the directed subnetwork involving the set of compounds and reactions identified at the scope step, starting from the source compound of interest.

2. Enumerate elementary flux modes. An elementary mode corresponds to a minimal unique set of reactions that (i) verified the stoichiometric constraints of the network and (ii) is able to carry nonzero fluxes at the system's steady state (Schuster, Fell, & Dandekar, 2000). Only enumerated flux modes linking source compounds to the sink compound are kept in order to form the final list of pathways.
3. Enumerate the pathways from the elementary flux modes. We provide at https://github.com/brsynth/rp2paths a separate utility program "RP2paths" allowing one to enumerate pathways from the results generated by RetroPath2.0.

3.3 Use Case: 1,4-Butanediol Pathways Prediction Using RetroPath2.0

1,4-Butanediol is an important commodity chemical used as a starting point for the synthesis of other chemicals and polymers such as the polybutylene terephthalate, a unique engineering plastic. While most of the production of 1,4-butanediol is performed by chemical synthesis and is still making use of petroleum-based feedstock, a bioprocess alternative has been first reported in Yim et al. (2011).

Here we showcase the usefulness of RetroPath2.0 in order to predict pathways enabling the bioproduction of 1,4-butanediol in *Escherichia coli*.

3.3.1.1 Materials and Methods

To build the retrosynthetic graph and enumerate the pathways, we use the procedure below:

- 1,4-Butanediol is considered as the source compound.
- Sink compounds were extracted from the iJO1366 *E. coli* whole-cell model (Orth et al., 2011) and MetaNetX cross-references (Moretti et al., 2016).
- Reaction information is collected from the MetaNetX database and reaction rules are generated following the aforementioned guidelines. The generated set contains from 6900 to 19,000 unique rules, depending on the diameter considered.
- RetroPath2.0 is applied with a maximum of four retrosynthesis iterations, keeping a maximum of 100 compounds for the next iteration, and reaction diameters ranging from 12 to 8.
- Pathways are enumerated thanks to the RP2paths software.

Using RetroPath2.0 we successfully retrieved the bioproduction pathway reported by Yim et al. (2011), as well as five alternatives enabling the production of BDA (depicted in Fig. 5). Four pathways, including the one reported by Yim et al. (2011), propose a common strategy where the coenzyme A to be attached to 4-hydroxybutyrate is supplied by different CoA-related chemicals (namely CoA, acetyl-CoA, succinyl-CoA, and butanoyl-CoA).

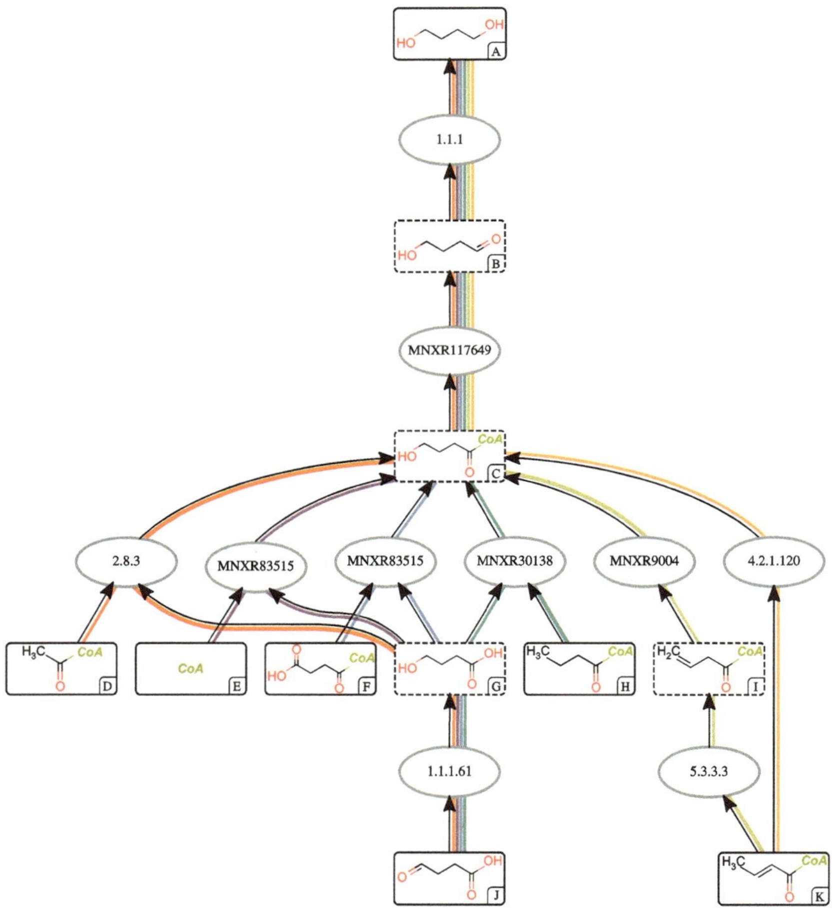

Fig. 5 Enumerated pathways for the production of styrene. Each pathway is depicted by a distinct color. Pathway referenced in Yim et al. (2011) is in *red* (J-G-D-C-B-A). Compounds are represented by their structures, and reactions by their EC numbers when known (else by the MetaNetX reaction ID). 1,4-Butanediol and sink compounds are surrounded by a *solid line*, intermediates by a *dashed line*. *A*, 1,4-butanediol; *B*, 4-hydroxybutanal; *C*, 4-hydroxybutyryl-CoA; *D*, acetyl-CoA; *E*, CoA; *F*, succinyl-CoA; *G*, 4-hydroxybutyrate; *H*, butanoyl-CoA; *I*, vinylacetyl-CoA; *J*, succinate semialdehyde; *K*, crotonoyl-CoA. Cofactors have been removed for clarity.

3.4 Other Applications of the Protocol

The choice of rule sets, sink, and source depends on the application.

1. For instance, if one wishes to find all producing pathways for a given compound, the source will be the target, the sink the metabolites of the chassis strain, and the rules the reversed form of all known metabolic reactions, as was presented here.
2. To degrade a given xenobiotic, the rule set can be the same metabolic reactions in the forward direction, the sink will be the metabolites of the chassis strain, and the source is the compound to degrade.
3. Another choice of settings allows for another application of interest: choosing a set of known detectable compounds as sink, a set of target compounds one wishes to detect as source, and the set of forward rules, one can design sensing-enabling pathways (Delépine et al., 2016).

The versatility of this tool allows us to use it for various applications, as showcased in Koch, Duigou, Carbonell, and Faulon (2017).

1. Using a set of rules for isomer generation or fragment exchange, instead of enzymatic rules, we can generate all structural isomers of a given compound or virtually screen the chemical space around a set of molecules of interest.
2. Moreover, the way we encode our rules using diameter allows us to tune the promiscuity we allow in our compounds' generation. Therefore, this workflow can be used for metabolome completion; with novel molecules generated using promiscuous enzymatic reaction rules.

4. SUMMARY AND CONCLUSION

We have covered in this chapter some possible scenarios involving enzyme discovery and pathway design for synthetic biology and metabolic engineering applications. For a given biochemical transformation, we can select enzymes sequences through the Selenzyme (Carbonell et al., 2018) protocol. Starting from a desired target compound, we can identify production pathways through the RetroPath2.0 (Delépine et al., 2018) protocol. As shown in Fig. 1, such discovery process can be in some cases reversed, for instance when an interesting pathway is known, but not its product. A tool such as antiSMASH (Weber et al., 2015) allows for the annotation of the transformation steps involved in the pathway starting from the gene cluster, as well as allowing integration of predicted biosynthesis pathways for secondary metabolites with genome-scale models of metabolism.

Once a putative biochemical pathway has been inferred from a gene cluster, antiSMASH and RetroPath provide means to predict and enumerate its main products and side products of enzyme promiscuity. The tools and protocols described in this chapter and the resulting integrated pipeline offer a rich synthetic biology toolbox for enzyme discovery.

ACKNOWLEDGMENTS

The authors would like to acknowledge funding support from the French National Research Agency (ANR-15-CE1-0008), the Biotechnology and Biological Sciences Research Council, Centre for Synthetic Biology of Fine and Speciality Chemicals (BB/M017702/1), and Synthetic Biology Applications for Protective Materials (EP/N025504/1). M.K. acknowledges funding provided by the DGA and École Polytechnique.

REFERENCES

Altman, T., Travers, M., Kothari, A., Caspi, R., & Karp, P. D. (2013). A systematic comparison of the MetaCyc and KEGG pathway databases. *BMC Bioinformatics*, *14*(1), 112. https://doi.org/10.1186/1471-2105-14-112.

Campodonico, M. A., Andrews, B. A., Asenjo, J. A., Palsson, B. Ø., & Feist, A. M. (2014). Generation of an atlas for commodity chemical production in Escherichia coli and a novel pathway prediction algorithm, GEM-path. *Metabolic Engineering*, *25*, 140–158. https://doi.org/10.1016/j.ymben.2014.07.009.

Carbonell, P., Currin, A., Dunstan, M., Fellows, D., Jervis, A., Rattray, N. J. W., et al. (2016). SYNBIOCHEM—A SynBio foundry for the biosynthesis and sustainable production of fine and speciality chemicals. *Biochemical Society Transactions*, *44*(3), 675–677. https://doi.org/10.1042/BST20160009.

Carbonell, P., Currin, A., Jervis, A. J. A. J., Rattray, N. J. W., Swainston, N., Yan, C., et al. (2016). Bioinformatics for the synthetic biology of natural products: Integrating across the Design-Build-Test cycle. *Natural Product Reports*, *33*(8), 925–932. https://doi.org/10.1039/c6np00018e.

Carbonell, P., Fichera, D., Pandit, S. B., & Faulon, J. L. (2012). Enumerating metabolic pathways for the production of heterologous target chemicals in chassis organisms. *BMC Systems Biology*, *6*(1), 10. https://doi.org/10.1186/1752-0509-6-10.

Carbonell, P., Parutto, P., Baudier, C., Junot, C., & Faulon, J. L. (2014). Retropath: Automated pipeline for embedded metabolic circuits. *ACS Synthetic Biology*, *3*(8), 565–577. https://doi.org/10.1021/sb4001273.

Carbonell, P., Parutto, P., Herisson, J., Pandit, S. B., & Faulon, J. L. (2014). XTMS: Pathway design in an eXTended metabolic space. *Nucleic Acids Research*, *42*(W1), W389–W394. https://doi.org/10.1093/nar/gku362.

Carbonell, P., Planson, A. G., Fichera, D., & Faulon, J. L. (2011). A retrosynthetic biology approach to metabolic pathway design for therapeutic production. *BMC Systems Biology*, *5*(1), 122. https://doi.org/10.1186/1752-0509-5-122.

Carbonell, P., & Trosset, J.-Y. (2015). Computational protein design methods for synthetic biology. *Methods in Molecular Biology*, *1244*, 3–21. https://doi.org/10.1007/978-1-4939-1878-2_1.

Carbonell, P., Wong, J., Swainston, N., Takano, E., Turner, N. J., Scrutton, N. S., et al. (2018). Selenzyme: Enzyme selection tool for pathway design. *Bioinformatics*, 188979. https://doi.org/10.1093/bioinformatics/bty065. Epub ahead of print.

Caspi, R., Altman, T., Dreher, K., Fulcher, C. A., Subhraveti, P., Keseler, I. M., et al. (2012). The MetaCyc database of metabolic pathways and enzymes and the BioCyc collection of pathway/genome databases. *Nucleic Acids Research*, *40*(D1), D742–D753. https://doi.org/10.1093/nar/gkr1014.

Chang, A., Schomburg, I., Placzek, S., Jeske, L., Ulbrich, M., Xiao, M., et al. (2014). BRENDA in 2015: Exciting developments in its 25th year of existence. *Nucleic Acids Research*, *43*(D1), D439–D446. https://doi.org/10.1093/nar/gku1068.

Cho, A., Yun, H., Park, J. H., Lee, S. Y., & Park, S. (2010). Prediction of novel synthetic pathways for the production of desired chemicals. *BMC Systems Biology*, *4*(1), 35. https://doi.org/10.1186/1752-0509-4-35.

Delépine, B., Duigou, T., Carbonell, P., & Faulon, J. L. (2018). RetroPath2.0: A retrosynthesis workflow for metabolic engineers. *Metabolic Engineering*, *45*, 158–170. https://doi.org/10.1016/j.ymben.2017.12.002.

Delépine, B., Libis, V., Carbonell, P., & Faulon, J. L. (2016). SensiPath: Computer-aided design of sensing-enabling metabolic pathways. *Nucleic Acids Research*, *44*(W1), W226–W231. https://doi.org/10.1093/nar/gkw305.

Feist, A. M., Henry, C. S., Reed, J. L., Krummenacker, M., Joyce, A. R., Karp, P. D., et al. (2007). A genome-scale metabolic reconstruction for Escherichia coli K-12 MG1655 that accounts for 1260 ORFs and thermodynamic information. *Molecular Systems Biology*, *3*(1), 121. https://doi.org/10.1038/msb4100155.

Fillbrunn, A., Dietz, C., Pfeuffer, J., Rahn, R., Landrum, G. A., & Berthold, M. R. (2017). KNIME for reproducible cross-domain analysis of life science data. *Journal of Biotechnology*, *261*, 149–156. https://doi.org/10.1016/j.jbiotec.2017.07.028.

Hadadi, N., Hafner, J., Soh, K. C., & Hatzimanikatis, V. (2016). Reconstruction of biological pathways and metabolic networks from in silico labeled metabolites. *Biotechnology Journal*, *12*(1), 1600464. https://doi.org/10.1002/biot.201600464.

Hadadi, N., & Hatzimanikatis, V. (2015). Design of computational retrobiosynthesis tools for the design of de novo synthetic pathways. *Current Opinion in Chemical Biology*, *28*, 99–104. https://doi.org/10.1016/j.cbpa.2015.06.025.

Henry, C. S., Broadbelt, L. J., & Hatzimanikatis, V. (2010). Discovery and analysis of novel metabolic pathways for the biosynthesis of industrial chemicals: 3-hydroxypropanoate. *Biotechnology and Bioengineering*, *106*(3), 462–473. https://doi.org/10.1002/bit.22673.

Jankowski, M. D., Henry, C. S., Broadbelt, L. J., & Hatzimanikatis, V. (2008). Group contribution method for thermodynamic analysis of complex metabolic networks. *Biophysical Journal*, *95*(3), 1487–1499. https://doi.org/10.1529/biophysj.107.124784.

Jeffryes, J. G., Colastani, R. L., Elbadawi-Sidhu, M., Kind, T., Niehaus, T. D., Broadbelt, L. J., et al. (2015). MINEs: Open access databases of computationally predicted enzyme promiscuity products for untargeted metabolomics. *Journal of Cheminformatics*, 7(1), 44. https://doi.org/10.1186/s13321-015-0087-1.

Kawashima, S., & Kanehisa, M. (2000). AAindex: Amino acid index database. *Nucleic Acids Research*, *28*(1), 374. https://doi.org/10.1093/nar/28.1.374.

Koch, M., Duigou, T., Carbonell, P., & Faulon, J.-L. (2017). Molecular structures enumeration and virtual screening in the chemical space with RetroPath2.0. *Journal of Cheminformatics*, *9*(1), 64. https://doi.org/10.1186/s13321-017-0252-9.

Landrum, G. (2017). *RDKit documentation*. http://www.rdkit.org. Accessed 1 September 2017. Retrieved September 1, 2017, from http://www.rdkit.org.

Lee, S. Y., & Kim, H. U. (2015). Systems strategies for developing industrial microbial strains. *Nature Biotechnology*, *33*(10), 1061–1072. https://doi.org/10.1038/nbt.3365.

Li, W., & Godzik, A. (2006). Cd-hit: A fast program for clustering and comparing large sets of protein or nucleotide sequences. *Bioinformatics*, *22*(13), 1658–1659. https://doi.org/10.1093/bioinformatics/btl158.

Li, Y., Wang, S., Umarov, R., Xie, B., Fan, M., Li, L., et al. (2018). DEEPre: Sequence-based enzyme EC number prediction by deep learning. *Bioinformatics*, *34*(5), 760–769. https://doi.org/10.1093/bioinformatics/btx680.

Liu, M., Bienfait, B., Sacher, O., Gasteiger, J., Siezen, R. J., Nauta, A., et al. (2014). Combining chemoinformatics with bioinformatics: In silico prediction of bacterial flavor-forming pathways by a chemical systems biology approach "Reverse Pathway Engineering" *PLoS One*, *9*(1), e84769. https://doi.org/10.1371/journal.pone.0084769.

Maggiora, G. M., & Shanmugasundaram, V. (2011). Molecular similarity measures. *Methods in Molecular Biology (Clifton, N.J.)*, *672*, 39–100. https://doi.org/10.1007/978-1-60761-839-3_2.

Medema, M. H., van Raaphorst, R., Takano, E., & Breitling, R. (2012). Computational tools for the synthetic design of biochemical pathways. *Nature Reviews Microbiology*, *10*(3), 191–202. https://doi.org/10.1038/nrmicro2717.

Mellor, J., Grigoras, I., Carbonell, P., & Faulon, J. L. (2016). Semisupervised Gaussian process for automated enzyme search. *ACS Synthetic Biology*, *5*(6), 518–528. https://doi.org/10.1021/acssynbio.5b00294.

Moretti, S., Martin, O., Van Du Tran, T., Bridge, A., Morgat, A., et al. (2016). MetaNetX/MNXref reconciliation of metabolites and biochemical reactions to bring together genome-scale metabolic networks. *Nucleic Acids Research*, *44*(D1), D523–D526. https://doi.org/10.1093/nar/gkv1117.

Moriya, Y., Shigemizu, D., Hattori, M., Tokimatsu, T., Kotera, M., Goto, S., et al. (2010). PathPred: An enzyme-catalyzed metabolic pathway prediction server. *Nucleic Acids Research*, *38*(Web Server issue), W138–W143. https://doi.org/10.1093/nar/gkq318.

O'Boyle, N. M., Banck, M., James, C. A., Morley, C., Vandermeersch, T., & Hutchison, G. R. (2011). Open Babel: An open chemical toolbox. *Journal of Cheminformatics*, *3*(10), 33. https://doi.org/10.1186/1758-2946-3-33.

Orth, J. D., Conrad, T. M., Na, J., Lerman, J. A., Nam, H., Feist, A. M., et al. (2011). A comprehensive genome-scale reconstruction of Escherichia coli metabolism-2011. *Molecular Systems Biology*, 7, 53. https://doi.org/10.1038/msb.2011.65.

Paricharak, S., Klenka, T., Augustin, M., Patel, U., & Bender, A. (2013). Are phylogenetic trees suitable for chemogenomics analyses of bioactivity data sets: The importance of shared active compounds and choosing a suitable data embedding method, as exemplified on kinases. *Journal of Cheminformatics*, *5*(1), 49. https://doi.org/10.1186/1758-2946-5-49.

Planson, A. G., Carbonell, P., Paillard, E., Pollet, N., & Faulon, J. L. (2012). Compound toxicity screening and structure-activity relationship modeling in Escherichia coli. *Biotechnology and Bioengineering*, *109*(3), 846–850. https://doi.org/10.1002/bit.24356.

Rahman, S. A., Cuesta, S. M., Furnham, N., Holliday, G. L., & Thornton, J. M. (2014). EC-BLAST: A tool to automatically search and compare enzyme reactions. *Nature Methods*, *11*(2), 171–174. https://doi.org/10.1038/nmeth.2803.

Rahman, S. A., Torrance, G., Baldacci, L., Martínez Cuesta, S., Fenninger, F., Gopal, N., et al. (2016). Reaction decoder tool (RDT): Extracting features from chemical reactions. *Bioinformatics*, *32*(13), 2065–2066. https://doi.org/10.1093/bioinformatics/btw096.

Rice, P., Longden, I., & Bleasby, A. (2000). EMBOSS: The European molecular biology open software suite. *Trends in Genetics*, *16*(6), 276–277.

Rodrigo, G., Carrera, J., Prather, K. J., & Jaramillo, A. (2008). DESHARKY: Automatic design of metabolic pathways for optimal cell growth. *Bioinformatics (Oxford, England)*, *24*(21), 2554–2556. https://doi.org/10.1093/bioinformatics/btn471.

Schuster, S., Fell, D. A., & Dandekar, T. (2000). A general definition of metabolic pathways useful for systematic organization and analysis of complex metabolic networks. *Nature Biotechnology*, *18*(3), 326–332. https://doi.org/10.1038/73786.

Swainston, N., Batista-Navarro, R., Carbonell, P., Dobson, P. D., Dunstan, M., Jervis, A. J., et al. (2017). biochem4j: Integrated and extensible biochemical knowledge through graph databases. *PLoS One*, *12*(7), e0179130. https://doi.org/10.1371/journal.pone.0179130.

Taly, J. F., Magis, C., Bussotti, G., Chang, J.-M., Di Tommaso, P., Erb, I., et al. (2011). Using the T-Coffee package to build multiple sequence alignments of protein, RNA, DNA sequences and 3D structures. *Nature Protocols*, *6*(11), 1669–1682. https://doi.org/10.1038/nprot.2011.393.

Weber, T., Blin, K., Duddela, S., Krug, D., Kim, H. U., Bruccoleri, R., et al. (2015). antiSMASH 3.0—A comprehensive resource for the genome mining of biosynthetic gene clusters. *Nucleic Acids Research*, *43*(W1), W237–W243. https://doi.org/10.1093/nar/gkv437.

Yachdav, G., Wilzbach, S., Rauscher, B., Sheridan, R., Sillitoe, I., Procter, J., et al. (2016). MSAViewer: Interactive JavaScript visualization of multiple sequence alignments. *Bioinformatics*, *32*(22), 3501–3503. https://doi.org/10.1093/bioinformatics/btw474.

Yamanishi, Y., Hattori, M., Kotera, M., Goto, S., & Kanehisa, M. (2009). E-zyme: Predicting potential EC numbers from the chemical transformation pattern of substrate-product pairs. *Bioinformatics (Oxford, England)*, *25*(12), i179–i186. https://doi.org/10.1093/bioinformatics/btp223.

Yim, H., Haselbeck, R., Niu, W., Pujol-Baxley, C., Burgard, A., Boldt, J., et al. (2011). Metabolic engineering of Escherichia coli for direct production of 1,4-butanediol. *Nature Chemical Biology*, 7(7), 445–452. https://doi.org/10.1038/nchembio.580.

SECTION II

Enzyme Engineering Tools and Platforms

CHAPTER TWO

Cell-Free Synthetic Biology for Pathway Prototyping

Ashty S. Karim[*,†,‡], **Michael C. Jewett**[*,†,‡,§,¶,1]
[*]Department of Chemical and Biological Engineering, Northwestern University, Evanston, IL, United States
[†]Chemistry of Life Processes Institute, Northwestern University, Evanston, IL, United States
[‡]Center for Synthetic Biology, Northwestern University, Evanston, IL, United States
[§]Robert H. Lurie Comprehensive Cancer Center, Northwestern University, Chicago, IL, United States
[¶]Simpson Querrey Institute, Northwestern University, Chicago, IL, United States
[1]Corresponding author: e-mail address: m-jewett@northwestern.edu

Contents

Abstract

Engineering biological systems for the production of biofuels and bioproducts holds great potential to transform the bioeconomy, but often requires laborious, time-consuming design-build-test cycles. For decades cell-free systems have offered quick and facile approaches to study enzymes with hopes of informing cellular processes, mainly in the form of purified single-enzyme activity assays. Over the past 20 years, cell-free systems have grown to include multienzymatic systems, both purified and crude. By decoupling cellular growth objectives from enzyme pathway engineering objectives, cell-free systems provide a controllable environment to direct substrates toward a single, desired product. Cell-free approaches are being developed for prototyping and for biomanufacturing. In prototyping applications, the idea is to use

Methods in Enzymology, Volume 608
ISSN 0076-6879
https://doi.org/10.1016/bs.mie.2018.04.029

cell-free systems to test and optimize biosynthetic pathways before implementation in live cells and scale-up. We present a detailed method for the generation of crude lysates for cell-free pathway prototyping, mix-and-match cell-free metabolic engineering using preenriched lysates, and cell-free protein synthesis driven cell-free metabolic engineering. The cell-free synthetic biology methods described herein are generalizable to any biosynthetic pathway of interest and provide a powerful approach to building pathways in crude lysates for the purpose of prototyping. The foundational principle of the presented approach is that we can construct discrete metabolic pathways through modular assembly of cell-free lysates containing enzyme components produced by overexpression in the lysate chassis strain or by cell-free protein synthesis (in vitro production). Overall, the modular and cell-free nature of our pathway prototyping framework is poised to facilitate multiplexed, automated study of biosynthetic pathways to inform systems-level cellular design.

1. INTRODUCTION

Increased demands for energy, climate change concerns, and reliance on petrochemicals as the source of 95% of today's chemicals and materials have intensified the need for sustainable, low-cost biofuels and bioproducts production (Sheldon, 2007; Werpy et al., 2004). Microbial cell factories offer one of the most attractive approaches for addressing this need (Curran & Alper, 2012). However, long research and development timelines (10–100s of person years) motivate the need for new methods to accelerate the design and optimization of biological systems (Bornscheuer et al., 2012; Curran & Alper, 2012; Erickson, Nelson, & Winters, 2012; Fritz, Timmerman, Daringer, Leonard, & Jewett, 2010; Nielsen et al., 2014; Rollié, Mangold, & Sundmacher, 2012).

1.1 The State of Metabolic Engineering

Often, biologically produced small molecules are insufficient for production at commercially relevant titers, rates, or yields because natural sources are difficult to optimize and to scale. Thus, engineers seek to design enzymatic reaction schemes in model microorganisms to meet manufacturing criteria. Success in these endeavors depends upon identifying sets of enzymes that can convert readily available molecules (e.g., glucose) to high-value products (e.g., medicines), with each enzyme performing one of a series of chemical modifications. For example, introducing heterologous pathways into model microbes (such as baker's yeast and *Escherichia coli*) and engineering them to maximize biochemical production has led to large-scale production of

1,3-propanediol, farnesene, and artemisinin with many more on their way to market (Hodgman & Jewett, 2012; Nielsen et al., 2014). Unfortunately, this is difficult because design-build-test (DBT) cycles—iterations of reengineering organisms to test new sets of enzymes—can be detrimentally slow due to the constraints of cell growth (Nielsen & Keasling, 2016). As a result, a typical project today might only explore dozens of variants of an enzymatic reaction pathway. This is often insufficient to identify a commercially relevant solution because selecting productive enzymes using existing single-enzyme kinetic data has limited applicability in multienzyme pathways and consequently requires more DBT iterations. While techniques continue to develop to multiplex DBT cycles for rationally engineering cells (Smanski et al., 2014), in vitro systems show promise in speeding up DBT cycles because they bypass many in vivo limitations by having direct access to the cellular contents (Carlson, Gan, Hodgman, & Jewett, 2012; Hodgman & Jewett, 2012; Moore et al., 2018; Siegal-Gaskins, Tuza, Kim, Noireaux, & Murray, 2014; Sun, Yeung, Hayes, Noireaux, & Murray, 2014). In concert with a trend toward automation, in vitro systems could transform the way we engineer metabolic pathways.

1.2 Emerging Cell-Free Biotechnology

Cell-free systems complement traditional cellular systems. By decoupling cellular growth objectives from engineering enzyme utilization objectives, cell-free systems provide controllable and open environment to direct substrates toward a single, desired product (Dudley, Karim, & Jewett, 2015). Thus, cell-free synthetic biology methods are being developed for both pathway prototyping and for biomanufacturing. In prototyping applications, the idea is to use cell-free systems to test and optimize biosynthetic pathways before implementation in live cells and scale-up. Already, cell-free biosynthetic pathway building methods are already being used for pathway operation and debugging (Dudley et al., 2015). However, using cell-free systems to study metabolism and enzymatic pathways are still in its early stages. Nevertheless, cell-free systems provide advantages in controlling enzymes and the reaction environment. In biomanufacturing applications, the idea is to use cell-free systems for making the product itself. There is also potential to make products unavailable to cells due to toxicity limitations and to focus substrates to products with yields unattainable in cells (Korman, Opgenorth, & Bowie, 2017; Korman et al., 2014). Cell-free systems might also offer exciting new directions in the synthesis of hybrid biochemicals comprised of parts derived from organic syntheses and parts derived from biological syntheses (Swartz, 2012). The ability to not only use enzymes from multiple organisms but also unique metabolisms from across the phylogenic spectra is also on the horizon. As CFME emerges (Dudley et al., 2015; Guterl et al., 2012; Karim, Dudley, & Jewett, 2016; You & Zhang, 2013), two broad classes dominate in vitro small-molecule synthesis: purified enzymes and crude cell lysates (Fig. 1).

1.2.1 Purified Enzyme Systems

Purified enzyme approaches involve individual overexpression and purification of enzymes, which are then used as individual biocatalysts or recombined to assemble pathways of interest. The benefit of these systems is that the reaction network is explicitly defined, which gives exquisite control of reaction conditions and pathway fluxes. Indeed, the clarity of the biosynthetic pathway comes from eliminating unnecessary enzymes and cellular interferences (i.e., growth, other off-pathway metabolites). There are several examples of simple purified enzyme systems in industrial biocatalysis (Bruggink, Roos, & de Vroom, 1998; Jensen & Rugh, 1987). However, few industrial examples of synthetic enzymatic pathways exist, in part because of the high catalyst

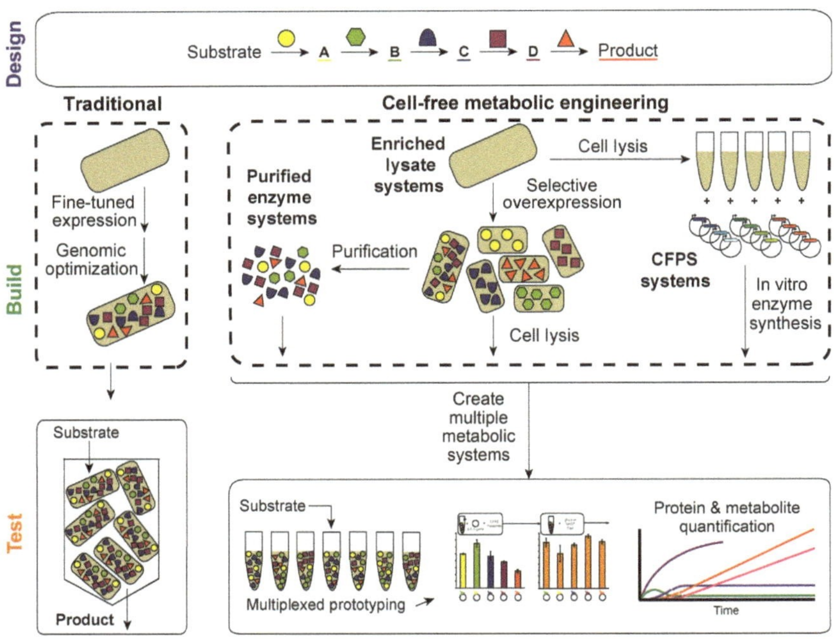

Fig. 1 Overview of cell-free synthetic biology methods for prototyping biosynthetic pathways.

and cofactor costs (as a result of enzyme purification and instability) and poor cofactor regeneration (Bujara & Panke, 2012). Despite these challenges, the majority of CFME research to date has utilized purified systems (Martin del Campo et al., 2013; Myung & Zhang, 2013; Toogood et al., 2015; Wang, Huang, Sathitsuksanoh, Zhu, & Zhang, 2011; Ye et al., 2009; Zhang, Evans, Mielenz, Hopkins, & Adams, 2007; Zhu, Kin Tam, Sun, You, & Percival Zhang, 2014). Recently, unique cofactor regeneration systems have been developed for purified systems that increase production capabilities and longevity of reactions (Opgenorth, Korman, & Bowie, 2014; Opgenorth, Korman, Iancu, & Bowie, 2017).

1.2.2 Crude Cell Lysate Systems

Crude extract-based systems rely on the ensemble of biocatalysts left after cell lysis. Upon centrifugation lipid membranes and genomic DNA are removed from cell lysates. Many groups have carried out extensive work in extract preparation and system optimization (Jewett, Calhoun, Voloshin, Wuu, & Swartz, 2008; Jewett & Swartz, 2004a, 2004b; Kwon & Jewett, 2015). There are several advantages to using crude lysates beyond their ease of preparation including lower system catalyst costs compared to purified counterparts, cofactor regeneration systems (Swartz, 2006, 2012), and the presence of native-like metabolism (Jewett et al., 2008; Jewett & Swartz, 2004a). Crude lysates thus allow for observations of metabolic interactions with biosynthetic pathways (Dudley, Anderson, & Jewett, 2016; Karim, Heggestad, Crowe, & Jewett, 2018; Kay & Jewett, 2015). There are a growing number of successes in using and characterizing crude lysate systems. A great example is the real-time monitoring and optimization of DHAP production (Bujara, Schumperli, Pellaux, Heinemann, & Panke, 2011; Hold, Billerbeck, & Panke, 2016). In addition, our group has shown that 2,3-butanediol (Kay & Jewett, 2015), mevalonate (Dudley et al., 2016), *n*-butanol (Karim et al., 2018; Karim & Jewett, 2016), and more complex products (Goering et al., 2016) can be constructed in crude lysates with high productivities (>g/L/h).

2. THE CELL-FREE METABOLIC ENGINEERING FRAMEWORK

The cell-free framework is a way to build pathways in the context of the DBT paradigm. The foundational principle of the cell-free metabolic engineering approach is that we can construct discrete metabolic pathways through modular assembly of cell-free lysates containing enzyme components produced by overexpression in the lysate chassis strain or by cell-free protein synthesis (CFPS; in vitro production) (Fig. 2). In addition, the open reaction environment allows for the supplementation of components such as cofactors and intermediates at any time during a cell-free reaction, which can be maintained at precise concentrations. In the Design phase, desired small-molecule products are selected, enzymes, homologs, and their stoichiometries are chosen, and likely beneficial conditions (i.e., substrates, cofactors, buffers, pH, and temperature) are selected. In the Build phase, the heart of this framework consists of assembling planned pathways for making the desired molecules from lysates, which can occur through two routes. One pathway construction route we term mix-and-match cell-free metabolic engineering (CFME) involves preparing multiple cell extracts for a selected pathway. One enzyme in the candidate pathway will be preenriched in each extract by overexpression in the host strain prior to lysis. Then, enriched extracts can be mixed in multiple, different ratios to build complete biosynthetic pathways. Another route termed CFPS-driven metabolic

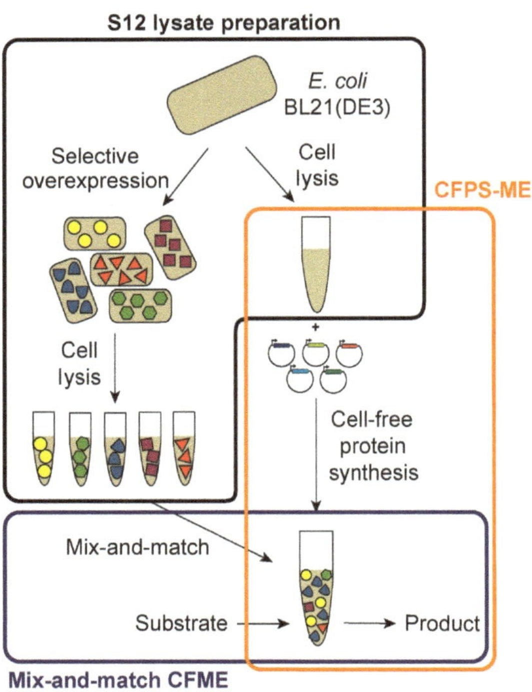

Fig. 2 Overview of methods for cell-free pathway prototyping. Lysate preparation, mix-and-match CFME, and CFPS-ME methods are outlined.

engineering (CFPS-ME) involves one-pot in vitro synthesis of biosynthetic enzymes and pathway operation. In this route, enzymes are made by CFPS in the extract postlysis by adding the DNA for each enzyme along with protein synthesis reagents followed by the addition of pathway operation reagents, faster than any previous approach (hours rather than days). Pathways can be built in 96- or 384-well plates, using liquid handling robotics, or in 1.5- or 2.0-mL microcentrifuge tubes (Karim et al., 2018). Significantly, this cell-free prototyping approach does not require the focus on flux balancing and delicate promoter tuning to maintain viability as is true for in vivo systems (Ajikumar et al., 2010; Alper & Stephanopoulos, 2007; Blazeck, Liu, Redden, & Alper, 2011; Du, Yuan, Si, Lian, & Zhao, 2012). Following the rapid and discrete construction of metabolic pathways, each pathway can be tested by using kinetically sampled batch reactions for fast DBT cycle times (Karim & Jewett, 2016).

3. S12 LYSATE PREPARATION FOR CELL-FREE METABOLIC ENGINEERING

The foundation of these cell-free prototyping strategies relies on crude lysates for pathway assembly. There are many ways to make *E. coli*-based crude lysates (Kigawa et al., 2004; Kim et al., 2006; Krinsky et al., 2016; Kwon & Jewett, 2015; Shrestha, Holland, & Bundy, 2012; Sun et al., 2013). In theory any of these methods would work to make lysates for pathway engineering, so our method focuses on one of the more high-throughput, high-yielding methods for crude lysate preparation (Kwon & Jewett, 2015). The following protocol is based on Kwon and Jewett's preparation with a few modifications for pathway prototyping strategies, also described in Karim and Jewett (2016).

3.1 Materials

3.1.1 Equipment

- 125 mL (small batch) and 4 L (large batch) baffled, Tunair shake flasks, or similar culture container; one per extract.
- 30°C and 37°C incubators with vigorous shaking at 250 rpm.
- Spectrophotometer, for cuvette-based OD_{600} measurements.
- Optically clear cuvettes for measuring optical density.
- Centrifuge capable of spinning 1-L centrifuge bottles at 5,000 × *g*, prechilled to 4°C.
- 1-L plastic centrifuge bottles and lids, prechilled to 4°C; one per extract.

- Table-top centrifuge capable of spinning 50-mL conical tubes at 12,000 × *g*, prechilled to 4°C.
- 50-mL conical tubes, prechilled to 4°C; one per extract.
- Weight scale to measure wet cell mass.
- Kim wipes.
- Bucket with ice.
- Beaker with ice water bath.
- Liquid nitrogen and dewar.
- −80°C freezer for storage.
- Q125 Sonicator (Qsonica, Newtown, CT), 3.175 mm diameter probe at frequency of 20 kHz.
- Heat block capable of heating to 90°C.
- XCell SureLock® Mini-Cell (Invitrogen) with power box.
- Gel Doc™ XR+ Gel Documentation System (BioRad).

3.1.2 Media

- Luria–Bertani (LB) broth: 1.0% (w/v) tryptone, 0.5% (w/v) yeast extract, 0.5% (w/v) NaCl, autoclaved.
- LB agar: add 1.5% (w/v) agar to LB prepared as above. Aliquot 30 mL of LB agar per 150 mm Petri dish into a plastic conical tube. Add appropriate antibiotics.
- 2 × YTPG broth: 1.6% (w/v) tryptone, 1.0% (w/v) yeast extract, 0.5% (w/v) NaCl, 1.8% (w/v) glucose, 0.7% (w/v) K_2HPO_4, 0.3% (w/v) KH_2PO_4, pH adjusted to 7.2 with 5N KOH, autoclaved. Add 40% (w/v) glucose, separately autoclaved, to broth prior use. Use 1 L per extract.

3.1.3 Media Supplements

- Carbenicillin (Carb) (100 μg/mL): To make a 1000 × stock, mix 1 g in 10 mL nanopure water, sterile filtered.
- Kanamycin (Kan) (50 μg/mL): To make a 1000 × stock, mix 0.5 g in 10 mL nanopure water, sterile filtered.
- Isopropyl-β-D-thiogalactopyranoside (IPTG) (0.1 m*M*): To make a 1000 × stock, mix 0.238 g in nanopure water to 10 mL, sterile filtered.

3.1.4 Bacterial Strains and Plasmids (See *Table 1*)

- *E. coli* DH5α (NEB).
- *E. coli* BL21(DE3) (NEB).

Table 1 A List of Strains and Plasmids Used to Carry Out Cell-Free Pathway Prototyping

Strain or Plasmid	Description	References
E. coli strains		
NEB Turbo™	F′ proA + B + lacIq Δ lacZM15/ fhuA2 Δ(lac-proAB) glnV galK16 galE15 R(zgb-210::Tn10)TetS endA1 thi-1 Δ(hsdS-mcrB)5—used for cloning purposes only	New England Biolabs
BL21 (DE3)	fhuA2 [lon] ompT gal (λ DE3) [dcm] Δ hsdS λ DE3 = λ s*Bam*HIo Δ *Eco*RI-B int::(lacI::PlacUV5::T7 gene1) i21 Δ nin5—used for enzyme overexpression and for lysate production	New England Biolabs
Plasmids		
pETBCS-rbsU-gene	Plasmid used for protein production and extract preparation	Karim et al. (2018) and Karim and Jewett (2016)
pJL1-gene	Plasmid used for CFPS	Karim et al. (2018), Karim and Jewett (2016), and Addgene #69496

- pETBCS-rbsU vector, used in previous studies (Karim et al., 2018; Karim & Jewett, 2016).
- pJL1 vector, used in previous studies (Karim et al., 2018; Karim & Jewett, 2016) (Addgene #69496).

3.1.5 Buffers and Reagents

- S30 buffer: 10 m*M* Tris–acetate (pH 8.2), 14 m*M* magnesium acetate, and 60 m*M* potassium glutamate, prechilled to 4°C.
- QuickStart™ Bradford Protein Assay Kit (BioRad Laboratories, Inc.).
- NuPAGE® LDS Sample Buffer (Invitrogen).
- 1 *M* Dithiothreitol (DTT).
- 4%–12% Bis-Tris Nu-PAGE gel (Invitrogen).
- 20 × NuPAGE MOPS SDS Running Buffer (Invitrogen).
- SeeBlue™ Plus2 Prestained Protein Standard (Thermo).
- SimplyBlue™ SafeStain (Thermo).

3.2 Procedure

3.2.1 Cell Preparation and Expression

Day 0 of procedure (about 16–18 h before starting Day 1):

1. Add 30 mL of LB media to a 125-mL baffled flask and sterilize. Add appropriate antibiotic if necessary following sterilization. Note: while you only need ~1 mL of culture per planned extract preparation for Day 1 of procedure, having additional liquid culture is suggested.
2. Inoculate media with desired *E. coli* strain from glycerol stock. Note: inoculating from a plate is also possible and should not make a substantial difference on lysate quality.
3. Incubate overnight at 37°C, shaking at 250 rpm.

Day 1 of procedure:

4. Add 1 L of 2 × YTPG media without antibiotic to an autoclaved 4-L Tunair flask. Note: we do not typically use antibiotics at this stage of the procedure. 200 mL of culture makes ~2 mL of extract.
5. Inoculate overnight culture (~1:100 dilution) into 2 × YTPG media so that the starting OD_{600} is within the range of 0.05–0.1. Note: the 1:100 dilution is not a strict guideline but undergoing several doublings before induction is ideal.
6. Incubate culture at 37°C, shaking at 250 rpm.
7. Check the OD_{600} about every 30 min. Note: there can often be a variable lag phase, so it is important to check the OD_{600} frequently. Use 2 × YTPG as a blank.
8. At OD_{600} of 0.5–0.8 (early exponential phase), induce recombinant protein overexpression by adding 0.1 m*M* (final concentration) IPTG. Note: IPTG concentrations can be tuned for optimal expression and other induction mechanisms can be used.
9. Incubate induced culture at 30°C, shaking at 250 rpm.
   - If enzyme overexpression is the objective, incubate in 30°C shaker for 4 h, or ~8 doublings, postinduction for protein expression. Note: this time it can be optimized to obtain the desired weight of cells and enzyme overexpression.
   - If cell-free protein synthesis is the objective, incubate in 30°C shaker to reach OD_{600} of 3. Check OD_{600} every ~30 min.
10. Following growth, immediately take the shake flask containing the culture out of the incubator and pour the contents into the prechilled 1-L centrifuge bottles on ice. Note: keeping cultures and subsequent handlings of the cultures on ice is imperative to maintain the quality of lysates produced.

11. Centrifuge samples in prechilled centrifuge bottles for 15 min at 5000 × *g* at 4°C.
12. Wipe the inside walls of the centrifuge bottle with large Kim wipes to remove as much of the leftover supernatant as possible before inverting the bottle back upright. Note: removing as much supernatant at each centrifugation steps reduces possible dilution of the extract.
13. Transfer cell pellet from centrifuge bottle to a prechilled 50-mL conical tube, using a spatula.
14. To obtain the remaining cells from the spatula and centrifuge bottle, add 2 mL of S30 buffer in the bottle and use the spatula to mix the cells into solution. Pipette the cell/buffer mixture into the 50-mL conical tube. Proceed with your remaining cell samples.
15. Place the 50-mL conical tube containing your cell sample submerged in ice.
16. Resuspend in 25 mL S30 buffer. Alternate between vortexing using a medium-high setting for 15 s and resting on ice for 15 s. Note: this can take 5–15 min or longer.
17. Once resuspended, centrifuge samples in 50-mL conical tubes at 5,000 × *g* for 10 min at 4°C to pellet cells.
18. Pour off supernatant into waste. Again, wipe the inside walls of the centrifuge bottle with Kim wipes to remove as much of the leftover supernatant as possible before inverting the bottle back upright.
19. Repeat steps 15–18.
20. Repeat step 16.
21. Once resuspended, centrifuge samples in 50-mL conical tubes at 7000 × *g* for 10 min at 4°C to pellet cells. Note: the change in centrifugation speed is to keep the cells somewhat tight to remove residual liquid but somewhat loose before freeze–thaw to increase ease of resuspension.
22. Completely remove supernatant by pouring into waste. Before inverting your tubes back upright, clean the residual supernatant on inside of tubes as much as possible using a Kim wipe.
23. Use a clean pipette tip to split the cell pellet in half to increase the surface area of the pellet. Note: this will increase the speed of pellet resuspension when making extract.
24. Measure pellet mass on weight scale and record weight on tube.
25. Flash freeze the pellets in liquid nitrogen and store at −80°C. Note: flash freezing in liquid nitrogen helps maintain the quality of cells before extract preparation compared to just placing pellets in −80°C storage.

3.2.2 Extract Preparation

1. Take out cell pellets from −80°C freezer. Let pellets thaw on ice (~1–2 h).
2. Resuspend pellets in 1 mL S30 buffer per g of weighed pellet via vortex. It is important that the cells are kept on ice. Vortex as necessary, but do not hold off ice for more than 15 s at a time. Note: this can take 15 min or more and the resuspension volume may be optimized for other strains.
3. Let samples rest on ice for suspension to settle.
4. Prepare a glass beaker with an ice water bath (90% ice, 10% water).
5. Transfer 1.4 mL of cell suspension into 1.5-mL microcentrifuge tubes. Note: avoid transferring bubbles to 1.5-mL microcentrifuge tubes.
6. Place microcentrifuge tube containing cell suspension in ice bath.
7. Turn on the Q125 sonicator with 3.175 mm diameter probe at frequency of 20 kHz. Note that this protocol is optimized for the Q125 sonicator and might need to be adjusted for other sonicator models and probe diameters.
8. Place the sonicator tip into the cell suspension just below center of the microcentrifuge tube. Note: tip should not rest against the sides of the tube. Moving tip around the cell suspension can improve energy transfer.
9. Set sonicator to 50% amplitude with pulsing at 10 s on, 10 s off. Note: this amplitude and pulsing time may be optimized for other strains.
10. Sonicate the 1.4-mL cell suspension with 820 J of sonication energy according to step 11 settings. Note: for a 1-mL suspension, the energy should be 530 J. For volumes between 1 and 1.4 mL, use linear interpolation to calculate sonication energy required. Refer to Kwon and Jewett for more information on sonication energies (Kwon & Jewett, 2015).
11. Different bacterial strains may require a preincubation step. Refer to Kwon and Jewett for more information on preincubation (Kwon & Jewett, 2015), noting that experiments have shown that variability at this step can lead to significant differences in extract performance. Note: for our previous pathway prototyping studies, preincubation was not needed.
12. Clean sonicator according to step 9 before moving to next sample. Repeat steps 9–13 for each sample.
13. Centrifuge samples at 12,000 × *g* for 10 min at 4°C.
14. Aliquot ~500–820 μL (if started with 1.4 mL extract mixture), top layer only, into a new tube. All samples from the original cell suspension should be pooled into the same new tube. Note: there may be three layers. Collect top layer only. In addition, the pellet after the first centrifugation tends to be loose. Therefore, immediately transfer

the supernatant of the first centrifugation (typically with a pipette) into another tube to avoid carrying over layers. Experiments have shown that multiple centrifugations yield more active extracts as compared to extracts performed with one centrifugation.

15. Pipet mix thoroughly but gently. Aliquot 100–200 μL each (makes ~20 aliquots) on ice in new 1.5-mL or 0.6-mL microcentrifuge tubes.
16. Flash freeze on liquid nitrogen and store at −80°C.

3.2.3 Extract Quantification of Total Protein by Bradford Assay

1. Remove the 1 × dye reagent (QuickStart™ Bradford Protein Assay Kit) and equilibrate to room temperature before use.
2. Refer to Table 2 for preparing the protein standards (use nanopure H_2O to dilute) Note: we use the bovine serum albumin standard that comes with the QuickStart™ Bradford Protein Assay Kit.

Table 2 Bradford Assay Sample Setup

Tube #	Source Volume (μL)	Source	Water Volume (μL)	[Protein] (μg/mL)
Standards				
1	10	2 mg/mL BSA stock	790	25
2	10	2 mg/mL BSA stock	990	20
3	6	2 mg/mL BSA stock	794	15
4	500	Tube 2	500	10
5	500	Tube 4	500	5
6	500	Tube 5	500	2.5
7	500	Tube 6	500	1.25
8	—	—	500	0
Extracts				
Dilute extract	10	Extract to measure	990	
E4x	6	Dilute extract	594	
E7x	8	Dilute extract	552	
E10x	16	Dilute extract	624	

The following is the experimental setup of samples for extract quantification.
This is adapted from the QuickStart™ Bradford Protein Assay Kit (BioRad Laboratories, Inc.) instruction manual.

3. Dilute extract samples (assuming extract concentration is ~40 mg/mL) by adding 10 μL extract to 990 μL water. This will be the "dilute extract" used to make further dilutions (See Table 2).
   - Make a 10,000× dilution by adding 6 μL "dilute extract" into 594 μL water.
   - Make a 7000× dilution by adding 8 μL "dilute extract" into 552 μL water.
   - Make a 4000× dilution by adding 16 μL "dilute extract" into 624 μL water.
4. Pipette 140 μL each standard and extract dilutions into separate microplate wells (refer to Fig. 3).
5. Use a multichannel pipette to add 140 μL 1× dye reagent to each well and carefully mix. Note: avoid bubbles as they can drastically alter Bradford assay readings.
6. Incubate at room temperature for ~5 min. Samples should not be incubated longer than 1 h at room temperature.
7. Measure the absorbance of the standards and extract samples at 595 nm on a plate reader. Note: the linear range of these assays for BSA is 1.25–10 μg/mL.
8. Create a standard curve of the BSA standards measured and use this curve to calculate the protein content of each extraction dilution. Multiple each dilution to get a 1× concentration of the measurement. Then, average the values together to get an extract total protein

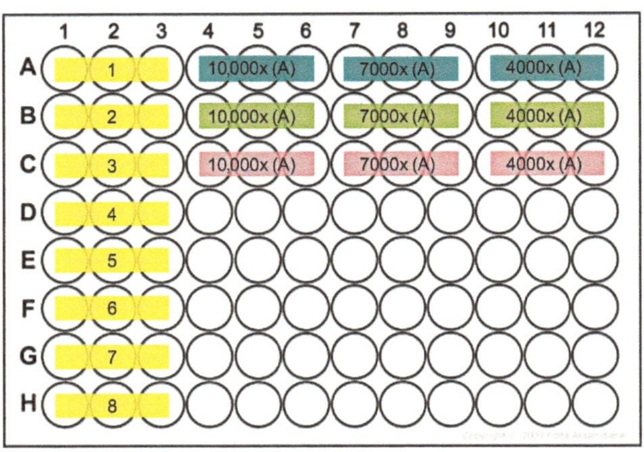

Fig. 3 Layout of Bradford assay plate. *Yellow* (columns 1–3) are the standards. *Blue*, *green*, and *purple* (rows 1, 2, and 3; columns 4–12) are three separate extracts with triplicates for each of the three dilutions: 10,000×, 7000×, and 4000×.

concentration. Note: total extract concentrations are typically between 35 and 45 mg total *E. coli* protein/mL. Refer to QuickStart™ Bradford Protein Assay Kit instructions.

3.2.4 Overexpressed Protein Quantification by Densitometry

1. Add ~2 μL of extract to 3 μL NuPAGE® LDS Sample Buffer (4×), 0.6 μL 1 *M* DTT, and water to make a total of 12 μL sample per extract being tested. Note: it is important for overexpressed protein quantification that a control sample of *E. coli* extract with no protein overexpressed is also made.
2. Heat each sample at 90°C for 10 min in a heat block.
3. Load 10 μL of each protein samples on a 4%–12% Bis-Tris Nu-PAGE gel with 1× MOPS buffer and protein standards (SeeBlue plus2 ladder). Note: other protein ladders can be used in place of the SeeBlue plus2 ladder.
4. Run gel at a constant 120 V for 105 min.
5. After the run, open gel container, and place gel in staining container.
6. Rinse 2× and shake in nanopure water for 5 min, then drain.
7. Stain with ~100 mL SimplyBlue™ SafeStain for 1 h on shaker. Note: any Coomassie stain can be used in place of SimplyBlue™ SafeStain.
8. Rinse 2× and shake in water for 2 h or overnight.
9. Image on white background with Gel Doc™ XR+ Gel Documentation System using a Coomassie stain filter.
10. Once imaged the gel can be used to loosely quantify what percentage of protein in the extract is the overexpressed protein of interest using ImageJ software and documentation (http://rsb.info.nih.gov/ij/index.html). See Fig. 4 for details.

4. MIX-AND-MATCH CELL-FREE METABOLIC ENGINEERING

One approach to building pathways is to construct them by mixing lysates together that separately contain heterologous enzymes catalyzing the chemical reactions to get to a product of interest. Rather than needing to engineer one organism to contain all enzymes of a pathway, we can overexpress single enzymes in chassis strains which can then be lysed and mixed to assemble the pathway. Provided are methods for biosynthetic reaction assembly.

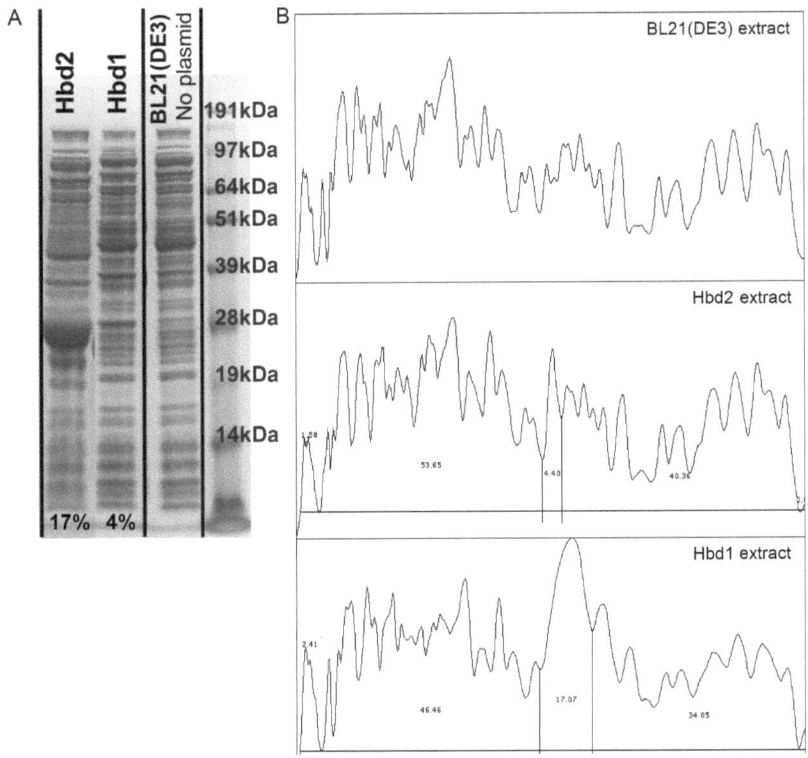

Fig. 4 An example of semiquantification overexpressed proteins in *E. coli* extracts. (A) An SDS-PAGE gel is shown for three extracts: BL21 (DE3) containing no plasmid, containing a plasmid expressing Hbd1, and containing a plasmid expressing Hbd2. The percent of overexpressed protein relative to total protein content is listed at the bottom of each lane. (B) ImageJ analysis described is demonstrated here. This analysis is performed on the SDS-PAGE gel in A.

4.1 Materials

4.1.1 Equipment

- 1.5-mL Eppendorf tubes
- 37°C incubator

4.1.2 Buffers and Reagents

- *E. coli* extracts individually expressing enzymes to assemble a biosynthetic pathway.
- 15 × Salt Solution: magnesium glutamate (150 m*M*), ammonium glutamate (150 m*M*), and potassium glutamate (2010 m*M*).
- Glucose (2.2 *M*).

- Dipotassium phosphate (1 *M*, pH 7.2).
- Bis-Tris (2 *M*).
- NAD (100 m*M*).
- ATP (100 m*M*).
- CoA (50 m*M*).
- DNAse/RNAse-free water.
- 10% (w/v) trichloroacetic acid (TCA), sterile filtered.

4.2 Procedure

4.2.1 Mix-and-Match Biosynthesis Reactions

1. Let reagents thaw on ice. Do not hand-thaw lysates.
2. For a first test of a pathway, label 21- to 1.5-mL Eppendorf tubes as "reagent mix," "extract mix," "reaction mix," and 0, 3, 6, 9, 18, and 24 h time points in triplicate. It is important to do a time course for new pathways.
3. Keep all tubes on ice.
4. Assemble the "reagent mix" by combining magnesium glutamate (8 m*M*), ammonium glutamate (10 m*M*), potassium glutamate (134 m*M*), glucose (200 m*M*), dipotassium phosphate (10 m*M*, pH 7.2), Bis-Tris (100 m*M*), NAD (1 m*M*), ATP (1 m*M*), and CoA (1 m*M*). It is helpful to make this mix at 1.2 × the volume needed for all reaction tubes (18 tubes, 25 μL in this example). Mix thoroughly by vortex and keep tube on ice.
5. Assemble the "extract mix" by combining each lysate preenriched with heterologous enzymes for the given pathway (in a five-enzyme pathway this would involve mixing five lysates). Start with each lysate at a final concentration of 2 mg/mL based on lysate quantification results. It is helpful to have this mix at 1.2 × the volume needed for all reaction tubes. Gently pipette mix and keep on ice.
6. To create the "reaction mix" combine the "extract mix" and "reagent mix" the volume in here should be at 1.1 × the volume needed to pipette each reaction tube at 25 μL each. Keep on ice.
7. Pipette mix and transfer 25 μL of "reaction mix" into each labeled reaction tube.
8. Immediately pull the 0 h time point (all replicates) and quench those reactions with 10% (w/v) trichloroacetic acid in a 1:1 ratio.
9. Incubate all other time points at 37°C.
10. Pull each reaction with replicates at time points. Terminate reactions by adding 10% (w/v) trichloroacetic acid in a 1:1 ratio.

11. Precipitate proteins by pelleting through centrifugation at 15,000 × *g* for 10 min.
12. The supernatant was stored at −80°C until analysis.

4.2.2 Biosynthesis Analysis

Metabolites can be quantified with current chromatography and mass spectroscopy techniques. In addition, chemical and enzymatic plate-based assays can be used when available (Dudley et al., 2016; Karim et al., 2018; Karim & Jewett, 2016; Kay & Jewett, 2015).

5. CELL-FREE PROTEIN SYNTHESIS-DRIVEN METABOLIC ENGINEERING

Using cell-free protein synthesis to enrich lysates with different enzymes for combinatorial assembly of different pathways enables parallelized pathway construction of combinatorial designs to accelerate DBT cycles.

5.1 Materials

5.1.1 Equipment

- 1.5-mL Eppendorf tubes.
- 37°C incubator.
- Microbeta scintillation detection instrument, PerkinElmer.
- Typhoon 7000 Imager (GE Healthcare Life Sciences, Pittsburgh, PA).
- Waterman chromatography paper.
- 12-in square cellophane sheets.
- Autoradiography cassettes.

5.1.2 Buffers and Reagents

- *E. coli* BL21 Star (DE3) extracts with or without individually expressing enzymes as part of a biosynthetic pathway.
- DNA encoding each enzyme in the biosynthetic pathway.
- 15 × Salt Solution: magnesium glutamate (150 m*M*), ammonium glutamate (150 m*M*), and potassium glutamate (2010 m*M*).
- 15 × Nucleotide master mix: ATP (18 m*M*), GTP (12.75 m*M*), UTP (12.75 m*M*), CTP (12.75 m*M*), folinic acid (0.51 mg/mL), and 2.559 mg/mL tRNA.
- Amino acid solution containing all 20 amino acids at 50 m*M*.
- Phosphoenolpyruvate (PEP) at 1 *M*.

- Putrescine (250 m*M*).
- Spermidine (250 m*M*).
- HEPES Buffer (1 *M*, pH 7.2).
- Glucose (2.2 *M*)
- Dipotassium phosphate (1 *M*, pH 7.2)
- NAD (100 m*M*)
- ATP (100 m*M*)
- CoA (50 m*M*)
- DNAse/RNAse-free water
- ^{14}C-leucine (10 μ*M*)
- 0.1 *M* sodium hydroxide.
- 5% (w/v) TCA.
- 10% (w/v) TCA, sterile filtered.
- Cytoscint™-ES liquid scintillation cocktail.
- Split-GFP Buffer: 50 m*M* Tris pH 7.4, 0.1 *M* NaCl, 10% glycerol (TNG buffer) (Cabantous, Terwilliger, & Waldo, 2005).

5.2 Procedure

CFPS reactions were performed to express enzymes involved in *n*-butanol production prior to starting the CFME portion of the reactions using a modified PANOx-SP system (Jewett & Swartz, 2004a).

5.2.1 CFPS-ME Reactions

1. Let reagents thaw on ice. Do not hand-thaw lysates.
2. For a first test of a pathway, label 21 1.5-mL Eppendorf tubes as "reagent mix," "extract mix," "reaction mix," and 0, 3, 6, 9, 18, and 24 h time points in triplicate. It is important to do a time course for new pathways.
3. Keep all tubes on ice.
4. Assemble the "CFPS Master Mix" by combining 15 × Salt Solution (magnesium glutamate (8 m*M*), ammonium glutamate (10 m*M*), potassium glutamate (134 m*M*)), 15 × nucleotide mix (ATP (1.2 m*M*); GTP, UTP, and CTP (0.85 m*M* each); folinic acid (34.0 μg/mL); *E. coli* tRNA mixture (170.0 μg/mL)), 20 standard amino acids (2 m*M* each), NAD (0.33 mm), CoA (0.27 m*M*), spermidine (1.5 m*M*), putrescine (1 m*M*), and PEP (33 m*M*). It is helpful to make this mix at 1.2 × the volume needed for all reaction tubes (18 tubes, 22.5 μL in this example). Mix thoroughly by vortex and keep tube on ice.

5. Assemble the "extract mix" by using plain *E. coli* extract with no enzyme enrichment or by combining each lysate preenriched with heterologous enzymes for the given pathway (in a five-enzyme pathway this would involve mixing five lysates). Start with each lysate at a final concentration of 2 mg/mL based on lysate quantification results. The total lysate concentration should be 10 mg/mL. It is helpful to have this mix at 1.2 × the volume needed for all reaction tubes. Gently pipette mix and keep on ice.
6. To create the "reaction mix" combine the "extract mix," "reagent mix," and plasmid DNA encoding each enzyme (~13.3 ng/mL each). The volume in here should be at 1.1 × the volume needed to pipette each reaction tube at 22.5 μL each. Keep on ice.
7. Pipette mix and transfer 22.5 μL of "reaction mix" into each labeled reaction tube. Note: it is important that the reaction components are evenly mixed. Cell-free protein synthesis has been plagued by variable results. One cause for this variability in our laboratory has been the lack of complete mixing of the reaction components. Specifically, this can be a problem when setting up numerous reactions from the same premix. Air bubbles, especially in the extract can also be an issue. When reactions are well controlled, variability is low.
8. Incubate all reaction tubes at 30°C.
9. Assemble the "CFME Master Mix" by combining glucose (200 m*M*), NAD (0.67 m*M*), and CoA (0.73 m*M*). Mix thoroughly by vortex and keep tube on ice.
10. After 3 h, spike in 2.5 μL "CFME Master Mix" to each reaction tube to initiate biosynthetic activity.
11. Immediately pull the "0 h" time point (all replicates) and quench those reactions with 10% (w/v) TCA in a 1:1 ratio.
12. Incubate all other time points at 30°C.
13. Pull each reaction with replicates at time points. Terminate reactions by adding 10% (w/v) TCA in a 1:1 ratio.

5.2.2 Quantification of Protein Produced In Vitro via Radioactive Incorporation

1. Using the procedure for CFPS-ME Reactions in Section 5.2.1, the "CFPS Master Mix" can be made with radioactive ^{14}C-leucine (10 μ*M*) at step 4.
2. Samples are quenched with 100 μL of 0.1 *M* sodium hydroxide and incubated at 37°C for 20 min.

3. Quenched samples are then split in half ("washed" vs "unwashed") and pipetted onto 0.25-in by 1-in Waterman paper strips. The strips are then dried.
4. The "washed" half of samples are then washed three times with 5% (w/v) TCA to precipitate radioactive protein samples.
5. The "washed" half of samples are then washed with molecular biology grade ethanol.
6. Radioactivity of TCA-precipitated samples and "unwashed" were measured by liquid scintillation counting to then quantify the protein produced as previously reported (MicroBeta2; PerkinElmer)(Jewett et al., 2008; Jewett & Swartz, 2004a).
7. These reactions were also run on a Coomassie-stained SDS-PAGE gel as done in Section 3.2.4 steps 1–11.
8. The gels can then be dried between 2- and 12-in square cellophane sheets overnight.
9. The dried gel can now be exposed by autoradiography for 3 days.
10. Autoradiographs are imaged with a Typhoon 7000 Imager (GE Healthcare Life Sciences, Pittsburgh, PA).
11. Multiple proteins produced in vitro were further quantified by gel image intensity comparisons using ImageJ (NIH) similar to Section 3.2.4 (Fig. 5A).

5.2.3 Quantification of Protein Produced In Vitro via Split-GFP Construct

If radioactive ^{14}C-leucine is unavailable, other quantification methods can be used. Here, we present a method for quantification by split-GFP fluorescence. By adding a 20-amino acid "GFP11" tag onto the end of each protein and expressing the corresponding "GFP1–10" protein, the association of the two will elicit a quantifiable fluorescent signal.

1. DNA constructs encoding enzymes to prototype can be designed to include the following encoded amino acid sequence directly at the end of the coding sequence before the stop codon "DGGSGGGSTSR DHMVLHEYV."
2. Also, a DNA construct pJL1-split-GFP encoding the other portion of GFP can be made "MGGTSMSKGEELFTGVVPILVELDGDVNGH KFSVRGEGEGDATIGKLTLKFICTTGKLPVPWPTLVTTLTYGV QCFSRYPDHMKRHDFFKSAMPEGYVQERTISFKDDGKYKTR AVVKFEGDTLVNRIELKGTDFKEDGNILGHKLEYNFNSHNVY ITADKQKNGIKANFTVRHNVEDGSVQLADHYQQNTPIGDGP VLLPDNHYLSTQTVLSKDPNEK" (Cabantous et al., 2005).

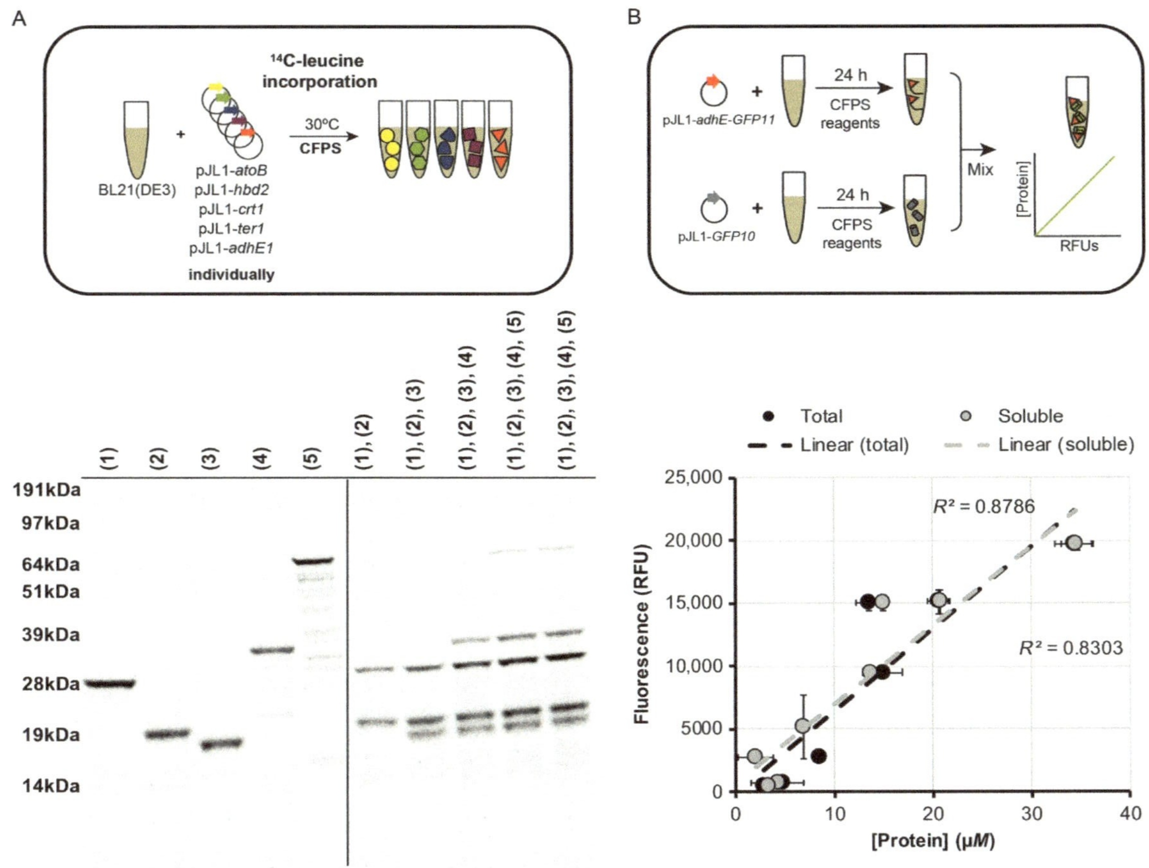

Fig. 5 Quantification of cell-free protein synthesis examples. (A) Data are representative of quantification by radioactivity. SDS-PAGE autoradiogram shows each protein produced by cell-free protein synthesis. Each column is a separate cell-free reaction producing one of five proteins or a combination of the five proteins. (B) Data are representative of quantification by splitGFP reporter. Each *circle* represents a concentration of protein measured by radioactive ^{14}C-leucine incorporation and the corresponding fluorescent readout.

3. Perform CFPS reactions of pJL1-gene-gfp according to Section 5.2.1, steps 1–9 (CFPS RXN 1).
4. Separately, perform CFPS reaction of pJL1-split-GFP (CFPS RXN 2).
5. Mix 10 μL of CFPS RXN 1, 5 μL of CFPS RXN 2, and 5 μL Split-GFP Buffer.
6. Measure fluorescence over a 20-h period (Fig. 5B).

5.2.4 Metabolite Quantification

Metabolites can be quantified with current chromatography and mass spectroscopy techniques (Goering et al., 2016; Karim et al., 2018). In addition, chemical and enzymatic plate-based assays can be used when available through commercially available kits or documented in the literature.

6. SUMMARY AND CONCLUSIONS

Cell-free synthetic biology provides powerful tools to prototype biosynthetic pathways, providing an unprecedented capability to test hundreds to thousands of pathways by avoiding inherent limitations of cell growth. The methods provided here describe multiple ways of constructing cell-free enzymatic pathways to iterate through DBT cycles at speeds 10 × faster than traditional approaches. Coupling cell-free protein synthesis, in particular, to the construction of a metabolic pathway in tandem with high-end metabolomics will offer a high degree of flexibility to model the kinetics and stability of individual enzymes, measure metabolite fluxes in multistep pathways, and experimentally isolate many other parameters confounded in living organisms. Overall, the modular and cell-free nature of our framework is poised to facilitate multiplexed, automated study of biosynthetic pathways to inform systems-level cellular design.

ACKNOWLEDGMENTS

We would like to thank Quentin M. Dudley for conversations that have led to development of these methods. In addition, we thank the Department of Energy (BER Grant: DE-SC0018249), the Air Force Research Laboratory Center of Excellence Grant (FA8650-15-2-5518), the David and Lucile Packard Foundation (2011-37152), and the Dreyfus Teacher-Scholar Program for funding and support. A.S.K. is an NSF Graduate Research Fellow. The U.S. Government is authorized to reproduce and distribute reprints for Governmental purposes notwithstanding any copyright notation thereon. The views and conclusions contained herein are those of the authors and should not be interpreted as necessarily representing the official policies or endorsements, either expressed or implied, of Air Force Research Laboratory or the U.S. Government.

REFERENCES

Ajikumar, P. K., Xiao, W., Tyo, K. E. J., Wang, Y., Simeon, F., Leonard, E., et al. (2010). Isoprenoid pathway optimization for taxol precursor overproduction in *Escherichia coli*. *Science*, *330*(6000), 70–74.

Alper, H. S., & Stephanopoulos, G. (2007). Global transcription machinery engineering: A new approach for improving cellular phenotype. *Metabolic Engineering*, *9*(3), 258–267.

Blazeck, J., Liu, L., Redden, H., & Alper, H. S. (2011). Tuning gene expression in *Yarrowia lipolytica* by a hybrid promoter approach. *Applied and Environmental Microbiology*, 77(22), 7905–7914.

Bornscheuer, U. T., Huisman, G. W., Kazlauskas, R. J., Lutz, S., Moore, J. C., & Robins, K. (2012). Engineering the third wave of biocatalysis. *Nature*, *485*(7397), 185–194. https://doi.org/10.1038/nature11117.

Bruggink, A., Roos, E. C., & de Vroom, E. (1998). Penicillin Acylase in the industrial production of β-lactam antibiotics. *Organic Process Research & Development*, *2*(2), 128–133. https://doi.org/10.1021/op9700643.

Bujara, M., & Panke, S. (2012). *In silico* assessment of cell-free systems. *Biotechnology and Bioengineering*, *109*(10), 2620–2629. https://doi.org/10.1002/bit.24534.

Bujara, M., Schumperli, M., Pellaux, R., Heinemann, M., & Panke, S. (2011). Optimization of a blueprint for *in vitro* glycolysis by metabolic real-time analysis. *Nature Chemical Biology*, 7(5), 271–277. https://doi.org/10.1038/nchembio.541.

Cabantous, S., Terwilliger, T. C., & Waldo, G. S. (2005). Protein tagging and detection with engineered self-assembling fragments of green fluorescent protein. *Nature Biotechnology*, *23*(1), 102–107. https://doi.org/10.1038/nbt1044.

Carlson, E. D., Gan, R., Hodgman, C. E., & Jewett, M. C. (2012). Cell-free protein synthesis: Applications come of age. *Biotechnology Advances*, *30*(5), 1185–1194. https://doi.org/10.1016/j.biotechadv.2011.09.016.

Curran, K. A., & Alper, H. S. (2012). Expanding the chemical palate of cells by combining systems biology and metabolic engineering. *Metabolic Engineering*, *14*(4), 289–297. https://doi.org/10.1016/j.ymben.2012.04.006.

Du, J., Yuan, Y., Si, T., Lian, J., & Zhao, H. (2012). Customized optimization of metabolic pathways by combinatorial transcriptional engineering. *Nucleic Acids Research*, *40*(18), e142.

Dudley, Q. M., Anderson, K. C., & Jewett, M. C. (2016). Cell-free mixing of *Escherichia coli* crude extracts to prototype and rationally engineer high-titer mevalonate synthesis. *ACS Synthetic Biology*, *5*(12), 1578–1588. https://doi.org/10.1021/acssynbio.6b00154.

Dudley, Q. M., Karim, A. S., & Jewett, M. C. (2015). Cell-free metabolic engineering: Biomanufacturing beyond the cell. *Biotechnology Journal*, *10*(1), 69–82. https://doi.org/10.1002/biot.201400330.

Erickson, B., Nelson, J. E., & Winters, P. (2012). Perspective on opportunities in industrial biotechnology in renewable chemicals. *Biotechnology Journal*, 7(2), 176–185. https://doi.org/10.1002/biot.201100069.

Fritz, B. R., Timmerman, L. E., Daringer, N. M., Leonard, J. N., & Jewett, M. C. (2010). Biology by design: From top to bottom and back. *Journal of Biomedicine & Biotechnology*, *2010*, 232016. https://doi.org/10.1155/2010/232016.

Goering, A. W., Li, J., McClure, R. A., Thomson, R. J., Jewett, M. C., & Kelleher, N. L. (2016). In vitro reconstruction of nonribosomal peptide biosynthesis directly from DNA using cell-free protein synthesis. *ACS Synthetic Biology*. https://doi.org/10.1021/acssynbio.6b00160.

Guterl, J. K., Garbe, D., Carsten, J., Steffler, F., Sommer, B., Reisse, S., et al. (2012). Cell-free metabolic engineering: Production of chemicals by minimized reaction cascades. *ChemSusChem*, *5*(11), 2165–2172. https://doi.org/10.1002/cssc.201200365.

Hodgman, C. E., & Jewett, M. C. (2012). Cell-free synthetic biology: Thinking outside the cell. *Metabolic Engineering*, *14*(3), 261–269. https://doi.org/10.1016/j.ymben.2011.09.002.

Hold, C., Billerbeck, S., & Panke, S. (2016). Forward design of a complex enzyme cascade reaction. *Nature Communications*, *7*, 12971. https://doi.org/10.1038/ncomms12971.

Jensen, V. J., & Rugh, S. (1987). Industrial-scale production and application of immobilized glucose isomerase. *Methods in Enzymology*, *136*, 356–370. https://doi.org/10.1016/s0076-6879(87)36035-5.

Jewett, M. C., Calhoun, K. A., Voloshin, A., Wuu, J. J., & Swartz, J. R. (2008). An integrated cell-free metabolic platform for protein production and synthetic biology. *Molecular Systems Biology*, *4*, 220. https://doi.org/10.1038/msb.2008.57.

Jewett, M. C., & Swartz, J. R. (2004a). Mimicking the *Escherichia coli* cytoplasmic environment activates long-lived and efficient cell-free protein synthesis. *Biotechnology and Bioengineering*, *86*(1), 19–26. https://doi.org/10.1002/bit.20026.

Jewett, M. C., & Swartz, J. R. (2004b). Substrate replenishment extends protein synthesis with an *in vitro* translation system designed to mimic the cytoplasm. *Biotechnology and Bioengineering*, *87*(4), 465–471. https://doi.org/10.1002/bit.20139.

Karim, A. S., Dudley, Q. M., & Jewett, M. C. (2016). Cell-free synthetic systems for metabolic engineering and biosynthetic pathway prototyping. In C. Wittmann & J. C. Liao (Eds.), *Industrial biotechnology: Microorganisms, Vol. 1*. (pp. 125–148). Wiley-VCH Verlag.

Karim, A. S., Heggestad, J. T., Crowe, S. A., & Jewett, M. C. (2018). Controlling cell-free metabolism through physiochemical perturbations. *Metabolic Engineering*, *45*, 86–94. https://doi.org/10.1016/j.ymben.2017.11.005.

Karim, A. S., & Jewett, M. C. (2016). A cell-free framework for rapid biosynthetic pathway prototyping and enzyme discovery. *Metabolic Engineering*, *36*, 116–126. https://doi.org/10.1016/j.ymben.2016.03.002.

Kay, J. E., & Jewett, M. C. (2015). Lysate of engineered *Escherichia coli* supports high-level conversion of glucose to 2,3-butanediol. *Metabolic Engineering*, *32*, 133–142. https://doi.org/10.1016/j.ymben.2015.09.015.

Kigawa, T., Yabuki, T., Matsuda, N., Matsuda, T., Nakajima, R., Tanaka, A., et al. (2004). Preparation of *Escherichia coli* cell extract for highly productive cell-free protein expression. *Journal of Structural and Functional Genomics*, *5*(1–2), 63–68. https://doi.org/10.1023/B:JSFG.0000029204.57846.7d.

Kim, T. W., Keum, J. W., Oh, I. S., Choi, C. Y., Park, C. G., & Kim, D. M. (2006). Simple procedures for the construction of a robust and cost-effective cell-free protein synthesis system. *Journal of Biotechnology*, *126*(4), 554–561. https://doi.org/10.1016/j.jbiotec.2006.05.014.

Korman, T. P., Opgenorth, P. H., & Bowie, J. U. (2017). A synthetic biochemistry platform for cell free production of monoterpenes from glucose. *Nature Communications*, *8*, 15526. https://doi.org/10.1038/ncomms15526.

Korman, T. P., Sahachartsiri, B., Li, D., Vinokur, J. M., Eisenberg, D., & Bowie, J. U. (2014). A synthetic biochemistry system for the in vitro production of isoprene from glycolysis intermediates. *Protein Science*, *23*(5), 576–585. https://doi.org/10.1002/pro.2436.

Krinsky, N., Kaduri, M., Shainsky-Roitman, J., Goldfeder, M., Ivanir, E., Benhar, I., et al. (2016). A simple and rapid method for preparing a cell-free bacterial lysate for protein synthesis. *PLoS One*, *11*(10), e0165137. https://doi.org/10.1371/journal.pone.0165137.

Kwon, Y. C., & Jewett, M. C. (2015). High-throughput preparation methods of crude extract for robust cell-free protein synthesis. *Scientific Reports*, *5*, 8663. https://doi.org/10.1038/srep08663.

Martin del Campo, J. S., Rollin, J., Myung, S., Chun, Y., Chandrayan, S., Patino, R., et al. (2013). High-yield production of dihydrogen from xylose by using a synthetic enzyme cascade in a cell-free system. *Angewandte Chemie (International Ed. in English)*, *52*(17), 4587–4590. https://doi.org/10.1002/anie.201300766.

Moore, S. J., MacDonald, J. T., Wienecke, S., Ishwarbhai, A., Tsipa, A., Aw, R., et al. (2018). Rapid acquisition and model-based analysis of cell-free transcription–Translation reactions from nonmodel bacteria. *Proceedings of the National Academy of Sciences of the United States of America*, *115*(19), E4340–E4349. https://doi.org/10.1073/pnas.1715806115.

Myung, S., & Zhang, Y. H. (2013). Non-complexed four cascade enzyme mixture: Simple purification and synergetic co-stabilization. *PLoS One*, *8*(4), e61500. https://doi.org/10.1371/journal.pone.0061500.

Nielsen, J., Fussenegger, M., Keasling, J., Lee, S. Y., Liao, J. C., Prather, K., et al. (2014). Engineering synergy in biotechnology. *Nature Chemical Biology*, *10*(5), 319–322. https://doi.org/10.1038/nchembio.1519.

Nielsen, J., & Keasling, J. D. (2016). Engineering cellular metabolism. *Cell*, *164*(6), 1185–1197. https://doi.org/10.1016/j.cell.2016.02.004.

Opgenorth, P. H., Korman, T. P., & Bowie, J. U. (2014). A synthetic biochemistry molecular purge valve module that maintains redox balance. *Nature Communications*, *5*, 4113. https://doi.org/10.1038/ncomms5113.

Opgenorth, P. H., Korman, T. P., Iancu, L., & Bowie, J. U. (2017). A molecular rheostat maintains ATP levels to drive a synthetic biochemistry system. *Nature Chemical Biology*, *13*(9), 938–942. https://doi.org/10.1038/nchembio.2418.

Rollié, S., Mangold, M., & Sundmacher, K. (2012). Designing biological systems: Systems engineering meets synthetic biology. *Chemical Engineering Science*, *69*(1), 1–29. https://doi.org/10.1016/j.ces.2011.10.068.

Sheldon, R. A. (2007). The E factor: Fifteen years on. *Green Chemistry*, *9*(12), 1273. https://doi.org/10.1039/b713736m.

Shrestha, P., Holland, T. M., & Bundy, B. C. (2012). Streamlined extract preparation for *Escherichia coli*-based cell-free protein synthesis by sonication or bead vortex mixing. *BioTechniques*, *53*(3), 163–174. https://doi.org/10.2144/0000113924.

Siegal-Gaskins, D., Tuza, Z. A., Kim, J., Noireaux, V., & Murray, R. M. (2014). Gene circuit performance characterization and resource usage in a cell-free "breadboard" *ACS Synthetic Biology*, *3*(6), 416–425. https://doi.org/10.1021/sb400203p.

Smanski, M. J., Bhatia, S., Zhao, D., Park, Y., Woodruff, L. B. A., Giannoukos, G., et al. (2014). Functional optimization of gene clusters by combinatorial design and assembly. *Nature Biotechnology*, *32*(12), 1241–1249. https://doi.org/10.1038/nbt.3063.

Sun, Z. Z., Hayes, C. A., Shin, J., Caschera, F., Murray, R. M., & Noireaux, V. (2013). Protocols for implementing an *Escherichia coli* based TX-TL cell-free expression system for synthetic biology. *Journal of Visualized Experiments* *79*, e50762. https://doi.org/10.3791/50762.

Sun, Z. Z., Yeung, E., Hayes, C. A., Noireaux, V., & Murray, R. M. (2014). Linear DNA for rapid prototyping of synthetic biological circuits in an *Escherichia coli* based TX-TL cell-free system. *ACS Synthetic Biology*, *3*(6), 387–397. https://doi.org/10.1021/sb400131a.

Swartz, J. R. (2006). Developing cell-free biology for industrial applications. *Journal of Industrial Microbiology & Biotechnology*, *33*(7), 476–485. https://doi.org/10.1007/s10295-006-0127-y.

Swartz, J. R. (2012). Transforming biochemical engineering with cell-free biology. *AICHE Journal*, *58*(1), 5–13. https://doi.org/10.1002/aic.13701.

Toogood, H. S., Ni Cheallaigh, A., Tait, S., Mansell, D. J., Jervis, A., Lygidakis, A., et al. (2015). Enzymatic menthol production: One-pot approach using engineered *Escherichia coli*. *ACS Synthetic Biology*, *4*(10), 1112–1123. https://doi.org/10.1021/acssynbio.5b00092.

Wang, Y., Huang, W., Sathitsuksanoh, N., Zhu, Z., & Zhang, Y. H. (2011). Biohydrogenation from biomass sugar mediated by in vitro synthetic enzymatic pathways. *Chemistry & Biology*, *18*(3), 372–380. https://doi.org/10.1016/j.chembiol.2010.12.019.

Werpy, T., Petersen, G. R., Aden, A., Bozell, J., Holladay, J., White, J., et al. (2004). Top value added chemicals from biomass. Volume 1—Results of screening for potential candidates from sugars and synthesis gas. *DTIC Document*, 1. https://www.nrel.gov/docs/fy04osti/35523.pdf.

Ye, X., Wang, Y., Hopkins, R. C., Adams, M. W., Evans, B. R., Mielenz, J. R., et al. (2009). Spontaneous high-yield production of hydrogen from cellulosic materials and water catalyzed by enzyme cocktails. *ChemSusChem*, *2*(2), 149–152. https://doi.org/10.1002/cssc.200900017.

You, C., & Zhang, Y. H. (2013). Cell-free biosystems for biomanufacturing. *Advances in Biochemical Engineering/Biotechnology*, *131*, 89–119. https://doi.org/10.1007/10_2012_159.

Zhang, Y. H., Evans, B. R., Mielenz, J. R., Hopkins, R. C., & Adams, M. W. (2007). High-yield hydrogen production from starch and water by a synthetic enzymatic pathway. *PLoS One*, *2*(5), e456. https://doi.org/10.1371/journal.pone.0000456.

Zhu, Z., Kin Tam, T., Sun, F., You, C., & Percival Zhang, Y. H. (2014). A high-energy-density sugar biobattery based on a synthetic enzymatic pathway. *Nature Communications*, *5*, 3026. https://doi.org/10.1038/ncomms4026.

CHAPTER THREE

Fast and Flexible Synthesis of Combinatorial Libraries for Directed Evolution

Joanna C. Sadler, Lucy Green, Neil Swainston, Douglas B. Kell, Andrew Currin[1]

School of Chemistry, Faculty of Science and Engineering, Manchester Synthetic Biology Research Centre for Fine and Speciality Chemicals (SYNBIOCHEM), Manchester Institute of Biotechnology, University of Manchester, Manchester, United Kingdom

[1]Corresponding author: e-mail address: andrew.currin@manchester.ac.uk

Contents

Abstract

Directed evolution (DE) is a powerful tool for optimizing an enzyme's properties toward a particular objective, such as broader substrate scope, greater thermostability, or increased k_{cat}. A successful DE project requires the generation of genetic diversity and subsequent screening or selection to identify variants with improved fitness. In contrast to random methods (error-prone PCR or DNA shuffling), site-directed mutagenesis enables the rational design of variant libraries and provides control over the nature and frequency of the encoded mutations. Knowledge of protein structure, dynamics, enzyme mechanisms, and natural evolution demonstrates that multiple

Methods in Enzymology, Volume 608
ISSN 0076-6879
https://doi.org/10.1016/bs.mie.2018.04.006

(combinatorial) mutations are required to discover the most improved variants. To this end, we describe an experimentally straightforward and low-cost method for the preparation of combinatorial variant libraries. Our approach employs a two-step PCR protocol, first producing mutagenic megaprimers, which can then be combined in a "mix-and-match" fashion to generate diverse sets of combinatorial variant libraries both quickly and accurately.

1. INTRODUCTION

Modern biotechnology frequently requires enzymes to operate beyond the bounds of what nature has provided; these requirements include activity toward nonnatural substrates, higher k_{cat}, thermostability, and/or organic solvent tolerance. Over the past couple of decades, directed evolution (DE) has emerged as an invaluable tool to engineer enzymes to meet these demands and is now a widely adopted technology both in academic and industrial settings (Denard, Ren, & Zhao, 2015; Turner, 2009). Indeed, there are numerous examples of the large-scale manufacture of high-value small molecules using evolved enzymes. For example, a Baeyer–Villiger monooxygenase has been engineered for the production of enantiopure esomeprazole, a proton pump inhibitor used for the treatment of stomach acid overproduction (Wilson, 2015). More recently there have also been significant advances in evolving enzymes toward reactions unprecedented in nature, including silylations (Kan, Lewis, Chen, & Arnold, 2016) and borylation (Kan, Huang, Gumulya, Chen, & Arnold, 2017).

In any DE project, the search for a suitable sequence is a vast and complex combinatorial search problem (Kell, 2012) and as such intelligent search strategies are ideally placed to tackle it (Currin, Swainston, Day, & Kell, 2015). A typical DE workflow comprises three main steps: (1) selection of a suitable parent enzyme and chassis, (2) introduction of sequence variation by DNA mutagenesis, and (3) screening and/or selection to identify variants with the desired property or properties (Fig. 1). We briefly discuss (1) and (3) but focus on step (2), which is the topic of this chapter.

In most cases, a parent enzyme is selected which displays at least some activity of interest, or, in the case of engineering stability toward organic solvents or elevated temperatures, activity under ambient conditions. There are many online tools available to aid this process, with databases such as KEGG (Kanehisa, Sato, Kawashima, Furumichi, & Tanabe, 2016), BRENDA (Placzek et al., 2017), SABIO-RK (Wittig et al., 2012), and biochem4j

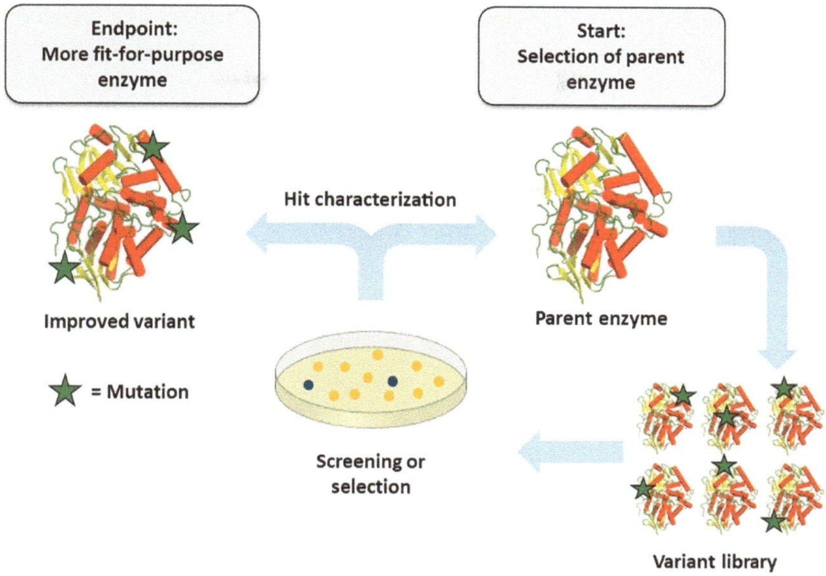

Fig. 1 Workflow of a typical DE project. A parent enzyme is selected, upon which iterative cycles of mutation and screening or selection are performed until a variant with the required fitness is obtained.

(Swainston, Batista-Navarro, et al., 2017, Swainston, Currin, et al., 2017) being popular first port-of-calls for known or closely related reactions. For more challenging reactions with no reported precedent in nature, a number of computational tools are available to aid selection of a suitable parent enzyme to evolve toward the desired function. These include Selenzyme (Carbonell et al., 2017), ATLAS of Biochemistry (Hadadi, Hafner, Shajkofci, Zisaki, & Hatzimanikatis, 2016), EC-BLAST (Rahman, Cuesta, Furnham, Holliday, & Thornton, 2014), and the CanOE Strategy (Smith, Belda, Viari, Medigue, & Vallenet, 2012), which are described in detail in a series of review articles (Kim, Peña, Moll, Bennett, & Kavraki, 2017; Wang, Dash, Ng, & Maranas, 2017). Additionally, enzymes with 3D structures or homology modeling available through the protein data bank are useful for rational engineering strategies (e.g., active-site modification).

Having identified a suitable parent enzyme, a chassis (host organism) or cell-free expression system is identified (Dodevski, Markou, & Sarkar, 2015; Pourmir & Johannes, 2012; Villarreal & Tan, 2017). Next, mutagenesis of the gene of interest (GOI) is carried out and variants with improved fitness relative to the parent enzyme are identified by screening or selection (Acevedo-Rocha, Agudo, & Reetz, 2014; Badran & Liu, 2015;

Tizei, Csibra, Torres, & Pinheiro, 2016; Xiao, Bao, & Zhao, 2015). Numerous techniques have been described, including many high- and ultrahigh-throughput methods (Chandrasekaran et al., 2013; Chen et al., 2016; Colin et al., 2015; Ostafe, Prodanovic, Nazor, & Fischer, 2014; Seitz et al., 2010; Yang & Withers, 2009; Yoo, Pogson, Iverson, & Georgiou, 2012) for the screening of large libraries (Ahn et al., 2008; Leemhuis, Kelly, & Dijkhuizen, 2009; Soumillion & Fastrez, 2001; Weiß, Pavlidis, Vickers, Hohne, & Bornscheuer, 2014; Willies, White, & Turner, 2012; Xiao et al., 2015; Yang & Withers, 2009). This process of mutagenesis followed by screening or selection is performed iteratively until a protein with the required fitness is obtained.

1.1 Mutagenesis: Introducing Sequence Variation

With a suitable parent enzyme, expression system, and screen or selection methodology in place, the next stage is to introduce sequence variation into the GOI to produce a library of variant sequences. Whereas early DE projects used an error-prone polymerase chain reaction (epPCR) and/or gene shuffling to introduce sequence variation randomly, recently there has been a move toward more rational engineering strategies, in which specific residues or regions of the GOI are targeted using site-directed mutagenesis (SDM) or gene synthesis (Currin, Swainston, Day, & Kell, 2014, 2017). Mutations are typically encoded by DNA oligonucleotides, and multiple amino acids can be encoded at a defined position using degenerate codons (Jacobs, Yumerefendi, Kuhlman, & Leaver-Fay, 2015; Patrick & Firth, 2005; Swainston, Batista-Navarro, et al., 2017, Swainston, Currin, et al., 2017). However, deciding upon which mutations to create is nontrivial. The number of possible permutations of a protein sequence (its "sequence space," namely, 20^n where n is the length of the protein chain of interest) is many orders of magnitude greater than the number of mutants that can be tested experimentally (even using ultrahigh-throughput screening methods). Consequently, successful DE projects must navigate this landscape sufficiently to identify variants with suitably improved fitness (Currin et al., 2015). In practice, library design depends upon which parameter one wishes to evolve the protein toward. For example, targeting first and second shell active-site residues has been demonstrated to be effective for the evolution of an enzyme toward activity with a different substrate (Kolev, Zaengle, Ravikumar, & Fasan, 2014; Kumar, Scott, Liu, & Halpert, 2003; Li et al., 2017; Manna & Mazumdar, 2010; Porter et al., 2015; Toogood et al., 2011), while regions with a high B-factor (Yuan, Bailey, & Teasdale, 2005) can be mutated to improve the protein's thermostability (Reetz, Carballeira, & Vogel, 2006; Wijma et al., 2014). For DE toward a higher k_{cat}, it is often necessary to create libraries that sample a broader area of sequence space, with successful DE projects in this area often identifying beneficial mutations distal to the active site (Fox et al., 2007; Jiménez-Osés et al., 2014; Suzuki et al., 2010). It should be noted that simple selection for the single best enzyme (when mutation rates are low) guarantees discovery of a local optimum and "escape" from this is impossible (Currin et al., 2015).

1.2 Combinatorial Libraries

"Epistasis" describes the set of phenomena in which the influence of one residue on a protein's fitness is dependent on the identity of other residues. Despite our lack of any detailed understanding of the complexity of sequence space, the fundamental importance of epistatic interactions to protein function is recognized (de Visser & Krug, 2014; Storz, 2018). As such, nonadditive increases (or decreases) in fitness are frequently observed when two mutations are combined (Fig. 2) (Currin et al., 2015; Miton & Tokuriki, 2016; Noor et al., 2012; Sato et al., 2016; Starr & Thornton, 2016; Storz, 2018). For DE, it is therefore appreciated that single residue mutations are often insufficient to yield significant improvement.

A pertinent example of exploiting epistasis in DE is Combinatorial Active Site Testing (CASTing), which has proved particularly successful

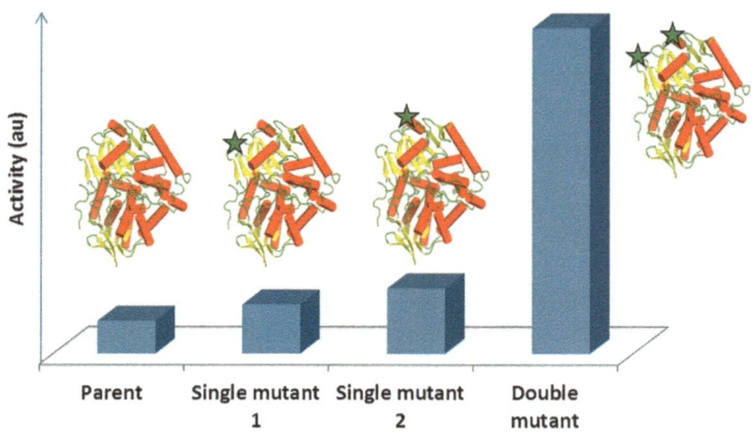

Fig. 2 A demonstration of epistasis, in which a nonadditive relationship between two mutations yields a significant increase in fitness (activity). Alternatively, mutant 2 could also be worse than the parent enzyme.

for engineering a protein toward nonnatural substrates (Reetz & Carballeira, 2007; Reetz, Carballeira, Peyralans, et al., 2006). CASTing involves the covariation of two residues that are close in 3D space to produce a combinatorial library, which is then screened for improvements in activity. Performing multiple rounds of CASTing iteratively can also be particularly successful (Reetz, Carballeira, & Vogel, 2006; Reetz, Wang, & Bocola, 2006). When target mutations are not in a contiguous sequence, combinatorial mutations at multiple separate positions must be created. While traditional SDM (e.g., using a QuikChange-based protocol; Braman, Papworth, & Greener, 1996; Zheng, Baumann, & Reymond, 2004) can be utilized for a single library, this method can become laborious if a library targeting a number of different regions of variation is sought. Several methods for the preparation of combinatorial libraries have been developed, including cassette mutagenesis (Seyfang & Huaqian Jin, 2004; Wells, Vasser, & Powers, 1985), PFunkel (Firnberg & Ostermeier, 2012), Darwin Assembly, (Cozens & Pinheiro, 2018), OmniChange (Dennig, Shivange, Marienhagen, & Schwaneberg, 2011), plasmid recombineering (Bratulic & Badran, 2017; Higgins, Ouonkap, & Savage, 2017), and combinatorial codon mutagenesis (Belsare et al., 2016). However, these often involve complex protocols and utilize large numbers of oligonucleotides or longer DNA fragments. Herein, we describe a fast and flexible method to introduce combinatorial mutations to a DNA template of interest using standard PCR-based techniques and reagents, which has been used for a number of studies to date (Currin, Korovin, et al., 2017; Currin et al., 2018). The protocol described here is divided into two main steps: asymmetric PCR to produce mutagenic megaprimers and assembly of the full-length gene using any combination of the mutagenic megaprimers. The method is described for the covariation of two regions in a gene simultaneously; however, the methodology could be extended for the covariation of more regions as required.

2. METHODS

The overall process for the preparation of combinatorial libraries comprises two main PCR steps: (1) preparation of mutagenic megaprimers and (2) combination of two mutagenic megaprimers to assemble the full-length gene (Fig. 3). PCR (2) can utilize any combination of mutagenic megaprimers in a "mix-and-match" fashion to produce different combinatorial libraries. To generate the megaprimer, a primer (typically forward direction) encoding mutations (termed "mutagenic primer") is employed in an

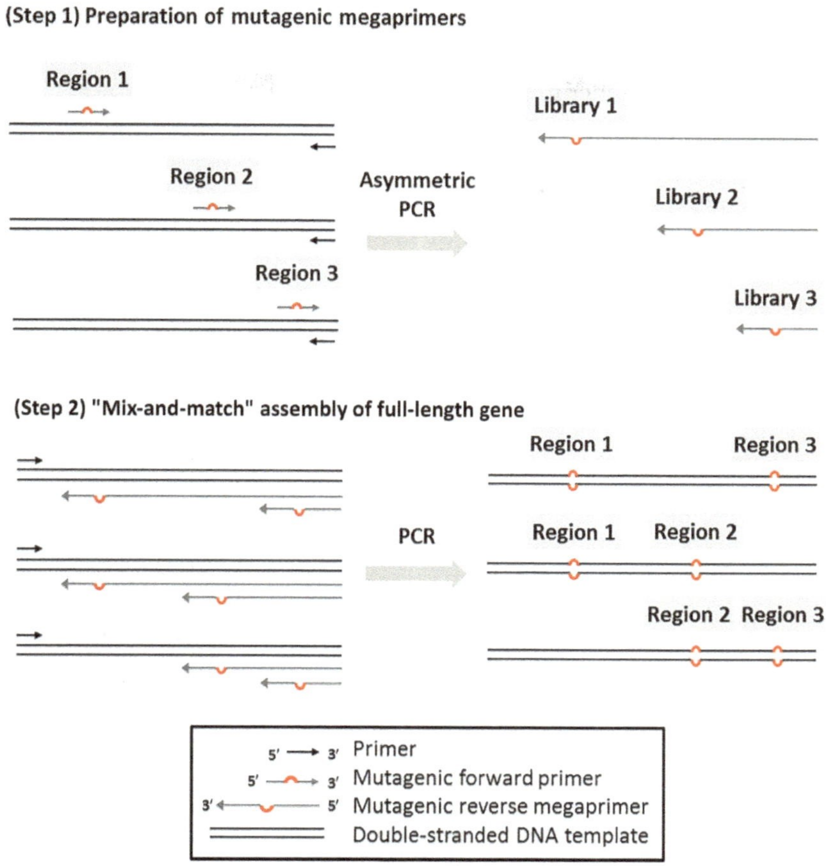

Fig. 3 Overview of the method for preparation of combinatorial libraries. Arrows denote the direction of amplification during PCR.

asymmetric PCR. Multiple reactions are performed in parallel to cover each of the target regions to be mutated (termed "target regions"). This generates mutagenic megaprimers (reverse direction) for each region (Fig. 3, step 1). In the second PCR, two megaprimers are pooled together, which combine to assemble a full-length gene library encoding mutations at both target regions (Fig. 3, step 2).

To obtain mutations in both regions in step 2, it is necessary to control the amount of each megaprimer that is added. During development work, a higher mutational frequency at the region at the 5′ end of the gene was consistently observed when equal amounts of both megaprimers were used. Two factors were hypothesized to cause this: (1) the longer megaprimer binds preferentially during PCR and/or (2) if mutation at the 3′ end of the gene has been incorporated in a gene, it will be "overwritten" if the longer megaprimer binds in a subsequent cycle of the PCR. Both of these situations result in a higher mutational frequency at the 5′ end of the gene. To address this effective bias, the ratio of long to short megaprimers is adjusted according to their relative lengths. It was found that increasing the ratio of the short megaprimer to the long megaprimer exponentially as the difference in length increased resulted in the most mutations at both positions (Fig. 4). This empirically determined relationship has been exploited extensively in our laboratory to provide combinatorial libraries. For straightforward implementation, we have developed CombiGenie, a ratio calculator which is available for download at http://combi.synbiochem.co.uk/.

2.1 Primer Design

Separate mutagenic primers need to be designed for each target region. For instance, in the example shown in Fig. 3, a total of three mutagenic forward primers would be designed. Design mutagenic primers according to the following guidelines:

1. A mutation can be a degenerate codon for a library or a specific codon if a particular residue is required. For example, the NNK codon will code for all 20 amino acids and one stop codon, while reduced degeneracy codons code for a subset of amino acids such as only nonpolar residues (NTN) or only residues with large, charged side chains (NAN). Optimized degenerate codons encoding these and any other combination of amino acids can be calculated using the CodonGenie tool, which is

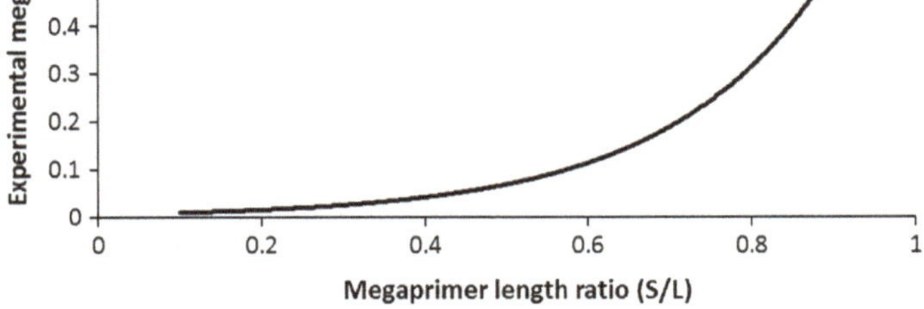

Fig. 4 A theoretical relationship between megaprimer lengths and the experimental ratio of megaprimers required for high levels of mutational efficiency in both positions.

available at http://codon.synbiochem.co.uk/ (Currin et al., 2015; Kille et al., 2013; Swainston, Batista-Navarro, et al., 2017, Swainston, Currin, et al., 2017).

2. Flanking sequences 5′ and 3′ to the mutations are next created, such that each has a suitable T_m (Fig. 5, see Notes 1–3).

In addition, two nonmutagenic primers (forward and reverse direction) are also designed. These bind the start and end of the GOI, respectively, such that when used together they amplify the full-length GOI. Primer parameters should match that of the mutagenic primer flanking regions.

2.1.1 Notes and Troubleshooting

Note 1. Primers should generally be between 45 and 65 nucleotides long, ideally with a G/C at both 5′ and 3′ termini. Where possible, avoid consecutive occurrences of four or more of the same base, e.g., -GAAAAT-. The mutagenic codon should fall near the center of the sequence.

Note 2. Mutagenic primers are routinely designed containing flanking sequences with a T_m of 60°C. If a T_m of 60°C for each of the regions flanking the mutation is not possible, design the required primer and alter the annealing temperature of the PCR to the new T_m.

Note 3. Unfavorable primer features like homodimerization and secondary structure can be identified using online tools (e.g., OligoAnalyzer from Integrated DNA Technologies (https://eu.idtdna.com/calc/analyzer) or QuikFold from the DINAMet Web server (Markham & Zuker, 2005, 2008) (http://unafold.rna.albany.edu/?q=DINAMelt/Quickfold)). Primers containing such undesired features should be redesigned.

2.2 Synthesis of Mutagenic Megaprimers by Asymmetric PCR

Asymmetric PCR is a well-established method for the production of megaprimers and has been used extensively for mutagenesis (Ke & Madison, 1997;

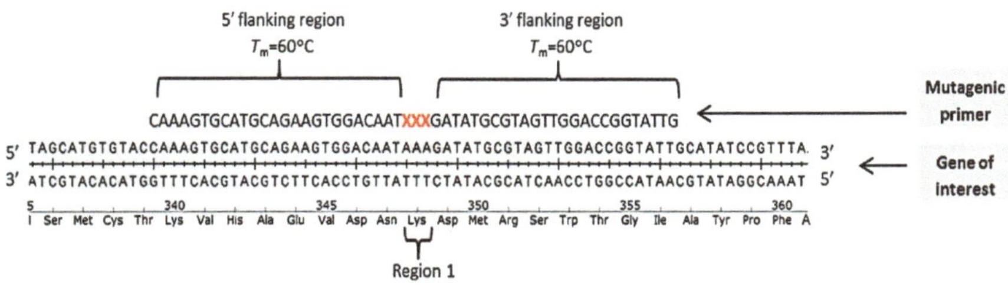

Fig. 5 Primer design for mutagenic primers. XXX = mutated codon (degenerate codon for a library or a single codon as required).

Tyagi, Lai, & Duggleby, 2004; Wang & Malcolm, 2002; Wooddell & Burgess, 1996; Xu, Colosimo, & Gruenert, 2003). Here, mutagenic megaprimers are prepared using a 20-fold molar excess of the antisense strand nonmutagenic end primer (reverse direction), relative to the sense strand mutagenic primer in the PCR reaction.

2.2.1 Equipment

- PCR thermocycler
- Spectrophotometer for DNA quantification
- Thin-walled PCR tubes

2.2.2 Reagents and Buffers

- Nonmutagenic end primer (reverse)
- Mutagenic primers
- dNTPs
- High-fidelity HotStart DNA polymerase and reaction buffer (see Note 4)
- DNase-free water
- Template DNA (plasmid or linear)

2.2.3 Protocol

1. Prepare primer stock solutions in DNase-free water at the following concentrations:
   a. Nonmutagenic end primer (reverse): 10 μ*M* (Stock 1a)
   b. Mutagenic primers (forward): 10 μ*M* (Stock 1b)
   c. Mutagenic primers (forward): 1 μ*M* (Stock 1c)
   d. Nonmutagenic end primer (forward): 10 μ*M* (Stock 1d)
2. For each target region, set up a separate asymmetric PCR in a thin-walled PCR tube as described in Table 1.
3. Recommended: for each asymmetric PCR reaction, carryout a symmetric PCR using 10 μ*M* mutagenic primer as a positive control as described in Table 2 (see Note 6).
4. Perform PCR as in Table 3.
5. Analyze PCR products by gel electrophoresis. ssDNA products may appear smeared and/or have a different retention time to the symmetric PCR products. An example gel is shown in Fig. 6; if the symmetric control has been successful, proceed to the next step (see Note 7).

Table 1 Asymmetric PCR Setup

	Stock Concentration	Reaction Concentration	Volume for 50 μL PCR (μL)
Stock 1a	10 μ*M*	500 n*M*	2.5
Stock 1c	1 μ*M*	25 n*M*	1.25
Template (see Note 5)	0.5 ng/μL plasmid or 0.05 ng/μL linear template	0.01 ng/μL plasmid or 0.001 ng/μL linear template	1
dNTPs	2 m*M*	200 μ*M*	5
Buffer	Variable	1 ×	Variable
DNA polymerase	Variable	Variable	Variable
DNase-free water	—	—	To 50 μL

Table 2 Symmetric PCR Setup (See Note 2)

	Stock Concentration	Reaction Concentration	Volume for 50 μL PCR (μL)
Stock 1a	10 μ*M*	500 n*M*	2.5
Stock 1b	10 μ*M*	500 n*M*	2.5
Template (see Note 5)	0.5 ng/μL plasmid or 0.05 ng/μL linear template	0.01 ng/μL plasmid or 0.001 ng/μL linear template	1
dNTPs	2 m*M*	200 μ*M*	5
Buffer	Variable	1 ×	Variable
DNA polymerase	Variable	Variable	Variable
DNase-free water	—	—	To 50 μL

Table 3 Suggested PCR Thermocycling Parameters (See Note 4)

Denaturation	98°C	2 min	
Denaturation	98°C	30 s	25 × cycles
Annealing	60°C	20 s	
Extension	72°C	Variable	
Hold	4°C		

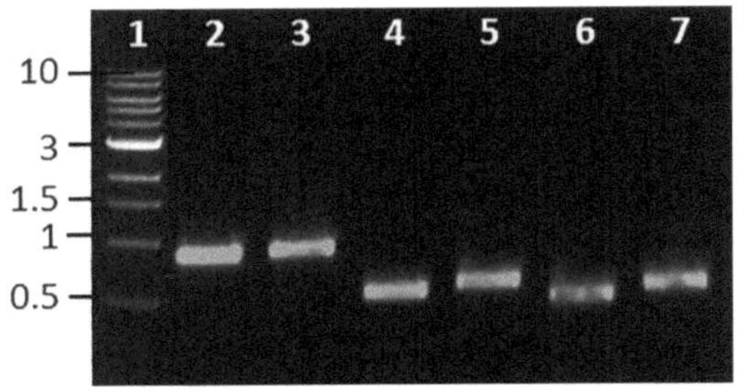

Fig. 6 Example gel from analysis of symmetric and asymmetric PCR products. Lanes: 1, ladder (size in kb); 2, sample 1 symmetric; 3, sample 1 asymmetric; 4, sample 2 symmetric; 5, sample 2 asymmetric; 6, sample 3 symmetric; 7, sample 3 asymmetric. Theoretical length of megaprimers: 1136, 454, and 443 bases for samples 1, 2, and 3, respectively.

6. If using methylated DNA as template (e.g., isolated from *E. coli*), perform a *Dpn*I digest of the PCR products (following manufacturer's protocol) to minimize carryover of the PCR template into the second stage. If using nonmethylated DNA as template (e.g., from PCR amplification), proceed directly to step 7.
7. Purify PCR products using a PCR cleanup kit. Measure the concentration of ssDNA for each of the megaprimers prepared using a spectrophotometer.

2.2.4 Notes and Troubleshooting

Note 4. We use Q5 HotStart DNA polymerase (New England Biolabs); however, other high-fidelity HotStart DNA polymerases may also be used (adjust PCR conditions accordingly).

Note 5. A recommended starting template concentration is provided. Template concentration may require optimization if the symmetric positive control PCR is unsuccessful.

Note 6. For an easier setup, prepare a master mix as for Table 1 and for the asymmetric PCR add 1.25 μL mutagenic primer mix and top up with 1.25 μL DNase-free water.

Note 7. If bands corresponding to the desired PCR product are not observed, try varying the template concentration and/or increasing the extension time during PCR.

Note 8. If very low yields of asymmetric product are observed, try increasing the ratio of end primer (reverse direction) to mutagenic primer, ranging from 20:1 to 100:1. Increasing the number of PCR cycles can also result in a higher level of single-stranded product; however, increasing this above 35 cycles is not recommended.

2.3 Assembly of Full-Length Gene Using Mutagenic Megaprimers

2.3.1 Equipment

- Spectrophotometer for DNA quantification
- Thermocycler
- Thin-walled PCR tubes

2.3.2 Reagents and Buffers

- Nonmutagenic end primer (forward)
- Mutagenic megaprimers prepared in Section 2.2
- dNTPs
- High-fidelity HotStart DNA polymerase and reaction buffer (see Note 4)
- DNase-free water
- Template DNA (plasmid or linear)

2.3.3 Protocol

1. Calculate volume of each megaprimer required in each library using CombiGenie (available at http://combi.synbiochem.co.uk/), by inputting the length of each megaprimer and the measured concentration of each megaprimer (Section 2.2.3, step 7).
2. For each library, set up a PCR according to Table 4.
3. Perform PCR in thin-walled PCR tubes. Suggested thermocycling parameters are provided in Table 5.

Table 4 PCR to Assemble Full-Length Combinatorial Libraries

	Stock Concentration	Final Concentration	Volume for 50 μL PCR (μL)
Stock 1d	10 μ*M*	500 n*M*	2.5
Mutagenic megaprimers (reverse)	100 n*M*	50 n*M*	Variable
Template	0.5 ng/μL plasmid or 0.05 ng/μL linear ds template	0.01 ng/μL plasmid or 0.001 ng/μL linear template	1
dNTPs	2 m*M*	200 μ*M*	5
Buffer	Variable	1 ×	Variable
DNA polymerase	Variable	Variable	Variable
DNase-free water	—	—	To 50 μL

Table 5 PCR Thermocycling Parameters for Preparation of Combinatorial Libraries Using Two Mutagenic Megaprimers

Denaturation	98°C	2 min	
Denaturation	98°C	30 s	25 × cycles
Annealing	60°C	20 s	
Extension	72°C	Variable	
Hold	4°C		

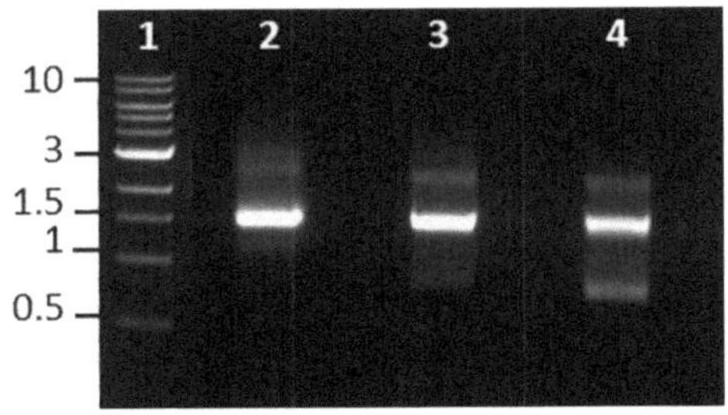

Fig. 7 Preparation of full-length gene libraries using mutagenic megaprimers. The theoretical length of the desired PCR product is ~1.5 kb.

4. Analyze PCR products by gel electrophoresis (Fig. 7).
5. If the correct PCR product is present perform a gel extraction to isolate the full-length product.

2.3.4 Notes and Troubleshooting

Note 9. For an easier setup, prepare a master mix containing the end primer (Stock 1d), template, dNTPs, DNA polymerase, reaction buffer, and DNase-free water to a final volume of 25 μL × no. PCR reactions. Add 25 μL mutagenic megaprimers to each PCR tube and 25 μL master mix.

Note 10. If PCR products corresponding to theoretical gene length are not observed, repeat the PCR with 0.05, 0.1, 1, and 2.5 ng plasmid template or 0.005, 0.01, 0.1, and 0.25 ng linear template. If still unsuccessful, optimize annealing and extension times of the PCR.

3. LIBRARY LIGATION, TRANSFORMATION, AND QUALITY CONTROL

A library validation step using Sanger DNA sequencing is highly recommended, to ensure the desired mutations have been introduced into the intended regions of the target gene.

3.1 Buffers and Reagents

- Cloning vector, ideally high-copy number
- Competent cells, such as 5-alpha (NEB) or Top10 (ThermoFisher) *E. coli*
- Sequencing primers

3.2 Protocol

1. Ligate each combinatorial library into an appropriate vector using any preferred ligation kit (see Note 11).
2. Transform the ligation reaction into competent cells following the manufacturer's protocol.
3. Perform colony PCR on ~10 clones from each library using a suitable DNA polymerase, according to the manufacturer's protocol (see Note 12). Analyze PCR products by gel electrophoresis. If the PCR reaction is successful, purify PCR products using a PCR cleanup kit and submit the resulting DNA fragments for sequencing following the service provider's protocol.
4. Assess the mutational frequency at the targeted regions of each clone.

3.3 Notes and Troubleshooting

Note 11. We use the In-Fusion cloning system (Takara Bio) for library ligation, which requires a 15-bp overlap between the insert and linearized vector. However, any reliable cloning method can be utilized for this step.

Note 12. We use KAPA HiFi DNA polymerase (Roche Diagnostics), which is recommended for colony PCR; however, other suitable DNA polymerases may also be employed.

Note 13. If a bias for mutation in one of the two regions is observed, it may be necessary to alter the ratio of the megaprimers in the second PCR step. If higher mutational frequency is observed at the 5′ region, decrease the ratio of long:short megaprimers. Conversely, if more mutations are found at the 3′ region, increase the ratio of long:short megaprimers (Fig. 8).

Note 14. If high levels of WT are observed, the use of ssDNA template during the second PCR step is recommended. This prevents the carryover of WT DNA into the final PCR products, as there is no WT antisense strand available for amplification and consequently only mutated DNA strands can be propagated (Fig. 9).

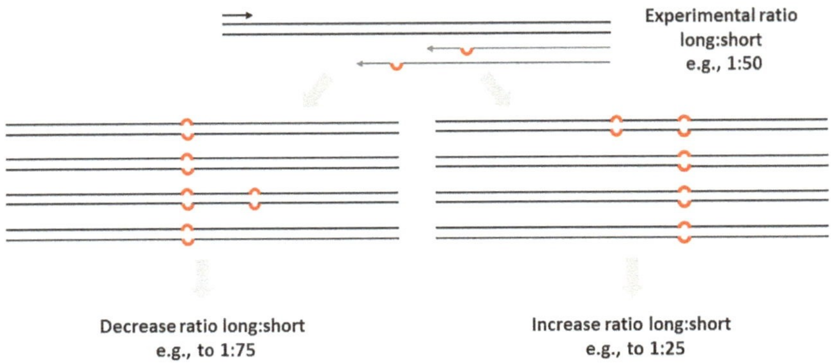

Fig. 8 Adjustment of the ratio of long:short megaprimers for preparation of combinatorial libraries.

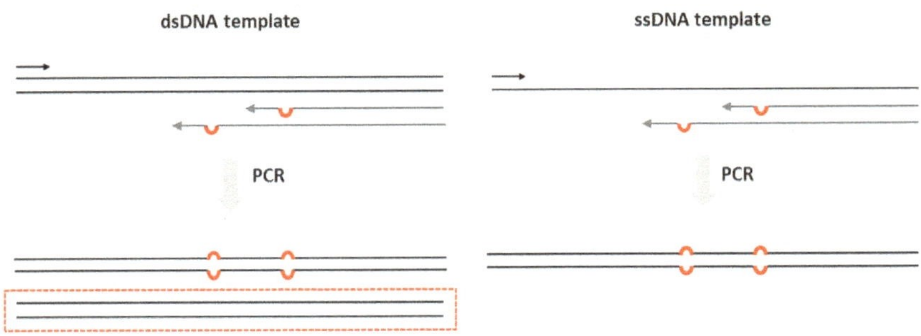

Fig. 9 Full-length gene PCR using an ssDNA template. Carryover of template DNA (*dashed box*) can be significantly reduced using an ssDNA template.

Note 15. ssDNA template DNA can be prepared using a 5′-phosphorylated reverse primer (Stock 1a) in a PCR (with Stock 1d) to produce a full-length gene product with a 5′-phosphorylated reverse (antisense) strand. Digestion of the resulting 5′-phosphorylated antisense strand with Lambda exonuclease (New England Biolabs) affords the ssDNA template, according to the manufacturer's protocol (Mitsis & Kwagh, 1999). If using ssDNA template, a much lower template concentration (0.0001–0.00001 ng/μL) is recommended for the PCR to prepare the combinatorial libraries.

4. SUMMARY AND CONCLUSION

The preparation of combinatorial libraries for the DE of biocatalysts is necessary for a sufficient exploration of sequence space to identify novel mutants with enhanced fitness. In particular, it is generally accepted that dramatic increases in fitness may be observed as a result of epistatic interactions. In this chapter, we have presented a novel method for the preparation of high quality, targeted combinatorial libraries in which two regions per clone are covaried. The two-step protocol is particularly useful if one wishes to investigate mutating a number of different regions in different combinations. A particular advantage with this approach is the "mix-and-match" nature of our methodology, as regions of interest can be interrogated in different combinations in a fast and flexible manner. This method can be applied to any gene to generate genetic diversity and to explore epistatic interactions during a DE project.

ACKNOWLEDGMENTS

The authors acknowledge the funding from the Biotechnology and Biological Sciences Research Council (BBSRC) under Grants BB/M017702/1, "Centre for Synthetic Biology of Fine and Speciality Chemicals (SYNBIOCHEM)" and a proof of concept Grant from BIOCATNET NIBB BBSRC Grant BB/L013649/1. This is a contribution from the Manchester Centre for Synthetic Biology of Fine and Speciality Chemicals (SYNBIOCHEM).

REFERENCES

Acevedo-Rocha, C. G., Agudo, R., & Reetz, M. T. (2014). Directed evolution of stereoselective enzymes based on genetic selection as opposed to screening systems. *Journal of Biotechnology*, *191*, 3–10.

Ahn, Y., Ko, S. B., Kim, M. J., Park, J., Alexeeva, M., Enright, A., et al. (2008). Deracemization of α-methylbenzylamine using an enzyme obtained by in vitro evolution. *Tetrahedron Letters*, *43*(4), 1247–1250.

Badran, A. H., & Liu, D. R. (2015). In vivo continuous directed evolution. *Current Opinion in Chemical Biology*, *24*, 1–10.

Belsare, K. D., Andorfer, M. C., Cardenas, F. S., Chael, J. R., Park, H. J., & Lewis, J. C. (2016). A simple combinatorial codon mutagenesis method for targeted protein engineering. *ACS Synthetic Biology*, *17*(3), 416–420.

Braman, J., Papworth, C., & Greener, A. (1996). Site-directed mutagenesis using double-stranded plasmid DNA templates. *In vitro mutagenesis protocols: Vol. 57*. (pp. 31–44). New Jersey: Humana Press.

Bratulic, S., & Badran, A. H. (2017). Modern methods for laboratory diversification of biomolecules. *Current Opinion in Chemical Biology*, *41*, 50–60.

Carbonell, P., Wong, J., Swainston, N., Takano, E., Turner, N. J., Scrutton, N. S., et al. (2017). Selenzyme: Enzyme selection tool for pathway design. *bioRxiv*, 188979.

Chandrasekaran, A., Deng, K., Koh, C.-Y., Takasuka, T., Bergeman, L. F., Fox, B. G., et al. (2013). A universal flow cytometry assay for screening carbohydrate-active enzymes using glycan microspheres. *Chemical Communications*, *49*(49), 5441–5443.

Chen, B., Lim, S., Kannan, A., Alford, S. C., Sunden, F., Herschlag, D., et al. (2016). High-throughput analysis and protein engineering using microcapillary arrays. *Nature Chemical Biology*, *12*(2), 76–81.

Colin, P.-Y., Kintses, B., Gielen, F., Miton, C. M., Fischer, G., Mohamed, M. F., et al. (2015). Ultrahigh-throughput discovery of promiscuous enzymes by picodroplet functional metagenomics. *Nature Communications, 6*, 10008.

Cozens, C., & Pinheiro, V. B. (2018). Darwin assembly: Fast, efficient, multi-site bespoke mutagenesis. *Nucleic Acids Research*. https://doi.org/10.1093/nar/gky067. [Epub ahead of print].

Currin, A., Dunstan, M. S., Johannissen, L. O., Hollywood, K. A., Vinaixa, M., Jervis, A. J., et al. (2018). Engineering the "missing link" in biosynthetic (−)-menthol production: Bacterial isopulegone isomerase. *ACS Catalysis, 8*, 2012–2020.

Currin, A., Korovin, K., Ababi, M., Roper, K., Kell, D. B., Day, P. J., et al. (2017). Computing exponentially faster: Implementing a non-deterministic universal Turing machine using DNA. *Journal of the Royal Society, Interface, 14*(128), 20160990.

Currin, A., Swainston, N., Day, P. J., & Kell, D. B. (2014). SpeedyGenes: An improved gene synthesis method for the efficient production of error-corrected, synthetic protein libraries for directed evolution. *Protein Engineering, Design and Selection, 27*(9), 273–280.

Currin, A., Swainston, N., Day, P. J., & Kell, D. B. (2015). Synthetic biology for the directed evolution of protein biocatalysts: Navigating sequence space intelligently. *Chemical Society Reviews, 44*(5), 1172–1239.

Currin, A., Swainston, N., Day, P. J., & Kell, D. B. (2017). SpeedyGenes: Exploiting an improved gene synthesis method for the efficient production of synthetic protein libraries for directed evolution. In R. A. Hughes (Ed.), *Methods in molecular biology: Vol. 1472*. (pp. 63–78). Clifton, NJ: Humana Press.

de Visser, J. A. G. M., & Krug, J. (2014). Empirical fitness landscapes and the predictability of evolution. *Nature Reviews. Genetics, 15*(7), 480–490.

Denard, C. A., Ren, H., & Zhao, H. (2015). Improving and repurposing biocatalysts via directed evolution. *Current Opinion in Chemical Biology, 25*, 55–64.

Dennig, A., Shivange, A. V., Marienhagen, J., & Schwaneberg, U. (2011). OmniChange: The sequence independent method for simultaneous site-saturation of five codons. *PLoS ONE, 6*(10), e26222.

Dodevski, I., Markou, G. C., & Sarkar, C. A. (2015). Conceptual and methodological advances in cell-free directed evolution. *Current Opinion in Structural Biology, 33*, 1–7.

Firnberg, E., & Ostermeier, M. (2012). PFunkel: Efficient, expansive, user-defined mutagenesis. *PLoS ONE, 7*(12), e52031.

Fox, R. J., Davis, S. C., Mundorff, E. C., Newman, L. M., Gavrilovic, V., Ma, S. K., et al. (2007). Improving catalytic function by ProSAR-driven enzyme evolution. *Nature Biotechnology, 25*(3), 338–344.

Hadadi, N., Hafner, J., Shajkofci, A., Zisaki, A., & Hatzimanikatis, V. (2016). ATLAS of biochemistry: A repository of all possible biochemical reactions for synthetic biology and metabolic engineering studies. *ACS Synthetic Biology, 5*(10), 1155–1166.

Higgins, S. A., Ouonkap, S. V. Y., & Savage, D. F. (2017). Rapid and programmable protein mutagenesis using plasmid recombineering. *ACS Synthetic Biology, 6*(10), 1825–1833.

Jacobs, T. M., Yumerefendi, H., Kuhlman, B., & Leaver-Fay, A. (2015). SwiftLib: Rapid degenerate-codon-library optimization through dynamic programming. *Nucleic Acids Research, 43*(5), e34.

Jiménez-Osés, G., Osuna, S., Gao, X., Sawaya, M. R., Gilson, L., Collier, S. J., et al. (2014). The role of distant mutations and allosteric regulation on LovD active site dynamics. *Nature Chemical Biology, 10*(6), 431–436.

Kan, S. B. J., Huang, X., Gumulya, Y., Chen, K., & Arnold, F. H. (2017). Genetically programmed chiral organoborane synthesis. *Nature, 552*(7683), 132.

Kan, S. B. J., Lewis, R. D., Chen, K., & Arnold, F. H. (2016). Directed evolution of cytochrome c for carbon–silicon bond formation: Bringing silicon to life. *Science, 354*(6315), 1048–1051.

Kanehisa, M., Sato, Y., Kawashima, M., Furumichi, M., & Tanabe, M. (2016). KEGG as a reference resource for gene and protein annotation. *Nucleic Acids Research, 44*, 457–462.

Ke, S. H., & Madison, E. L. (1997). Rapid and efficient site-directed mutagenesis by single-tube "megaprimer" PCR method. *Nucleic Acids Research, 25*(16), 3371–3372.

Kell, D. B. (2012). Scientific discovery as a combinatorial optimisation problem: How best to navigate the landscape of possible experiments? *BioEssays: News and Reviews in Molecular, Cellular and Developmental Biology, 34*(3), 236–244.

Kille, S., Acevedo-Rocha, C. G., Parra, L. P., Zhang, Z.-G., Opperman, D. J., Reetz, M. T., et al. (2013). Reducing codon redundancy and screening effort of combinatorial protein libraries created by saturation mutagenesis. *ACS Synthetic Biology, 2*(2), 83–92.

Kim, S. M., Peña, M. I., Moll, M., Bennett, G. N., & Kavraki, L. E. (2017). A review of parameters and heuristics for guiding metabolic pathfinding. *Journal of Cheminformatics, 9*(1), 51.

Kolev, J. N., Zaengle, J. M., Ravikumar, R., & Fasan, R. (2014). Enhancing the efficiency and regioselectivity of P450 oxidation catalysts by unnatural amino acid mutagenesis. *ChemBioChem, 15*(7), 1001–1010.

Kumar, S., Scott, E. E., Liu, H., & Halpert, J. R. (2003). A rational approach to re-engineer cytochrome P450 2B1 regioselectivity based on the crystal structure of cytochrome P450 2C5. *The Journal of Biological Chemistry, 278*(19), 17178–17184.

Leemhuis, H., Kelly, R. M., & Dijkhuizen, L. (2009). Directed evolution of enzymes: Library screening strategies. *IUBMB Life, 61*(3), 222–228.

Li, G., Yao, P., Gong, R., Li, J., Liu, P., Lonsdale, R., et al. (2017). Simultaneous engineering of an enzyme's entrance tunnel and active site: The case of monoamine oxidase MAO-N. *Chemical Science, 8*(5), 4093–4099.

Manna, S. K., & Mazumdar, S. (2010). Tuning the substrate specificity by engineering the active site of cytochrome P450cam: A rational approach. *Dalton Transactions, 39*(12), 3115.

Markham, N. R., & Zuker, M. (2005). DINAMelt web server for nucleic acid melting prediction. *Nucleic Acids Research, 33*(Web Server), W577–W581.

Markham, N. R., & Zuker, M. (2008). UNAFold: Software for nucleic acid folding and hybridization. *Methods in Molecular Biology, 453*, 3–31.

Miton, C. M., & Tokuriki, N. (2016). How mutational epistasis impairs predictability in protein evolution and design. *Protein Science, 25*(7), 1260–1272.

Mitsis, P. G., & Kwagh, J. G. (1999). Characterization of the interaction of lambda exonuclease with the ends of DNA. *Nucleic Acids Research, 27*(15), 3057–3063.

Noor, S., Taylor, M. C., Russell, R. J., Jermiin, L. S., Jackson, C. J., Oakeshott, J. G., et al. (2012). Intramolecular epistasis and the evolution of a new enzymatic function. *PLoS ONE, 7*(6), e39822.

Ostafe, R., Prodanovic, R., Nazor, J., & Fischer, R. (2014). Ultra-high-throughput screening method for the directed evolution of glucose oxidase. *Chemistry and Biology, 21*(3), 414–421.

Patrick, W. M., & Firth, A. E. (2005). Strategies and computational tools for improving randomized protein libraries. *Biomolecular Engineering, 22*(4), 105–112.

Placzek, S., Schomburg, I., Chang, A., Jeske, L., Ulbrich, M., Tillack, J., et al. (2017). BRENDA in 2017: New perspectives and new tools in BRENDA. *Nucleic Acids Research, 45*, D380–D388.

Porter, J. L., Boon, P. L. S., Murray, T. P., Huber, T., Collyer, C. A., & Ollis, D. L. (2015). Directed evolution of new and improved enzyme functions using an evolutionary intermediate and multi-directional search. *ACS Chemical Biology, 10*(2), 611–621.

Pourmir, A., & Johannes, T. W. (2012). Directed evolution: Selection of the host organism. *Computational and Structural Biotechnology Journal, 2*, e201209012.

Rahman, S. A., Cuesta, S. M., Furnham, N., Holliday, G. L., & Thornton, J. M. (2014). EC-BLAST: A tool to automatically search and compare enzyme reactions. *Nature Methods*, *11*(2), 171–174.

Reetz, M. T., & Carballeira, J. D. (2007). Iterative saturation mutagenesis (ISM) for rapid directed evolution of functional enzymes. *Nature Protocols*, *2*(4), 891–903.

Reetz, M. T., Carballeira, J. D., Peyralans, J., Höbenreich, H., Maichele, A., & Vogel, A. (2006). Expanding the substrate scope of enzymes: Combining mutations obtained by CASTing. *Chemistry: A European Journal*, *12*(23), 6031–6038.

Reetz, M. T., Carballeira, J. D., & Vogel, A. (2006). Iterative saturation mutagenesis on the basis of B factors as a strategy for increasing protein thermostability. *Angewandte Chemie International Edition*, *45*(46), 7745–7751.

Reetz, M. T., Wang, L. W., & Bocola, M. (2006). Directed evolution of enantioselective enzymes: Iterative cycles of CASTing for probing protein-sequence space. *Angewandte Chemie International Edition*, *45*(8), 1236–1241.

Sato, T. K., Tremaine, M., Parreiras, L. S., Hebert, A. S., Myers, K. S., Higbee, A. J., et al. (2016). Directed evolution reveals unexpected epistatic interactions that alter metabolic regulation and enable anaerobic xylose use by Saccharomyces cerevisiae. *PLoS Genetics*, *12*(10), e1006372.

Seitz, T., Thoma, R., Schoch, G. A., Stihle, M., Benz, J., Wiget, A., et al. (2010). Enhancing the stability and solubility of the glucocorticoid receptor ligand-binding domain by high-throughput library screening. *Journal of Molecular Biology*, *403*, 562–577.

Seyfang, A., & Huaqian Jin, J. (2004). Multiple site-directed mutagenesis of more than 10 sites simultaneously and in a single round. *Analytical Biochemistry*, *324*(2), 285–291.

Smith, A. A. T., Belda, E., Viari, A., Medigue, C., & Vallenet, D. (2012). The CanOE strategy: Integrating genomic and metabolic contexts across multiple prokaryote genomes to find candidate genes for orphan enzymes. *PLoS Computational Biology*, *8*(5), e1002540.

Soumillion, P., & Fastrez, J. (2001). Novel concepts for selection of catalytic activity. *Current Opinion in Biotechnology*, *12*(4), 387–394.

Starr, T. N., & Thornton, J. W. (2016). Epistasis in protein evolution. *Protein Science*, *25*(7), 1204–1218.

Storz, J. F. (2018). Compensatory mutations and epistasis for protein function. *Current Opinion in Structural Biology*, *50*, 18–25.

Suzuki, Y., Asada, K., Miyazaki, J., Tomita, T., Kuzuyama, T., & Nishiyama, M. (2010). Enhancement of the latent 3-isopropylmalate dehydrogenase activity of promiscuous homoisocitrate dehydrogenase by directed evolution. *The Biochemical Journal*, *431*(3), 401–410.

Swainston, N., Batista-Navarro, R., Carbonell, P., Dobson, P. D., Dunstan, M., Jervis, A. J., et al. (2017). biochem4j: Integrated and extensible biochemical knowledge through graph databases. *PLoS ONE*, *12*(7), e0179130.

Swainston, N., Currin, A., Green, L., Breitling, R., Day, P. J., & Kell, D. B. (2017). CodonGenie: Optimised ambiguous codon design tools. *PeerJ Computer Science*, *3*, e120.

Tizei, P. A. G., Csibra, E., Torres, L., & Pinheiro, V. B. (2016). Selection platforms for directed evolution in synthetic biology. *Biochemical Society Transactions*, *44*(4), 1165–1175.

Toogood, H. S., Fryszkowska, A., Hulley, M., Sakuma, M., Mansell, D., Stephens, G. M., et al. (2011). A site-saturated mutagenesis study of pentaerythritol tetranitrate reductase reveals that residues 181 and 184 influence ligand binding, stereochemistry and reactivity. *ChemBioChem*, *12*(5), 738–749.

Turner, N. J. (2009). Directed evolution drives the next generation of biocatalysts. *Nature Chemical Biology*, *5*(8), 567–573.

Tyagi, R., Lai, R., & Duggleby, R. G. (2004). A new approach to "megaprimer" polymerase chain reaction mutagenesis without an intermediate gel purification step. *BMC Biotechnology*, *4*, 2.

Villarreal, F., & Tan, C. (2017). Cell-free systems in the new age of synthetic biology. *Frontiers of Chemical Science and Engineering*, *11*(1), 58–65.

Wang, L., Dash, S., Ng, C. Y., & Maranas, C. D. (2017). A review of computational tools for design and reconstruction of metabolic pathways. *Synthetic and Systems Biotechnology*, *2*(4), 243–252.

Wang, W., & Malcolm, B. A. (2002). Two-stage polymerase chain reaction protocol allowing introduction of multiple mutations, deletions, and insertions, using QuikChangeTM site-directed mutagenesis. In *In vitro mutagenesis protocols* (pp. 37–43). Totowa, NJ: Humana Press.

Weiß, M. S., Pavlidis, I. V., Vickers, C., Hohne, M., & Bornscheuer, U. T. (2014). Glycine oxidase based high-throughput solid-phase assay for substrate profiling and directed evolution of (R)- and (S)-selective amine transaminases. *Analytical Chemistry*, *86*(23), 11847–11853.

Wells, J. A., Vasser, M., & Powers, D. B. (1985). Cassette mutagenesis: An efficient method for generation of multiple mutations at defined sites. *Gene*, *34*(2–3), 315–323.

Wijma, H. J., Floor, R. J., Jekel, P. A., Baker, D., Marrink, S. J., & Janssen, D. B. (2014). Computationally designed libraries for rapid enzyme stabilization. *Protein Engineering, Design and Selection*, *27*(2), 49–58.

Willies, S. C., White, J. L., & Turner, N. J. (2012). Development of a high-throughput screening method for racemase activity and its application to the identification of alanine racemase variants with activity towards L-arginine. *Tetrahedron*, *68*(37), 7564–7567.

Wilson, R. (2015). Evolving BVMOs for increasing the yield and purity of enzyme catalysed oxidation. *Chimica Oggi/Chemistry Today*, *33*(2), 50–52.

Wittig, U., Kania, R., Golebiewski, M., Rey, M., Shi, L., Jong, L., et al. (2012). SABIO-RK—Database for biochemical reaction kinetics. *Nucleic Acids Research*, *40*(Database issue), D790–6.

Wooddell, C. I., & Burgess, R. R. (1996). Use of asymmetric PCR to generate long primers and single-stranded DNA for incorporating cross-linking analogs into specific sites in a DNA probe. *Genome Research*, *6*(9), 886–892.

Xiao, H., Bao, Z., & Zhao, H. (2015). High throughput screening and selection methods for directed enzyme evolution. *Industrial and Engineering Chemistry Research*, *54*(16), 4011–4020.

Xu, Z., Colosimo, A., & Gruenert, D. C. (2003). Site-directed mutagenesis using the megaprimer method. In *E. coli plasmid vectors* (pp. 203–208). New Jersey: Humana Press.

Yang, G., & Withers, S. G. (2009). Ultrahigh-throughput FACS-based screening for directed enzyme evolution. *ChemBioChem*, *10*(17), 2704–2715.

Yoo, T. H., Pogson, M., Iverson, B. L., & Georgiou, G. (2012). Directed evolution of highly selective proteases by using a novel FACS-based screen that capitalizes on the p53 regulator MDM2. *ChemBioChem*, *13*(5), 649–653.

Yuan, Z., Bailey, T. L., & Teasdale, R. D. (2005). Prediction of protein B-factor profiles. *Proteins: Structure, Function, and Bioinformatics*, *58*(4), 905–912.

Zheng, L., Baumann, U., & Reymond, J.-L. (2004). An efficient one-step site-directed and site-saturation mutagenesis protocol. *Nucleic Acids Research*, *32*(14), e115.

SECTION III

Enzyme Families: Parts and Platforms for Chemical Diversity

CHAPTER FOUR

Sesquiterpene Synthase-Catalyzed Conversion of a Farnesyl Diphosphate Analogue to a Nonnatural Terpenoid Ether

Florence Huynh, David J. Miller, Rudolf K. Allemann[1]
School of Chemistry, Cardiff University, Cardiff, United Kingdom
[1]Corresponding author: e-mail address: allemannrk@cardiff.ac.uk

Contents

Abstract

Sesquiterpene synthases catalyze the conversion of farnesyl diphosphate to more than 300 different hydrocarbon and alcohol natural products which often contain multiple fused rings and stereocenters. Recent work has taken advantage of the exquisite stereospecificity of these enzymes to synthesize complex novel sesquiterpenoids using nonnatural substrates. In this chapter, we describe the expression, purification, and use of one such synthase to convert a nonnatural substrate to a novel cyclic ether, thereby expanding the terpenome.

1. INTRODUCTION

Synthetic biology aims to engineer enzymes for a broad range of applications such as finding a biological pathway to produce complex molecules with multiple stereocenters that traditional synthetic chemistry cannot achieve in an easy and cost-effective manner. Terpenes are currently an

Methods in Enzymology, Volume 608
ISSN 0076-6879
https://doi.org/10.1016/bs.mie.2018.05.004

underexploited class of natural products representing a significant synthetic challenge due to their enormous variety of chemical structures. They comprise the biggest family of natural products with more than 80,000 known members (Buckingham, Cooper, & Purchase, 2016) and have a wide variety of applications in medicine, agriculture, and the food industry and as semiochemicals or fragrances. Well-known examples include the anticancer agent paclitaxel, the antimalarial artemisinin, the semiochemical germacrene D, or steroid hormones (Croteau, Ketchum, Long, Kaspera, & Wildung, 2006; Croteau, Kutchan, & Lewis, 2000; Pickett, Allemann, & Birkett, 2013). The biosynthesis of terpenoids starts from linear isoprenyl diphosphates that are converted to a diverse range of natural products often with multiple fused rings and stereocenters by terpene synthases. These structurally complex hydrocarbons or alcohols are formed through a Mg^{2+}-mediated cleavage of diphosphate to generate an initial carbocation that then undergoes a cascade of electrophilic ring closures, hydride, and methyl shifts culminating in proton loss and/or nucleophilic capture of a final carbocation by water (Christianson, 2017). For instance, farnesyl diphosphate (FDP), the C15 linear isoprenyl diphosphate precursor of sesquiterpenes, is converted to more than 300 different products by sesquiterpene synthases (Miller & Allemann, 2012). Most sesquiterpene synthases give multiple products from one substrate, but recent engineering has enabled us and others to control and modify their products. For example, δ-cadinene synthase has been converted into germacradien-4-ol synthase by only a single amino acid mutation (Loizzi, González, Miller, & Allemann, 2018). Yoshikuni et al. have also successfully modified the highly promiscuous cyclase γ-humulene synthase to give a series of selective mutant cyclases. These produced sesquiterpenes such as β-bisabolene and longifolene with a minimum number of mutations to change product distribution toward a specific sesquiterpene while suppressing side reaction products (Yoshikuni, Ferrin, & Keasling, 2006). Another strategy to create further terpenoid diversity is to take advantage of terpene synthase plasticity whereby FDP analogues with bulkier alkyl groups than the natural substrate and/or containing heteroatoms are often tolerated by the active site, generating nonnatural terpenoids. A successful use of this property is the production of dihydroartemisinic aldehyde (DHAA), a key intermediate in the synthesis of artemisinin from 12-hydroxyfarnesyl diphosphate (**3**) by amorphadiene synthase (ADS; Fig. 1). ADS catalyzes the conversion of FDP (**1**) to amorpha-4,11-diene (**2**) which can then be converted to DHAA in three chemical steps. By starting from this oxygenated FDP analogue (**3**),

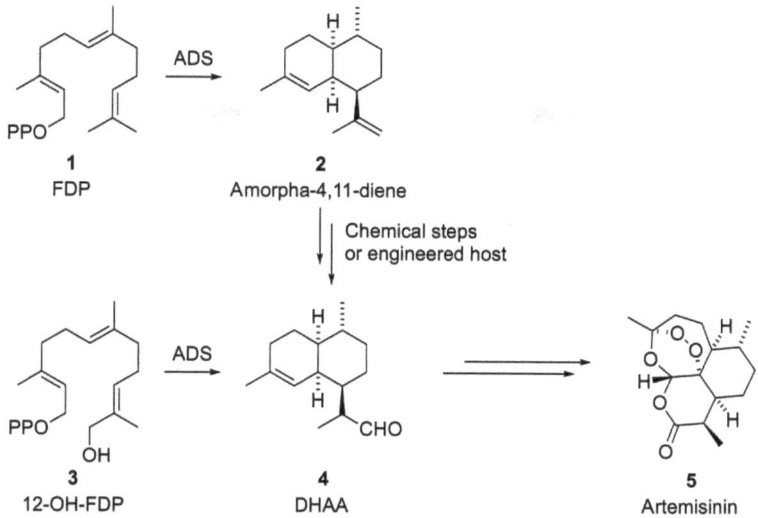

Fig. 1 Schematic synthesis of artemisinin from ADS-catalyzed cyclization of FDP and 12-OH FDP.

DHAA (**4**) was generated in a single step through ADS catalysis (Demiray, Tang, Wirth, Faraldos, & Allemann, 2017; Tang, Demiray, Wirth, & Allemann, 2018). Completely novel structures have also been synthesized under terpene synthase catalysis such as a 13-membered macrocyclic paracyclophane from anilinogeranyl diphosphate by 5-*epi*-aristolochene synthase (Rising et al., 2015).

Over decades, a deeper understanding of terpene synthases has also permitted enzyme engineering for unnatural substrate acceptance. Novel terpenoids have been successfully synthesized from substrate analogues as by-products from mechanistic investigation (Faraldos et al., 2012; Miller, Yu, & Allemann, 2007; Miller, Yu, Knight, & Allemann, 2009) or for biological applications such as (*S*)-14,15-dimethylgermacrene D by (*S*)-germacrene D synthase, a powerful aphid attractant (Cascón et al., 2012; Touchet et al., 2015). A pheromone used by destructive pine beetles, (+)-verbenone, was also synthesized in one pot from geranyl pyrophosphate (Yoosuf-Aly, Faraldos, Miller, & Allemann, 2012).

Here we describe how to express and purify one of the numerous sesquiterpene synthases and then employ it to catalyze the formation of a new sesquiterpenoid. We chose germacradiene-4-ol synthase (GdolS) from *Streptomyces citricolor* to illustrate the procedure. GdolS is a 315 amino acid bacterial sesquiterpene cyclase that catalyzes the conversion of FDP to

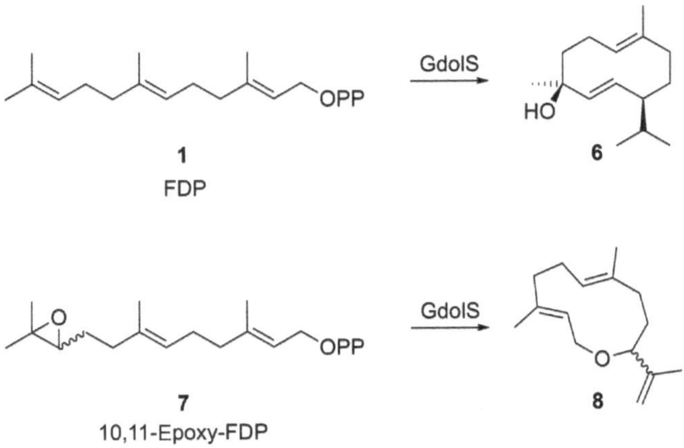

Fig. 2 GdolS-catalyzed cyclization of FDP (**1**) and 10,11-epoxy-FDP (**7**) to germacradien-4-ol (**6**) and cyclic ether **8**.

germacradien-4-ol (**6**) as the sole product (Grundy et al., 2016). Recently, we have shown that GdolS also accepts other substrates such as 10,11-epoxy-FDP (**7**), which it turns over to the 11-membered cyclic ether **8** (Fig. 2) (Huynh, Grundy, Jenkins, Miller, & Allemann, accepted).

2. SESQUITERPENES PRODUCTION AND PURIFICATION

2.1 Recombinant Enzyme Production and Purification

The gene encoding GdolS is included in the pET16b-SC1 expression vector under the control of a T7 promoter (a generous gift from Prof. Yasuo Ohnishi, University of Tokyo). It contains a C-terminal 10xHis tag (Grundy, 2015) and is readily expressed in *Escherichia coli* BL21-(DE3) and the resulting protein purified from the cell lysate.

Equipment needed:

- Temperature-controlled oven
- Temperature-controlled water bath
- Temperature-controlled rotary shaker with clamp for 2L Erlenmeyer flask
- Autoclave
- High-speed centrifuge and rotors for 1L and 50mL samples
- Ni^{2+}–NTA column (5mL for 1L expression)
- SDS-PAGE equipment, 12% SDS-PAGE gels
- Dialysis tubing (10- to 12-kDa cutoff, 3cm flat width)
- EMD Millipore Amicon™ Bioseparations stirred cells with ultrafiltration membrane 30kDa cutoff

- Nitrogen cylinder
- pH meter
- Shimadzu UV-2401 PC spectrophotometer
- Quartz cuvette 1 mL

Materials required:

- LB plate (200 μg/mL ampicillin)
- LB medium
- Stock solution of 1 *M* isopropyl β-D-1-thiogalactopyranoside (IPTG)
- Ampicillin 100 mg/mL stock filter sterilized
- Resuspension buffer: 50 m*M* Tris (pH 8.0), 100 m*M* NaCl
- Elution buffer 50 m*M* Tris (pH 8.0), 100 m*M* NaCl, 500 m*M* imidazole
- Dialysis buffer 10 m*M* Tris (pH 8.0), 50 m*M* NaCl
- Bovine serum albumin (BSA)
- Bradford reagent (can be purchased or prepared as described in the literature (Bradford, 1976))

(1) Grow an overnight culture of BL21 (DE3), transformed with pET16b-SC1 harboring GdolS gene at 37°C with shaking in 100 mL LB medium with 100 μg/mL ampicillin.

(2) Inoculate 750 mL LB medium + 100 μg/mL ampicillin with 5 mL of the overnight stationary-phase culture from step 1. Grow the culture at 37°C with shaking at 250 rpm until OD_{600} is 0.6. Induce gene expression by addition of 150 μL of IPTG stock solution and continue the growth at 37°C for 3 h.

(3) Harvest the cells by centrifugation at 3400 × *g* at 4°C for 15 min and resuspend the resultant pellet in the resuspension buffer (50 m*M* Tris pH 8.0, 100 m*M* NaCl). Typically 20 mL of buffer is used per 750 mL culture.

Tip: Cell pellets may be stored at −20°C until required.

(4) Lyse the cells by sonication for 3 min with bursts of 5 s on and 10 s off at 37% amplitude with the cell suspension kept cold in an ice-water bath.

(5) Centrifuge at 38,000 × *g* for 30 min at 4°C and load the supernatant onto a 5-mL Ni^{2+}–NTA column preequilibrated with 5 column volumes (CV) of resuspension buffer. Wash the resin with a gradient of diluted elution buffer (5, 20, 60, 100, 300 m*M*). For each concentration 10 CV is applied.

(6) Use SDS-PAGE analysis (12% gel) to determine which fractions contain protein with the expected molecular weight for GdolS (molecular weight 38,687 Da)

(7) Pool the fractions into dialysis tubing and dialyze against dialysis buffer (10 m*M* Tris (pH 8.0), 50 m*M* NaCl) at 4°C overnight. Typically use a volume of dialysis buffer at least 100-fold greater than the sample volume.

(8) Concentrate the dialyzed sample to 10 mL using Amicon ultrafiltration (1.5 bars pressure). Assess purity using 12% SDS-PAGE.

Tips: GdolS can be stored at −80°C in 20% glycerol until further use.

(9) Determine the total protein concentration using the method of Bradford (Bradford, 1976; Zor & Selinger, 1996). The total protein concentration is determined by comparison to a calibration curve of different BSA concentrations (0–100 μg/mL). The calibration curve is obtained by plotting the concentration of BSA against the ratio of absorbance at 595–450 nm.

2.2 Determination of Optimum Catalysis Conditions

Several parameters influence the catalytic efficiency of sesquiterpene synthases. The optimum pH, enzyme, magnesium ion, and substrate concentrations are determined from a radiolabeled assay using [1-^{3}H]-(*E*,*E*)-FDP (Cane, Wu, Proctor, & Hohn, 1993).

Equipment needed:

- Scintillation counter (Packard 2500 TR™)
- Mini vortex mixer
- Scintillation vial with cap (20 mL)
- Plugged small Pasteur pipette with silica gel (pore size 60 Å, 230–400 mesh)
- Dry block heating and cooling shaker for Eppendorf tubes

Materials needed:

- Incubation buffer I (20 mL): 50 m*M* HEPES (pH 8.0), 5 m*M* β-mercaptoethanol
- Incubation buffer IIa (20 mL): 50 m*M* MES (pH 6.0), 5 m*M* β-mercaptoethanol, 2.5 m*M* $MgCl_2$
- Incubation buffer IIb (20 mL): 50 m*M* MES (pH 6.5), 5 m*M* β-mercaptoethanol, 2.5 m*M* $MgCl_2$
- Incubation buffer IIc (20 mL): 50 m*M* HEPES (pH 7.0), 5 m*M* β-mercaptoethanol, 2.5 m*M* $MgCl_2$
- Incubation buffer IId (20 mL): 50 m*M* HEPES (pH 7.5), 5 m*M* β-mercaptoethanol, 2.5 m*M* $MgCl_2$
- Incubation buffer IIe (20 mL): 50 m*M* HEPES (pH 8.0), 5 m*M* β-mercaptoethanol, 2.5 m*M* $MgCl_2$

- Incubation buffer IIf (20 mL): 50 mM HEPES (pH 8.5), 5 mM β-mercaptoethanol, 2.5 mM $MgCl_2$
- Incubation buffer IIg (20 mL): 50 mM CHES (pH 9.0), 5 mM β-mercaptoethanol, 2.5 mM $MgCl_2$
- Incubation buffer IIh (20 mL): 50 mM CHES (pH 9.5), 5 mM β-mercaptoethanol, 2.5 mM $MgCl_2$
- Incubation buffer IIi (20 mL): 50 mM CHES (pH 10.0), 5 mM β-mercaptoethanol, 2.5 mM $MgCl_2$
- 0.5 M EDTA
- Stock solution of 1 M $MgCl_2$
- 11:1 hexane/diethyl ether
- Stock solution of 500 mM [1-^{3}H]-FDP (15,000–30,000 dpm nmol^{-1})
- Ecoscint™ A fluid (National Diagnostics)

2.2.1 *Magnesium Ion Concentration*

(1) Assays are carried out in a total volume of 250 μL. Incubation solutions are prepared in 1.5 mL Eppendorf tubes at 0°C by successive addition of incubation buffer, magnesium stock solution, and tritium-labeled FDP. The final volume is completed by addition of different magnesium ion concentrations from 0.5 to 15 mM and fixed enzyme concentration (5 nM), hot-FDP (4 μM), and incubation buffer I.

(2) Overlay reactions with 1 mL hexane.

(3) Incubate at 37°C for 10 min with shaking and then cool on ice immediately.

(4) Add 50 μL EDTA stock solution (0.5 M) to each incubation and vortex for 30 s.

(5) Pass the organic layer through a small plug of silica into 15 mL Ecoscint™ A fluid.

(6) Add 750 μL 11:1 hexane/diethyl ether to the remaining aqueous layer and vortex 10 s. Repeat step 5. Then repeat this step a second time.

(7) Wash the small plug of silica with 750 μL 11:1 hexane/diethyl ether and add to the Ecoscint™ fluid.

(8) Analyze the radioactivity of each sample of Ecoscint™ fluid containing the combined organic extract on a scintillation counter in ^{3}H mode (4 min per sample).

Tips: Always determine background activity by repeating all the incubations but this time without enzyme. For accuracy, repeat the experiment at least three times.

(9) Fit the data into the Michaelis–Menten equation using the Systat Sigmaplot package (Systat Sigmaplot 10.0, 2007) (Sigmaplot for

Windows Version 10.0, Build 10.0.0.54, 2006, Systat Software Inc. 1735, Technology Drive, Ste. 430, San Jose, CA 95110, USA. Molecular Operating Environment (MOE 2004.03) Chemical Computing Group ULC, 1010 Sherbooke St. West, Suite #910, Montreal, QC, Canada, H3A 2R7, 2018).

(10) Plot the concentration of magnesium ions against V_{max}. The rate is typically proportional to the magnesium ion concentration until saturating conditions are reached. Therefore, the concentration of magnesium ions giving a rate of half V_{max} was selected for all future assays.

2.2.2 Optimum pH

(1) Perform small incubations of 250 μL total volume in 1.5 mL Eppendorf tubes at 0°C varying pH by using incubation buffer IIa–IIi with a fixed hot-FDP concentration (4 μ*M*) and fixed enzyme concentration (5 n*M*).

(2) Repeat steps 2 to 10 from Section 2.2.1

(3) Plot the pH against V_{max}. The optimum pH is found close to the plateau before the point where the rate decreases.

2.2.3 Enzyme Concentration

(1) Perform small incubations of 250 μL total volume in 1.5 mL Eppendorf tubes at 0°C using a range of different enzyme concentrations from 5 n*M* to 5 μ*M* and fixed hot-FDP concentration (4 μ*M*), incubation buffer at the optimal pH, and magnesium ion concentration, in this example (50 m*M* HEPES, 5 m*M* $MgCl_2$, 5 m*M* βME, pH 8.0).

(2) Repeat steps 2–10 from Section 2.2.1

(3) Plot the enzyme concentration against V_{max}. Optimum concentration is found at half V_{max}. The rate will plateau as sesquiterpenes aggregate at high concentrations.

3. ENZYMATIC SYNTHESIS OF CYCLIC ETHER

After the successful expression of the sesquiterpenes synthase, the recombinant enzyme can be used to catalyze the conversion of a selected substrate to a sesquiterpenoid. Conversion of 10,11-epoxy-FDP by GdolS will be illustrated in this chapter. When using an unnatural substrate, it is advised to always test the activity of the enzyme against its natural substrate. This procedure is described in Section 3.1.

3.1 Enzyme Activity Assay

Equipment needed:

- Shaker with small vial holder
- Mini vortex mixer
- Gas chromatography–mass spectrometer: Perkin Elmer Elite-1 capillary column, 30 m × 0.25 mm, 0.25 µm, 100% dimethyl polysiloxane

Materials needed:

- FDP (can be purchased or synthesized as described in the literature (Davisson et al., 1986))
- Incubation buffer 50 m*M* HEPES (pH 8.0), 5 m*M* β-mercaptoethanol, 5 m*M* $MgCl_2$
- Pentane (HPLC grade)

(1) Dissolve 4.2 mg of FPP in 1 mL deionized water. This is the stock solution.

Tips: Aliquots of 250 µL are usually stored at −80°C until further use.

(2) Add, in a 2 mL screw cap glass vial, the incubation buffer (200 µL), the purified enzyme (1 µ*M*), and then FDP (200 µ*M*).

(3) Overlay the sample with 800 µL pentane.

(4) Incubate the reaction on a shaker at room temperature for 12 h with shaking 200 rpm.

(5) Stop the enzymatic reaction by mixing the two phases with a mini vortex mixer and allow the two phases to separate.

(6) Transfer the organic layer using a glass pipette, equipped with a rubber pipette bulb, into a glass vial.

(7) Analyze the sample using gas chromatography–mass spectrometry with the following column: Perkin Elmer Elite-1 capillary column, 30 m × 0.25 mm, 0.25 µm, 100% dimethyl polysiloxane. The following GC settings are used: injection volume 1 µL, injector temperature 100°C, injector split 5:1, oven heating program: initial hold of 80°C for 1 min, followed by a ramp to 180°C (4°C min^{-1}), and final hold at 180°C for 2 min.

Tips: A chiral column may be used as chiral products have been formed.

3.2 Preparative Scale Incubation

Equipment needed:

- Magnetic stirrer
- Polygon magnetic stirring bar (*l* = 60 mm, diameter = 10 mm)
- 250 mL separating funnel
- Rotary evaporator

- TLC tank
- UV lamp (365 nm)

Materials needed:

- Pentane (HPLC grade)
- Incubation buffer 50 m*M* HEPES (pH 8.0), 5 m*M* β-mercaptoethanol, 5 m*M* $MgCl_2$
- Magnesium chloride stock solution in water (1 *M*)
- Dichloromethane
- Sodium sulfate anhydrous
- Florisol®, 60–100 mesh
- TLC silica gel $60F_{254}$, aluminum sheets 20 × 20 cm
- 0.1% berberine hydrochloride in ethanol
- Ether (HPLC grade)
- 10,11-Epoxy-FDP stock solution (10 m*M*)

(1) Add the enzyme (3 μ*M*) and substrate (0.35 m*M*) into a 500-mL flask and make up the total volume to 100 mL with buffer. Add additional magnesium chloride to adjust the final magnesium concentration to 5 m*M* as the enzyme and substrate solutions lack magnesium.

(2) Add a polygon magnetic stirring bar (l = 60 mm, diameter = 10 mm)

(3) Overlay the solution with 200 mL of pentane.

(4) Leave the reaction to stir at ~25 rpm for 12 h at room temperature.

(5) Remove the pentane layer using a glass pipette and keep it in a sealed glass bottle.

(6) Add 200 mL of fresh pentane and leave the reaction to stir again at ~25 rpm for 12 h at room temperature.

(7) Pour the two phases into a separating funnel and separate the two layers.

Tips: The enzyme can be used again: The pentane layer should be removed with a glass pipette, then the substrate can be added, and steps 3 and 4 are repeated.

(8) Extract the aqueous layer with dichloromethane (3 × 150 mL)

Tips: The biphasic system can be centrifuged to break the emulsion.

(9) Combine all the organic layers, dry (Na_2SO_4), and filter.

(10) Remove the solvent under reduced pressure (400 mbar, 30°C)

Tips: A splash head adapter is recommended to avoid loss of compound.

(11) Dilute the residue with approx. 0.5 mL of diethyl ether.

(12) Pass the sample through a short plug of Florisil®.

(13) Purify the products using a preparative TLC plate in 5% ether/pentane on a berberine prestained silica plate. The visualization is performed using a UV lamp at (365 nm).

Tips: If the desired band is not well separated, repeat the run. Optimal purification of a novel terpenoid can require optimization and specific changes of this protocol. A concentration of more than 5% ether generally leads to a single band on the plate. When a separation is difficult to achieve the TLC plate can be impregnated beforehand with 1%–20% silver nitrate in acetonitrile (these plates cannot be stored over a long time as silver ions are photoreactive). It has been shown that cyclic or acyclic terpenoids containing an exocyclic or terminal alkene are able to bind a silver oleofin complex but those containing one or two internal alkenes (unless cis conjugated) do not (Lawrence, 1968). Some terpenoids are acid sensitive, therefore the plate can be deactivated by soaking in 1% triethylamine in dichloromethane and drying it before use.

(14) The product band is scraped off the TLC plate using a clean spatula. Extract the resulting silica with 5 mL pentane by vigorous shaking. Filter the mixture and concentrate the purified compound.

Tips: Sesquiterpenes are quite volatile and have a high vapor pressure; it is advised to avoid leaving the compound under high vacuum or reduced pressure for prolonged periods. Always test for the presence of the compound in the rotary evaporator by collecting from the flask and analyzing GC–MS.

(15) Characterize the isolated compound by nuclear magnetic resonance spectroscopy as well as mass spectrometry.

4. CONCLUSION

Over the last decade, numerous novel sesquiterpenoids have been synthesized using chemoenzymatic processes with modified substrates. FDP substrates modified with hydroxy, fluoro, or methyl groups not only result in new compounds but can also provide access to novel activities, as seen with (14,15-dimethyl)-GDS (Touchet et al., 2015). Often chemoenzymatic approaches to terpenoids allow a shortening of the synthetic pathway as was the case for the production of dihydroartemisinic aldehyde, a key intermediate for the generation of artemisinin (Demiray et al., 2017). The heterologous expression of germacradien-4-ol synthase in *E. coli* allowed us to screen a variety of modified substrates. One of these generated an unusual cyclic ether, a linkage currently not known to be formed as a direct product of terpene cyclases in nature; it represents an entirely novel expansion of the terpenome. The use of unnatural substrates with sesquiterpene synthases is an elegant route to create new structures in a cost-effective and ecologically friendly manner. In our hands, this method has already been applied to aristolochene synthase, germacrene A synthase, and δ-cadinene synthase to yield ethereal products. This methodology may prove useful for discovering novel biologically active compounds and for easy access to chiral building blocks.

REFERENCES

Bradford, M. M. (1976). A rapid and sensitive method for the quantitation of microgram quantities of protein utilizing the principle of protein-dye binding. *Analytical Biochemistry, 72*, 248–254.

Buckingham, J., Cooper, C., & Purchase, R. (2016). *Natural products desk reference.* Boca Raton: CRC Press, Taylor & Francis Group.

Cane, D. E., Wu, Z., Proctor, R. H., & Hohn, T. M. (1993). Overexpression in Escherichia coli of soluble aristolochene synthase from Penicillium roqueforti. *Archives of Biochemistry and Biophysics, 304*, 415–419.

Cascón, O., Touchet, S., Miller, D. J., González, V., Faraldos, J. A., & Allemann, R. K. (2012). Chemoenzymatic preparation of germacrene analogues. *Chemical Communications, 48*, 9702–9704.

Christianson, D. W. (2017). Structural and chemical biology of terpenoid cyclases. *Chemical Reviews, 117*, 11570–11648.

Croteau, R., Ketchum, R. E. B., Long, R. M., Kaspera, R., & Wildung, M. R. (2006). Taxol biosynthesis and molecular genetics. *Phytochemistry Reviews, 5*, 75–97.

Croteau, R., Kutchan, T. M., & Lewis, N. G. (2000). Natural products (secondary metabolites). In R. J. B. Buchanan & W. Gruissem (Eds.), *Biochemistry & molecular biology of plant* (pp. 1250–1318). American Society of Plant Physiologists.

Davisson, V. J., Woodside, A. B., Neal, T. R., Stremler, K. E., Muehlbacher, M., & Poulter, C. D. (1986). Phosphorylation of isoprenoid alcohols. *The Journal of Organic Chemistry, 51*, 4768–4779.

Demiray, M., Tang, X., Wirth, T., Faraldos, J. A., & Allemann, R. K. (2017). An efficient chemoenzymatic synthesis of dihydroartemisinic aldehyde. *Angew. Chem. Int. Ed., 56*, 4347–4350.

Faraldos, J. A., Miller, D. J., González, V., Yoosuf-Aly, Z., Cascón, O., Li, A., et al. (2012). A 1,6-ring closure mechanism for (+)-δ-cadinene synthase? *Journal of the American Chemical Society, 134*, 5900–5908.

Grundy, D. J. (2015). *Expanding the terpenome: Complementary approaches to novel terpenoids.* Cardiff University, PhD Thesis.

Grundy, D. J., Chen, M., González, V., Leoni, S., Miller, D. J., Christianson, D. W., et al. (2016). Mechanism of germacradien-4-ol synthase-controlled water capture. *Biochemistry, 55*, 2112–2121.

Huynh F., Grundy D.J., Jenkins R.L., Miller D.J., & Allemann R.K., Sesquiterpene synthase catalysed formation of a novel medium sized cyclic terpenoid ether from farnesyl diphosphate analogues, *Chembiochem*, accepted.

Lawrence, B. M. (1968). The use of silver nitrate impregnated silica gel layers in the separation of monoterpene hydrocarbons. *Journal of Chromatography A, 38*, 535–537.

Loizzi, M., González, V., Miller, D. J., & Allemann, R. K. (2018). Nucleophilic water capture or proton loss: Single amino acid switch converts δ-cadinene synthase into germacradien-4-ol synthase. *Chembiochem, 19*, 100–105.

Miller, D. J., & Allemann, R. K. (2012). Sesquiterpene synthases: Passive catalysts or active players? *Natural Product Reports*, *29*, 60–71.

Miller, D. J., Yu, F., & Allemann, R. K. (2007). Aristolochene synthase-catalyzed cyclization of 2-fluorofarnesyl-diphosphate to 2-fluorogermacrene A. *Chembiochem*, *8*, 1819–1825.

Miller, D. J., Yu, F., Knight, D. W., & Allemann, R. K. (2009). 6- and 14-Fluoro farnesyl diphosphate: Mechanistic probes for the reaction catalysed by aristolochene synthase. *Organic & Biomolecular Chemistry*, *7*, 962–975.

Pickett, J. A., Allemann, R. K., & Birkett, M. A. (2013). The semiochemistry of aphids. *Natural Product Reports*, *30*, 1277–1283.

Rising, K. A., Crenshaw, C. M., Koo, H. J., Subramanian, T., Chehade, K. A. H., Starks, C., et al. (2015). Formation of a novel macrocyclic alkaloid from the unnatural farnesyl diphosphate analogue anilinogeranyl diphosphate by 5-epi-aristolochene synthase. *ACS Chemical Biology*, *10*, 1729–1736.

Tang, X., Demiray, M., Wirth, T., & Allemann, R. K. (2018). Concise synthesis of artemisinin from a farnesyl diphosphate analogue. *Bioorganic & Medicinal Chemistry*, *26*(7), 1314–1319.

Touchet, S., Chamberlain, K., Woodcock, C. M., Miller, D. J., Birkett, M. A., Pickett, J. A., et al. (2015). Novel olfactory ligands via terpene synthases. *Chemical Communications*, *51*, 7550–7553.

Yoosuf-Aly, Z., Faraldos, J. A., Miller, D. J., & Allemann, R. K. (2012). Chemoenzymatic synthesis of the alarm pheromone (+)-verbenone from geranyl diphosphate. *Chemical Communications*, *48*, 7040–7042.

Yoshikuni, Y., Ferrin, T. E., & Keasling, J. D. (2006). Designed divergent evolution of enzyme function. *Nature*, *440*, 1078–1082.

Zor, T., & Selinger, Z. (1996). Linearization of the Bradford protein assay increases its sensitivity: Theoretical and experimental studies. *Analytical Biochemistry*, *236*, 302–308.

CHAPTER FIVE

In Vivo Platforms for Terpenoid Overproduction and the Generation of Chemical Diversity

Guangkai Bian*, Tian Ma*, Tiangang Liu*,†,1

*Key Laboratory of Combinatorial Biosynthesis and Drug Discovery, Ministry of Education and School of Pharmaceutical Sciences, Wuhan University, Wuhan, PR China

†Hubei Engineering Laboratory for Synthetic Microbiology, Wuhan Institute of Biotechnology, Wuhan, PR China

[1]Corresponding author: e-mail address: liutg@whu.edu.cn

Contents

Abstract

Terpenoids represent a highly diverse group of natural products with wide applications. Engineering approaches have been used to increase titers of many value-added terpenoids, such as farnesene, taxadiene, lycopene, and astaxanthin. In this chapter, we review the in vitro reconstitution-based targeted engineering of terpenoids, as well as approaches for the mining of terpene cyclases and for increasing the chemical diversity. Information gained from in vitro reconstitution extends our understanding of the mechanisms underlying terpenoid biosynthesis, the contributions of enzymes and cofactors, and key enzymes and rate-limiting steps for the development of an ideal biosynthetic production system. The in vitro reconstitution-based targeted engineering strategy provides a rational and accurate engineering approach for terpenoid

Methods in Enzymology, Volume 608
ISSN 0076-6879
https://doi.org/10.1016/bs.mie.2018.04.025

overproduction with high efficiency. Furthermore, an efficient terpenoid overproduction platform can accelerate the entire process for the mining of terpene cyclases and the discovery of novel terpenoids and can substantially increase the chemical diversity of these kinds of terpenoids.

1. INTRODUCTION

Terpenoids are widely distributed in plants, microorganisms, insects, and various marine organisms. They represent the most diverse group of natural products. It has been estimated that more than 80,000 terpenoids have been structurally characterized (Pemberton et al., 2017).

Two distinct pathways, the methylerythritol phosphate (MEP) pathway and the mevalonate (MVA) pathway, in nature can produce the two universal 5-carbon isoprenoid building blocks, isopentenyl diphosphate (IPP) and dimethylallyl diphosphate (DMAPP), for the production of terpenoids. Subsequently, prenyltransferase assembles DMAPP and IPP to generate polyisoprenoid pyrophosphates with different lengths (C_{10}–C_{25}), such as geranyl diphosphate (GPP), farnesyl diphosphate (FPP), geranylgeranyl diphosphate (GGPP), and geranylfarnesyl diphosphate (GFPP). Terpenoids are then produced via terpene synthases using the polyisoprenoid pyrophosphates as substrates, resulting in various terpene skeletons (mono-, sesqui-, di-, sester-, tri-, tetra-, and polyterpenes). Finally, terpene skeletons are modified by tailor enzymes, such as cytochrome P450s, glycosyltransferase, and halogenase, to generate terpenoids with high structural diversity and complexity. Owing to their structural diversity, terpenoids have a wide range of applications; for example, they are used in food products, as antitumor and antimalaria agents, and as new jet fuel (Paddon et al., 2013; Wani, Taylor, Wall, Coggon, & McPhail, 1971; Zhu et al., 2015, 2014). However, the practical application of these value-added terpenoids is significantly hindered by low productivity and high cost.

Recent developments in metabolic engineering and synthetic biology provide alternative methods for producing value-added terpenoids in *Escherichia coli*, *Saccharomyces cerevisiae*, and filamentous fungi, resulting in several remarkable breakthroughs. The efficient overproduction of terpenoids, such as taxadiene (Ajikumar et al., 2010) and artemisinic acid (Paddon et al., 2013), has been realized by optimizing the precursor supply, balancing upstream and downstream pathways, balancing redox reactions, and introducing a dynamic regulatory system. To improve the convenience and efficiency of engineering, the full reconstitution of the MVA pathway has been investigated using a systematic kinetic analysis. An in vitro reconstitution-based targeted metabolic engineering approach was developed for rationally and accurately engineering terpenoid overproduction with high efficiency (Zhu et al., 2014). Directed by this strategy, based on the well-engineered MVA pathway, high titers of farnesene, lycopene, taxadiene, and astaxanthin have been obtained using *E. coli* (Fig. 1), demonstrating the high efficiency and extensive applicability of the targeted engineering strategy (Bian, Yuan, Tao, Shi, & Liu, 2017; Ma et al., 2016; Zhu et al., 2015, 2014).

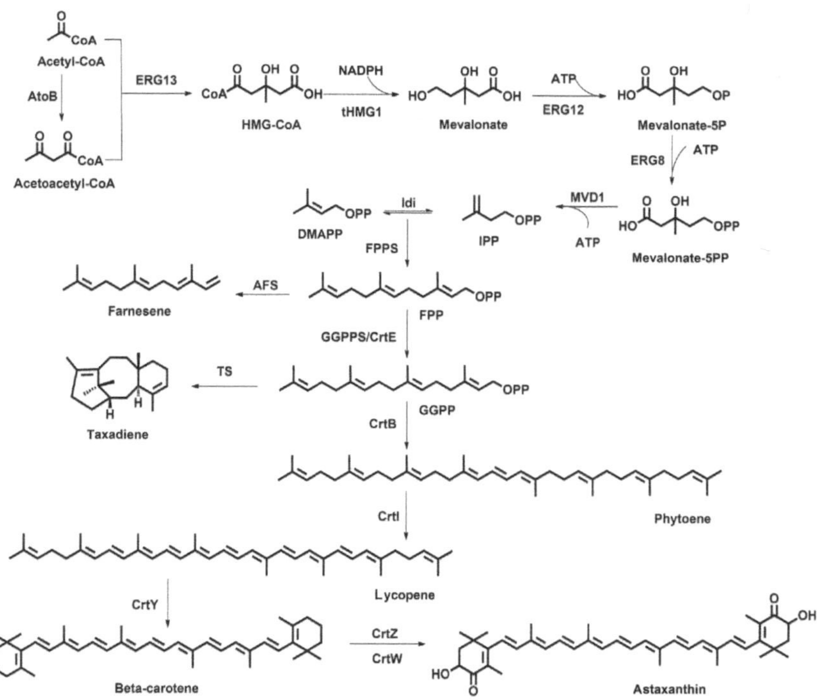

Fig. 1 Biosynthesis of farnesene, taxadiene, lycopene, and astaxanthin via the MVA pathway. *AFS*, α-farnesene synthase; *AtoB*, acetyl-CoA acetyltransferase; *CrtB*, phytoene synthase; *CrtI*, phytoene desaturase; *CrtW*, beta-carotene ketolase; CrtY, lycopene cyclase; *CrtZ*, beta-carotene hydroxylase; *DMAPP*, dimethylallyl pyrophosphate; *ERG8*, phosphomevalonate kinase; *ERG12*, mevalonate kinase; *ERG13*, HMG-CoA synthase; *FPP*, farnesyl pyrophosphate; *FPPS*, FPP synthase; *GGPP*, geranylgeranyl diphosphate; *GGPPS/CrtE*, GGPP synthase; HMG-CoA, (*S*)-3-hydroxy-3-methylglutaryl-CoA; *Idi*, IPP isomerase; *IPP*, isopentenyl diphosphate; *IspA*, FPPS from *E. coli*; *MVD1*, mevalonate pyrophosphate decarboxylase; *tHMG1*, the truncated 3-hydroxy-3-methylglutaryl-CoA reductase; *TS*, taxadiene synthase.

A large number of terpenoids have been structurally characterized, and it is unclear how many terpenoids exist in nature that are yet to be discovered. Various terpenoids as well as their biosynthetic elements have been characterized from diverse sources, such as the sesquiterpene synthase from *Streptomyces pristinaespiralis* (Klapschinski, Rabe, & Dickschat, 2016), the diterpene cyclases from *Hericium erinaceum* (Yang et al., 2017), and the sesterterpene synthase from plants and filamentous fungi (Bian, Han, Hou, Yuan, & Liu, 2017; Huang et al., 2017; Qin et al., 2015; Ye et al., 2015). The discovery of new terpene cyclases and new terpenoids provides a basis for understanding the biosynthetic mechanism, enriching the structural diversity, and screening for new bioactive compounds. An efficient terpenoid overproduction platform can accelerate the entire process for the mining of terpene cyclases and the discovery of terpenoids. Based on a well-developed terpenoid overproduction platform, the functions of promiscuous terpene cyclases FgMS (*Fusarium graminearum* mangicdiene synthase) and FgGS (*F. graminearum* GJ1012 synthase) were fully exploited to produce seven novel diterpenes and sesterterpenes with three unusual skeletons (Bian, Han, et al., 2017). Furthermore, by using well-engineered precursor production platforms, combined with a combinatorial biosynthesis strategy, a large number of terpenoids have been generated after the combination of class I and class II diterpene synthases from various sources (Andersen-Ranberg et al., 2016; Jia, Potter, & Peters, 2016). Consequently, a metabolic engineering-based terpenoid discovery strategy may accelerate the process of novel terpene skeletons identification and provide bioactive or high value-added compounds for drug discovery and industrial applications. In this chapter, we outline a targeted engineering strategy for the overproduction of terpenoids and the mining of terpene cyclases to increase chemical diversity.

2. IN VITRO RECONSTITUTION OF THE MVA PATHWAY AND TARGETED ENGINEERING OF VALUE-ADDED TERPENOIDS

2.1 In Vitro Reconstitution of the MVA Pathway

In vitro reconstitution is a powerful method to monitor the detailed information regarding enzymatic reaction mechanisms, steady-state kinetics, biochemical parameters, and the accumulation of intermediates (Lowry, Walsh, & Khosla, 2015). The information obtained by this method can improve our understanding of the mechanism underlying terpenoid synthesis, the contributions of particular enzymes and cofactors, and the key enzymes and rate-limiting steps for an ideal biosynthetic production system. Directed by this strategy, researchers have realized the efficient overproduction of fatty acids and terpenoids efficiently in vitro and in vivo (Korman, Opgenorth, & Bowie, 2017; Liu et al., 2017; Yu, Liu, Zhu, & Khosla, 2011; Zhu et al., 2014).

For the in vitro reconstitution of the MVA pathway (Fig. 2), farnesene was selected as an indicator to evaluate the efficiency of genes related to this pathway. The *atoB*, *idi*, and *ispA* genes of *E. coli*, the *erg13*, *thmg1* (Polakowski, Stahl, & Lang, 1998), *erg12*, *erg8*, and *mvd1* (also known as *erg19*) of *S. cerevisiae*, and codon-optimized *afs* from *Malus × domestica* (Pechous & Whitaker, 2004) were selected and cloned into pET28a(+) for protein overexpression and purification (Zhu et al., 2014). Then, the plasmids for protein expression were individually transformed into *E. coli* BL21 (DE3). The proteins were purified and maintained in storage buffer (100 m*M* phosphate buffer, 10% glycerol, pH 7.6).

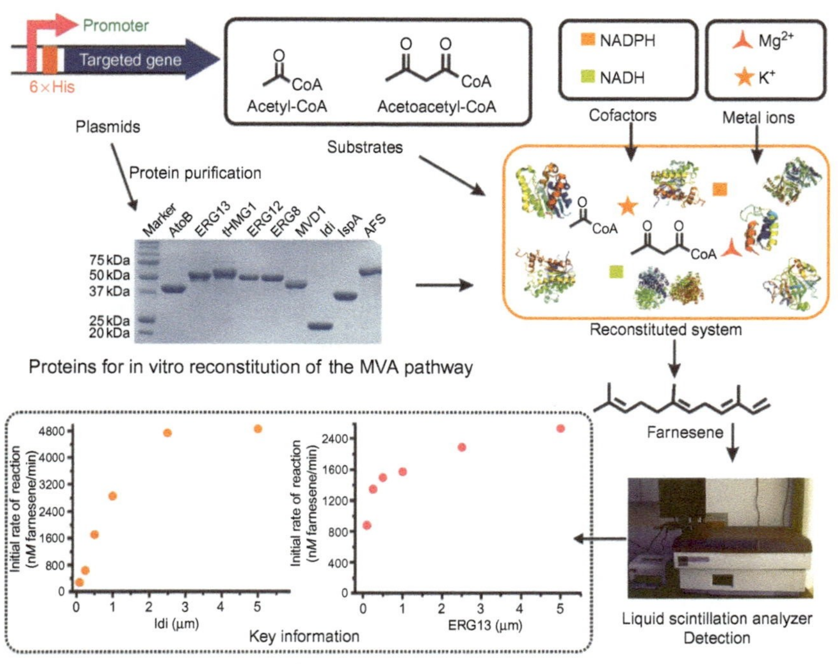

Fig. 2 In vitro reconstitution of MVA pathway.

An in vitro reconstitution assay was performed and the production of farnesene was detected by GC/MS and radioactive thin-layer chromatography. Based on a kinetic analysis, increasing the concentrations of ERG13, ERG8, IspA, and AFS from 0.1 to 5 μ*M* can mildly (1.5- to 2.7-fold) promote the production of farnesene. Increasing the concentration of Idi (0.1–5 μ*M*) can substantially increase the initial rate of farnesene production by 17-fold (Fig. 2), indicating the key role of Idi in terpenoid engineering. Additionally, an analysis of synergistic effects showed that simultaneously increasing the concentrations of Idi together with ERG13, ERG8, and AFS can increase the production of farnesene by fivefold. Based on the results of titration studies, the molar ratio of AtoB:ERG13:tHMG1:ERG12:ERG8:MVD1:Idi:IspA:AFS was adjusted slightly to 1:10:2:5:5:2:5:2:2 (Zhu et al., 2014). These previous studies provide an ultimate goal for metabolic engineering to realize the overproduction of terpenoids with high efficiency.

Instead of constructing a series of mutants to evaluate the contributions of enzymes to terpenoid yield, key information from in vitro assays provides direction for the engineering of terpenoid production in microorganisms. Thus, this approach provides a fantastic strategy for the realization of the overproduction of value-added natural products in a convenient and efficient manner.

2.1.1 Protein Expression and Purification for In Vitro Reconstitution

The procedures for protein expression and purification for in vitro reconstitution can be summarized as follows.

1. Plasmids related to the expression of each enzyme in the MVA pathway and the downstream farnesene production pathway are transformed into *E. coli* BL21 (DE3) for protein purification.
2. Mutants are cultivated in 2-L flasks containing 1 L of LB medium at 37°C with 50 mg/L kanamycin (Kan).
3. When the OD_{600} reaches 0.6–0.8, 0.1 m*M* IPTG is added to the culture, followed by cultivation for an additional 18 h at 16°C.
4. Cells are harvested and resuspended in 40 mL of buffer A (50 m*M* Tris–Cl, 300 m*M* NaCl, 4 m*M* β-mercaptoethanol, pH 7.6).
5. A high-pressure homogenizer is used to lyse the cell suspensions, followed by centrifugation at 30,000 × *g* for 1 h.

6. The supernatant is filtered and supplemented with 6% buffer B (500 m*M* imidazole in buffer A) to adjust the supernatant to 30 m*M* imidazole to reduce nonspecific binding.
7. The samples are subjected to metal affinity chromatography using a Ni-NTA column.
8. The sample is supplemented with 8% buffer B to rinse unbound and nonspecifically bound proteins.
9. Buffer B is linearly increased to 60% in 12 CVs (column volumes) and then to 100% for an additional 4 CVs.
10. SDS-PAGE is used to detect the targeted proteins and their purity.
11. Amicon Ultra-15 centrifugal filter devices are used to concentrate the purified target protein to a final volume of 2.5 mL.
12. The PD-10 column is equilibrated with storage buffer (PB buffer: 100 m*M* phosphate buffer, 10% glycerol, pH 7.4), and the concentrated protein is loaded to change buffer.
13. A Pierce BCA Protein Assay Kit and EnSpire Multimode Plate Reader are used to measure the protein concentration.
14. Finally, proteins are stored at −80°C after flash freezing in liquid nitrogen.

2.1.2 Extraction and Detection of Farnesene

To detect the production of farnesene, the reaction mixture was extracted by the following procedures:

1. Extract the reaction mixture with an equal volume of hexane.
2. Centrifuge the hexane layer at 10,000 × *g* for 5 min and then collect the organic layer for GC/MS analysis.
3. Inject the samples into a TRACE TR-5MS volume (30 m × 0.25 mm × 0.25 μm). Set the oven temperature at 80°C for 1 min, increased to 220°C at a rate of 10°C/min, and held at 220°C for 15 min. Maintain the injector and transfer lines at 230°C and 240°C, respectively.

2.2 Targeted Engineering of Farnesene

Farnesene is one of the most important sesquiterpenes with wide applications, e.g., as a precursor of next-generation jet fuel, lubricants, solvents, polymers, and vitamin E (https://amyris.com/products/polymers/).

For the targeted engineering of farnesene, different from the traditional metabolic engineering strategy that requires the construction of a large number of mutants, information from the in vitro reconstitution system can indicate the rate-limiting steps in the whole pathway and can be used to establish target yields. Directed by this information, and together with targeted proteomic analyses and mass spectrometry-based intermediate analyses, inefficiencies in the concentrations of ERG13, ERG8, Idi as well as the accumulation of ATP, NADPH, and the intermediates Acetyl-CoA, MVA, IPP/DMAPP, and FPP were clearly identified, providing a basis for further metabolic engineering research (Zhu et al., 2014). Based on these findings, the expression levels of ERG13, ERG8, and Idi were increased. As a result, the production of farnesene increased predictably stepwise to 1.1 g/L in strain F4 (2000-fold) at the shake-flask scale (Fig. 3) (Zhu et al., 2014).

2.2.1 Plasmids for the MVA Pathway and Farnesene Overexpression

To realize the overproduction of the MVA pathway, the plasmids pMH1 and pFZ81 with genes in the MVA pathway and plasmid pFZ71 with the downstream farnesene production pathway were constructed by Simple Cloning (You, Zhang, & Zhang, 2012) and Gibson method (Gibson et al., 2009). A detailed description of the construction procedure is as follows.

For plasmid pMH1 construction, the origin of pBBR1MCS-1 was replaced with p15A before the insertion of the indicated genes. The primers are listed in Table 1. The p15A origin was amplified from pMSD15 using the primer pair P1/P2, and pBBR1MCS1 was simultaneously amplified using the primer pair P3/P4. The two PCR fragments were combined using the Simple Cloning method to obtain the plasmid pBBR1MCS1/p15A. Gibson method was then used to assemble the indicated genes into pBBR1MCS1/p15A to yield pMH1. First, the three primer pairs P5/P6, P7/P8, and P9/P10 were used to amplify *atoB*, *erg13*, and *thmg1*, respectively, and the primers P11/P12 were used to amplify the pBBR1MCS1/p15A plasmid backbone. Finally, the pBBR1MCS1/p15A PCR fragment, *atoB*, *erg13*, and *thmg1* were assembled by the Gibson method to generate pMH1. For plasmid pFZ81 construction, *erg12*, *erg8*, *mvd1*, and *idi* were amplified from the expression vector using the primers P13/P14, P15/P16, P17/P18, and P19/P20, respectively. pBBR1MCS2 was amplified

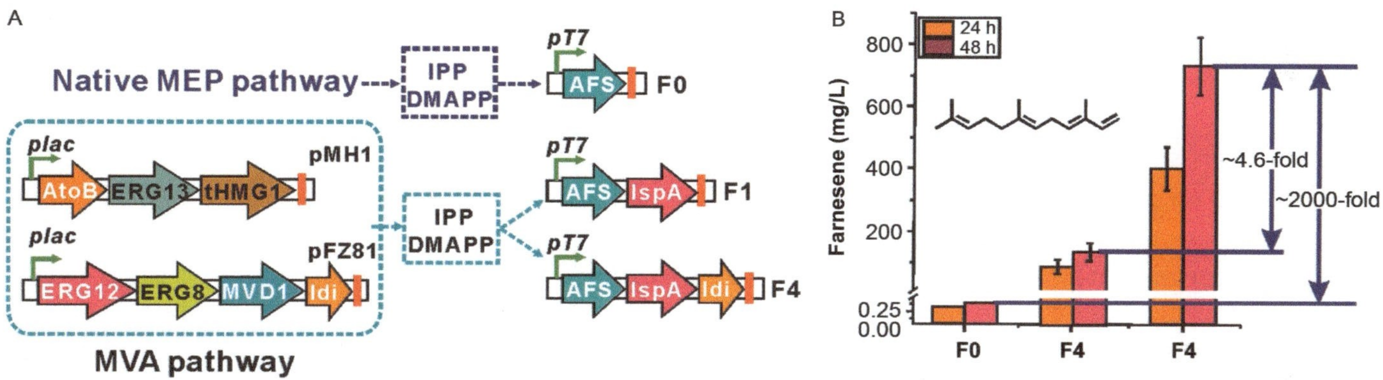

Fig. 3 Overproduction of farnesene in *E. coli* (Zhu et al., 2014). (A) Production of α-farnesene by the endogenous MEP pathway or synthetic MVA pathways and depiction of the synthetic operons used in this study. (B) Production of farnesene in engineered strains.

Table 1 Primers for In Vitro Reconstitution of the Mevalonate Pathway, Terpenoid Overproduction, and Functional Characterization of Terpene Cyclases

Primers	Sequence 5′–3′
P1	GCATTGCCGCCGTGGGTTTCTCGAGCGGGGCGGAGATTTCCTGGAAGATG
P2	GGGGCCACTTTTTGCCGGAGCTCGAGAAATATTTTATCTGATTAATAAGATG
P3	CATCTTATTAATCAGATAAAATATTTCTCGAGCTCCGGCAAAAAGTGGCCCC
P4	CATCTTCCAGGAAATCTCCGCCCCGCTCGAGAAACCCACGGCGGCAATGC
P5	CCCGGGGATCCTCTAGAGTCGACTAGGAGGAATATAAAATGAAAAATTGTGTCATCGTC
P6	GTTGAGAGTTTCATTTAGCTGTCCTCCTTAATTCAACCGTTCAATCACC
P7	ACGGTTGAATTAAGGAGGACAGCTAAATGAAACTCTCAACTAAACT
P8	TGGCTGCTGCCCATAGTGTAATCCTCCTTATTTTTTAACATCGTAAGAT
P9	ATGTTAAAAAATAAGGAGGATTACACTATGGGCAGCAGCCATCATCA
P10	TTAGGATTTAATGCAGGTGACGG
P11	TTTGAAAGATGGGTCCGTCACCTGCATTAAATCCTAAGGATCCACTAGTTCTAGA
P12	TTTTATATTCCTCCTAGTCGACTCTAGAGGATCCCCGGGCTGCAGGAATTCGATATCAA
P13	CTGCAGTAGGAGGAATTAACCATGTCATTACCGTTCTTAACT
P14	CTCAACTCTGACATTTGATCTGCCTCCTATGAAGTCCATGGTAA
P15	TTACCATGGACTTCATAGGAGGCAGATCAAATGTCAGAGTTGAGAGCCTTC
P16	GATGCTGTGTAAACGGTCATGAGTATTACCTCCTATTTATCAAGATAAGTTTC
P17	ATCTTGATAAATAGGAGGTAATACTCATGACCGTTTACACAGCATCCGTT
P18	TGCCCATATAGTAATCCTCCTCCCGGGCTGCAGTTATTCCTTTGGTAGACCAGTC
P19	GGAATAACTGCAGCCCGGGAGGAGGATTACTATATGGGCAGCAGCCATCAT
P20	TTATTTAAGCTGGGTAAATGCAGA
P21	CAGAAAACGATTATCTGCATTTACCCAGCTTAAATAAGAGCTCCAATTCGCCCTATAGT
P22	TTAAGAACGGTAATGACATGGTTAATTCCTCCTACTGCAGGAATTCGATATCAAG
P23	ACCCAGCTTAAATAAAGGAGGATTACACTATGCCGCG GTATGATCTGATTC
P24	GCATTCCAAATCCACAACATATAGTAATCCTCCTTCAT TGCATCGCCTGTTGAC
P25	CAGGCGATGCAATGAAGGAGGATTACTATATGTTGTG GATTTGGAATGCCCTGA
P26	CCACTGCGGCGGTCATTACTCATTCCTCCTTTACTTCC CGGGTGGCGCGTC
P27	CGCCACCCGGGAAGTAAAGGAGGAATGAGTAATGAC CGCCGCAGTGGCAGAG
P28	GCGGTTTCTTTACCAGACTCGAGTTAGCTCTCACCACG CCATAAG
P29	GAAATAATTTTGTTTAACTTTAAGAAGGAGATATACC ATGACCGTGTGTGCGAAAAAAC
P30	TCACCGTGGTCGGTTTCATGGTTAATTCCTCCTTTACG ACACCGCTGCCAG
P31	CTGGCAGCGGTGTCGTAAAGGAGGAATTAACCATGAA ACCGACCACGGTGA
P32	TCAGCAGGGACGGATTATTCATGAGTATTACCTCCTTT AAATCAGGTCTTCCAGCATC
P33	GATGCTGGAAGACCTGATTTAAAGGAGGTAATACTCATGAATAATCCGTCCCTGCTGA

Table 1 Primers for In Vitro Reconstitution of the Mevalonate Pathway, Terpenoid Overproduction, and Functional Characterization of Terpene Cyclases—cont'd

Primers	Sequence 5′–3′
P34	TTGTCGACCTGCAGGCGCGCCGAGCTCGAATTCACTA GTTAAACGGGGCGCTGCCAGAG
P35	TTGCGTTTCGTCAGAACTGAGTCGACAAGCTTGCGGCCG
P36	GTACCCGTAGAGCTAGACATGGTATATCTCCTTCTTAAAG
P37	CTTTAAGAAGGAGATATACCATGTCTAGCTCTACGGGTAC
P38	ACATATGTATATCTCCTTCTTTAGACCTGGATTGGATCGAT
P39	ATCGATCCAATCCAGGTCTAAAGAAGGAGATATACATATGT
P40	CGGCCGCAAGCTTGTCGACTCAGTTCTGACGAAACGCAA
P41	ATATCATATGGATTTCACCTACCGTTATAG
P42	ATATCTCGAGACTAGTTAGTCGATACGCAGCATCTC
P43	GCACGAGCTCATGGATCCCTACAGTGAAACATCAG
P44	AGTCAAGCTTACTAGTCAACCCTCGTGCTGACAAGC
P45	ATATGGATCCATGGACTTTCCGCAGCAACTC
P46	ATATGAATTCACTAGTTTATTTATTACGCTGGATGATG
P47	ATCGCATATGCAAACGGAACACGTCATTTTATTG
P48	ATATCTCGAGACTAGTTATTTAAGCTGGGTAAATGC
P49	CTTTAAGAAGGAGATATACCATGGATTTCACCTACCGTTATAG
P50	GCCATATGTATATCTCCTTCTTTAGTCGATACGCAGCATCTC
P51	GAGATGCTGCGTATCGACTAAAGAAGGAGATATACATATGGCGCAGCTGAGCG
P52	CTGTTCGACTTAAGCATTATGCGGCCGCAAGCTTGTCGACTTAGTGGTCACGCGCCGCAACC
P53	GTCGACAAGCTTGCGGCCGCATAATGCTTAAGTCGAACAG
P54	ATAACGGTAGGTGAAATCCATGGTATATCTCCTTCTTAAAG
P55	CTTTAAGAAGGAGATATACCATGGATCCCTACAGTGAAAC
P56	CATATGTATATCTCCTTCTTCAACCCTCGTGCTGACAAGCC
P57	GCTTGTCAGCACGAGGGTTGAAGAAGGAGATATACATATGTTTGATTTCAATGAATATATG
P58	CTGTTCGACTTAAGCATTATGCGGCCGCAAGCTTGTCGACTCAGTTCTGACGAAACGCAATG
P59	GTTTCACTGTAGGGATCCATGGTATATCTCCTTCTTAAAG

by PCR using the primers P21/P22. Finally, *erg12*, *erg8*, *mvd1*, *idi*, and pBBR1MCS2 were assembled by the Gibson method to obtain pFZ81. For farnesene overproduction, the *Xba*I–*Xho*I fragment of *afs* from pFZ31 was inserted into pET21a(+) to generate pFZ35, and the *Xba*I–*Xho*I fragment of *ispA* from pFZ32 was ligated into the *Spe*I–*Xho*I fragment of pFZ35 to generate pFZ38. An additional copy of *idi* was introduced by inserting the *Xba*I–*Xho*I fragment of *idi* into the *Spe*I–*Xho*I fragment of pFZ38 to generate pFZ71 under the control of the T7 promoter. Plasmids are listed in Table 2.

Table 2 Strains and Plasmids Used for the In Vitro Reconstitution of the Mevalonate Pathway, Terpenoid Overproduction, and Functional Characterization of Terpene Cyclases

Strains	Relevant Genotype	References
BL21 (DE3)	*E. coli B* F^- *dcm ompT hsdSB*(rB^- mB^-) *gal*	Invitrogen
K12 MG1655 (DE3)	F^- λ^- ilvG-rfb-50 rph-1 (DE3)	Invitrogen
F4	*E. coli* BL21:: pMH1, pFZ81, pFZ71	Zhu et al. (2014)
L3	*E. coli* BL21:: pMH1, pFZ81, pFZ111	Zhu et al. (2015)
A1	*E. coli* K12 MG1655: pMH1, pFZ81, pFZ153	Ma et al. (2016)
T4	*E. coli* BL21:: pMH1, pFZ81, pFZ131	Bian, Yuan, et al. (2017)
T7	*E. coli* BL21:: pMH1, pFZ81, pGB309	Bian, Han, et al. (2017)
T11	*E. coli* BL21:: pMH1, pFZ81, pGB313	Bian, Han, et al. (2017)
T12	*E. coli* BL21:: pMH1, pFZ81, pGB314	Bian, Han, et al. (2017)

Plasmids	Origin	Description	References
pGZI	pBR322	pT7: N-terminal His-tagged Idi, Kan^+	Zhu et al. (2014)
pJW1	pBR322	pT7: N-terminal His-tagged AtoB, Kan^+	Zhu et al. (2014)
pFZ12	pBR322	pT7: N-terminal His-tagged ERG13, Kan^+	Zhu et al. (2014)
pJW2	pBR322	pT7: N-terminal His-tagged tHMG1, Kan^+	Zhu et al. (2014)
pFZ13	pBR322	pT7: N-terminal His-tagged ERG12, Kan^+	Zhu et al. (2014)
pFZ14	pBR322	pT7: N-terminal His-tagged ERG8, Kan^+	Zhu et al. (2014)
pFZ15	pBR322	pT7: N-terminal His-tagged MVD1, Kan^+	Zhu et al. (2014)

Table 2 Strains and Plasmids Used for the In Vitro Reconstitution of the Mevalonate Pathway, Terpenoid Overproduction, and Functional Characterization of Terpene Cyclases—cont'd

Plasmids	Origin	Description	References
pFZ31	pBR322	pT7: N-terminal His-tagged AFS, Kan^+	Zhu et al. (2014)
pFZ32	pBR322	pT7: N-terminal His-tagged IspA, Kan^+	Zhu et al. (2014)
pFZ35	pBR322	pT7: AFS, Amp^+	Zhu et al. (2014)
pFZ36	pBR322	pT7: N-terminal His-tagged IspA, Kan^+	Zhu et al. (2014)
pFZ38	pBR322	pT7: AFS, N-terminal His-tagged IspA, Amp^+	Zhu et al. (2014)
pFZ71	pBR322	pT7: AFS, N-terminal His-tagged IspA and Idi, Amp^+	Zhu et al. (2014)
pFZ81	pBBR1MCS	plac: ERG12, ERG8, MVD1 and N-terminal His-tagged Idi, Kan^+	Zhu et al. (2014)
pMH1	p15A	plac: AtoB, ERG13 and N-terminal His-tagged tHMG1, Chl^+	Zhu et al. (2014)
pFZ21	pBR322	pT7: N-terminal His-tagged CrtE, Kan^+	Zhu et al. (2015)
pFZ22	pBR322	pT7: N-terminal His-tagged CrtB, Kan^+	Zhu et al. (2015)
pFZ23	pBR322	pT7: N-terminal His-tagged CrtI, Kan^+	Zhu et al. (2015)
pFZ27	pBR322	pT7: N-terminal His-tagged CrtE, CrtB and CrtI, Kan^+	Zhu et al. (2015)
pFZ110	pBR322	pT7: N-terminal His-tagged CrtE, CrtB and CrtI, Amp^+	Zhu et al. (2015)
pFZ111	pBR322	pT7: N-terminal His-tagged CrtE, CrtB, CrtI and Idi, Amp^+	Ma et al. (2016)
pFZ112	pBR322	pT7: CrtE, CrtI, CrtB, Amp^+	Ma et al. (2016)
pFZ153	pBR322	pT7: CrtE, CrtI, CrtB, pT7: Idi, CrtY, CrtZ, CrtW, Amp^+	Ma et al. (2016)

Continued

Table 2 Strains and Plasmids Used for the In Vitro Reconstitution of the Mevalonate Pathway, Terpenoid Overproduction, and Functional Characterization of Terpene Cyclases—cont'd

Plasmids	Origin	Description	References
pFZ131	pBR322	pT7: TS, GGPPS, Amp^+	Bian, Yuan, et al. (2017)
pMH1	p15A	plac: AtoB, ERG13, tHMG1, Chl^+	Zhu et al. (2014)
pFZ81	pBBR1MCS	plac: ERG12, ERG8, MVD1, Idi, Kan^+	Zhu et al. (2014)
pGB309	pBR322	pT7: FgMS, GFPPS; pT7: Idi, Amp^+	Bian, Han, et al. (2017)
pGB313	pBR322	pT7: FgGS, GGPPS; pT7: Idi, Amp^+	Bian, Han, et al. (2017)
pGB314	pBR322	pT7: FgGS, FPPS, Idi, Amp^+	Bian, Han, et al. (2017)

2.2.2 Shake-Flask Fermentation and Analysis of α-Farnesene Production in Engineered Strains

To assess farnesene production in shake flasks, the engineered *E. coli* F4 (Table 2) was cultured in 500-mL flasks containing 150 mL of 2× TY medium at 30°C with 2% glycerol, 100 mg/mL ampicillin (Amp), 50 mg/mL Kan, and 34 mg/mL chloramphenicol (Chl). When the OD_{600} reached 0.6–0.8, IPTG was added to a final concentration of 0.1 m*M*. After 3 h of induction, the medium was overlaid with 30 mL of decane. Samples containing 1 mL of decane and 5 mL of culture were withdrawn at designated time points. After centrifugation, the decane phase was subjected to GC/MS according to the method described in Section 2.1.2. The production α-farnesene was quantified by GC/MS; trans-β-farnesene in decane was used as a calibration standard as described previously (Wang et al., 2011).

2.3 Targeted Engineering of Lycopene

Lycopene is a well-known tetraterpenoid with attractive antioxidative, anticancer, and antiinflammatory activities; it has been used as a pharmaceutical compound, nutraceutical, functional food, and cosmetic additive

(Zhu et al., 2015). Extensive efforts have focused on increasing the production efficiency of lycopene in *E. coli* and *S. cerevisiae* (Coussement, Bauwens, Maertens, & De Mey, 2016; Sun et al., 2014; Xie, Lv, Ye, Zhou, & Yu, 2015). The biosynthesis of lycopene begins with the condensation of IPP and DMAPP, which are catalyzed by CrtE to form GGPP, and then GGPP is catalyzed by CrtB to produce phytoene, and phytoene is catalyzed by CrtI to produce lycopene (Fig. 4).

With a clear biochemical basis and few rate-limiting steps, the MVA pathway is widely used for carotenoid overproduction in *E. coli*. The above-mentioned targeted engineering strategy has been employed to accelerate the engineering process and efficiently produce lycopene. By taking advantage of the efficient precursor-producing platform (*E. coli* BL21 (DE3)/pMH1/pFZ81) constructed for farnesene production together with the downstream lycopene-forming pathway (pFZ111), the titer of lycopene in strain *E. coli* L3 was increased to 34.3 mg/g DCW (dry cell weight) (1.23 g/L) in a 100-L fermenter (Zhu et al., 2015). This indicates that the information obtained from the targeted engineering strategy is universal and can be utilized to realize the overproduction of other terpenoids with limited effort in a short period of time.

2.3.1 Plasmids for Lycopene Overproduction

To produce lycopene in *E. coli*, *crtE*, *crtB*, and *crtI* from *Pantoea ananatis* (Misawa et al., 1990) were codon-optimized, synthesized, and inserted into pET28a(+) to produce pFZ21, pFZ22, and pFZ23, respectively. Subsequently, the *Xba*I–*Xho*I fragments of *crtB* and *crtI* were sequentially inserted into pFZ21 to produce pFZ27. The *Xba*I–*Xho*I fragment of pFZ27 was inserted into pET21a(+) to produce pFZ110. An additional copy of *idi* was introduced into pFZ110 by inserting the *Xba*I–*Xho*I fragment of *idi* into the *Spe*I–*Xho*I fragment of pFZ110 to produce pFZ111.

2.3.2 Analysis of Lycopene Titers in Engineered Strains

Overproduction of lycopene is a fascinating work, and many researchers have realized the high efficiency production of lycopene in *E. coli*, for example, Zhu et al. reporting a concentration of 34.3 mg/g DCW (Zhu et al., 2015) and Sun et al. reporting 50.6 mg/g DCW (Sun et al., 2014). However, different quantitative assessment method used can result in significant differences in the yield of lycopene and may mislead the metabolic engineering process. A closer examination of the study by Coussement

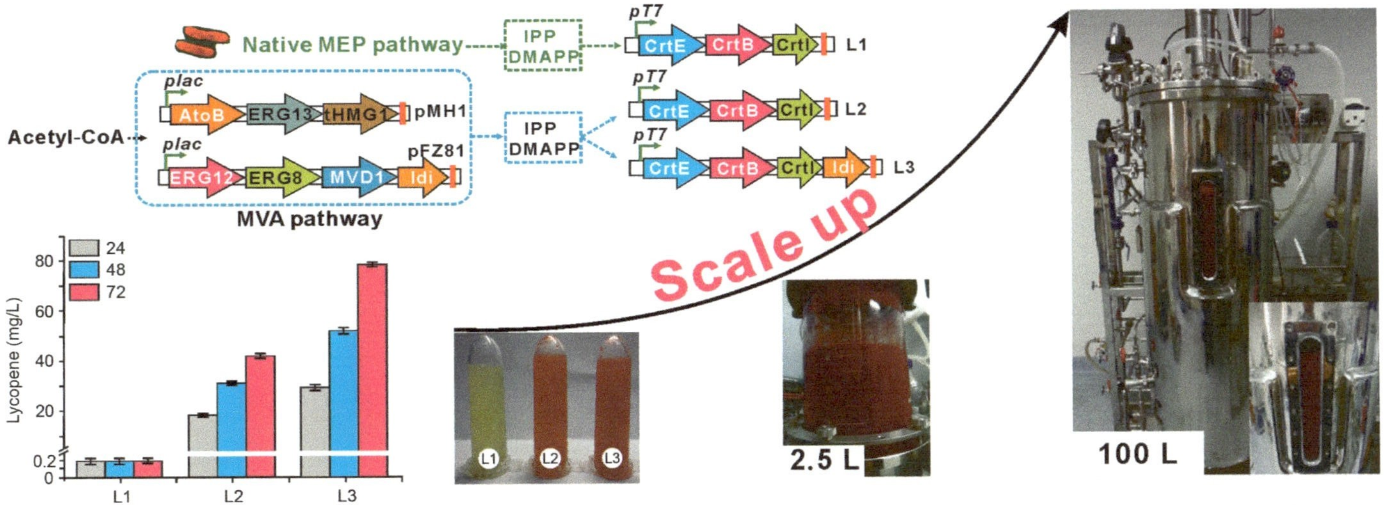

Fig. 4 Targeted engineering of lycopene overproduction in *E. coli*. Production of lycopene via endogenous MEP pathway or synthetic MVA pathway and depiction of the synthetic operons used in this study. Production of lycopene in engineered strains L1–L3 at shake-flask level was detected, and the fed-batch fermentation of strain L3 was scaled up to a 100-L fermenter.

et al. (2016), which yielded 448 mg/g DCW, revealed several notable details: the reference strain should not produce any lycopene, but it already showed a high production level; after incubating the 96-well plate containing the extraction solution at 50°C for 10 min, the total solvent decreased to <100μL, and although this volume would not meet the requirements for the next operation, there were no details on how to avoid reagent volatilization. Most importantly, though, no detailed genetic information was provided regarding the highest production strain, which has prevented other groups from repeating this experiment and building on the research. The significant differences in quantitative assessment of lycopene production, raises questions regarding how assessments are made and the best way to compare different strategies.

The extraction and titration method we recommend is as follows:

1. Harvest 1 mL cell culture of engineered strain and centrifuge at 5000 × *g* for 5 min, remove the supernatants completely.
2. Vortex to disperse the cell pellet, then use acetone (with 1% butylated hydroxytoluene, BHT) to extract the pellets at 55°C for 15 min in the dark, with vortexing for 1 min every 5 min.
3. Centrifuge the solution at 13,000 × *g* for 10 min.
4. Transfer the supernatant to 96-well polypropylene plates for direct UV absorbance measurements or to sample vials for HPLC measurements.
5. Calculate the titer of lycopene using a standard curve with appropriate dilution factor and averaged for three replicate analyses as described previously (Alper, Jin, Moxley, & Stephanopoulos, 2005).
6. Count the cell mass by OD_{600} with a coefficient of 0.3 g DCW/OD_{600}.

For lycopene standard calibration, the content of lycopene in the standard was corrected with an extinction coefficient of 3450 ($A^{1\%}=3450$) at 472 nm (hexane as solvent), as described previously (Joint FAO/WHO Expert Committee on Food Additives, 2006). The details are as follows: ~5 mg of the standard was weighed in a 100-mL volumetric flask, dissolved in 5 mL of tetrahydrofuran (0.025% BHT), and then supplement with hexane to a final volume of 100 mL. Next, 5.0 mL of the solution was transferred into a 100-mL volumetric flask, and then supplement with hexane to a final volume of 100 mL, producing the standard solution for UV detection. The absorbances of the standard and candidate solutions were measured in 1-cm cuvettes at the wavelength of maximum absorption (approximately 472 nm). Hexane was used as the blank.

$$\text{Calculations}: C_{\text{St}}\,(\text{mg/L}) = A_{472} \times 10,000/3450$$

where C_{St} is the lycopene concentration, A_{472} is the absorbance at the wavelength of lycopene maximum absorption in hexane, and 3450 is the specific absorbance $A^{1\%}$ of lycopene in hexane. The scaling factor is 10,000.

$$\text{Lycopene}\,\% = C_{St} \times 20 \times 0.1/W$$

where lycopene % is the percentage of lycopene in the standard, 20 is the dilution ratio, 0.1 is the solvent volume, and W is the weight of the sample.

There are two methods for lycopene titer analyzing. For UV detection method, detection of lycopene can be carried out using the Perkin Elmer EnSpire Multimode plate reader at 474 nm (acetone as solvent) with 150 μL acetone extraction solution in polypropylene 96-well plates. For HPLC detection method, an HPLC system (Thermo Scientific, Dionex, UltiMate 3000) equipped with a C18 column (4.6 × 150 mm, 5 μm, Agilent Technologies, Inc.) was used to detect the 474 nm signals at 30°C. Samples were eluted with solvent A [90% aqueous acetonitrile (HPLC grade)] and solvent B [methyl alcohol-isopropyl alcohol (3:2, v/v, HPLC grade)] according to the following procedure with flow rate set as 1.0 mL/min: 0–15 min, 100%–10% solvent A, 0%–90% solvent B; 15–30 min, 10% solvent A, 90% solvent B; 30–35 min, 10%–100% solvent A, 90%–0% solvent B (Zhou, Ye, Xie, Lv, & Yu, 2015). Both detection methods are recommended; however, UV is more suitable for high-throughput lycopene detection and HPLC is preferred for quantification of lycopene content in mixture.

2.4 Targeted Engineering of Astaxanthin

Astaxanthin is a kind of cartenoid with various beneficial effects on human health; it has widely used in nutritional supplements, pharmaceuticals, cosmetics, food colorants, and animal feed additives (Fraser & Bramley, 2004; Tominaga, Hongo, Karato, & Yamashita, 2012). Astaxanthin can be produced by using the lycopene production pathway in addition to the downstream astaxanthin production pathway consisting of CrtY, CrtZ, and CrtW. CrtY can catalyze the transformation of lycopene to beta-carotene, and CrtZ together with CrtW can catalyze the formation of astaxanthin by using β-carotene as a substrate (Fig. 1). Consequently, similar to the above-mentioned overproduction of farnesene and lycopene, an efficient precursor production pathway plays a significant role in the overproduction of astaxanthin.

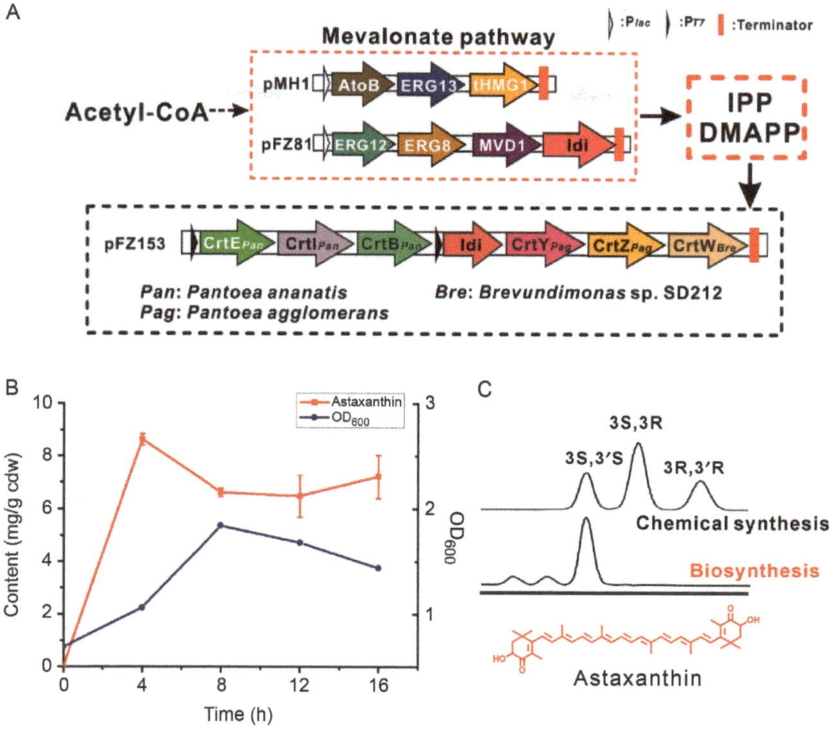

Fig. 5 Targeted engineering of astaxanthin overproduction in *E. coli*. (A) *E. coli* A1 with pMH1, pFZ81, and pFZ153 for astaxanthin overproduction. (B) Content of astaxanthin produced by *E. coli* A1. (C) Configuration of astaxanthin produced by chemical synthesis and biosynthesis in our system.

Using the targeted engineering strategy, an efficient astaxanthin overproduction mutant *E. coli* A1 (Table 2) was constructed by employing the same farnesene precursor-providing platform, together with *CrtEIB*, *CrtYZ*, *CrtW*, and an additional copy of *Idi* from *E. coli*. The specificity production of 3S, 3′S-astaxanthin was realized (3S, 3′S configuration has the highest biological activity), and the highest yield of 8.64 mg/g DCW was obtained in *E. coli* (Fig. 5) (Ma et al., 2016).

2.4.1 Plasmid Construction for Astaxanthin Overproduction

For astaxanthin overproduction, a plasmid harboring the downstream astaxanthin-forming pathway was constructed. Primers used for astaxanthin overproduction are listed in Table 1. The *idi* gene from pGZI (Zhu et al., 2014) was digested by *Nde*I and *Xho*I and cloned into the plasmid pETDuet-1 to generate pFZ87. The *crtY* and *crtZ* genes were amplified individually from *Pantoea agglomerans* (Hundle et al., 1994; Schnurr, Schmidt, & Sandmann, 1991) using the primer pairs P23/P24 and P25/P26. The *crtW* gene from *Brevundimonas* sp. SD212 (Nishida et al., 2005) was codon optimized, synthesized, and amplified using the primer pair P27/P28. The four fragments containing *idi*, *crtY*, *crtZ*, and *crtW* were assembled by the Gibson method to obtain the plasmid pFZ152. The *crtE*, *crtI*, and *crtB* genes were separately amplified from pFZ21, pFZ22, and pFZ23 with the primer pairs P29/P30, P31/P32, and P33/P34, respectively (Zhu et al., 2015). These three fragments were assembled by the Gibson method to generate pFZ112. Finally, the *Nde*I–*Eco*RI digested fragment of pFZ112 (including *crtE*, *crtI*, and *crtB*) was inserted into pFZ152 to produce pFZ153. The constructed plasmids are listed in Table 2.

2.4.2 Fermentation and Quantification of Astaxanthin Produced by E. coli

E. coli K-12 MG1655 (DE3) was selected as a host for astaxanthin overproduction. During the shake-flask fermentation period, *E. coli* A1 (*E. coli* K12 MG1655/pMH1/pFZ81/pFZ153) was cultured and induced by IPTG as described in Section 2.2.2. The production of astaxanthin was achieved and detected according to the following procedure.

1. The cell culture (2 mL) is harvested by centrifugation at 10,000 $\times g$ for 5 min.
2. The cell pellet is dispersed by vortexing, and methanol and acetone (1:4, v:v) are used to extract the pellets in the dark until the residues become colorless.
3. The supernatant is pooled, centrifuged, and transferred to sample vials for HPLC detection.
4. Samples are detected by the HPLC system according to the procedure described in Section 2.3.2.

2.5 Targeted Engineering of Taxadiene

As described earlier, the targeted engineering of terpenoids by the MVA pathway has been successfully conducted in *E. coli* for the production of the sesquiterpene product farnesene and the tetraterpene compounds lycopene and astaxanthin (Ma et al., 2016; Zhu et al., 2015, 2014). The efficiency of this platform for diterpene overproduction in *E. coli* was also evaluated. Taxadiene, a precursor of one of the most important anticancer drugs, contains a C20 backbone derived from the condensation and cyclization of the universal C5 units IPP and DMAPP. To test the efficiency of a targeted engineering strategy for diterpene overproduction in *E. coli*, the downstream GGPPS, the taxadiene synthase, and an additional copy of Idi were introduced to generate the pFZ131 plasmid for taxadiene production (Fig. 6) (Bian, Yuan, et al., 2017).

2.5.1 Plasmid Construction for Taxadiene Overproduction

For plasmid pFZ131 construction, the backbone of the pETDuet-1 fragment was amplified using the primer pair P35/P36 (Table 1), and the *ts* and *ggpps* genes were synthesized as described by Ajikumar and amplified by P37/P38 and P39/P40, respectively (Ajikumar et al., 2010). The three fragments were assembled by the Gibson method to generate pFZ131 (Gibson et al., 2009).

2.5.2 Overproduction and Detection of Taxadiene in E. coli

For taxadiene overproduction in *E. coli*, strain *E. coli* T4 (Table 2) was constructed and cultivated in 500-mL shake flasks according to the procedure described in Section 2.2.2, with slight modifications, including cultivation at 28°C for an additional 3 days after induction with IPTG. Production of taxadiene was detected by GC–MS following the method described in Section 2.1.2. The cell concentration was measured by OD_{600} with a coefficient of 0.3 g DCW/OD_{600}. As a result, taxadiene was produced by *E. coli* T4, with a titer of 11.3 mg/L (13.2 g/g DCW) at the shaking flask level (Bian, Yuan, et al., 2017). This demonstrates the wide applicability of the targeted engineering strategy.

3. GENOME MINING OF TERPENE CYCLASES AND THE GENERATION OF CHEMICAL DIVERSITY

The development of genome sequencing and various bioinformatics tools, such as antiSMASH and SMURF (Khaldi et al., 2010; Weber et al., 2015), have enabled the characterization of a large number of terpene cyclases and have substantially increased the chemical diversity of terpenoids (Fig. 7) (Bian, Han, et al., 2017; Bian et al., 2018; Yamada et al., 2015). However, there are still many terpene cyclases were unknown functions. The well-established terpenoid overproduction platform makes it possible to efficiently characterize the functions of terpene cyclases, irrespective of the organism and type of terpenoid. This allows us to avoid the negative

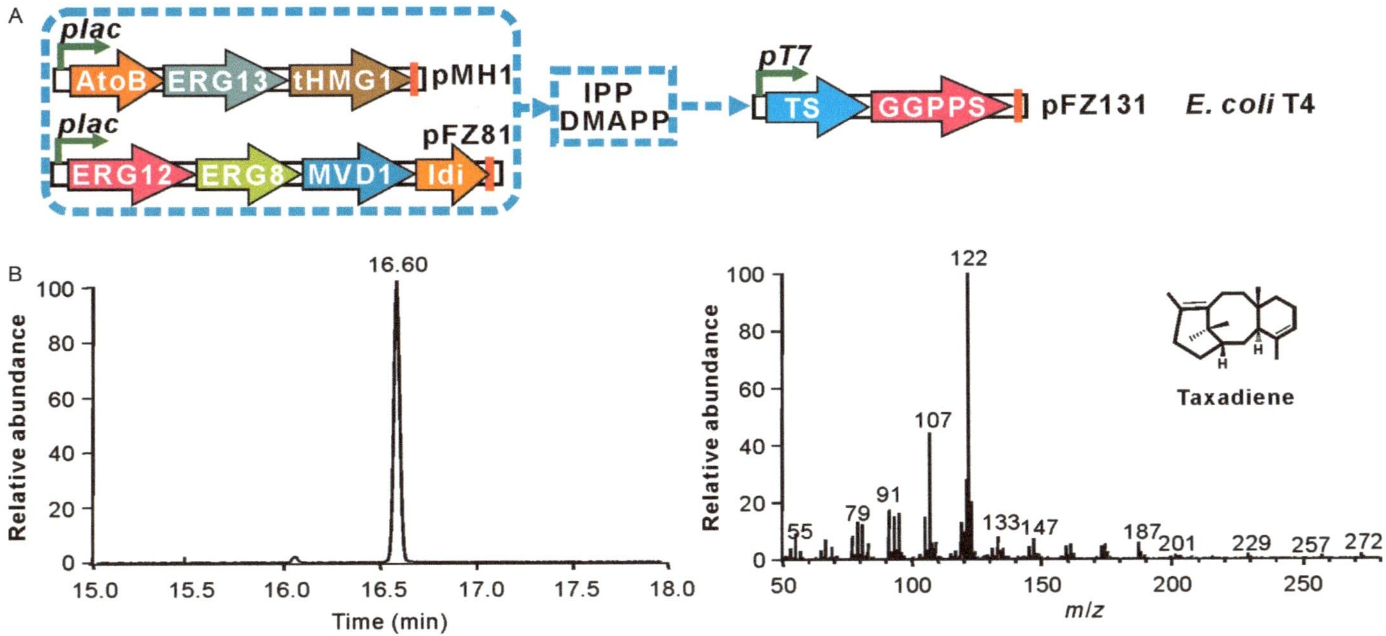

Fig. 6 Production of taxadiene in *E. coli*. (A) *E. coli* T4 for taxadiene overproduction. (B) Detection of taxadiene by GC–MS.

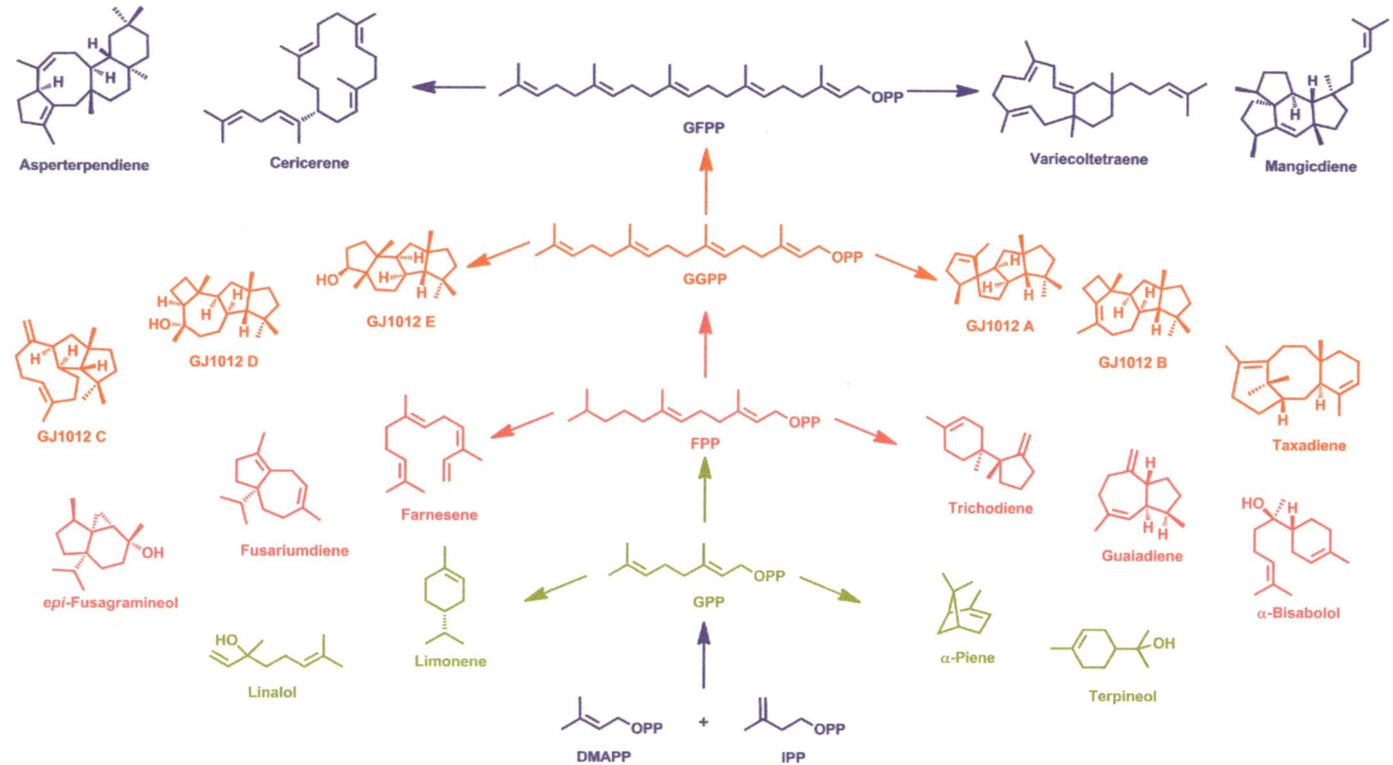

Fig. 7 Diversity of terpenoid biosynthesis. Examples of monoterpenes (C10), sesquiterpenes (C15), diterpenes (C20), and sesterterpenes (C25) are shown.

effects of gene silencing, complex regulatory systems, or low yields in the native host, accelerating the entire process for the mining of terpene cyclases and the discovery of terpenoids (Bian, Han, et al., 2017; Bian et al., 2018; Jia et al., 2016; Shao et al., 2017). Consequently, a metabolic engineering-based terpenoid discovery strategy may accelerate the process of genome mining for terpene cyclases and increase the chemical diversity.

3.1 Genome Mining of Terpene Cyclases and Exploration of the Chemical Diversity of Terpenoids

Two kinds of terpene cyclases exist in nature; class I terpene synthases initiate carbocation formation by substrate ionization and class II terpene cyclases initiate carbocation formation by substrate protonation (Meguro, Tomita, Nishiyama, & Kuzuyama, 2013). Class I terpene cyclases FgMS and FgGS were predicted from *F. graminearum* J1-012 with the highly conserved aspartate-rich motif **DD**VIE and the NSE/DTE triad **ND**LY**S**YDK**E**, which chelate Mg^{2+} ions used to bind to the diphosphate moiety of the substrate and required for the ionization of the allylic diphosphate ester bond (Meguro et al., 2013; Shishova, Di Costanzo, Cane, & Christianson, 2007). A bioinformatic analysis showed that FgMS is a chimeric terpene cyclase with the prenyltransferase and the terpene cyclase domain and can utilize IPP and DMAPP as substrates to produce terpene skeletons. In vitro assays have shown that both of these enzymes could directly catalyze the conversion GPP, FPP, GGPP, and GFPP to generate multiple products, exhibiting exceptionally broad substrate specificities for polyisoprenoid diphosphates (Bian, Han, et al., 2017).

Directed by the combinatorial biosynthesis approach, a platform consisting of the terpene-overproducing chassis, the prenyltransferase (FPPS, GGPPS, and GFPPS) library, and the enzymes FgMS and FgGS was built to realize the full potential of these terpene synthases. First, the above-mentioned efficient precursor production platform was employed to provide sufficient IPP and DMAPP for terpenoid overproduction (Zhu et al., 2014). Second, efficient prenyltransferases (FPPS, GGPPS, and GFPPS) were employed to construct a library to produce FPP, GGPP, and GFPP for terpene cyclase to produce sesqui-, di-, and sesterterpenes, respectively. Third, the terpene cyclase (FgMS and FgGS) was combined with appropriate prenyltransferases to generate relevant mutants. In this section, we describe the construction of three mutants for the characterization of FgMS and FgGS and the production of sesqui-, di-, and sesterterpenes.

3.1.1 Identification of Class I Terpene Synthases

Class I terpene synthases were identified by screening all predicted proteins against a set of hidden Markov models constructed for each terpene synthase family. Proteins with the highly conserved aspartate-rich motif **DD**xxD/**E** and the NSE/DTE triad **(N/H)D**xx**S**xxx**E** were selected for subsequent analyses. Based on sequence alignments, both FgMS and FgGS contain the highly conserved aspartate-rich motif **DD**xxD/**E** and NSE/DTE triad (Fig. 8), which make it possible to produce terpene skeletons using an appropriate polyisoprenoid diphosphate as a substrate.

3.1.2 Construction of Plasmids

Primers used for plasmid construction are listed in Table 1. Codon-optimized FgMS was amplified using the primer pair P41/P42 and subcloned into pET28a (+) to generate pGB301. The coding sequence of FgGS was assembled and then amplified using the primer pair P43/P44 and subcloned into pET28a (+) and pET21a (+) to generate pGB303 and pGB304, respectively. *FPPS* was amplified using the primer pair P45/P46 and subcloned into pET21a to generate pGB305. Next, *idi* was amplified using the primers P47/P48 and subcloned into pET21a (+) and pETduet-1 to generate pGB306 and pGB307, respectively. Then, pGB307 was digested with *Xba*I/*Xho*I, and the *idi* fragment was ligated into pGB305, which was digested by *Spe*I/*Xho*I, to generate pGB308. The *Xba*I/*Xho*I digested fragment (*FPPS-Idi*) of pGB308 was then inserted into the *Spe*I/*Xho*I digested plasmid pGB304 to generate pGB314. FgMS, GFPPS, and the backbone of pGB307 were amplified using the primer pairs P49/50, P51/P52, and P53/P54, respectively, and these three fragments were then assembled by the Gibson method to generate pGB309 (Gibson et al., 2009). Similarly, FgGS, GGPPS, and the backbone of pGB307 were amplified using the primer pairs P55/56, P57/P58, and P53/P59 and then assembled to generate pGB313.

3.1.3 In Vitro Assays of Terpene Cyclases

To test the in vitro activities of FgMS and FgGS, the following reactions were conducted.

1. First, 10 μ*M* purified proteins, 100 μ*M* substrates (GPP, FPP, GGPP, or GFPP), and 2 m*M* Mg^{2+} in 200 μL of 50 m*M* PB buffer (pH 7.6) with 10% glycerol were incubated at 30°C overnight.

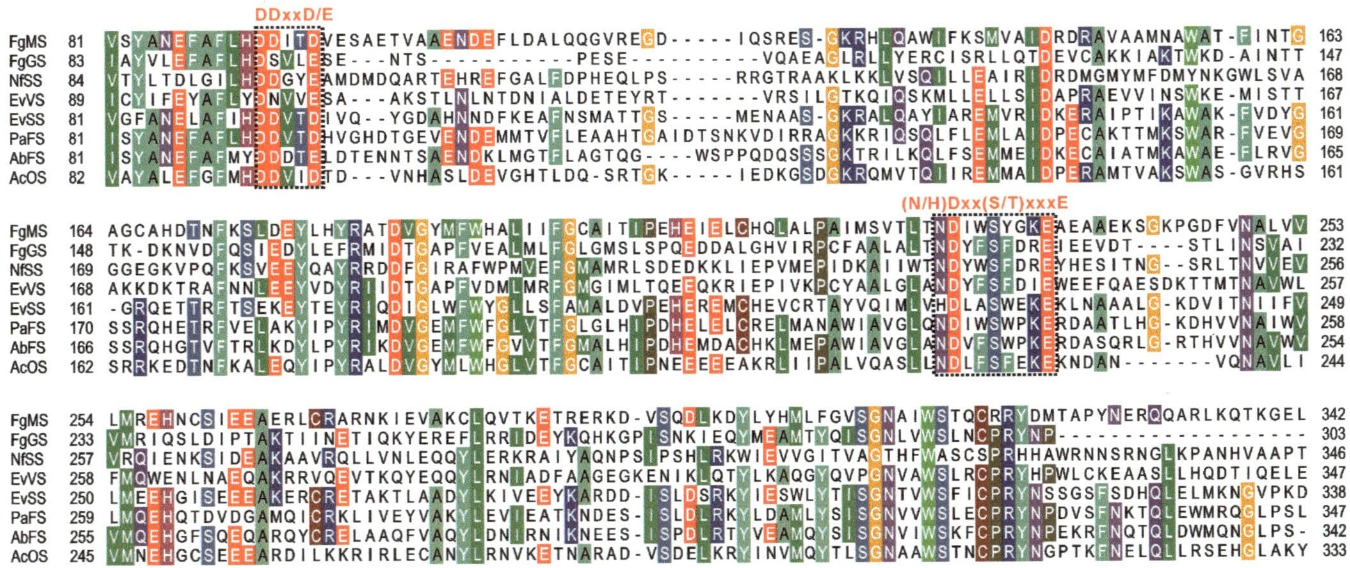

Fig. 8 Sequence alignment of FgMS, FgGS, and the class I terpene synthase of filamentous fungi. The *black boxes* represent the conserved "DDXXD/E" and "NSE/DTE" motifs.

2. The samples were extracted with 200 μL hexane twice and centrifuged at 1000 rpm for 5 min; the organic layer was collected.
3. The production of terpenes was detected GC–MS using the method described in Section 2.1.2.

In the absence of IPP, both FgMS and FgGS could directly catalyze the conversion GPP, FPP, GGPP, and GFPP to generate 51 sesqui-, di-, and sesterterpenes in vitro, as detected by GC–MS, exhibiting exceptionally broad substrate specificities for polyisoprenoid diphosphates (Fig. 9).

3.1.4 Fermentation and Purification of Terpenoids

E. coli T7 was constructed for the production of sesterterpenes by FgMS. *E. coli* T11 and T12 were constructed for the production of diterpenes and sesquiterpenes produced by FgGS (Table 2).

The fermentation and purification of terpenoids produced by FgMS and FgGS were carried out as follows.

1. *E. coli* T7, T11, and T12 were cultivated in 2-L flasks containing 1 L of LB medium according to the description in Section 2.2.2.
2. The cell cultures were extracted twice with an equal volume of hexane and detected by GC/MS according to the description in Section 2.1.2.
3. The organic layer was concentrated with rotary evaporators.
4. The extract was resolved in methanol (with an appropriate ratio DMSO, about 20%), and separation and purification were performed using the XBridge™ Prep C18 column (Waters, 10 × 250 mm, 5 μm) and prep-HPLC system with acetonitrile and water as the mobile phase.
5. Structures of sesqui-, di-, and sesterterpenes were determined by comparison with authentic standards, comparisons with mass spectra data from the NIST library, and NMR analyses.

The fermentation data showed that, in accordance with the in vitro data, 1 sesquiterpene, 11 diterpenes, and 8 sesterterpenes were produced by *E. coli* T7, T11, and T12, respectively (Table 2). Two sesterterpenes and five novel diterpenes with three unusual skeletons were structurally characterized from three mutants (Fig. 10). Accordingly, the terpenoid overproduction platform and a combinatorial biosynthesis strategy can accelerate the entire process for the discovery and creation of terpenoids and thereby can expand the chemical diversity of terpene skeletons. This strategy can also be extended to other systems to realize the characterization of terpene cyclases with high efficiency (Bian et al., 2018).

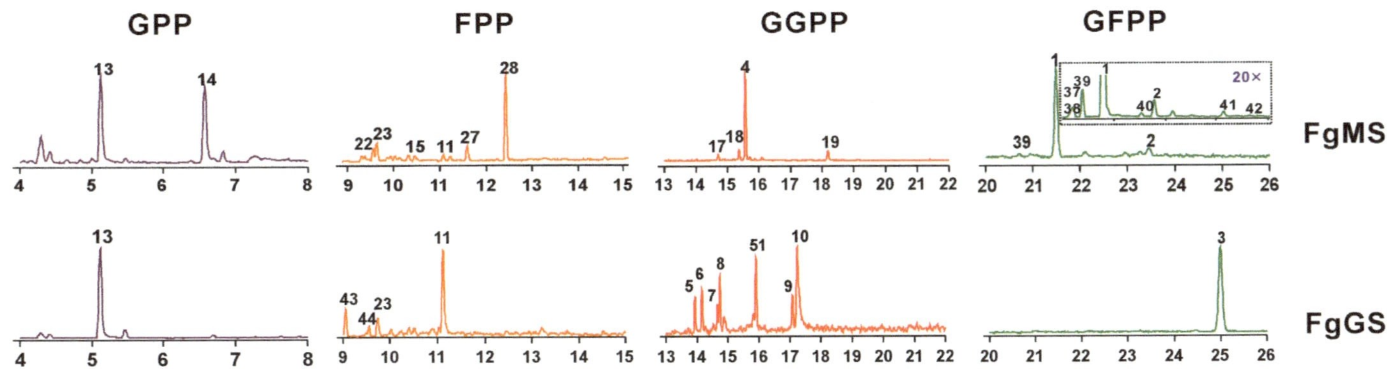

Fig. 9 In vitro characterization of FgMS and FgGS (Bian, Han, et al., 2017). Chromatograms showing the GC–MS analysis of terpenoid products from in vitro assays of FgMS (*top*) and FgGS (*bottom*) with GPP, FPP, GGPP, and GFPP in the absence of IPP.

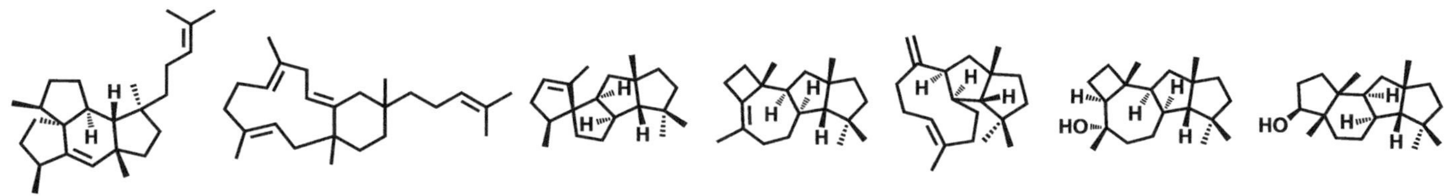

Mangicdiene **Variecoltetraene** **GJ1012 A** **GJ1012 B** **GJ1012 C** **GJ1012 D** **GJ1012 E**

Fig. 10 Structural characterization of seven new terpenoids with three unusual skeletons.

4. CONCLUSION

Terpenoids represent a highly diverse group of natural products with wide applications. The overproduction of value-added terpenoids, the efficient mining of new terpene cyclases, and increased chemical diversity are major goals in the fields of metabolic engineering and natural product discovery. In vitro reconstitution assays provide key information regarding the rate-limiting steps and mechanism of action of the MVA pathway. This information (e.g., the concentration of Idi and the optimal ratio of each component) provides specific direction for engineering research. Directed by the results of these assays, an efficient precursor production platform was constructed, providing a solid foundation for the subsequent overproduction of terpenoids, thereby enabling the overproduction of various kinds of terpenoids, such as farnesene, lycopene, astaxanthin, and taxadiene, with minimal effort and high efficiency.

Metabolic engineering-based genome mining of terpene cyclases makes it possible for researchers to utilize the potential power of terpene cyclases and acquire sufficient quantities of products for structural characterization in a short period of time. In many cases, for promiscuous terpene cyclases that can produce a series of terpenoids, it is only possible to characterize the main products by traditional heterologous or homologous expression, and in vitro feeding experiments are both time-consuming and expensive. The targeted engineering-based precursor production platform can provide a sufficient amount of precursor for the production of mono-, sesqui-, di-, and sesterterpenes with high yields. In this way, the combinatorial biosynthesis approach can release the potential power of terpene cyclases and enable the characterization of all resulting products. As observed in the case of FgGS, some by-products were new terpenes with unusual skeletons, demonstrating that this approach may contribute to increasing the chemical diversity of terpenoids. Owing to the large number of functionally uncharacterized terpene cyclases, an efficient terpenoid precursor-providing platform, the combinatorial biosynthesis approach, the process for genome mining, and the extension of chemical diversity will enable significant advances. This approach is expected to provide a large number of bioactive or high value-added compounds for drug discovery and industrial applications.

ACKNOWLEDGMENTS

This work was financially supported by the funding from J1 Biotech Co., Ltd. Grants from the Young Talents Program of National High-level Personnel of Special Support Program (The "Ten Thousand Talent Program") and the China Postdoctoral Science Foundation (Grant 2017M622507, 2018T110793).

REFERENCES

Ajikumar, P. K., Xiao, W. H., Tyo, K. E. J., Wang, Y., Simeon, F., Leonard, E., et al. (2010). Isoprenoid pathway optimization for Taxol precursor overproduction in *Escherichia coli*. *Science*, *330*, 70–74.

Alper, H., Jin, Y.-S., Moxley, J., & Stephanopoulos, G. (2005). Identifying gene targets for the metabolic engineering of lycopene biosynthesis in *Escherichia coli*. *Metabolic Engineering*, *7*, 155–164.

Andersen-Ranberg, J., Kongstad, K. T., Nielsen, M. T., Jensen, N. B., Pateraki, I., Bach, S. S., et al. (2016). Expanding the landscape of diterpene structural diversity through stereochemically controlled combinatorial biosynthesis. *Angewandte Chemie International Edition*, *55*, 2142–2146.

Bian, G., Han, Y., Hou, A., Yuan, Y., Liu, X., Deng, Z., et al. (2017). Releasing the potential power of terpene synthases by a robust precursor supply platform. *Metabolic Engineering*, *42*, 1–8.

Bian, G., Hou, A., Yuan, Y., Hu, B., Cheng, S., Ye, Z., et al. (2018). Metabolic engineering-based rapid characterization of a sesquiterpene cyclase and the skeletons of Fusariumdiene and Fusagramineol from *Fusarium graminearum*. *Organic Letters*, *20*, 1626–1629.

Bian, G., Yuan, Y., Tao, H., Shi, X., Zhong, X., Han, Y., et al. (2017). Production of taxadiene by engineering of mevalonate pathway in *Escherichia coli* and endophytic fungus *Alternaria alternata* TPF6. *Biotechnology Journal*, *12*, 1600697.

Coussement, P., Bauwens, D., Maertens, J., & De Mey, M. (2016). Direct combinatorial pathway optimization. *ACS Synthetic Biology*, *6*, 224–232.

Fraser, P. D., & Bramley, P. M. (2004). The biosynthesis and nutritional uses of carotenoids. *Progress in Lipid Research*, *43*, 228–265.

Gibson, D. G., Young, L., Chuang, R. Y., Venter, J. C., Hutchison, C. A., & Smith, H. O. (2009). Enzymatic assembly of DNA molecules up to several hundred kilobases. *Nature Methods*, *6*, 343–345.

Huang, A. C., Kautsar, S. A., Hong, Y. J., Medema, M. H., Bond, A. D., Tantillo, D. J., et al. (2017). Unearthing a sesterterpene biosynthetic repertoire in the *Brassicaceae* through genome mining reveals convergent evolution. *Proceedings of the National Academy of Sciences of the United States of America*, *114*, 201705567. https://doi.org/10.1073/pnas.1705567114.

Hundle, B., Alberti, M., Nievelstein, V., Beyer, P., Kleinig, H., Armstrong, G., et al. (1994). Functional assignment of *Erwinia herbicola* Eho10 carotenoid genes expressed in *Escherichia coli*. *Molecular and General Genetics*, *245*, 406–416.

Jia, M., Potter, K. C., & Peters, R. J. (2016). Extreme promiscuity of a bacterial and a plant diterpene synthase enables combinatorial biosynthesis. *Metabolic Engineering*, *37*, 24–34.

Joint FAO/WHO Expert Committee on Food Additives. (2006). *Compendium of food additive specifications: Joint FAO/WHO Expert Committee on Food Additives, 67th meeting 2006*. Food and Agriculture Organization.

Khaldi, N., Seifuddin, F. T., Turner, G., Haft, D., Nierman, W. C., Wolfe, K. H., et al. (2010). SMURF: Genomic mapping of fungal secondary metabolite clusters. *Fungal Genetics and Biology*, *47*, 736–741.

Klapschinski, T. A., Rabe, P., & Dickschat, J. S. (2016). Pristinol, a sesquiterpene alcohol with an unusual skeleton from *Streptomyces pristinaespiralis*. *Angewandte Chemie International Edition*, *55*, 10141–10144.

Korman, T. P., Opgenorth, P. H., & Bowie, J. U. (2017). A synthetic biochemistry platform for cell free production of monoterpenes from glucose. *Nature Communications*, *8*, 15526.

Liu, Z., Zhang, Y., Jia, X., Hu, M., Deng, Z., Xu, Y., et al. (2017). In vitro reconstitution and optimization of the entire pathway to convert glucose into fatty acid. *ACS Synthetic Biology*, *6*, 701–709.

Lowry, B., Walsh, C. T., & Khosla, C. (2015). In vitro reconstitution of metabolic pathways: Insights into nature's chemical logic. *Synlett*, *26*, 1008–1025.

Ma, T., Zhou, Y., Li, X., Zhu, F., Cheng, Y., Liu, Y., et al. (2016). Genome mining of astaxanthin biosynthetic genes from *Sphingomonas* sp. ATCC 55669 for heterologous overproduction in *Escherichia coli*. *Biotechnology Journal*, *11*, 228–237.
Meguro, A., Tomita, T., Nishiyama, M., & Kuzuyama, T. (2013). Identification and characterization of bacterial diterpene cyclases that synthesize the cembrane skeleton. *Chembiochem*, *14*, 316–321.
Misawa, N., Nakagawa, M., Kobayashi, K., Yamano, S., Izawa, Y., Nakamura, K., et al. (1990). Elucidation of the *Erwinia uredovora* carotenoid biosynthetic pathway by functional analysis of gene products expressed in *Escherichia coli*. *Journal of Bacteriology*, *172*, 6704–6712.
Nishida, Y., Adachi, K., Kasai, H., Shizuri, Y., Shindo, K., Sawabe, A., et al. (2005). Elucidation of a carotenoid biosynthesis gene cluster encoding a novel enzyme, 2, 2′-β-hydroxylase, from *Brevundimonas* sp. strain SD212 and combinatorial biosynthesis of new or rare xanthophylls. *Applied and Environmental Microbiology*, *71*, 4286–4296.
Paddon, C. J., Westfall, P. J., Pitera, D. J., Benjamin, K., Fisher, K., McPhee, D., et al. (2013). High-level semi-synthetic production of the potent antimalarial artemisinin. *Nature*, *496*, 528–532.
Pechous, S. W., & Whitaker, B. D. (2004). Cloning and functional expression of an (*E*, *E*)-α-farnesene synthase cDNA from peel tissue of apple fruit. *Planta*, *219*, 84–94.
Pemberton, T. A., Chen, M., Harris, G. G., Chou, W. K., Duan, L., Köksal, M., et al. (2017). Exploring the influence of domain architecture on the catalytic function of diterpene synthases. *Biochemistry*, *56*, 2010–2023.
Polakowski, T., Stahl, U., & Lang, C. (1998). Overexpression of a cytosolic hydroxymethylglutaryl-CoA reductase leads to squalene accumulation in yeast. *Applied Microbiology and Biotechnology*, *49*, 66–71.
Qin, B., Matsuda, Y., Mori, T., Okada, M., Quan, Z., Mitsuhashi, T., et al. (2015). An unusual chimeric diterpene synthase from *Emericella variecolor* and its functional conversion into a sesterterpene synthase by domain swapping. *Angewandte Chemie International Edition*, *54*, 1–5.
Schnurr, G., Schmidt, A., & Sandmann, G. (1991). Mapping of a carotenogenic gene cluster from *Erwinia herbicola* and functional identification of six genes. *FEMS Microbiology Letters*, *78*, 157–161.
Shao, J., Chen, Q.-W., Lv, H.-J., He, J., Liu, Z.-F., Lu, Y.-N., et al. (2017). (+)-Thalianatriene and (−)-retigeranin B catalyzed by sesterterpene synthases from *Arabidopsis thaliana*. *Organic Letters*, *19*, 1816–1819.
Shishova, E. Y., Di Costanzo, L., Cane, D. E., & Christianson, D. W. (2007). X-ray crystal structure of aristolochene synthase from *Aspergillus terreus* and evolution of templates for the cyclization of farnesyl diphosphate. *Biochemistry*, *46*, 1941–1951.
Sun, T., Miao, L., Li, Q., Dai, G., Lu, F., Liu, T., et al. (2014). Production of lycopene by metabolically-engineered *Escherichia coli*. *Biotechnology Letters*, *36*, 1515–1522.
Tominaga, K., Hongo, N., Karato, M., & Yamashita, E. (2012). Cosmetic benefits of astaxanthin on humans subjects. *Acta Biochimica Polonica*, *59*.
Wang, C., Yoon, S.-H., Jang, H.-J., Chung, Y.-R., Kim, J.-Y., Choi, E.-S., et al. (2011). Metabolic engineering of *Escherichia coli* for α-farnesene production. *Metabolic Engineering*, *13*, 648–655.
Wani, M. C., Taylor, H. L., Wall, M. E., Coggon, P., & McPhail, A. T. (1971). Plant antitumor agents. VI. Isolation and structure of taxol, a novel antileukemic and antitumor agent from *Taxus brevifolia*. *Journal of the American Chemical Society*, *93*, 2325–2327.
Weber, T., Blin, K., Duddela, S., Krug, D., Kim, H. U., Bruccoleri, R., et al. (2015). antiSMASH 3.0—A comprehensive resource for the genome mining of biosynthetic gene clusters. *Nucleic Acids Research*, *43*, 237–243.

Xie, W., Lv, X., Ye, L., Zhou, P., & Yu, H. (2015). Construction of lycopene-overproducing *Saccharomyces cerevisiae* by combining directed evolution and metabolic engineering. *Metabolic Engineering*, *30*, 69–78.

Yamada, Y., Kuzuyama, T., Komatsu, M., Shin-ya, K., Omura, S., Cane, D. E., et al. (2015). Terpene synthases are widely distributed in bacteria. *Proceedings of the National Academy of Sciences of the United States of America*, *112*, 857–862.

Yang, Y. l., Zhang, S., Ma, K., Xu, Y., Tao, Q., Chen, Y., et al. (2017). Discovery and characterization of a new family of diterpene cyclases in bacteria and fungi. *Angewandte Chemie International Edition*, *56*, 4749–4752.

Ye, Y., Minami, A., Mandi, A., Liu, C., Taniguchi, T., Kuzuyama, T., et al. (2015). Genome mining for sesterterpenes using bifunctional terpene synthases reveals a unified intermediate of di/sesterterpenes. *Journal of the American Chemical Society*, *137*, 11846–11853.

You, C., Zhang, X. Z., & Zhang, Y. H. P. (2012). Simple cloning via direct transformation of PCR product (DNA Multimer) to *Escherichia coli* and *Bacillus subtilis*. *Applied and Environmental Microbiology*, *78*, 1593–1595.

Yu, X., Liu, T., Zhu, F., & Khosla, C. (2011). In vitro reconstitution and steady-state analysis of the fatty acid synthase from *Escherichia coli*. *Proceedings of the National Academy of Sciences of the United States of America*, *108*, 18643–18648.

Zhou, P., Ye, L., Xie, W., Lv, X., & Yu, H. (2015). Highly efficient biosynthesis of astaxanthin in *Saccharomyces cerevisiae* by integration and tuning of algal crtZ and bkt. *Applied Microbiology and Biotechnology*, *99*, 8419–8428.

Zhu, F., Lu, L., Fu, S., Zhong, X., Hu, M., Deng, Z., et al. (2015). Targeted engineering and scale up of lycopene overproduction in *Escherichia coli*. *Process Biochemistry*, *50*, 341–346.

Zhu, F., Zhong, X., Hu, M., Lu, L., Deng, Z., & Liu, T. (2014). In vitro reconstitution of mevalonate pathway and targeted engineering of farnesene overproduction in *Escherichia coli*. *Biotechnology and Bioengineering*, *111*, 1396–1405.

CHAPTER SIX

Imine Reductases, Reductive Aminases, and Amine Oxidases for the Synthesis of Chiral Amines: Discovery, Characterization, and Synthetic Applications

Sebastian C. Cosgrove, Agata Brzezniak, Scott P. France, Jeremy I. Ramsden, Juan Mangas-Sanchez, Sarah L. Montgomery, Rachel S. Heath, Nicholas J. Turner[1]

Manchester Institute of Biotechnology, School of Chemistry, University of Manchester, Manchester, United Kingdom

[1]Corresponding author: e-mail address: nicholas.turner@manchester.ac.uk

Contents

Abstract

Synthesis of the chiral amine moiety is a key challenge for synthetic organic chemistry due to its prevalence in many biologically active molecules. Imine reductase and amine oxidase enzymes have enabled the biocatalytic synthesis of a host of chiral amine compounds. In this chapter, procedures for the synthesis of chiral amines using imine

Methods in Enzymology, Volume 608
ISSN 0076-6879
https://doi.org/10.1016/bs.mie.2018.04.022

reductases (IREDs), the recently discovered IRED homologues reductive aminases, and amine oxidases (AOs) are described. Amine oxidases have been the subject of mutagenesis approaches for improvement of substrate scope. The high-throughput screening method for determining active variants in amine oxidase libraries is illustrated. Finally, in an approach which takes inspiration from nature, many enzymes can be combined with each other in cascade reactions. The incorporation of imine reductase and monoamine oxidase biocatalysts into several cascade reactions, both in vitro and in vivo (where the approach moves toward synthetic biology), is reported.

1. INTRODUCTION

The amine moiety is present in many bioactive compounds and therefore the development of methodologies and strategies for the synthesis of amines is of great interest (Vitaku, Smith, & Njardarson, 2014). Alongside traditional synthetic chemistry, biocatalysis offers an extensive toolbox of enzymes to be used either as sole catalysts or in combination with other enzymes to mimic biosynthetic pathways to prepare this family of compounds in a more effective manner. In this chapter, we focus on two different types of enzymes that have already shown a huge potential for synthetic biology purposes.

1.1 Imine Reductases and Reductive Aminases

Imine reductases (IREDs) are NADPH-dependent oxidoreductases that catalyze the asymmetric reduction of imines to form the corresponding amines. Imine reductions can be found in different biosynthetic pathways such as folate (Posner et al., 1996), pipecolate (Muramatsu et al., 2005), siderophore (Meneely et al., 2016), morphine (Winzer et al., 2015), and argimycin P alkaloid (Ye et al., 2018) biosynthesis. The enzymes involved in these pathways are very diverse and often unrelated by sequence. They have been extensively studied although, due to their narrow substrate scope, limited synthetic applications have been developed to date. Ever since Mitsukura and coworkers discovered imine reductase activity in *Streptomyces* sp. (Mitsukura et al., 2011, 2013), we and others have contributed to the discovery and application of these enzymes in organic chemistry and now they constitute a valuable instrument in the enzymatic toolbox for the synthesis of chiral amines (Lenz et al., 2017; Mangas-Sanchez et al., 2017). Recently, we discovered an IRED homologue from *Aspergillus oryzae* that displays remarkable activity toward reductive amination (Fig. 1), prompting

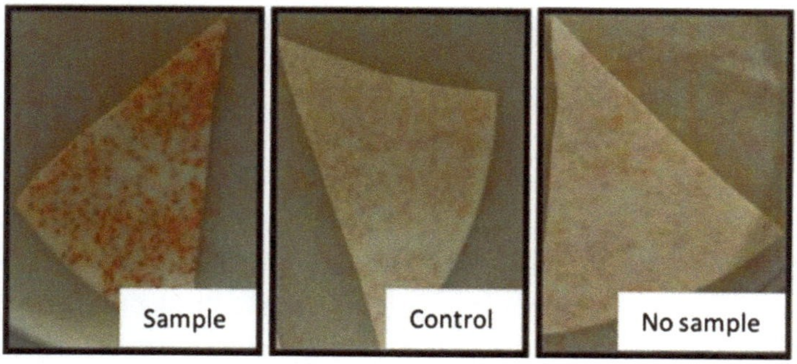

Fig. 1 Example of a screen of MAO-N activity using HRP-DAB assay. The *red/brown* spots indicate positives. The other lighter colored colonies were inactive on the substrate. The pictures alongside are a negative control and no sample.

us to name this enzyme a reductive aminase (RedAm) (Aleku et al., 2017). This finding opens a new horizon in the field, as biocatalytic intermolecular reductive amination reactions between carbonyl compounds and amines are now possible. This reductive aminase (*Asp*RedAm) is not the only example, as other enzymes have since been found to have the same activity (Roiban et al., 2017). Considering the synthetic applications of IREDs and RedAms, we envision that the integration of these biocatalysts into either in vivo or in vitro cascades will have a large impact for the preparation of industrially relevant amines.

1.2 Amine Oxidases

Monoamine oxidase (MAO) belongs to the flavin adenine dinucleotide (FAD)-dependent amine-oxidoreductase enzyme class and catalyzes the oxygen-driven conversion of amines to their corresponding imines. In nature MAOs are responsible for the oxidative deamination of amines, producing the derivative carbonyl and amine compounds (Sablin et al., 1997). The enzyme that has seen the most application in synthetic chemistry is a fungal variant isolated from *Aspergillus niger*. First discovered in 1995 and termed MAO-N, it exists as a homotetramer with a molecular mass of 55.6 kDa per subunit, with FAD noncovalently bound. MAO-N is highly selective for chiral molecules that have (*S*)-orientation at the chiral carbon atom in the alpha-position to the nitrogen atom (Atkin et al., 2008; Ghislieri et al., 2013). As well as MAO-N, there are other MAO enzymes that have found applications as biocatalysts. Cyclohexylamine oxidase (CHAO), isolated from *Brevibacterium oxydans*, is also an (*S*)-selective oxidoreductase.

6-Hydroxy-D-nicotine oxidase (6-HDNO) from *Arthrobacter nicotinovorans* is involved in the metabolism of nicotine and is enantiocomplimentary to (*S*)-selective MAO-N and CHAO, in that it demonstrates selectivity for (*R*)-enantiomers (Heath et al., 2014; Li et al., 2014). The oxygen-dependent mechanism of MAO enzymes proceeds through oxidation of the amine via abstraction of the alpha-hydride to afford the imine, with hydrogen peroxide being produced as a by-product. Both CHAO and 6-HDNO are also FAD dependent and require oxygen to work. Production of hydrogen peroxide during the reaction has been exploited by many in the development of activity screens that use horseradish peroxidase (HRP) in combination with colorless substrates that will react in the presence of hydrogen peroxide and HRP to give a colored compound. The most important synthetic application of the multiple MAO-N mutants is the chemoenzymatic deracemization reactions of α-substituted chiral amines, which will be discussed later in the chapter.

This chapter will discuss the methods that are used to produce and purify IRED and MAO enzymes, and describe some of the ways they have been applied in chemical biology and in synthetic organic chemistry. In particular, discussion of cascade reactions using these enzymes will be discussed. This chapter serves to highlight the potential power of IRED and MAO biocatalysts in the context of synthetic biology and also show how some of the enzymatic pathways we have designed mimic those found in nature.

2. PROTEIN ENGINEERING

2.1 Introduction

The use of enzymes in organic synthesis offers many advantages; however, most wild-type biocatalysts are not inherently suitable to perform under industrially relevant conditions in terms of activity, stability, and substrate tolerance. Fortunately, many of these issues can be addressed through protein engineering, i.e., the modification of the amino acid sequence of a protein to increase its performance or to study its catalytic properties. There are different approaches to protein engineering depending on the level of knowledge known about the enzyme's structure and these can be divided into three different groups: random, rational, or semirational (Crameri et al., 1998; Gupta & Tawfik, 2008; Reetz & Carballeira, 2007). As an example, our approach to improve the activity of MAO-N will be discussed in this section. For IRED/RedAm we have created mutations using standard site-directed mutagenesis methods and screened using NADPH depletion assays (Aleku et al., 2017).

2.2 Amine Oxidase Libraries and Screening

2.2.1 Introduction

Due to lack of sequence diversity in the families of amine oxidases (AOs), the substrate scope of these enzymes has been altered by mutagenesis. Alongside the oxidation of the amine, hydrogen peroxide is produced as a by-product of the reoxidation of the FAD cofactor by molecular oxygen. This is easily detected with HRP and a dye, which is a noncolored compound that is a substrate for HRP. The dye molecules are reduced in the presence of HRP and hydrogen peroxide and polymerize to form a colored compound (Scheme 1). Thousands of colonies expressing different variants of the enzyme can be screened by this method. A version of this assay using a soluble dye can be used for activity assays (kinetics and specific activity measurements) (Heath et al., 2014).

The color change is due to the reaction of hydrogen peroxide with 3,3′-diaminobenzidine, which is catalyzed by the HRP. The reaction, shown below, produces a polymeric compound which appears as a pink/red color (Scheme 2).

2.2.2 Library Generation

We have generated libraries of AO variants for screening using random and semirational methods. Random methods are useful when there is no crystal structure available, or they could be used to improve stability, for example. Once the crystal structures were known we took a semirational approach, for example, confining our sites of mutation to the active site or entry tunnel but including all possible amino acids in the library (using NNS or NNK

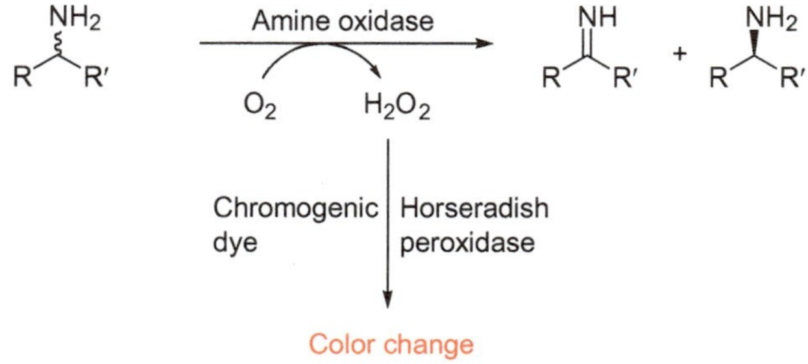

Scheme 1 Method for screening amine oxidase variants.

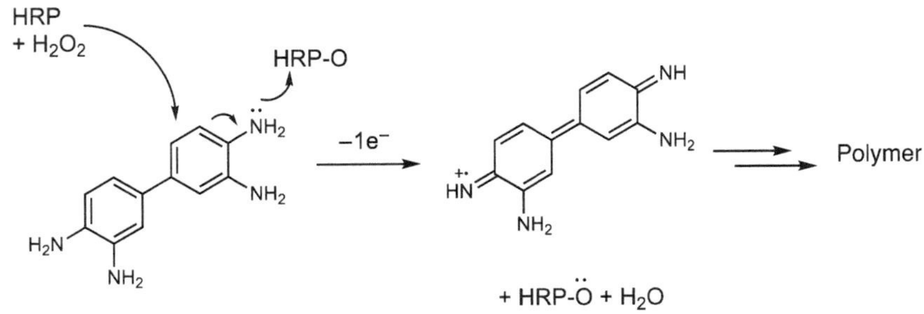

Scheme 2 HRP-catalyzed polymerization of 3,3′-diaminobenzidine with hydrogen peroxide.

codons in our primer design). To generate the libraries, we followed the manufacturers' protocols for the kits listed below:

Random

1. Follow *E. coli* XL1-Red mutator strain (Agilent) protocol
2. Follow GeneMorph II EZClone Domain Mutagenesis Kit (Agilent) protocol

Semirational

1. Follow QuikChange II site-directed mutagenesis kit (Agilent) protocol

Notes

Using the GeneMorph kit you can use primers to select the portion of the gene over which you want the mutations to be.

We replaced the QuikChange II kit components with our own Phusion™ polymerase, and dNTPs from an alternative source (e.g., NEB). (The extension temperature was adjusted to 72°C which is the optimum for Phusion™ polymerase.)

2.2.3 Solid-Phase Screen

Equipment

Standing incubator

Shaking incubator

Reagents

150-mm Petri dishes

150-mm Petri dishes containing LB agar

150-mm Petri dishes containing LB agar supplemented with 100 μg/mL ampicillin and 1 m*M* isopropyl β-D-1-thiogalactopyranoside (IPTG)

HRP solution: 0.1 mg/mL HRP (Type VI) (Sigma) in 100 m*M* potassium phosphate buffer, pH 8.0

SigmaFast™ 3,3′-Diaminobenzidine (DAB) tablets

Amine substrate
Filter paper, cut to fit inside 150-mm Petri dishes
E. coli BL21 (DE3) cells
Plasmid library
HyBond-N membrane (RPN303N from GE Healthcare, cut to fit inside 150-mm Petri dishes)
LB supplemented with 100 μg/mL ampicillin

Protocol

1. Transform your library into *E. coli* BL21 (DE3) following the manufacturer's protocol.
2. Place a membrane (HyBond) on top of LB/amp agar and spread the entire transformation reaction over the top.
3. Incubate overnight at 30°C.
4. Transfer the membrane to an LB/amp/IPTG plate and induce for 4 h at 30°C.
5. Transfer the membrane into an empty Petri dish and store at −20°C until ready to use.
6. Freeze thaw the membrane 4 × (freeze in liquid N_2, allow to thaw, and repeat).
7. Place the filter paper inside an empty Petri dish, soak with HRP solution, and pour off the excess.
8. Place the membrane on top of the filter paper, cover, and leave for 1 h at room temperature to remove any cellular H_2O_2 that may be present.
9. Prepare the assay mix by dissolving the DAB tablet in HRP solution (1 tablet per 15 mL) and add 20 m*M* of substrate.
10. Place a new filter paper inside an empty Petri dish, soak with assay mix, and pour off the excess.
11. Transfer the membrane to the filter paper, cover, and leave for at least 30 min or overnight.
12. Pick colonies that have changed to dark red/brown (see Fig. 1 as an example of a screen) and inoculate into LB/amp.
13. Grow overnight at 37°C and 250 rpm, extract the plasmid using your method of choice, and DNA sequence to determine the mutation(s).

Notes

The DAB is slow to dissolve.

5 mL assay mix is required per membrane.

Other membranes may work, but the HyBond-N is the one we have found to be the best. 1500–2000 colonies per membrane is a typical yield and so for a library in which two sites are mutated to all possible amino acids, carry out two transformation reactions so that at least 3000 colonies are screened (Reetz, Kahakeaw, & Lohmer, 2008).

For random libraries more may need to be screened depending on the mutagenic frequency and the length of the gene targeted.

Single colonies are required for the screening, so serial dilutions may be required to get the right colony density.

Growth at 30°C rather than 37°C gives smaller colonies but the density of color is better.

3. BIOTRANSFORMATIONS, INTENSIFICATION, AND SCALE-UP

3.1 RedAms and IREDs

3.1.1 Analytical-Scale Biotransformations

The form of the biocatalyst preparation determines the setup of biotransformation reactions.

3.1.1.1 Whole Cell

As IREDs and RedAms are NADPH-dependent enzymes, reactions require the presence of this cofactor and a means by which it can be recycled when only present in substoichiometric quantities. When IREDs are employed as resting whole-cell biocatalysts, no addition of cofactor is required as it is already present within the cells. Glucose is typically added for cell metabolism to ensure sufficient cofactor turnover in these cases.

3.1.1.2 Cell-Free Extracts and Purified Enzyme

With crude cell extracts or purified enzyme preparations, NADPH has to be supplied. The cofactor is prohibitively expensive for stoichiometric use and therefore a recycling system is required. A glucose/glucose dehydrogenase system is the method of choice in this case in which the oxidation of glucose produces the reduced cofactor that is used by the IRED reaction. However, it is worth noting that the by-product of this recycling system is ultimately gluconic acid which can cause the pH of the reaction to drop significantly. This effect is more pronounced at high substrate concentration in which a correspondingly greater amount of gluconic acid will be produced. This can be controlled in a number of ways such as by increasing the buffer strength of the reaction or titrating in NaOH solution to maintain a constant pH. For analytical-scale reactions (<500 μL reaction volume), monitoring and controlling pH is not trivial; however if substrate concentration is low, then it will be within the buffering capacity of the reaction medium. Other methods of recycling the cofactor such as using sacrificial IRED cosubstrates have also been demonstrated, although is not widely used (Gand et al., 2016).

Equipment

Shaking incubator

Analytical instrumentation such as GC-FID or HPLC

Reagents

Substrate imine (imine reduction mode)

Substrate ketone (reductive amination mode)

Substrate amine (reductive amination mode)

IRED enzyme in whole-cell *E. coli* or lysate or purified enzyme

D-glucose

$NADP^+$

DMSO

100 m*M* sodium phosphate buffer, pH 7.0 (imine reduction mode)

100 m*M* Tris–HCl buffer, pH 9.0 (reductive amination mode)

Protocol

1. *For imine reduction with whole-cell biocatalyst*: add 5 μL of a 500 m*M* stock solution of substrate imine in DMSO (final reaction concentration 5 m*M*) to a reaction mixture consisting of whole-cell biocatalyst (50 mg/mL) and D-glucose (50 m*M*) in pH 7.0, 100 m*M* sodium phosphate buffer (500 μL final volume).

   For imine reduction with purified enzyme or lysate: add 5 μL of a 500 m*M* stock solution of substrate imine in DMSO (final reaction concentration 5 m*M*) to a reaction mixture consisting of purified enzyme (0.5 $mg\,mL^{-1}$) or cell lysate (20 vol%), GDH (0.3 $mg\,mL^{-1}$), $NADP^+$ (1 m*M*), D-glucose (50 m*M*) in pH 7.0, 100 m*M* sodium phosphate buffer (500 μL final volume).

   For reductive amination with whole-cell biocatalyst: add 5 μL of a 500 m*M* stock solution of substrate ketone in DMSO (final reaction concentration 5 m*M*) to a reaction mixture consisting of whole-cell biocatalyst (50 mg/mL), D-glucose (50 m*M*), and amine substrate (1–20 molar equivalents w.r.t. ketone concentration) in 100 m*M* Tris–HCl buffer (500 μL final volume) adjusted to pH 8.0.

   For reductive amination with purified enzyme or lysate: add 5 μL of a 500 m*M* stock solution of substrate ketone in DMSO (final reaction concentration 5 m*M*) to a reaction mixture consisting of purified enzyme (0.5 $mg\,mL^{-1}$) or cell lysate (20 vol%), GDH (0.3 $mg\,mL^{-1}$), $NADP^+$ (1 m*M*), D-glucose (50 m*M*), and amine substrate (1–20 molar equivalents w.r.t. ketone concentration) in 100 m*M* Tris buffer (500 μL final volume) adjusted to pH 8.0.

2. Incubate the biotransformation at 30°C and 250 rpm for 18 h.
3. Quench the reaction by adding 5 *M* NaOH (80 μL) and ethyl acetate (1 mL).
4. Vortex the mixture and then centrifuge at 13,000 rpm for 1 min.
5. Separate the organic layer and dry over $MgSO_4$.
6. Analyze using appropriate analytical method (GC-FID, GC-MS, HPLC).

Notes

Often, an entirely aqueous reaction medium is problematic for substrates which are poorly water-soluble and therefore there is considerable scope for reaction optimization by employing organic cosolvents. Water-miscible DMSO and methanol, in small quantities (up to 10% v/v), are typically well tolerated in biotransformations and are frequently used to prepare stock solutions of substrate to be added to reactions. Water-immiscible solvents such as cyclohexane and heptane have been successfully employed with IREDs without significant attenuation of enzyme activity.

3.1.2 Intensification and Scale-Up

When testing a potential new IRED or RedAm substrate it is important to first identify the appropriate enzyme for the transformation that confers high conversion and desired enantioselectivity. Screening a diverse panel of enzymes on an analytical scale is required to select the best candidate for intensification studies in which the main goal is to maximize substrate concentration and minimize enzyme loading to make the process more scalable. Here, design of experiment (DoE) can be put to good use for the investigation of multiple reaction parameters simultaneously. Once suitable scale-up conditions have been found, the process can then be implemented on scale.

3.1.2.1 Typical Procedure for Preparative-Scale Reductive Amination Exemplified With Cyclohexanone and Allylamine

1. Add allyl amine (575 μL, 7.59 mmol) to 45 mL of Tris–HCl buffer (pH 8.0, 100 m*M*) in a 100 mL EasyMax reactor equipped with pH probe, temperature probe, and overhead stirrer
2. Adjust the pH of the solution to 9.0 with hydrochloric acid
3. Add D-glucose (1 g, 5.5 mmol), $NADP^+$ (30 mg, 0.037 mmol), GDH (CDX 901, 30 mg), and 5 mL of enzyme lysate (from a stock solution of 200 mg/mL based on wet cell weight)

4. Add cyclohexanone (500 mg, 5.1 mmol)
5. Stir the reaction at 30°C for 24 h with titration of 2 *M* sodium hydroxide to maintain a pH of 8.0
6. After 24 h, adjust the pH to 12.5 with 5 *M* sodium hydroxide solution and extract the product with methyl *tert*-butyl ether (1 × 35 mL) and 2-methyltetrahydrofuran (1 × 15 mL)
7. Add methanol (5 mL) then centrifuge to break the emulsion to improve phase separation
8. Combine the organic layers, dry over anhydrous Na_2SO_4, and carefully concentrate under reduced pressure to give the product

3.2 Amine Oxidase-Catalyzed Deracemization

AOs can be used to deracemize amines via selective oxidation of one enantiomer of a racemic amine mixture which is converted to the imine. With the addition of a nonselective reducing agent, the imine is reduced back to racemic amine but the unoxidized enantiomer is still present, giving a 25:75 ratio of the enantiomers. These cycles continue until a single enantiomer is accumulated in >99% *ee* (Scheme 3). The protocols below are described for MAO-N but are equally applicable to 6-HDNO and CHAO.

Reagents

E. coli cells containing MAO-N from *A. niger*
Potassium phosphate buffer (1 *M*, pH 7.8)
Amine substrate stock solution (1 *M* in DMSO)
Reducing agent stock solution (1 *M* $NH_3{\cdot}BH_3$ in 1 *M* potassium phosphate, pH 7.8 or 3 *M* $NaBH(OAc)_3$ in 1 *M* potassium phosphate, pH 7.8)

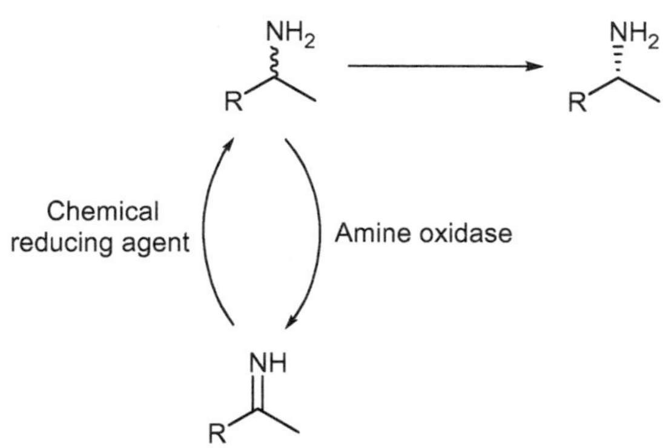

Scheme 3 Deracemization of amines using amine oxidases and a chemical reducing agent.

10 *M* aqueous NaOH solution
CH_2Cl_2

3.2.1 Analytical-Scale Deracemization Using Whole-Cell MAO-N

Protocol

1. Thaw frozen cells of *E. coli* containing MAO-N from *A. niger* (30 mg mL^{-1}) and suspend in reaction buffer (279 µL, 1 *M* potassium phosphate, pH 7.8)
2. Add the amine substrate (3 µL of 1 *M* solution in DMSO)
3. Add the reducing agent (18 µL of 1 *M* $NH_3 \cdot BH_3$ solution in 1 *M* potassium phosphate, pH 7.8 or 18 µL of 3 *M* $NaBH(OAc)_3$ solution in 1 *M* potassium phosphate, pH 7.8; final reaction volume of 300 µL)
4. Incubate the reaction mixture at 37°C and 250 rpm
5. Upon completion, basify the reaction by addition of 10 *M* NaOH (40 µL) and extract into CH_2Cl_2 (1 mL)
6. Vortex the mixture and then centrifuge at 13,200 rpm for 2 min
7. Separate the organic layer and dry over $MgSO_4$

Notes

More cells can be added. Formation of ketone by-products may be reduced by addition of organic cosolvents, such as hexane or toluene, in a 50% v/v ratio. For substrates with low water solubility, water-miscible organic cosolvents (DMSO, methanol) can be added up to 10% v/v. A variety of borane reducing agents may be used for the chemical reduction step (e.g., $BH_3 \cdot (CH_3)_2NH$, $BH_3 \cdot (CH_3)_3N$, $BH_3 \cdot t\text{-}BuNH_2$, $BH_3 \cdot C_4H_9NO$); however, $NH_3 \cdot BH_3$ and $NaBH(OAc)_3$ have been shown to be most effective. The addition of the chemical reducing agent can be omitted if only the enzymatic oxidation is required. Due to the oxygen dependency of the enzymes, the reactions are usually run in vessels that contain a head space of 75%–80% (for example, 500 µL reactions are run in 2-mL Eppendorf tubes).

3.2.2 Analytical-Scale Deracemization Using Purified MAO-N or MAO-N Cell Lysate

Protocol

1. To a mixture of purified MAO-N (0.5 mg/mL) or MAO-N cell lysate (20 vol%) add the racemic amine substrate (10 m*M* final concentration from a 1 *M* stock in DMSO)
2. Repeat steps 3–7 described in the preceding protocol

3.2.3 Preparative-Scale Deracemization Using Whole-Cell MAO-N

Protocol

1. In a 500-mL conical flask or glass bottle, dissolve the racemic amine (10 m*M*) and the reducing agent (50 m*M*, 5 equiv. of $NH_3{\cdot}BH_3$ or 100 m*M*, 10 equiv. of $NaBH(OAc)_3$) in 1 *M* potassium phosphate buffer (165 mL, pH 7.8)
2. Add *E. coli* cells containing MAO-N from *A. niger* (5 g)
3. Incubate the reaction mixture at 37°C and 250 rpm for 24 h or longer if required
4. Upon completion, basify the reaction by addition of 10 *M* NaOH (2 mL) and extract with CH_2Cl_2 (150 mL)
5. Vortex the mixture and then centrifuge at 4000 rpm for 5 min to separate the layers
6. Extract again with CH_2Cl_2 (2 × 150 mL) and repeat step 5
7. Combine the organic layers and dry over $MgSO_4$
8. Concentrate the organic extracts in vacuo to obtain the product

4. IN VITRO AND IN VIVO CASCADES

Enzyme cascades allow multiple biocatalysts to be employed simultaneously to carry out a series of chemical transformations, either in vitro with the component enzymes being added to a cell-free solution, or in vivo where the required biocatalysts are coexpressed in a living organism.

The development of in vitro cascades can be thought of as a "bottom-up" approach to synthetic biology, where individual biocatalytic steps are combined under controlled conditions and applied to chemical synthesis. In some cases, researchers attempt to directly replicate natural systems in a controlled environment. However, the Turner group typically applies biocatalytic retrosynthesis to a synthetic target, often identifying new applications for known enzyme families (Turner & O'Reilly, 2013).

While building cascades in vitro is beneficial for studying individual parameters outside the complex cellular metabolism, ultimately it is more practical to apply a single whole cell for the transformation (France et al., 2017). This requires coexpression of the enzymes on a single vector, for example, by using a BioBrick® system (Shetty, Endy, & Knight, 2008), or the cotransformation of multiple expression vectors (Jia & Jeon, 2016).

The efficiency of a cascade can be improved by combining two opposing redox processes with a common intermediate, so that the chemical transformations occur in the same pot with no additional oxidant or reductant required (see Section 4.2). However, the chemistry of a process will not always favor a simple cofactor exchange. For example, AOs cannot recycle the IRED cofactor in a cascade for amine deracemization. Nonetheless, our group created a working system by applying IRED whole cells supplied with glucose to drive endogenous regeneration of NADPH (see Section 4.3). Ultimately, it is possible to build up sequences of three or more biocatalytic transformations to convert simple precursors into valuable functionalized products. Our group has demonstrated this in vitro with the biocatalytic conversion of cycloalkanes into secondary amines (Tavanti et al., 2017) and in vivo with both the stereoselective C–H amination of 4-substituted ethylbenzenes (Both et al., 2016) and the synthesis of chiral cyclic amines from simple ketoacids (see Section 4.3).

4.1 Biocatalytic Hydrogen Borrowing

"Borrowing hydrogen" is a phrase used to describe a reductive process being driven by oxidation of the starting material in the same pot (Nixon, Whittlesey, & Williams, 2009). This has clear parallels with the biochemistry of living cells, where a limited number of cofactors act as electron shuttles to drive concurrent processes. However, "closed-loop cofactor recycling" is more typically used to describe the balance of redox chemistry across a larger network of enzymes (Chen, Li, & Liu, 2014).

Several cascades have been published which utilize an alcohol dehydrogenase (ADH) to oxidize an alcohol and thus form the aldehyde or ketone substrate for an enzyme which produces chiral amines. Amine dehydrogenases (AmDHs) were combined with two enantiocomplementary NAD-dependent ADHs to produce chiral primary amines from their racemic alcohol precursors with no additional cofactor regeneration (Mutti et al., 2015). Subsequently, a mutant ADH from *Thermoanaerobacter ethanolicus* (TeSADH W110A) was engineered to switch its cofactor dependency from NADPH to NADH, resulting in a more efficient cascade which required only two enzymes instead of three (Thompson & Turner, 2017). Most recently, the reductive aminase from *A. oryzae* was combined with either TesADH W110A or one of two other ADHs in a cascade for the direct bioalkylation of amines with primary and secondary alcohols (Montgomery et al., 2017). This is the first example of an in vitro biocatalytic cascade to produce secondary amines from an alcohol precursor in a single step (Fig. 2).

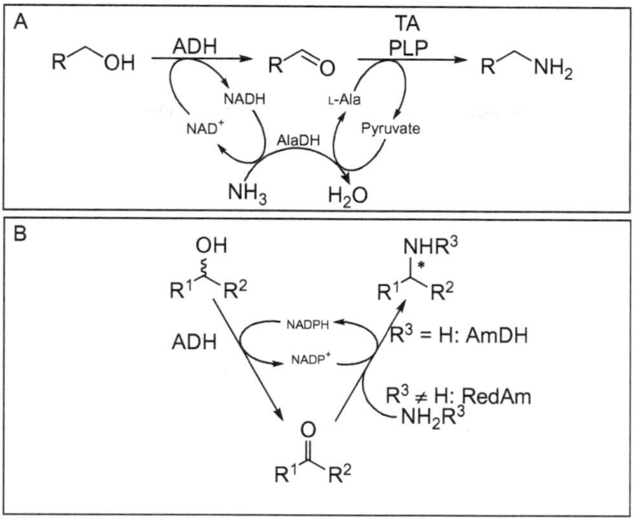

Fig. 2 (A) Biocatalyst network for the amination of primary alcohols; *TA*, transaminase. (B) Two-enzyme cascade for the amination of alcohols.

4.2 Amine Oxidase/IRED Cascade for Enantioselective Amine Synthesis

The imine product from an AO reaction is often a substrate for an IRED. Thus, AOs and imine reductases can be combined in one pot to perform a deracemization of a racemic amine without the need for a chemical reducing agent (Scheme 4). The reaction is performed by combining the whole-cell biocatalysts in buffer with glucose to enable in vivo recycling of the nicotinamide cofactor. Coexpression of the AO with an IRED was successful but would require further optimization in order to give full conversions (Heath et al., 2016).

4.3 Carboxylic Acid Reductase, Transaminase, and IRED Cascade Reactions

Due to their stability in aqueous media, cyclic imines constitute the major substrate scope of imine reductases. Despite the fact that these substrates may be complex or not commercially available, biocatalytic retrosynthetic analysis allows for many to be broken down into more simple precursors. For example, monosubstituted piperidines and pyrrolidines may be synthesized in a three-step enzymatic sequence from the corresponding ketoacids. This involves an initial reduction to the ketoaldehyde with a carboxylic acid

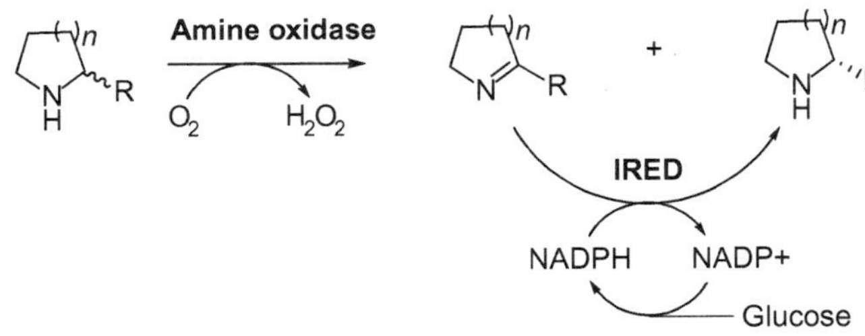

Scheme 4 Combined amine oxidase and IRED cascade for the deracemization of amines.

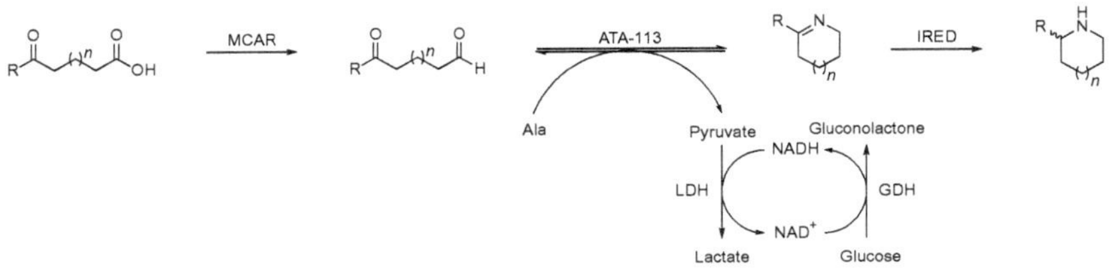

Scheme 5 A three-step biocatalytic cascade delivers chiral substituted heterocycles from simple ketoacid precursors (France et al., 2016).

reductase (CAR), followed by the regioselective installation of amine functionality by a transaminase, and (following a spontaneous cyclization) stereoselective reduction of the cyclic imine using an imine reductase (France et al., 2016) (Scheme 5).

As CAR must activate substrates through adenylation prior to reduction, ATP is required for enzyme function as well as the NADPH redox cofactor. Because of these demands, CAR was applied as a whole-cell biocatalyst in this work to exploit the internal cofactor regeneration systems of *E. coli*. Although the IRED was also used whole cell, a commercially sourced transaminase was employed as an isolated enzyme alongside lactose dehydrogenase and GDH to shift the equilibrium toward the amine. Ultimately, this hybrid system delivers an extremely useful chemical transformation but demands the use of four separate cofactors. This made this cascade a promising target for a synthetic biology approach where the transformation may be achieved within a single cell. Through the coexpression of the three biocatalysts in a single recombinant organism, this de novo pathway was delivered with conversions comparable to the original work with a key advantage of the in vivo cascade being no requirement for the addition of exogenous cofactors, as these were supplied by glucose metabolism (Hepworth et al., 2017). Although, when conceived, this biocatalytic pathway to piperidine derivatives was considered nonnatural, it later emerged the same route is employed in the biosynthesis of the antibiotic indolizomycin (Peng et al., 2016). While most natural piperidine moieties are derived from the cyclization of lysine, here a thioester linked to a polyketide synthase, as opposed to the adenylation domain of CAR, is reduced to the ketoaldehyde before transamination, cyclization, and reduction.

5. CONCLUSIONS

IREDs, RedAms, and AOs are all valuable biocatalysts for the synthesis of chiral amines in enantiomerically enriched form. Together they constitute a toolbox of enzymes that use complementary substrates and give access to a range of different products. By applying biocatalytic retrosynthesis, we are not only able to devise new routes to important molecules including APIs and natural products, but "invent" biological pathways that are hitherto undiscovered, as demonstrated by our CAR/TA/IRED cascade. Increasingly these biocatalysts are finding application not just for single-step transformations but also in multienzyme processes with great potential and scope for wider implementation in synthetic biology in the near future.

REFERENCES

Aleku, G. A., et al. (2017). A reductive aminase from Aspergillus oryzae. *Nature Chemistry*, *9*(10), 961–969. Nature Publishing Group. https://doi.org/10.1038/nchem.2782.

Atkin, K. E., et al. (2008). The structure of monoamine oxidase from Aspergillus niger provides a molecular context for improvements in activity obtained by directed evolution. *Journal of Molecular Biology*, *384*(5), 1218–1231. https://doi.org/10.1016/j.jmb.2008.09.090.

Both, P., et al. (2016). Whole-cell biocatalysts for stereoselective C-H amination reactions. *Angewandte Chemie, International Edition*, *55*(4), 1511–1513. Wiley-Blackwell. https://doi.org/10.1002/anie.201510028.

Chen, X., Li, S., & Liu, L. (2014). Engineering redox balance through cofactor systems. *Trends in Biotechnology*, *32*(6), 337–343. Elsevier. https://doi.org/10.1016/j.tibtech.2014.04.003.

Crameri, A., et al. (1998). DNA shuffling of a family of genes from diverse species accelerates directed evolution. *Nature*, *391*(6664), 288–291. Nature Publishing Group. https://doi.org/10.1038/34663.

France, S. P., et al. (2016). One-pot cascade synthesis of mono- and disubstituted piperidines and pyrrolidines using carboxylic acid reductase (CAR), ω-transaminase (ω-TA), and imine reductase (IRED) biocatalysts. *ACS Catalysis*, *6*(6), 3753–3759. American Chemical Society. https://doi.org/10.1021/acscatal.6b00855.

France, S. P., et al. (2017). Constructing biocatalytic cascades: In vitro and in vivo approaches to de novo multi-enzyme pathways. *ACS Catalysis*, 7(1), 710–724. American Chemical Society. https://doi.org/10.1021/acscatal.6b02979.

Gand, M., et al. (2016). A NADH-accepting imine reductase variant: Immobilization and cofactor regeneration by oxidative deamination. *Journal of Biotechnology*, *230*, 11–18. Elsevier Ltd. https://doi.org/10.1016/j.jbiotec.2016.05.006.

Ghislieri, D., et al. (2013). Engineering an enantioselective amine oxidase for the synthesis of pharmaceutical building blocks and alkaloid natural products. *Journal of the American Chemical Society*, *135*(29), 10863–10869. American Chemical Society. https://doi.org/10.1021/ja4051235.

Gupta, R. D., & Tawfik, D. S. (2008). Directed enzyme evolution via small and effective neutral drift libraries. *Nature Methods*, *5*(11), 939–942. Nature Publishing Group. https://doi.org/10.1038/nmeth.1262.

Heath, R. S., et al. (2014). Development of an R-selective amine oxidase with broad substrate specificity and high enantioselectivity. *ChemCatChem*, *6*(4), 996–1002. WILEY-VCH Verlag. https://doi.org/10.1002/cctc.201301008.

Heath, R. S., et al. (2016). Combined imine reductase and amine oxidase catalyzed deracemization of nitrogen heterocycles. *ChemCatChem*, *8*(1), 117–120. https://doi.org/10.1002/cctc.201500822.

Hepworth, L. J., et al. (2017). Enzyme cascades in whole cells for the synthesis of chiral cyclic amines. *ACS Catalysis*, 7(4), 2920–2925. American Chemical Society. https://doi.org/10.1021/acscatal.7b00513.

Jia, B., & Jeon, C. O. (2016). High-throughput recombinant protein expression in *Escherichia coli*: Current status and future perspectives. *Open Biology*, *6*(8), 160196. Royal Society Journals. https://doi.org/10.1098/rsob.160196.

Lenz, M., et al. (2017). Recent advances in imine reductase-catalyzed reactions. *World Journal of Microbiology and Biotechnology*, *33*(11), 1–10. Springer Netherlands. https://doi.org/10.1007/s11274-017-2365-8.

Li, G., et al. (2014). Substrate profiling of cyclohexylamine oxidase and its mutants reveals new biocatalytic potential in deracemization of racemic amines. *Applied Microbiology and Biotechnology*, *98*(4), 1681–1689. Springer Berlin Heidelberg. https://doi.org/10.1007/s00253-013-5028-1.

Mangas-Sanchez, J., et al. (2017). Imine reductases (IREDs). *Current Opinion in Chemical Biology*, *37*, 19–25. Elsevier Ltd. https://doi.org/10.1016/j.cbpa.2016.11.022.

Meneely, K. M., et al. (2016). Holo structure and steady state kinetics of the thiazolinyl imine reductases for siderophore biosynthesis. *Biochemistry*, *55*(38), 5423–5433. https://doi.org/10.1021/acs.biochem.6b00735.

Mitsukura, K., et al. (2011). Purification and characterization of a novel (R)-imine reductase from Streptomyces sp. GF3587. *Bioscience, Biotechnology, and Biochemistry*, *75*(9), 1778–1782. https://doi.org/10.1271/bbb.110303.

Mitsukura, K., et al. (2013). A NADPH-dependent (S)-imine reductase (SIR) from Streptomyces sp. GF3546 for asymmetric synthesis of optically active amines: Purification, characterization, gene cloning, and expression. *Applied Microbiology and Biotechnology*, *97*(18), 8079–8086. https://doi.org/10.1007/s00253-012-4629-4.

Montgomery, S. L., et al. (2017). Direct alkylation of amines with primary and secondary alcohols through biocatalytic hydrogen borrowing. *Angewandte Chemie, International Edition*, *56*(35), 10491–10494. Wiley-Blackwell. https://doi.org/10.1002/anie.201705848.

Muramatsu, H., et al. (2005). The putative malate/lactate dehydrogenase from Pseudomonas putida is an NADPH-dependent Δ1-piperideine-2-carboxylate/Δ1-pyrroline-2-carboxylate reductase involved in the catabolism of D-lysine and D-proline. *Journal of Biological Chemistry*, *280*(7), 5329–5335. https://doi.org/10.1074/jbc.M411918200.

Mutti, F. G., et al. (2015). Conversion of alcohols to enantiopure amines through dual-enzyme hydrogen-borrowing cascades. *Science*, *349*(6255), 1525–1529. https://doi.org/10.1126/science.aac9283.

Nixon, T. D., Whittlesey, M. K., & Williams, J. M. J. (2009). Transition metal catalysed reactions of alcohols using borrowing hydrogen methodology. *Dalton Transactions*, *5*, 753–762. The Royal Society of Chemistry. https://doi.org/10.1039/B813383B.

Peng, H., et al. (2016). Deciphering piperidine formation in polyketide-derived indolizidines reveals a thioester reduction, transamination, and unusual imine reduction process. *ACS Chemical Biology*, *11*(12), 3278–3283. American Chemical Society. https://doi.org/10.1021/acschembio.6b00875.

Posner, B. A., et al. (1996). Engineering specificity for folate into dihydrofolate reductase from Escherichia coli. *Biochemistry*, *35*(5), 1653–1663. https://doi.org/10.1021/bi9518095.

Reetz, M. T., & Carballeira, J. D. (2007). Iterative saturation mutagenesis (ISM) for rapid directed evolution of functional enzymes. *Nature Protocols*, *2*(4), 891–903. Nature Publishing Group. https://doi.org/10.1038/nprot.2007.72.

Reetz, M. T., Kahakeaw, D., & Lohmer, R. (2008). Addressing the numbers problem in directed evolution. *Chembiochem*, *9*(11), 1797–1804. Wiley-VCH Verlag. https://doi.org/10.1002/cbic.200800298.

Roiban, G. D., et al. (2017). Efficient biocatalytic reductive aminations by extending the imine reductase toolbox. *ChemCatChem*, *9*(24), 4475–4479. https://doi.org/10.1002/cctc.201701379.

Sablin, S. O., et al. (1997). Isolation and characterization of an evolutionary precursor of human MAO A and B. *European Journal of Biochemistry*, *43*(6), 455–456. Springer-Verlag. https://doi.org/10.1046/j.1432-1327.1998.2530270.x.

Shetty, R. P., Endy, D., & Knight, T. F. (2008). Engineering BioBrick vectors from BioBrick parts. *Journal of Biological Engineering*, *2*(1), 5. https://doi.org/10.1186/1754-1611-2-5.

Tavanti, M., et al. (2017). A biocatalytic cascade for the amination of unfunctionalised cycloalkanes. *Organic & Biomolecular Chemistry*, *15*(46), 9790–9793. The Royal Society of Chemistry. https://doi.org/10.1039/C7OB02569F.

Thompson, M. P., & Turner, N. J. (2017). Two-enzyme hydrogen-borrowing amination of alcohols enabled by a cofactor-switched alcohol dehydrogenase. *ChemCatChem*, *9*(20), 3833–3836. Wiley-Blackwell. https://doi.org/10.1002/cctc.201701092.

Turner, N. J., & O'Reilly, E. (2013). Biocatalytic retrosynthesis. *Nature Chemical Biology*, *9*(5), 285–288. Nature Publishing Group. https://doi.org/10.1038/nchembio.1235.

Vitaku, E., Smith, D. T., & Njardarson, J. T. (2014). Analysis of the structural diversity, substitution patterns, and frequency of nitrogen heterocycles among U.S. FDA approved pharmaceuticals. *Journal of Medicinal Chemistry*, *57*(24), 10257–10274. American Chemical Society. https://doi.org/10.1021/jm501100b.

Winzer, T., et al. (2015). Morphinan biosynthesis in opium poppy requires a P450-oxidoreductase fusion protein. *Science*, *349*(6245), 309–312.

Ye, S., et al. (2018). New insights into the biosynthesis pathway of polyketide alkaloid argimycins P in Streptomyces argillaceus. *Frontiers in Microbiology*, *9*(February), 1–12. https://doi.org/10.3389/fmicb.2018.00252.

FURTHER READING

Grogan, G., & Turner, N. J. (2016). InspIRED by nature: NADPH-dependent imine reductases (IREDs) as catalysts for the preparation of chiral amines. *Chemistry—A European Journal*, *22*(6), 1900–1907. https://doi.org/10.1002/chem.201503954.

CHAPTER SEVEN

Experimental Protocols for Generating Focused Mutant Libraries and Screening for Thermostable Proteins

Max J.L.J. Fürst, Caterina Martin, Nikola Lončar, Marco W. Fraaije[1]
Groningen Biomolecular Sciences and Biotechnology Institute, University of Groningen, Groningen, The Netherlands
[1]Corresponding author: e-mail address: m.w.fraaije@rug.nl

Contents

Abstract

Many proteins are rapidly deactivated when exposed to high or even ambient temperatures. This cannot only impede the study of a particular protein, but also is one of the major reasons why enzyme catalysis is still widely unable to compete with established chemical processes. Furthermore, differences in protein stability are a challenge in synthetic biology, when individual modules prove to be incompatible. The targeted stabilization of proteins can overcome these hurdles, and protein engineering techniques are more and more reliably supported by computational chemistry tools. Accordingly,

Methods in Enzymology, Volume 608
ISSN 0076-6879
https://doi.org/10.1016/bs.mie.2018.04.007

algorithms to predict the differences in folding energy of a mutant compared to the wild-type, $\Delta\Delta G^{fold}$, are used in the highly successful FRESCO workflow. The resulting single mutant prediction library consists typically of a few hundred amino acid exchanges, and after combining the most successful hits we so far obtained stabilized mutants which exhibited increases in apparent melting temperature of 20–35°C and showed vastly increased half-lives, as well as resistance to cosolvents. Here, we report a detailed protocol to generate these mutant libraries experimentally, covering the entire workflow from primer design, through mutagenesis, protein production and screening, to mutation combination strategies. The individual parts of the method are furthermore applicable to many other scenarios besides protein stabilization, and these protocols are valuable for any project requiring individual or semi high-throughput site-directed mutagenesis, protein expression and purification, or generation of mutant combination libraries.

ABBREVIATIONS

bp basepair(s)
o/n overnight
OD optical density
PCR polymerase chain reaction
ss secondary structure
T_m melting temperature

1. INTRODUCTION

Synthetic biology builds on the modular aspects by which cells function and can be programmed. Crucial elements in cellular functions are enzymes. With the current level of knowledge in enzymology, which ranges from detailed insights into how enzymes are produced and modified by posttranslational processes to the availability of databases that can be queried for known and characterized enzymes, design of novel metabolic pathways has become possible. By establishing the expression of a complementary set of enzymes, it has become feasible to create recombinant organisms capable of producing a desired target molecule. Such metabolic engineering efforts require enzymes to fulfill a wide range of criteria (Liu, Shin, Li, & Liu, 2015; Marcheschi, Gronenberg, & Liao, 2013). In fact, for generating effective metabolic pathways in synthetic biology, often the performance of the enzymes involved needs to be tuned. Targets of enzyme engineering often are optimizing kinetics, improved chemo-, regio-, and/or enantioselectivity, altering the substrate specificity, improving thermostability, and lowering inhibition or inactivation. Moreover, it is also crucial to optimize enzymes for many other biotechnological applications. While numerous approaches in enzyme engineering have emerged in the last few decades, it can still be a challenge to engineer an enzyme toward obeying the criteria set by the target application. From the wide range of different approaches explored, it seems a consensus in enzyme engineering has emerged in recent years: a hybrid approach of random mutagenesis combined with high-throughput screening (directed evolution) and knowledge-based directed mutagenesis. Such hybrid approaches typically make use of structural and/or mechanistic knowledge to design so-called clever libraries that still contain a fair amount of freedom by allowing several mutations at various positions in a protein sequence.

In this contribution, we provide experimental protocols that can be used for almost any protein engineering effort that involves the creation and screening of small-targeted mutant libraries. We have used these protocols for successfully generating proteins with improved (thermo)stability. The "knowledge-based" component to guide the preparation of the mutant libraries is based on a recently developed computational method which predicts mutations that are beneficial for enzyme stability: Framework for Rapid Enzyme Stabilization by Computational libraries (FRESCO). In our hands, the computational FRESCO method in combination with the experimental protocols as described here has delivered marked improvements in thermostability of 20–35°C in apparent melting temperature (T_m) for various proteins (Wijma et al., 2014). The FRESCO method proved to be useful not just in increasing the apparent T_m of the enzymes but also in giving a remarkable increase in resistance to cosolvents (Arabnejad et al., 2017). We have also used this approach to generate thermostable variants of flavoenzymes by using the ThermoFAD method for screening for improved mutants (Forneris, Orru, Bonivento, Chiarelli, & Mattevi, 2009).

The FRESCO in silico library can be created even by computationally inexperienced biochemists in a few weeks (Wijma, Fürst, & Janssen, 2018); however, the protocols provided here can also be used for the preparation and screening of mutant libraries based by other stabilization predictions methods or even entirely different design criteria. The provided protocols can be applied to any other small library and (flavo)enzyme engineering target. We set ourselves a target to provide the reader with comprehensive and detailed, yet straightforward protocols for semi high-throughput generation, expression, production, and verification of mutants in a 96-well format. Furthermore, we provide two experimental strategies for efficient combination of the best hits.

2. SINGLE MUTANTS GENERATION

Through FRESCO, other computational predictions tools or other structure-inspired input, it is often desirable to generate a large collection of site-directed mutant proteins. From the prediction generated through FRESCO, typically a library of a few hundred mutant proteins should be prepared. Also other methods often target such a library size, as it allows medium-throughput screening methods for establishing which mutant proteins perform best. We have optimized a protocol that allows the generation of such libraries, by combining established molecular biology methods to an efficient workflow, and by fine-tuning crucial steps to work seamlessly in the 96-well format (Fig. 1).

A well-established way to generate single mutants is the QuikChange™ mutagenesis method developed by Stratagene (now Agilent). In this method, a complete vector is polymerase chain reaction (PCR) amplified by two completely complementary primers that contain the desired mutation, and DpnI is used to digest the methylated template. The resulting linear PCR fragments contain complementary overhangs and thus circularize to a nicked plasmid, which becomes ligated by *E. coli*'s DNA repair machinery after transformation. Despite yielding relatively reliable results when using the original protocol, there was room for improvement. Limiting the failure rate in this initial experimental phase is essential to limit labor and material costs. For a typical library size of a few hundred mutants, primer synthesis, polymerase, and sequencing service become major cost factors of the method. Consequently, a reliable primer design strategy and a high-PCR success rate are critical.

The two most relevant limitations of the original QuikChange protocol are (i) favored primer–dimer formation over template annealing, particularly for an increased number of mismatches and (ii) the insufficient accumulation of product, due to the linear rather than exponential amplification of strands (Fig. 2). Adaptations first reported by Zheng et al. and further developed by Liu et al. elegantly overcame both limitations, while maintaining the simplicity and the general workflow. Accordingly, instead of using a completely complementary pair of primers, the overlapping region is confined to the 5′ end. The nonoverlapping 3′ end is designed to have a higher T_m than the

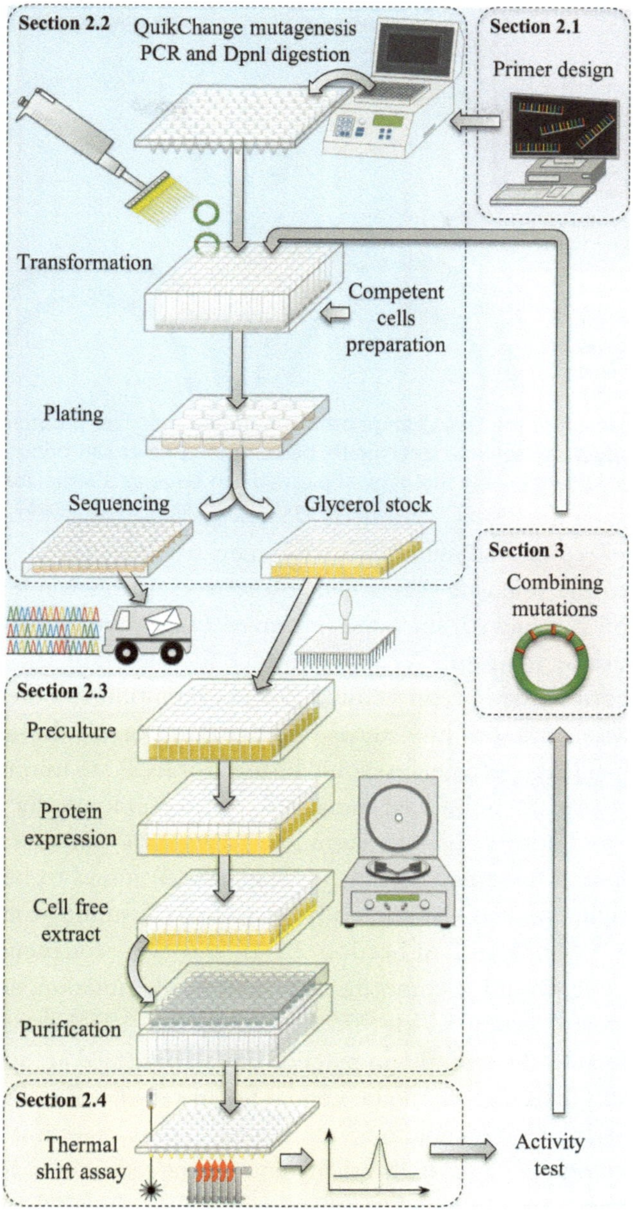

Fig. 1 Overview of the workflow for generating stabilized proteins. Individual steps and some of the main equipment required are depicted. Individual sections in this chapter where the corresponding procedures are described are also given.

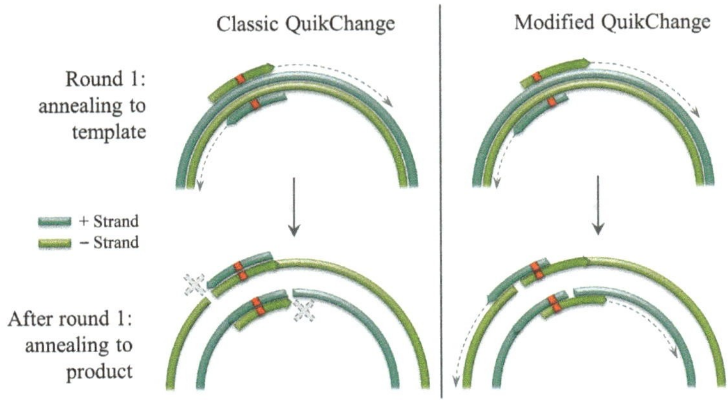

Fig. 2 Comparison of the QuikChange protocols. In the modified protocol, the linearized PCR product can serve as a template, because the primer can bridge the nick. In the classic QuikChange, only the original plasmid can serve as a template.

overlapping region, allowing primer–dimers to be melted, while the annealing to the template is retained. Moreover, in this way exponential amplification is achieved, since the linearized PCR product can serve as a template because the primer is able to bridge the nick (Fig. 2).

Besides primer design, one should also pay attention to other details of the (otherwise common) procedures in order to maximize the success rate. First, a high-fidelity polymerase should be used for PCR, to minimize undesired and potentially unnoticed mutations. As an economically interesting alternative to Phusion, we are commonly using *Pfu* polymerase, which we and others (McInerney, Adams, & Hadi, 2014) found to be similar in fidelity, despite what is commercially advertised. Assuming an error rate of 2.8×10^{-6} (see note 1 in Section 2.2.4), and two sequencing reads of 1 kb each, one would accumulate one unnoticed mutation per 96-well plate for a vector size of 5.7 kb ($[(5700-2000) \cdot 2.8 \cdot 10^{-6}]^{-1} \approx 96$), which we find tolerable. Second, the in practice often overdosed amount of template DNA in the PCR should be kept at the lower end of the polymerase manufacturer's recommendations, to facilitate DpnI digestion and avoid wild-type colonies. Third, high-quality competent cells should be prepared in order to counter a lowered transformation efficiency in 96-well plates.

2.1 Primer Design

Since reliably designing hundreds of primers manually is impractical, we recommend using automated software with adjustable parameters in order to satisfy both common primer rules, as well as guidelines specific to the partially overlapping design strategy discussed by Liu and Naismith. One example is AAscan (Sun et al., 2013), an open source tool run under Windows. Although the software has a batch mode to generate a list of primers on different positions, the input can currently not stem from a defined list of mutations, and is furthermore restricted to one amino acid at a time. There are two options to get to the desired list of primers: either by manual input of each position and mutant codon, or by generating primers for all positions and each amino acid, and subsequent filtering of the relevant primers. We describe the procedure for the latter option, including a small UNIX command line script that automatically filters the final list of primers, using as input a list of mutations generated in the computational part of the FRESCO procedure, or any other list that sticks to a common format (see note 1).

2.1.1 Equipment

- PC running Windows
- DNA sequence of the template plasmid
- AAscan, available from https://www.psi.ch/lbr/aascan
- A UNIX command line (shell), such as the terminal of MacOS or Linux, or the Windows 10 feature known as "Windows subsystem for Linux"

2.1.2 Procedure

1. Download AAscan and execute it; there is no install process needed (Fig. 3).
2. Copy and paste the unformatted sequence of the vector containing the wild-type gene in the DNA sequence field and enter the number

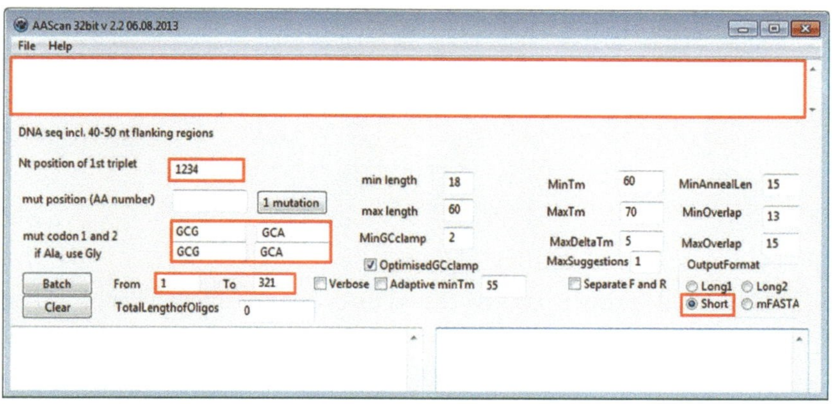

Fig. 3 Screenshot of AAscan. Fields requiring adaptation are marked in *red*.

corresponding to the sequence position of the gene's first nucleotide (usually the A of the ATG start codon).

3. Since the batch mode will be used, leave the "mut position" field empty.
4. Type the preferred codon for alanine into the "mut codon 1" field, as well as into the first "if Ala, use Gly" field below, and an alternative codon in the two fields on the right. It is important to enter the codon twice, otherwise, the software will mutate all alanines to glycines. Choose the codons with the help of an *E. coli* codon usage table (see note 2).
5. In the batch fields, type "1" in the first and the number of the last amino acid of your protein in the second field.
6. Choose the "Short" output format and keep the rest of the parameters unchanged or adapt according to specific requirements (see note 3).
7. Click "Batch" and copy the output in a plain text file, named with the one letter code of the amino acid (e.g., "A.txt" for alanine).
8. Repeat steps 4–7 for all 20 amino acids.
9. On a computer system with a UNIX command line, create a folder containing these 20 files as well as a plain text file (.txt) that contains a list of all mutants in an XNY format (e.g., G129S, see note 1) and give the file the name "mutations.txt."
10. To get only the primers for the mutations in this list, open a command line terminal, go to the folder containing the primer files and the mutations list, and run the following script either by copy and pasting into the terminal, or by creating and executing a script file:

```
while read -r line; do
    mut=$(echo "$line" | grep -o -E '[A-Z][0-9]+[A-Z]')
    num=$(echo "$mut" | sed 's/[a-zA-Z]//g')
    res=$(echo "$mut"| grep -o -E '[A-Z]$')
    fw=$(cat "$res".txt | grep ^"$num"_F | grep -o -E [agctACGT]+)
    rev=$(cat "$res".txt | grep ^"$num"_R | grep -o -E [agctACGT]+)
    printf "%s\t%s\t%s\t%s\n" "$mut"_fw "$fw" "$mut"_rev "$rev"
    done < mutations.txt > primers.txt
```

11. A file called "primers.txt" is created that contains the final list. Verify on a few examples that the primers are correct. Then order the primers from an oligonucleotide synthesis service. Ideally, order the primers in a 96-well plate and dissolved in water to 100 μ*M*.

2.1.3 Notes

1. During the visual inspection of the computational FRESCO procedure, an automatically generated spreadsheet is filled in with notes leading to acceptance or rejections of predicted stabilizing mutations. One of the columns in this file contains the mutations in the XNY format. A list compatible with the above script can be generated by simply copy and pasting the accepted mutations from this column to a plain text file (e.g., created when using Notepad under Windows).
2. Codon usage tables for *E. coli* can be found abundantly on the web. Considering that (i) the current understanding of the interrelation of DNA sequence (including codon usage) with protein expression is still far from fully understood (Boël et al., 2016) and (ii) one can assume that the impact on expression exerted by the change of a relatively small fraction of the gene's codons is minimal, in practice it is probably sufficient to just make sure to avoid *E. coli*'s very rare codons: CGA, AGA, AGG (Arg), ATA (Ile), and CTA (Leu).
3. If AAscan cannot produce a primer that fulfills the criteria, it will output "*******," which will result in an empty field for the primer, after the script. In such a case, the parameters should be adapted. Increasing the MaxTm field to 75°C can help with GC-rich DNA stretches, otherwise the MinTm can also be lowered, however, in that case the annealing temperature of the PCR should be adapted as well (see note 5 in Section 2.2.4).

2.2 QuikChange Library Creation

The protocol below describes the procedure for the usage of the *PfuUltra* II Hotstart Master Mix for the PCR, basically following the instructions of Agilent. The protocol for the preparation of chemically competent cells in a cation-mix buffer is based on the one from Green and Rogers (2013), enabled through the research of Hanahan, Jessee, and Bloom (1991). It is advantageous to transform the PCR product of the QuikChange reaction directly into the strain used for expression, in order to avoid the need of having to perform a second transformation step. However, many dedicated expression strains are not optimized for molecular biology applications, resulting in low transformation efficiencies. On the other hand, cloning strains are often unsuitable for protein overexpression (see note 3 in Section 2.3.4). We advise to investigate the feasibility of direct transformation for one's particular system. Test first a couple of mutants, using the high-throughput equipment, before applying the procedure to one or maximally two 96-well plates of mutants at a time. Most of the following steps can and should be performed using an (electronic) multichannel pipette, even when it is not explicitly stated.

2.2.1 Equipment

- Multichannel pipette (Sartorius, Biohit Picus electronic pipette 12-ch)
- 37°C shaking incubator for test tubes (Infors, Multitron Standard)
- 96-Well PCR plates with dedicated sealing (Bio-Rad, HSL9601)
- Sterile disposable or autoclavable 96-square deep well plates (Waters, 186002482)
- Sterile disposable or autoclavable 24-well plates (Greiner Bio One, 662102)
- Gas-permeable and water vapor retaining microtiter plate sealing (Excel Scientific, AeraSeal B-104)
- Adhesive aluminum sealing foils for microtiter plates (Greiner Bio One, 676090)
- Cooling centrifuge (Eppendorf, 5804R), rotors for tubes, and microtiter plates (Eppendorf, A-2-DWP)
- PCR machine (Peqlab, 96 Universal Gradient Peqstar)
- Sterile toothpicks
- Temperature controlled water bath (Julabo, TW20)
- Temperature controlled shaker, suitable for microtiter plates (see note 4 in Section 2.3.4)
- Sequencing service, ideally including plasmid preparation from clones (Eurofins Genomics)

2.2.2 Buffers and Reagents

- High-fidelity polymerase and reaction components (Agilent, *PfuUltra* II Hotstart PCR Master Mix 2×, see note 1)
- DpnI restriction enzyme (New England Biolabs, R0176S)
- *E. coli* cells (see introduction of this section and note 3 in Section 2.3.4)
- LB Miller's high-salt medium (10 g tryptone, 5 g yeast extract, 10 g NaCl in 1 L H_2O)
- LB agar plates with antibiotic
- SOC medium (20 g tryptone, 5 g yeast extract, 0.58 g NaCl, 0.19 g KCl, 2.4 g $MgSO_4$ in 1 L H_2O, pH adjusted to 7.5 with NaOH. After cooling medium to less than 50°C, add 20 mL filter sterilized 20% glucose solution).

- Buffer I for chemically competent cells: 100 mM RbCl, 50 mM $MnCl_2$, 30 mM potassium acetate, 10 mM $CaCl_2$, 15% w/v glycerol; pH adjusted to 5.8 with 0.2 M acetic acid. Sterilized by filtration with 0.45 μM pore size filter.
- Buffer II for chemically competent cells: 10 mM RbCl, 75 mM $CaCl_2$, 10 mM MOPS, 15% w/v glycerol; pH adjusted to 6.5 with 0.5 M KOH. Sterilized by filtration with 0.45 μM pore size filter.
- Sterile 70% glycerol

2.2.3 Procedure

1. Preparation of chemically competent cells. Perform all steps observing sterility.
   - **a.** Inoculate 5 mL LB in a sterile test tube with the desired *E. coli* strain (see note 3 in Section 2.3.4). Incubate overnight (o/n) at 37°C, shaking.
   - **b.** With this preculture, inoculate 165 mL LB (equivalent to one 96-well plate of competent cells) in a sterile culture flask and incubate shaking at 37°C until an OD_{600} of 0.25–0.5.
   - **c.** Transfer the culture to (a) sterile centrifuge tube(s) and place on ice for 20 min.
   - **d.** Centrifuge at 4000 × *g* in a precooled centrifuge at 4°C for 10 min. Discard the supernatant and resuspend the pellet in 1 mL of ice-cold buffer I. Fill to 60 mL with buffer I and keep on ice for 20 min.
   - **e.** Centrifuge at 4000 × *g* and 4°C for 10 min. Discard the supernatant and resuspend in 1 mL of ice-cold buffer II. Mix well, but carefully, fill to 6 mL and keep on ice for 15 min.
   - **f.** In a cold room, or with precooled equipment and keeping everything on ice as much as possible, aliquot 50 μL in each well of a 96-deep well plate (see note 2).
   - **g.** Use directly for transformation, or seal with adhesive aluminum sealing foil, flash-freeze in liquid nitrogen and store at −80°C.
2. PCR
   - **a.** Prepare a DNA template stock by diluting a fresh plasmid preparation of the wild-type gene-containing vector to a concentration of 1 ng/μL and store it at −20°C (see note 3).
   - **b.** If shipped otherwise, prepare primer stocks in a 96-well PCR plate by diluting all forward and reverse oligonucleotides to a concentration of 100 μM and store at −20°C.

c. Add 240 μL of sterile water to each well of a 96-well PCR plate with a multichannel pipette. Add 5 μL of forward primer stock and 5 μL of reverse primer stock to get a primer mix with 2 μ*M* of each primer.

d. In a 96-well PCR plate, mix the following reagents to a final volume of 25 μL (see note 4) (Table 1):

Table 1 PCR Recipe

Component, Stock Concentration	Final Concentration	Volume to Pipet (μL)
Template stock, 1 ng/μL	0.2 ng/μL	5
Primer mix, 2 μ*M* each	0.4 μ*M* each	5
PfuUltra II Hotstart PCR Master Mix 2 ×	1 ×	12.5
MiliQ water		2.5

e. Run a PCR on a thermal cycler with the following program (Table 2):

Table 2 PCR Program

Temperature (°C)	Time	Cycles
95	2 min	1
95	20 s	
55[a]	20 s	26
72	30 s/kb[b]	
72	10 min	1

[a]Adapt if the default parameters of AAscan were changed (see note 5).
[b]Doubled amplification time compared to manual, to increase yield (see note 6).

3. When the PCR is finished, add 10 U DpnI to each well and incubate o/n at 37°C.
4. Thaw a plate of competent cells on ice (or use freshly prepared cells). Add 5 μL of the DpnI-digested PCR products, but do not mix by pipetting up and down. Incubate on ice for 20 min.
5. Heat shock by placing the plate in a 42°C water bath for 40 s (see note 7).
6. Put the plate back on ice for 2 min, then add 500 μL of prewarmed SOC medium.

7. Incubate on a plate shaker for 1 h at 37°C and 1000 rpm.
8. Centrifuge at 2250 × *g* for 10 min and remove 430 μL of supernatant.
9. Resuspend the pellet with the remaining 70 μL and plate one-by-one on a 24-well plate with LB agar containing the proper antibiotic (see note 8). Incubate o/n at 37°C.
10. Use a sterile toothpick to pick a colony of each well and transfer it to two 96-well plates: one with LB-agar plus antibiotic to be sent for sequencing (see note 9) and one with liquid LB medium plus antibiotic to grow o/n.
11. Add sterile glycerol to a final concentration of 30% to the o/n liquid culture plate, mix well, seal with adhesive aluminum sealing foil, flash-freeze in liquid nitrogen, and store at −80°C.

2.2.4 Notes

1. The polymerase described here, *PfuUltra* II, is an engineered variant of the original *Pfu* polymerase, which—according to the supplier—is even higher in fidelity than the original *Pfu* which was compared to Phusion (McInerney et al., 2014).
2. Use precooled pipette tips and a precooled sterile or an autoclavable multichannel reservoir on ice.
3. Since multichannel pipettes tend to be less precise than single channel pipettes, and precision decreases with volume, we recommend the preparation of diluted stocks that require pipetting volumes of ≥5 μL.
4. If cost is an issue, the volume can be decreased up to 15 μL. Lower yields due to smaller volumes are compensated by high product yields due to efficient PCRs and high-quality competent cells. If a different polymerase is used, adapt both recipe as well as thermal cycling program according to the manufacturer's instructions.
5. The minimum primer T_m parameter in AAscan is per default set as 60°C. The annealing temperature should be 5°C below the lowest T_m.
6. Since amplification speed is one of the main competition parameters between polymerase manufacturers, we impute a possible exaggeration.
7. The strain's commercial supplier should be consulted to infer the strain dependent optimal heat shock time. We recommend the extension of this time by ~50% for transformation in polypropylene deep well plates, due to the impaired heat transfer compared to reaction tubes.
8. This is the only step where the 96-well format has to be abandoned. Use manual pipetting for this step. Plating can be performed by swirling motions or using sterile glass beads.

9. Several DNA sequencing companies offer the plasmid preparation step from colonies on 96-well plates. In our case, this extra cost is reasonably low to outcompete kit prices and labor of manual plasmid preparation of 96 samples. If retransformation into an expression strain is required, it is necessary to ensure that the sequencing company can send back the plasmid preparation. However, we recommend to request this from the company in any case.

2.3 Expression and Protein Purification in a 96-Well Plate

Expression and purification should ideally be performed in a 96-well format as well. For a reliable T_m measurement (see Section 2.4), 20 μL of a solution of 0.3 mg/mL of the target protein are minimally required (see note 1). It should be checked in advance, whether or not enough protein is obtained from the 1 mL plate culture in each individual case. If the target amount is not achieved, an effort should be made to improve expression yields, a topic which is beyond the scope of this protocol (see note 4 for tips on expression in 96-well plates). Otherwise, one can prepare two 96-well plates with identical clones, and combine the cell free extracts (CFEs), or switch to 24-well plates, or else express in 50 mL cultures using 250-mL flasks. Though laborious, such individual expression can be performed in a 1–2 weeks period for 50–100 mutants. When plates are used, it is important to include in each plate the wild-type as a reference, and a noninoculated negative control for contamination in one well, or in several wells diagonal across the plate.

If expression is very high, purification from the CFE might be omitted for some proteins, see Section 2.4 for details. Otherwise, many of the commonly applied purification methods are feasible in a 96-well format. We describe in this protocol the procedure for the widely applied polyhistidine-tag affinity chromatography purification using divalent nickel immobilized on beads, in our case Ni Sepharose®. Furthermore, the Tris/HCl buffer as well as the pH described here can be adapted to one's specific needs or preferences, as long as they are compatible with the purification method.

2.3.1 Equipment

- Temperature-controlled shaker, suitable for microtiter plates (see note 4 and Table 3)
- Cooling centrifuge and rotor for microtiter plates
- Sterile disposable or autoclavable 96-square deep well plates

Table 3 Shaking Frequencies for Instruments With Different Orbitals[a]

Orbital (mm)	50	25	12.5	6	3	1.5
Critical value freq. (rpm)	195	255	395	565	735	1070[b]
Maximum O_2 transfer freq. (rpm)	290	390	560	770	1000	1400[b]

[a]Values approximated from Hermann, Lehmann, and Büchs (2003), Figure 1, assuming a 2 × increase in O_2 transfer due to square wells.
[b]Extrapolated from fitting the data using an exponential function.

- 96-Well filter plates, glass polypropylene membrane (Whatman, Unifilter 2 mL, 7720-7236)
- 96-Well filter plates, polysulfone membrane (Pall, AcroPrep advance, PN8130)
- Disposable 96-well microtiter plate (Greiner, 650161)
- Multichannel pipettes
- Gas-permeable and water vapor retaining microtiter plate sealing
- Adhesive aluminum sealing foils for microtiter plates

2.3.2 Buffer and Reagents

- LB medium
- Ni Sepharose® resin (GE Healthcare, 17526802)
- Buffer A: 50 m*M* Tris/HCl, pH 7.5
- Buffer B: 50 m*M* Tris/HCl, pH 7.5, 5 m*M* imidazole
- Buffer C: 50 m*M* Tris/HCl, pH 7.5, 500 m*M* imidazole
- Lysis buffer: 50 m*M* Tris/HCl, pH 7.5, 10 m*M* $MgCl_2$, 0.5 mg/mL DNase I (NEB, M0303L), 1 mg/mL lysozyme (Sigma-Aldrich, L6876).

2.3.3 Procedure

1. Dispense 1 mL of LB medium with antibiotic in each well of a sterile 96-deep well plate and inoculate from the glycerol stock plate (see note 2). Grow the precultures in a suitable shaker at 37°C and use appropriate agitation (see note 4) o/n.
2. Inoculate the main culture plate from this preculture plate and express the protein according to your specific expression system. In order to achieve the high expression yields needed for the stability assay, consider the following guidelines:
   - Use high copy number plasmids, strongly overexpressing and inducible promoters such as P_{BAD} or T7 and induce expression after an initial growth (see note 3).

- If you have a well-established expression protocol, take into consideration that you might have to modify it for the small-scale expression. Typical adjustments are significantly larger inoculation volumes and higher OD_{600} induction.
- Use a high-cell density culture medium such as terrific broth (TB).
- Use square deep well plates and vigorous shaking for high aeration (see note 4).

3. Centrifuge 20 min at 2250 × *g* at 4°C and remove the supernatant by inverting the plate.
4. Thoroughly resuspend the pellet in 200 μL of lysis buffer (see note 5), cover the plate with aluminum seal, and incubate at 20°C with mild shaking for 30 min (see note 6).
5. Flash-freeze the plate in liquid nitrogen and thaw again in a water bath at room temperature.
6. Centrifuge the lysed cells in the plate for 45 min at 2250 × *g* at 4°C to remove cells debris.
7. Equilibrate 10 mL of Ni-Sepharose resin in a gravity flow column by washing with at least 50 mL of buffer A.
8. Resuspend the beads in an equal amount of buffer A and dispense 200 μL of the resin suspension (see note 7) in each well of a 96-well polysulfone membrane filter plate. Place the filter plate on top of a 96-deep well plate and remove excess buffer by centrifuging for 15 s at 200 × *g*.
9. Transfer the supernatant of the centrifuged lysed cells to a 96-well glass polypropylene membrane filter plate, placed on top of a 96 deep-well plate, and centrifuge at 200 × *g* for 1 min to remove residual cell debris.
10. Transfer the cell extracts to the filter plate containing the resin and resuspend very carefully. Close top and bottom with adhesive aluminum seal and incubate for 15–30 min at 4°C in a plate shaker with mild agitation.
11. Centrifuge 15 s at 200 × *g* and collect the flow through in a 96-well microtiter plate.
12. Wash the resin with 200 μL of buffer A by centrifuging 15 s at 200 × *g*.
13. Wash with 200 μL of buffer B by centrifuging 15 s at 200 × *g*.
14. Elute with 100 μL of buffer C by centrifuging 15 s at 200 × *g* two times and collect the fractions separately in a 96-well microtiter plate.
15. Identify the most concentrated fraction (by eye or using a NanoDrop) and transfer this fraction slowly to an equilibrated desalting plate.

16. Centrifuge 15 s at 200 × *g* to load the columns, and elute by adding 100 μL of buffer A and centrifuging for 15 s at 200 × *g* and collecting in a 96-well microtiter plate (or proceed according to the desalting plate manufacturer's instructions).
17. Proceed directly with the T_m measurement, and flash-freeze the remaining purified protein in liquid nitrogen and store at −80°C.

2.3.4 Notes

1. The minimum value varies slightly for different proteins and should for each case be determined with the wild-type beforehand.
2. This can be done with dedicated equipment (we use a cryoreplicator from EnzyScreen), or a multichannel pipette. Since this step bears a risk of cross-contamination, extra care is required, and the use of high concentrations of glycerol in the stock plate (step 11 in Section 2.2.3) avoids "jumping" frozen debris upon penetration.
3. One of our preferred expression systems is the pBAD vector (Guzman, Belin, Carson, & Beckwith, 1995), which contains the tightly controlled and L-arabinose-inducible P_{BAD} promoter, a pBR322 origin of replication and an ampicillin resistance marker. If successful expression is dependent on constant or very well-defined levels of arabinose (in practice seldom the case), the use of an *araABCD*⁻ strain such as top10 or NEB® 10-beta is obligatory. Otherwise, any strain with an intact arabinose import system (*araEFGH*⁺, virtually all lab strains) can be used, as long as high enough amounts of arabinose (usually 0.1%–2%) are added. In our experience, *E. coli* TOP10 is an excellent strain for both cloning/transformation, as well as expression (at least from pBAD), with the resulting advantage of avoiding the additional retransformation step.
4. A range of studies investigated expression conditions in 96-well plates (reviewed in Duetz, 2007), mainly focusing on oxygen limitation—the main challenge in these systems. General conclusions are as follows:
   - Square wells permit more effective volume than round wells, and shaking aeration is ~2 × higher, since the corners act as baffles to introduce turbulences (Duetz & Witholt, 2004).
   - Sufficient head space between culture surface and seal needs to be ensured to avoid splashes, causing gas-exchange-impairing droplets on the seal and cross-contaminations.
   - The larger the orbital diameter of the shaking device is, the lower needs to be the shaking frequency, in order to achieve the same oxygen transfer rates (Hermann et al., 2003).

- Oxygen transfer rates increase exponentially with the shaking frequency, but only beyond a critical value (Table 3), which is often higher than common shaking operating conditions (Hermann et al., 2003).

The practical problem is that the proficiency of the equipment that an individual researcher has at hand is influenced by many parameters: the plate (well dimensions and shape, material), the shaking device (orbital diameter, frequency range), and the cultivation conditions (culture volume, medium, organism, and strain). Since the number of variables impedes both the encounter of applicable literature and empirical determination, we recommend to approach highest productivity by using 96-square deep well plates with at least 2 mL well volume, and an instrument which permits a shaking frequency range as specified in Table 3. The ideal volume/frequency pair should be determined by either using a fixed frequency for maximum oxygen transfer (Table 3), or a fixed volume of 1 mL. In both cases, the second parameter can be determined by a stepwise increase in volume/frequency to a plate sealed with paper tissue, where a sample of an expression culture was added to some representative wells. Use the volume/frequency just before the occurrence of splashing (as inferred from tissue wetting) for the expression culture.

5. To resuspend the pellet more easily, combine circular motions with pipetting up and down.
6. Estimate lysis efficiency by SDS-PAGE. If necessary, increase incubation time, or—if the protein's stability is guaranteed—increase the incubation temperature to 37°C.
7. Be careful when pipetting pressure sensitive beads. We recommend the use of 1000 μL pipette tips where the thinnest part of the tip has been cut off.

2.4 Melting Temperature Screening

Several methods to measure the thermostability of proteins have been developed. However, the feature most commonly targeted is operational stability, i.e., the stability over time under working conditions. Developing a high-throughput screen sensitive to subtle changes for this feature is, however, notoriously difficult. As an attractive alternative, a protein's apparent T_m is a quantitative parameter requiring minimal protein amounts for its accurate determination. Furthermore, it can be measured in less than 1 h for one 96-well plate of mutant proteins. It is also easy to screen many variants for stability under different conditions, such as pH ranges and buffer additives. If dedicated equipment is not available, a real-time PCR (qPCR) machine, commonly available in many molecular biology laboratories, can be used. The T_m is then determined by following the change of the emission intensity of a fluorescent dye which is added in the assay (or by monitoring the fluorescence of a protein-bound ligand, such as a flavin cofactor). Upon unfolding, proteins expose otherwise buried hydrophobic patches to which these dyes can bind, effecting a shift of the dye's fluorescence spectrum (thus the name fluorescence-based thermal shift assay, trademarked as *Thermofluor*®). The bandpass filters with which qPCR instruments are equipped are adapted to the properties of the DNA-binding dyes most commonly used in real-time PCR, but are also perfectly suitable for many protein-binding dyes (e.g., SyproOrange), which have similar fluorescent properties. Conveniently, flavins also have fluorescence emission and excitation spectra that fall in the same ranges. In the case of most flavoproteins, the release of the cofactor results in a significant increase in fluorescence, as flavins are normally quenched when bound to the protein. Thus, flavin-containing proteins can be used for the thermal shift assay without the addition of dyes, and this method was named ThermoFAD (Forneris et al., 2009). Since this method is so specific for flavoproteins, purification from the CFE might even be omitted under certain circumstances. Besides the requirement for high-expression levels, one has to carefully check a relevant number of mutants and the wild type, to see whether or not the T_m measured in CFE and the T_m of purified protein is the same (or only marginally different). We have had cases of fairly similar proteins, where in one case the T_m of purified protein was very different from the T_m measured in CFE (unpublished data), while it was nearly identical in the other case (Martin, Ovalle Maqueo, Wijma, & Fraaije, 2018). Care has also to be taken since this approach can result in off-target peaks, potentially overlapping with the desired signal.

2.4.1 Equipment

- qPCR Machine (Bio-Rad, CFX96 C1000 Touch Thermal Cycler)
- 96-Well qPCR Plates (Bio-Rad, iQ High-Profile, 2239441)
- Adhesive qPCR Plate Sealing Film (Bio-Rad, Microseal "B," MSB1001)
- Multichannel pipette

2.4.2 Buffers and Reagents

- Solution of target protein with a concentration of at least 0.3 mg/mL
- Fluorescent dye (if the protein does not contain a flavin)

2.4.3 Procedure

1. In a qPCR machine, setup a melt curve program, where the temperature ramps from 20°C to 99°C with 0.5°C increments after a 10 s delay and the fluorescence is measured at each interval. Choose the emission filter such that it has the highest overlap with the emission spectrum of free flavin (peak at 524 nm) or the dye used (see note 1 and introduction of this section).
2. Transfer 20 μL of protein solution to a 96-well qPCR plate (see note 2). Unless you have indications that the protein concentration has a noticeable effect on the T_m, there is no need to correct for concentration variations.
3. For proteins not containing flavins, add a suitable dye (see introduction of this section).
4. Eliminate bubbles by centrifuging the plate on top of a 96 deep-well plate. Seal the plate with qPCR plate sealing film and use a sheet of lint-free tissue to fix the film air-tight and remove fingerprints and other residues.
5. Place the plate in the qPCR instrument and start the melt curve program.
6. Use the instruments software or manually determine the first derivative of the fluorescence signal as a function of temperature. The T_m is defined as the temperature where this curve has a maximum.

2.4.4 Notes

1. Bio-Rad has a very convenient chart with numerous commercial fluorophores and their excitation and emission data. Perform a web search for "Bio-Rad fluorophore reference guide" to retrieve the pdf.
2. It is necessary to use dedicated qPCR plates and sealing foils, permeable for the excitation and emission wavelength range.

3. COMBINING MUTATIONS

The most thermostable single mutants should be verified for preserved catalytic activity before proceeding with the combination of mutations aimed to obtain a final highly thermostable variant. With the FRESCO approach, typically 10–20 mutations are obtained that lead to an increased stability. As the next step in the engineering approach, one would like to combine the mutations in a highly stable mutant. Often, some of the individual mutations are not compatible, though. In order to identify efficiently the most optimal combination of mutations, several strategies can be followed.

One obvious approach is to directly generate the mutant containing all the mutations that were shown to increase the T_m on a single mutant level. This approach has proven successful in a few cases where we applied the FRESCO scheme. However, it often results in an inactive enzyme. If only relatively few (<5) mutations were identified, we recommend to introduce mutations by successive rounds of QuikChange (see Section 2.2), and to express, purify, and determine the T_m of all intermediates, in order to verify an additive effect of each mutation. We usually start with adding mutations that do not alter the charge of the protein, and generally go from highest to lowest effect on T_m. In many cases, it is found that one or several mutations give a problem, once combined with others. It may be that a mutation does not add to the T_m of an already more stable combined mutant, or that it decreases the T_m, or even leads to expression of insoluble protein. In these cases, it is preferable to omit that particular mutation in the final variant. However, it is possible that a mutation causes this problem only in a certain combination with other mutations, while it would be beneficial in another. If, for example, a highly stabilizing point mutant leads to insoluble protein after already having combined five other single mutants, one would like to know whether the insolubility is caused by a specific interaction with only one of the previous mutations. If that was the case, and if the effect on T_m of the first mutation was lower in comparison, it would be better to remove the earlier, rather than the newly added, mutation. To investigate this, however, one would have to remove one-by-one all the previous mutations, and express, purify, and compare the T_m again. If this occurs with several mutations, the effort and time spent becomes impractical.

As an alternative approach, we developed two strategies that aim to create a library of randomly combined mutations. This allows screening of all possible combinations for up to around eight mutated positions. Including more positions would result in a library size that is too big to be practicably screened in a 96-well format, unless a robotic system can be used (Reetz, Kahakeaw, & Lohmer, 2008). If more than eight positions are to be combined, the mutations can be split in sets, and the same or different strategies for combinations can be applied to each set.

The described methods are based on the generation of one DNA fragment per targeted position and subsequent cloning of a shuffled pool of wild type and mutant fragments. The methods vary with respect to the cloning techniques applied—Golden Gate and Gibson cloning—as well as the fragment generation, which is either PCR based, or through more and more affordable commercial gene synthesis. Although these techniques can in principle be combined indiscriminately, we present here two successfully applied work-flows as alternatives: Golden Gate cloning from two synthetic genes, as a currently more expensive but very easy and fast method once a suitable cloning vector is available, and Gibson cloning of PCR fragments, as a very affordable, albeit more work-intensive method.

3.1 Golden Gate Gene Shuffling

The first strategy proposed for randomly shuffled mutations is based on Golden Gate cloning methodology (Engler, Gruetzner, Kandzia, & Marillonnet, 2009). The method described and schematized in Fig. 4 is a one-pot reaction that requires two donor vectors, one containing the wild-type gene sequence and the other the mutated gene sequence, plus a destination vector. Previously described methods implicated the cloning of each module, flanked by BsaI sites in different vectors and addition of type IIS restriction enzyme sites by separate PCRs for each module (Engler et al., 2009; Zhang et al., 2011). We simplified this approach developed by involving two synthetic genes: one wild-type version and one with the selected mutations, each containing BsaI restriction sites flanking sections of the gene that contain the target mutation sites. The modules are flanked by two mirrored BsaI sites with four overhang nucleotides, thus determining the end of one module and the start of the next. Restriction and ligation leads to the excision of the BsaI sites, which leads to a scarless ligation. Having the BsaI restriction sites already inside the synthetic genes allows one to obtain a shuffled library in a single restriction–ligation reaction using only three vectors. The Golden Gate restriction–ligation proved to be very efficient: 97% of the colonies contained the correct restriction pattern and we obtained 65 uniquely shuffled clones from one 96-well plate. The main advantage of such a method is the rapidity, because the synthetic gene design takes less than a day and the library can be obtained in another day. On the other hand, although synthetic gene synthesis is becoming relatively cheap, the extra cost can remain a drawback.

3.1.1 Equipment

- Synthetic gene synthesis service (Genscript)
- PCR machine
- NanoDrop (Thermo Scientific, NanoDrop 1000)

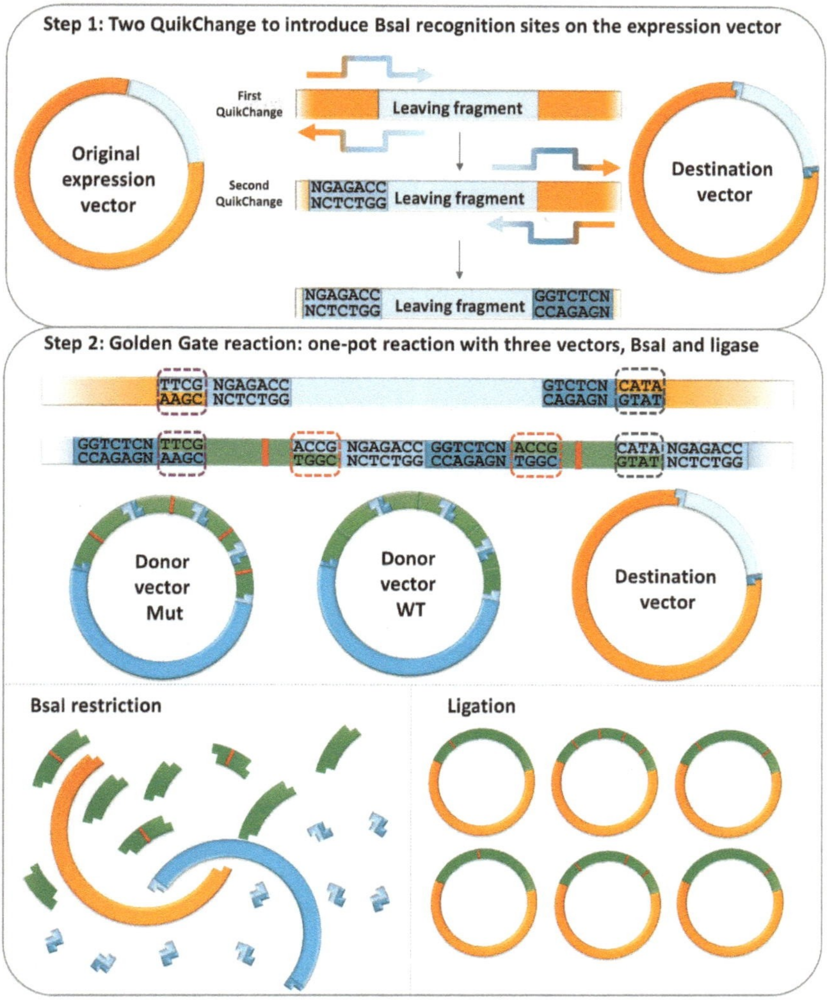

Fig. 4 Design of the Golden Gate gene shuffling (example with four mutations). *Top panel*: generation of the destination vector. Starting from the original expression vector, two BsaI sites flanking a leaving fragment are introduced by QuikChange mutagenesis. Additional QuikChange mutagenesis may be necessary to remove unwanted BsaI recognition sites occurring in the vector. *Bottom panel*: one-pot Golden Gate reaction. In the case of the donor vectors, the BsaI sticky ends are designed to pair with the sticky ends of the destination vector. Moreover, in the donor vectors the BsaI recognition sites between two modules are flanked and mirrored. The four nucleotide overhang at the end of one fragment is replicated at the beginning of the next one to avoid nucleotide loss after the ligation. The donor vectors have the same fragment arrangement. The *blue sections* (donor vectors backbone, BsaI sites and leaving fragment of the destination vector) are lost after the ligation. The shuffled library consists of destination vectors with a random combination and correct order of the four gene modules (with or without mutations).

3.1.2 Buffers and Reagents

- Plasmid isolation kit (Qiagen, QiaPrep, 27106)
- BsaI (NEB, R3535L)
- T4 ligase 30 WU (Thermo Scientific, EL0013)
- T4 ligase buffer (Thermo Scientific)
- Destination vector with leaving BsaI sites
- LB media with and without agar and antibiotics
- SOC medium
- Chemically competent *E. coli* cells (see Section 2.2.3)

3.1.3 Procedure

1. Design and construct the destination vector.
   a. The destination vector can be designed based on any vector desired for expression. First, it has to be ensured that there are no BsaI sites are in the vector. If they occur, they have to be removed by QuikChange, introducing silent mutations in coding regions, or otherwise mutations that do not alter the vector's properties.
   b. Introduce BsaI sites at the sites where the donor gene is supposed to enter. An unrelated placeholder (leaving fragment) sequence can be kept in between. The donor vectors and the destination vector should confer different antibiotic resistances. In this way the selection after the transformation will lead to the exclusion of clones with the donor vectors.
2. In silico design of the two synthetic genes.
   a. First identify the mutagenesis sites. The mutations must be separated by at least one codon (otherwise the sticky ends will not be the same).
   b. Design the two synthetic genes identically (one with mutations and one without).
   c. Add the 4 bp of the sticky ends of the destination vector at the beginning and at the end of your gene fragment. They will have to match with the sticky ends produced after BsaI cuts the destination vector.
   d. To separate all the mutation sites, select 4 freely chosen bp which will delimit the fragments, with the following precautions:
      i. The mutation sites must be inside of the modules, not on the selected 4 bp.
      ii. The sequence of 4 bp must be unique for at least 2 bp out of 4.
      iii. The 4 bp must not be palindromic.

- **e.** Once the fragments are identified, place the BsaI recognition sequence (see Fig. 4):
  - **i.** For the first and last fragment: add NGAGACC after the first 4 bp and GGTCTCN before the last 4 bp (see note 1).
  - **ii.** Next add the double BsaI recognition site NGAGACCGGTCTCN between the first and second module (and so on for the next modules). Replicate the chosen 4 bp that are before the NGAGACC after the GGTCTCN (this is to get a scarless ligation without basepair loss).

3. Order the two synthetic genes. If you order them already cloned in a vector (see note 2) they can be used directly in the one-pot reaction, otherwise a cloning step is needed.
4. Set the one-pot reaction with 20 μL of final volume and the following components (Table 4):

Table 4 Golden Gate Cloning Reaction

Component	Final Concentration
Destination vector	3.75 ng/μL
Donor vector wild type	2.5 ng/μL
Donor vector with mutations	2.5 ng/μL
T4 DNA ligase buffer	1 ×
T4 DNA ligase	1.5 U/μL
BsaI 1	1 U/μL

5. In order to directly create a fully mutated variant, another assembly can be performed without the wild-type donor plasmid.
6. The thermocycler program alternates between the optimal temperatures of BsaI and the ligase, followed by a final digestion step to cut all template and BsaI containing products, and an enzyme inactivation (Table 5):

Table 5 Golden Gate Cloning Program

Temperature (°C)	Time (min)	Cycles
37	5	50
16	10	
50	10	1
80	10	1

7. The restriction–ligation reaction (5 μL) can be used to transform 100 μL of chemically competent NEB 10β cells plated on LB agar plates with antibiotic.

3.1.4 Notes

1. The fragmentation of the two genes must be the same (BsaI cutting sites must be placed at the same position and the same sticky ends need to be used). To check the uniqueness of the sites a useful tool is provided by New England Biolabs in collaboration with Benchling (https://goldengate.neb.com/). NGAGACC and GGTCTCN represent BsaI recognition sites in different orientations (N is a random base pair that can be any of AGTC).
2. Companies like GenScript usually offer also a cloning service. Ordering the synthetic gene already cloned in a vector will save time, because it can be used directly for the one-pot reaction.

3.2 Gibson Shuffling

This protocol makes use of a multiple site-directed mutagenesis procedure developed by Mitchell et al. which allows simultaneous mutations at several positions on a plasmid (Mitchell et al., 2013). Based on this method, we have developed a protocol to construct a library of mutant combinations. First, the fully mutated plasmid is created by PCR-amplifying fragments with mutated primers and subsequent Gibson assembly. In the second step, PCR is used again to generate a mix of fragments with and without the mutations, and another Gibson assembly results in the shuffled library. Although this PCR-based method requires more work when compared with the Golden Gate-based protocol, it can be applied when a limited budget prohibits custom gene synthesis. Moreover, it is particularly useful if it is desired to randomize a position for more than one mutant variant. The FRESCO procedure commonly predicts several mutations for the same position, and if more than one variant proved to be stabilizing on single mutant level, this method allows the inclusion of all variations in the library, without requiring another full custom-made gene.

If two mutations are separated by 5–15 amino acids (or 15–45 bp), extra attention is required, because all of the individual steps (PCR, DNA purification, and cloning) will be less efficient for smaller fragments. The lowest limit for Gibson assembly can be considered as 150 bp—however, this length includes the length of the forward and reverse primers. Since the primer should not exceed 75 bp (the ideal is 30–50), and the mutation should always be at least 15 bp from the termini (the ideal is in the middle of the primer), the minimum separation of two mutations is approximately 11 amino acids, or 33 bp for the smallest possible fragment. If the two desired mutations are closer together they have to be included on the same primer. Due to the same limitations, the maximum separation of two mutations on one primer is ≈ 14 amino acids, or 42 bp. Thus, the method in principle covers any separation of two mutations.

The maximum amount of fragments is another limitation. However, the exponential increase of possible combinations will usually prohibit a library of more than eight target positions, and assembling eight fragments is still feasible. Efficiency does, however, decrease with more fragments, and the general considerations for highest success rates, as pointed out before, apply here as well (Fig. 5).

3.2.1 Equipment

- DNA sequence analysis software (Biomatters, Geneious R8)
- RNAfold software, part of the open source ViennaRNA package, web server available at http://rna.tbi.univie.ac.at/cgi-bin/RNAWebSuite/RNAfold.cgi (Lorenz et al., 2011)
- PCR machine
- Standard agarose gel electrophoresis equipment (Bio-Rad, Mini-Sub Cell GT)
- UV transilluminator or preferably blue-light variant (see note 2)
- NanoDrop

3.2.2 Buffers and Reagents

- High fidelity polymerase and reaction components
- 1% agarose gel
- Gel extraction kit, if necessary dedicated for small fragments (Qiagen, 28606)
- Gibson assembly master mix (NEB, E2611L, or see note 3)
- Chemically competent *E. coli* cells (see Section 2.2.3)

3.2.3 Procedure

1. Select sites for mutagenesis and verify they are in appropriate distance (see introduction).
2. Primer design is critical to the success of the method, so strict care is required.

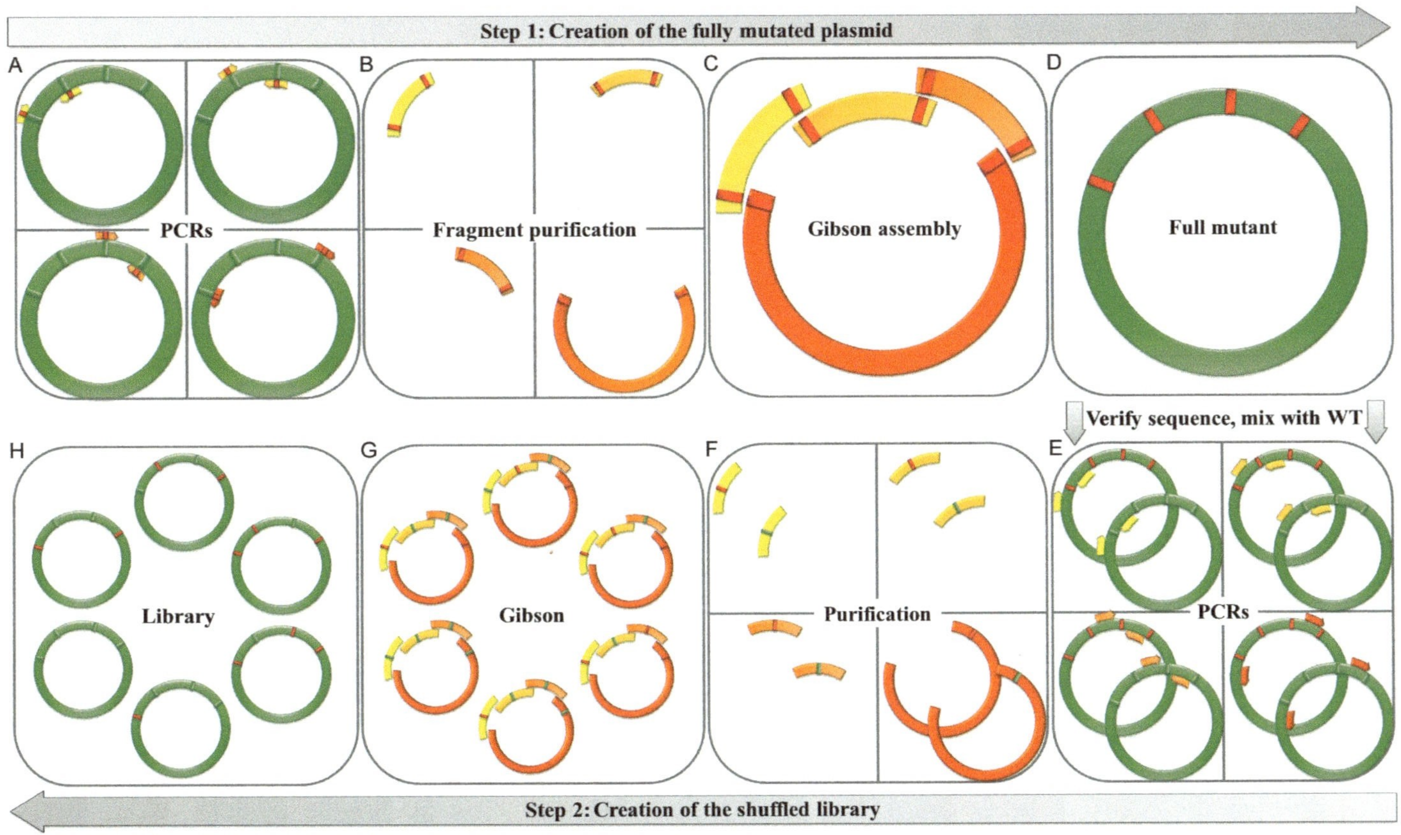

Fig. 5 Overview of the Gibson shuffling method (example with four mutations). By performing PCR on the WT plasmid with primers that contain the mutations (A), fragments are generated that overlap at the site of the mutation where the primer originally bound. After purifying the fragments (B), they are assembled in a Gibson cloning step (C) to result in the fully mutated plasmid (D). This plasmid is mixed with the original WT plasmid to serve as a template in a second PCR step (E) where primers bind in between the mutated regions. This creates pools of fragments that do or do not contain the mutation, which have to be purified (F) and assembled (G) again, to yield the shuffled library (H).

a. Choose the mutagenesis codon and select around 13–25 bp upstream and downstream, ensuring on both ends to apply the GC-clamp rule [A/T][A/T/C/G][C/G] (i.e., end on one or two G or C, but not more).

b. All primers should have a similar T_m of at least 55°C and a G/C content of 40%–60%.

c. Mutate the codon to the desired amino acid making sure to avoid rare codons (see note 2 in Section 2.1.3)

d. For successful Gibson assembly, it is critical to avoid secondary structures (ss) of the homologous regions (which corresponds to the primer). The most intuitive way of removing ss from the initial primer is by visual inspection, via ss prediction software, such as the RNAfold utility of the ViennaRNA package. This tool is freely available as a web server, but we commonly use its implementation in the bioinformatics software Geneious. Be sure to predict the ss at 50°C and inspect the result (Fig. 6).

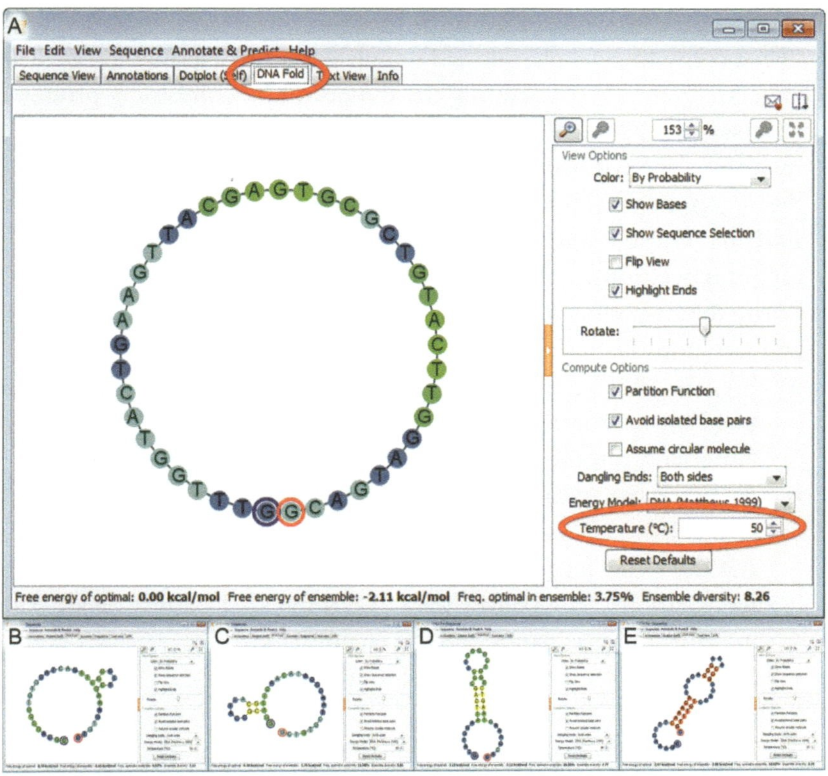

Fig. 6 Secondary structure of primers/homologous regions. In Geneious, go to the DNA-fold tab and select 50°C. Structure is (A) ideal, (B) good, (C) tolerable, (D) critical, and (E) unacceptable.

   e. Weak ss (probability <0.5, color code green/blue), especially when distant from the termini, can be accepted.
   f. Stronger ss (yellow/red), especially when close to the termini, should be removed by (i) change of mutant codon; (ii) introduction of (not more than a few) silent mutations for other codons, not too close to the termini; and/or (iii) removal/addition of nucleotides at the 5′ and/or 3′ end of the primer (keeping the GC-clamp).
   g. In this way, the best possible compromise has to be found between avoiding rare codons, removing strong ss and avoiding too many silent mutations, especially close to the termini.
   h. The reverse complement of the generated sequence serves as the reverse primer for the previous fragment.
   i. Repeat steps 2a–h for all mutant positions.
   j. Design a second set of primers which bind in between the mutated positions (Fig. 6E). They should again be 30–50 bp long, contain a GC-clamp on both sides and the exact position may be chosen freely with the intention to again keep the ss minimal. As before, the reverse complements act as primers for the adjacent fragments.
3. Using the wild-type plasmid as a template, perform individual PCRs to generate fragments containing the mutations by combining primers as color-coded in Fig. 6A: the forward primer of position 1 with the reverse primer of position 2, etc. Make 40 μL reactions and stick to the recommendations of the polymerase's manufacturer. When using *PfuUltra* II, see step 2 in Section 2.2.3 for recipe and program recommendations.
4. Purify the fragments by gel purification after agarose gel electrophoresis and (see note 1)
   a. Prepare a 1% agarose gel with pockets large enough to hold 40 μL. Load an appropriate ladder and the PCR products with loading buffer and run the gel at 85 V, 30 min.
   b. Avoid taking an image of the gel, to minimize the exposure time of the DNA to UV (see note 2). Verify the correct size of the fragments and the PCR success directly while excising the bands from the gel with a scalpel. Then purify the DNA using a gel-extraction kit. Elute in 20 μL (intense band) or 10 μL (faint band).
   c. Measure DNA concentration using a NanoDrop.
5. Gibson assembly reaction
   a. Thaw two 15 μL aliquots of Gibson assembly master mix (see note 3) on ice.

b. Calculate the correct amounts of vector to insert *molar ratios*, e.g., by using the NEB ligation calculator (http://nebiocalculator.neb.com/#!/ligation). This tool gives the required nanogram of insert for different ratios, upon input of a certain insert and vector length. For optimal assembly, use 100 ng of vector (i.e., the largest fragment). For fragments <200 bp, use 1:5, for 200–400 bp use 1:3, and for >400 bp use a 1:2 *M* ratio.

c. Next, calculate the volumes corresponding to this mass, considering the concentrations of your fragments. The total volume of the reaction mix is 20 μL, leaving 5 μL for fragments. If the sum of the volumes exceeds 5 μL, decrease the vector mass to 50 ng and use half of the fragments. If it still exceeds 5 μL, increase the total volume by using two Gibson MM aliquots, or try to get higher concentrations by optimizing PCR, extraction yield, and using lower elution volumes. Fill the Gibson assembly mix up to 20 μL with sterile MilliQ water.

d. As a negative control, prepare an identical sample excluding the smallest fragment.

e. Run the isothermal assembly at 50°C for 1 h.

6. For transformation, use 3.5 μL of the assembly mix (no clean-up required) and follow the recommendations outlined in steps 4–7 of Section 2.2.3.

7. Pick several colonies and grow them o/n in 5 mL LB medium. Make a plasmid preparation and send for sequencing.

8. Once verified that the full mutant was generated, perform another round of individual PCRs.

   a. For the template, create a mix of the wild-type plasmid with the fully mutated plasmid. Check the concentration with the NanoDrop and make a precisely 1:1 mixed stock solution. Volumes that require reasonable pipetting amounts (>10 μL) should be used for improved precision.

   b. Prepare PCR reactions with primer combinations as color-coded in Fig. 5E: the forward primer binding before position 1 with the reverse primer before position 2, etc. Make 40 μL reactions and stick to the recommendations of the polymerase's manufacturer. When using *PfuUltra* II, see step 2 in Section 2.2.3, for recipe and program recommendations.

9. Proceed in exactly the same way as described in steps 4–6 to purify and assemble the fragments, resulting in a shuffled library of mutant combinations.

3.2.4 Notes

1. In principle it is possible to replace this step with DpnI digestion. We recommend a gel extraction anyway, because (i) DpnI digestion is never 100% efficient in removing template, while gel purification is; (ii) it will immediately show success or failure of the PCR; (iii) a clean-up step is also required after DpnI digestion; (iv) even if yields are low, the required amounts in the assembly are very small; and (v) it is the faster method (gel 30 min + extraction 15 min vs ≥2 h DpnI digest + clean-up).
2. DNA damage occurs very rapidly under UV light. A blue-light transilluminator prevents UV damage and allows band excision without time pressure.
3. An alternative to the commercial Gibson mix is to prepare it in-house:
   a. Prepare first a 5 × reaction buffer containing 25% w/v PEG-8000, 500 m*M* Tris/HCl pH 7.5, 50 m*M* $MgCl_2$, 50 m*M* DTT, 1 m*M* dATP, 1 m*M* dCTP, 1 m*M* dGTP, 1 m*M* dTTP, and 5 m*M* NAD. Sterilize by filtration and freeze at −20°C in aliquots of 100 μL.
   b. Prepare next the Gibson assembly master mix containing 1 × reaction buffer, 0.0053 U/μL T5 exonuclease, 0.033 U/μL Phusion polymerase, and 5.33 U/μL Taq DNA ligase. Freeze at −20°C in aliquots of 15 μL.

3.3 Final Stabilized Mutant

After creating the shuffled library by either of the above methods, the success and shuffling efficiency should be verified by sequencing. In principle it is sufficient to sequence a mix of the entire library. In that case, individual clones are screened for thermostability without specific sequence information, and only the most stable variants are then sequenced again. A greater amount of information is gained if a statistically relevant portion of individual clones (e.g., 10% of the library size) or even the entire library is sequenced. In that case, one can draw more conclusions about the effect of different combinations and determine incompatible patterns as well as beneficial arrangements.

After determination of the mutant with the combination that yields the highest T_m, a verification of preserved catalytic activity is self-evident. This confirmation has to be adapted to one's specific requirements and is not described in this protocol. It is important to notice, however, that a common observation is an upshift of the temperature for optimal enzyme activity. Thus, even though one could observe a lowered activity at the assay temperature that is standard for the wild type, a higher activity is likely to be observed at a higher temperature. Nevertheless, if the activity is generally lowered, one can test the mutants next in rank with regard to the T_m to find the ideal compromise. Since the computational part of the FRESCO procedure already avoids mutations near the enzyme's active site, we usually do not observe an impaired catalytic activity for the thermostable mutants.

3.3.1 Equipment

- Sterile disposable or autoclavable 96-square deep well plates
- Sterile 15 mL tube (Greiner)
- Sterile toothpicks
- DNA sequence analysis software

3.3.2 Buffers and Reagents

- LB medium

3.3.3 Procedure

1. Analyze the transformed combination library generated by either of the two methods outlined in Sections 3.1 and 3.2. By using the highly competent cells (step 1 of Section 2.2.3), one should obtain a few hundred colonies. Transform more of the shuffled library assembly mix, if necessary.
2. Estimate the amount of clones that have to be analyzed to achieve a good coverage of the library. By rule of thumb, analyze three times the size of the library, e.g., for one mutant amino acid at seven positions, the number of possible combinations is $2^7 = 128$. Threefold oversampling gives 384 clones, which corresponds to exactly four 96-well plates.
3. Prepare the 96-well plates containing LB medium and appropriate antibiotics and a sterile 15-mL tube filled with 14 mL LB.
4. Pick individual colonies from the library plate with sterile toothpicks and place them in the 96-well plate to inoculate one well per colony. Keep some wells for the wild type, the fully mutated variant, and non-inoculated contamination controls.
5. Remove the toothpicks one by one, and dip each of them in the tube with LB medium before discarding them.
6. Mix the tube inoculated with all mutants, and grow 5 mL of culture o/n, to isolate the mix of plasmids and send it for Sanger sequencing.
7. Analyze the sequencing chromatogram, and verify that the peaks for the mutated nucleotides are approximately of the same intensity as the

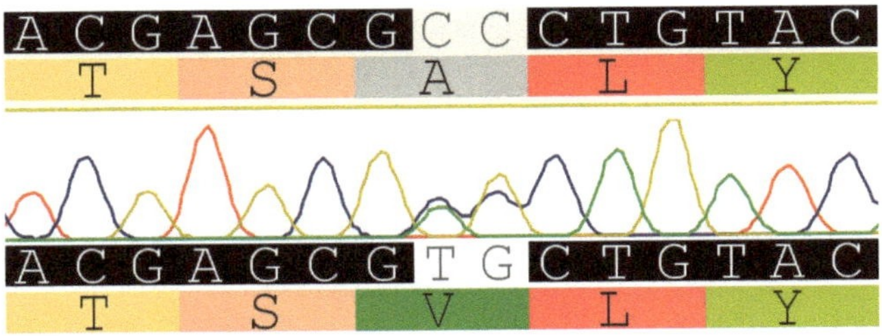

Fig. 7 Example of a sequencing chromatogram obtained from a successful mix of WT and mutant. The third amino acid position shows an even mix of peaks for GCC (Ala) and GTG (Val).

wild-type nucleotides (Fig. 7). If either WT or mutant signals seem to dominate, sequence a significant portion of individual clones to determine if there is a bias in the library. If that is the case, repeat the assembly reaction of the library generation, and pay extra care to using exactly the same amounts of wild type and mutant plasmid.

8. To produce and screen the individual clones of the library, proceed in exactly the same way as outlined in Sections 2.3 and 2.4.

4. SUMMARY AND CONCLUSIONS

The fourth wave of protein engineering is being announced as the synergy of directed evolution with rational and computational design to create more robust enzymes with better specificities or novel reactivities. Yet, one often still faces the quite tedious experimental work in generating and evaluating the corresponding libraries. With the protocols described earlier, we hope to contribute to efficient strategies to create improved enzymes.

After promising pioneering work done using FRESCO for thermostabilization of limonene epoxide hydrolase (T_m increase of 35°C) (Wijma et al., 2014), haloalkane dehalogenase (T_m increase of 23°C) (Floor et al., 2014), and halohydrin dehalogenase (T_m increase of 28°C) (Arabnejad et al., 2017), we have decided to further expand the application of FRESCO to flavin-containing enzymes. Thereby, the FRESCO approach was proven to be useful in engineering stability in a flavoenzyme: a more robust variant of 5-(hydroxymethyl)furfural oxidase (HMF oxidase) was engineered (Martin et al., 2018). As described in the protocol earlier, we could take advantage of the fluorescent properties of the flavin cofactor to rapidly identify improved HMF oxidase variants. HMF oxidase was shown to be able to convert HMF into the polymer building block FDCA (Dijkman, Groothuis, & Fraaije,

2014). By engineering an HMF oxidase that is efficient in converting HMF into FDCA and is thermostable, a valuable biocatalytic tool is now available to be used in strategies for producing renewable-based polymers. It demonstrates that the FRESCO methodology combined with the protocols described earlier is a powerful combination for engineering tuned (flavo)enzymes. Because many structures are available for various biocatalytically interesting flavoenzymes which display an astonishing catalytic flexibility, they are ideally suited in this methodology or other knowledge-based enzyme engineering approaches. There is a growing list of studies in which a flavoenzyme is a key for the design of a new metabolic pathway or biocatalytic cascade reaction (Wang et al., 2018). Examples of newly engineered or proposed metabolic pathways that include flavin-containing enzymes are (i) the recent illustrative example that employs a bacterial flavin-containing monooxygenase and a glucosyltransferase for a sustainable indigo dyeing strategy (Hsu et al., 2018), (ii) the whole-cell conversion of limonene into chiral carvolactone with the help of a flavin-containing Baeyer–Villiger monooxygenase (Oberleitner et al., 2017), (iii) the redesign of cyclohexanone monooxygenase to enable production of methyl propanoate (van Beek, Romero, & Fraaije, 2017), and (iv) the use of flavoprotein oxidases fused to a peroxidase for biocatalytic cascade reactions (Colpa, Lončar, Schmidt, & Fraaije, 2017). Except for FRESCO, the protocols described here can also be used as part of other (flavo)enzyme engineering efforts. Therefore, we hope that they will be adopted by other laboratories for the generation of improved proteins for biocatalysis.

ACKNOWLEDGMENTS

Part of the research for this work has received funding from the European Union (EU) project ROBOX (Grant agreement 635734) under EU's Horizon 2020 Programme Research and Innovation actions H2020-LEIT BIO-2014-1.

REFERENCES

Arabnejad, H., Dal Lago, M., Jekel, P. A., Floor, R. J., Thunnissen, A.-M. W. H., Terwisscha van Scheltinga, A. C., et al. (2017). A robust cosolvent-compatible halohydrin dehalogenase by computational library design. *Protein Engineering, Design and Selection*, *30*(3), 175–189.

Boël, G., Letso, R., Neely, H., Price, W. N., Wong, K.-H., Su, M., et al. (2016). Codon influence on protein expression in *E. coli* correlates with mRNA levels. *Nature*, *529*, 358–363.

Colpa, D. I., Lončar, N., Schmidt, M., & Fraaije, M. W. (2017). Creating oxidase–peroxidase fusion enzymes as a toolbox for Cascade reactions. *ChemBioChem*, *18*(22), 2226–2230.

Dijkman, W. P., Groothuis, D. E., & Fraaije, M. W. (2014). Enzyme-catalyzed oxidation of 5-hydroxymethylfurfural to furan-2,5-dicarboxylic acid. *Angewandte Chemie International Edition in English, 53*(25), 6515–6518.

Duetz, W. A. (2007). Microtiter plates as mini-bioreactors: Miniaturization of fermentation methods. *Trends in Microbiology, 15*(10), 469–475.

Duetz, W. A., & Witholt, B. (2004). Oxygen transfer by orbital shaking of square vessels and deepwell microtiter plates of various dimensions. *Biochemical Engineering Journal, 17*(3), 181–185.

Engler, C., Gruetzner, R., Kandzia, R., & Marillonnet, S. (2009). Golden gate shuffling: A one-pot DNA shuffling method based on type IIs restriction enzymes. *PLoS ONE, 4*(5), e5553.

Floor, R. J., Wijma, H. J., Colpa, D. I., Ramos-Silva, A., Jekel, P. A., Szymanski, W., et al. (2014). Computational library design for increasing haloalkane dehalogenase stability. *ChemBioChem, 15*(11), 1660–1672.

Forneris, F., Orru, R., Bonivento, D., Chiarelli, L. R., & Mattevi, A. (2009). ThermoFAD, a thermofluor-adapted flavin ad hoc detection system for protein folding and ligand binding. *FEBS Journal, 276*(10), 2833–2840.

Green, R., & Rogers, E. J. (2013). Transformation of chemically competent *E. coli*. *Changes, 29*, 329–336.

Guzman, L. M., Belin, D., Carson, M. J., & Beckwith, J. (1995). Tight regulation, modulation, and high-level expression by vectors containing the arabinose P_{BAD} promoter. *Journal of Bacteriology, 177*(14), 4121–4130.

Hanahan, D., Jessee, J., & Bloom, F. R. (1991). Plasmid transformation of *Escherichia coli* and other bacteria. In *Methods in enzymology: Vol. 204* (pp. 63–113). Academic Press.

Hermann, R., Lehmann, M., & Büchs, J. (2003). Characterization of gas–liquid mass transfer phenomena in microtiter plates. *Biotechnology and Bioengineering, 81*(2), 178–186.

Hsu, T. M., Welner, D. H., Russ, Z. N., Cervantes, B., Prathuri, R. L., Adams, P. D., et al. (2018). Employing a biochemical protecting group for a sustainable indigo dyeing strategy. *Nature Chemical Biology, 14*(3), 256–261.

Liu, Y., Shin, H.-d., Li, J., & Liu, L. (2015). Toward metabolic engineering in the context of system biology and synthetic biology: Advances and prospects. *Applied Microbiology and Biotechnology, 99*(3), 1109–1118.

Lorenz, R., Bernhart, S. H., Höner zu Siederdissen, C., Tafer, H., Flamm, C., Stadler, P. F., et al. (2011). ViennaRNA Package 2.0. *Algorithms for Molecular Biology, 6*(1), 26–40.

Marcheschi, R. J., Gronenberg, L. S., & Liao, J. C. (2013). Protein engineering for metabolic engineering: Current and next-generation tools. *Biotechnology Journal, 8*(5), 545–555.

Martin, C., Ovalle Maqueo, A., Wijma, H. J., & Fraaije, M. W. (2018). Creating a more robust 5-hydroxymethylfurfural oxidase by combining computational predictions with a novel effective library design. *Biotechnology for Biofuels, 11*(1), 56.

McInerney, P., Adams, P., & Hadi, M. Z. (2014). Error rate comparison during polymerase chain reaction by DNA polymerase. *Molecular Biology International, 2014*(287430), 1–8.

Mitchell, L. A., Cai, Y., Taylor, M., Noronha, A. M., Chuang, J., Dai, L., et al. (2013). Multichange isothermal mutagenesis: A new strategy for multiple site-directed mutations in plasmid DNA. *ACS Synthetic Biology, 2*(8), 473–477.

Oberleitner, N., Ressmann, A., Bica, K., Gärtner, P., Fraaije, M., Bornscheuer, U., et al. (2017). From waste to value-direct utilization of limonene from orange peel in a biocatalytic cascade reaction towards chiral carvolactone. *Green Chemistry, 19*(2), 367–371.

Reetz, M. T., Kahakeaw, D., & Lohmer, R. (2008). Addressing the numbers problem in directed evolution. *ChemBioChem, 9*(11), 1797–1804.

Sun, D., Ostermaier, M. K., Heydenreich, F. M., Mayer, D., Jaussi, R., Standfuss, J., et al. (2013). AAscan, PCRdesign and MutantChecker: A suite of programs for primer design and sequence analysis for high-throughput scanning mutagenesis. *PLoS ONE, 8*(10), e78878.

van Beek, H. L., Romero, E., & Fraaije, M. W. (2017). Engineering cyclohexanone monooxygenase for the production of methyl propanoate. *ACS Chemical Biology*, *12*(1), 291–299.

Wang, J., Yang, Y., Zhang, R., Shen, X., Chen, Z., Wang, J., et al. (2018). Microbial production of branched-chain dicarboxylate 2-methylsuccinic acid via enoate reductase-mediated bioreduction. *Metabolic Engineering*, *45*, 1–10.

Wijma, H. J., Floor, R. J., Jekel, P. A., Baker, D., Marrink, S. J., & Janssen, D. B. (2014). Computationally designed libraries for rapid enzyme stabilization. *Protein Engineering, Design and Selection*, *27*(2), 49–58.

Wijma, H. J., Fürst, M. J. L. J., & Janssen, D. B. (2018). A computational library design protocol for rapid improvement of protein stability: FRESCO. In U. T. Bornscheuer & M. Höhne (Eds.), *Protein engineering: Methods and protocols* (pp. 69–85). New York, NY: Springer New York.

Zhang, F., Cong, L., Lodato, S., Kosuri, S., Church, G. M., & Arlotta, P. (2011). Efficient construction of sequence-specific TAL effectors for modulating mammalian transcription. *Nature Biotechnology*, *29*(2), 149–154.

FURTHER READING

Zheng, L., Baumann, U., & Reymond, J.-L. (2004). An efficient one-step site-directed and site-saturation mutagenesis protocol. *Nucleic Acids Research*, *32*(14), e115.

CHAPTER EIGHT

Characterization of Cytochrome P450 Enzymes and Their Applications in Synthetic Biology

Laura N. Jeffreys, Hazel M. Girvan, Kirsty J. McLean, Andrew W. Munro[1]
Manchester Institute of Biotechnology, University of Manchester, Manchester, United Kingdom
[1]Corresponding author: e-mail address: andrew.munro@manchester.ac.uk

Contents

Methods in Enzymology, Volume 608
ISSN 0076-6879
https://doi.org/10.1016/bs.mie.2018.06.013

Abstract

The cytochrome P450 monooxygenase enzymes (P450s) catalyze a diverse array of chemical transformations, most originating from the insertion of an oxygen atom into a substrate that binds close to the P450 heme. The oxygen is delivered by a highly reactive heme iron-oxo species (compound I) and, according to the chemical nature of the substrate and its position in the active site, the P450 can catalyze a wide range of reactions including, e.g., hydroxylation, reduction, decarboxylation, sulfoxidation, *N*- and *O*-demethylation, epoxidation, deamination, C—C bond formation and breakage, nitration, and dehalogenation. In this chapter, we describe the structural, biochemical, and catalytic properties of the P450s, along with spectroscopic and analytical methods used to characterize P450 enzymes and their redox partners. Important uses of P450 enzymes are highlighted, including how various P450s have been exploited for applications in synthetic biology.

1. INTRODUCTION

The cytochromes P450 (abbreviated to P450s or CYPs) are a large enzyme class, predominantly formed of monooxygenase enzymes. They are found across nature from archaea and prokaryotes through to higher eukaryotes and man (Denisov, Makris, Sligar, & Schlichting, 2005; Munro, Girvan, Mason, Dunford, & McLean, 2013). The majority of these enzymes catalyze the reductive scission of dioxygen (O_2) and insert one of the oxygen atoms into a substrate that binds close to a reactive heme iron-oxo species (known as compound I) in the P450 active site. The other oxygen atom is reduced to water (Rittle & Green, 2010). Fig. 1 shows a catalytic cycle for a typical P450 enzyme, in this case describing the hydroxylation of the substrate, and following the general reaction scheme:

$$RH + O_2 + NAD(P)H + H^+ \rightarrow ROH + NADP^+ + H_2O$$

The P450s typically use redox partner enzymes to deliver electrons to the P450 from the cofactors NADPH or NADH. In prokaryotic systems, these are generally NAD(P)H dependent, flavin adenine dinucleotide (FAD)-binding ferredoxin reductases, which transfer electrons onto iron–sulfur cluster-binding ferredoxins. The ferredoxin in turn interacts with the P450 partner and reduces its heme iron from the ferric to the ferrous state (enabling oxygen to bind the ferrous iron), and then returns to provide a second electron that converts the ferric-superoxo species to the ferric-peroxo state in the cycle (McLean, Luciakova, Belcher, Tee, & Munro, 2015). In these cases, the P450 and redox partners are soluble, cytoplasmic enzymes, and are referred to as a class I redox partner system. A similar redox system is used for P450s located in the mitochondria of eukaryotes—with adrenodoxin reductase and adrenodoxin providing the electrons (from NADPH) to P450s involved in steroid biosynthesis (Ewen, Kleser, & Bernhardt, 2011). However, the majority of the P450s found in higher eukaryotes are located in the microsomes of the liver and other organs. The hepatic P450s are heavily involved in the oxidative metabolism, detoxification, and subsequent excretion of drugs and other xenobiotics (Guengerich, 2015). In these cases, a single reductase enzyme (the NADPH dependent, FAD- and FMN-binding cytochrome P450 reductase (CPR), a class II redox partner system) provide the electrons required for P450 catalysis. Both the CPR and its P450 partners are membrane-bound enzymes that possess a single transmembrane helix at their N-terminus (Munro, Girvan, & McLean, 2007). Structures of various redox partners for bacterial and eukaryotic P450s are shown in Fig. 2.

The P450 enzyme class derives its name from the absorbance maximum of the main (Soret) band of the heme in its ferrous/CO-bound complex at ~450 nm (Omura & Sato, 1964a, 1964b). This characteristic spectrum results from heme iron coordination in the proximal (fifth) position from a cysteine thiolate, with CO binding in the distal (sixth) position. An alternative form of the Fe^{II}–CO complex has its Soret spectral maximum at ~420 nm, which likely results from proximal coordination of the heme

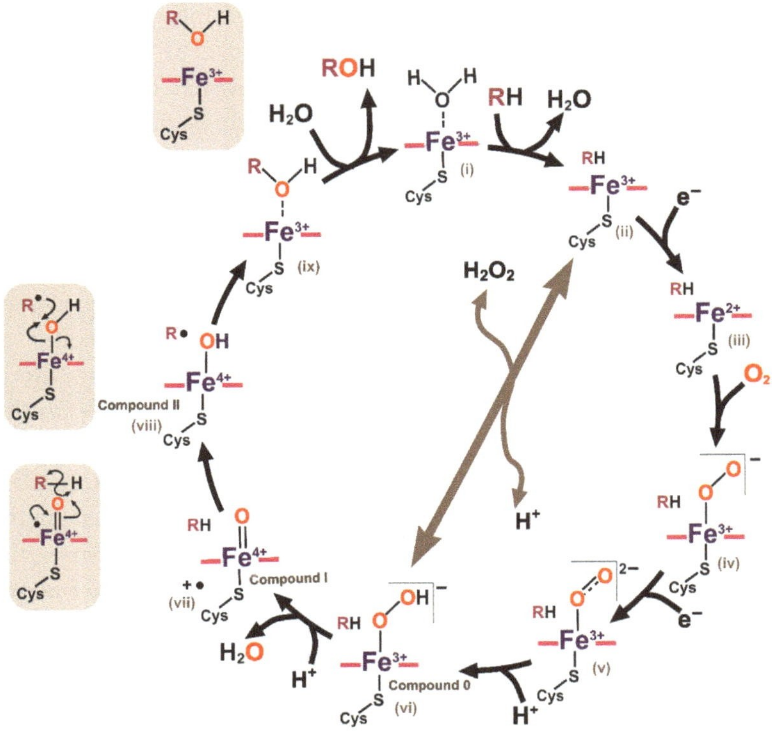

Fig. 1 The canonical cytochrome P450 catalytic cycle. The image shows a nine-step P450 catalytic cycle starting from the substrate-free resting state and finishing with the release of the oxidized product from the P450. The reaction steps are as follows: (i) the ferric resting state of the P450 is proximally coordinated by a cysteine in its thiolate form, with a distal water ligand in the resting state. Substrate (RH) binds to displace the water ligand and produces a substrate-bound P450 (ii). The pentacoordinate substrate-bound form undergoes electronic rearrangement in its 3d orbitals from a low-spin to a high-spin ferric state, with the latter form typically having a more positive heme iron reduction potential (favoring electron transfer from a redox partner to progress the catalytic process). A single electron transfer from a redox partner generates the ferrous species (iii), which then binds dioxygen (O_2) to form a species represented in (iv) as a ferric superoxy state. This species is in an equilibrium with the ferrous-oxy state. A further single electron transfer from the redox partner leads to the ferric peroxy state (v). This species is protonated to form the ferric hydroperoxy state (compound 0) (vi), with the proton (in most cases) delivered from bulk water using a proton shuttle system that involves an acid–alcohol amino acid pair (e.g., Asp-251 and Thr-252 in the well-studied P450cam enzyme) positioned to protonate the distal oxygen atom (Denisov et al., 2005). This reactive species undergoes a further protonation step (as described earlier) and then dehydrates to form the transient and highly reactive ferryl-oxo porphyrin radical cation state (compound I) (vii). Compound I abstracts a hydrogen atom from the substrate to form a ferryl-hydroxo species (compound II)

(Continued)

iron by a protonated cysteine thiol. P450s in their low-spin ferric, substrate-free state typically have their Soret maximum at ~417–419 nm (Miles et al., 1992). Many P450s undergo a substrate-induced heme spectral shift from low-spin to high-spin heme iron, with the high-spin Soret band usually located at ~390 nm. The high-spin shift is associated with the displacement of the distal (aqua) ligand and the formation of a pentacoordinate heme iron state. In most P450s, the conversion to the high-spin state is also accompanied by an increase in heme iron reduction potential, which in turn helps facilitate heme iron reduction by a redox partner protein (Daff et al., 1997). The extent of the Soret band shift can be variable according to aspects including the proximity of the substrate to the heme, its mobility in the active site and its binding affinity (K_d value).

While most P450s rely on interactions with NAD(P)H-dependent redox partners, some P450s use alternative strategies—including the peroxidase/P450 fusion enzyme PpoA from *Aspergillus nidulans*, in which a heme-containing peroxidase domain oxidizes linoleic acid to (8*R*)-hydroperoxy-octadecadienoic acid, and the P450 domain isomerizes the product to form dihydroxyoctadecadienoic acid in pathways for regulation of asexual and sexual life cycles. These psi (precocious sexual inducer) factors are produced by specific oxygenase catalysts (Brodhun & Feussner, 2011; Brodhun, Gobel, Hornung, & Feussner, 2009). Other examples include members of the peroxygenase (CYP152) P450 family, which have evolved to catalyze efficient hydroxylation and oxidative decarboxylation of fatty acids (forming terminal alkenes in the latter case) using hydrogen peroxide to form a reactive heme iron ferric-hydroperoxo species (compound 0) which rapidly forms the active oxidant P450 species compound I through successive protonation and dehydration reactions. Recent years have also seen the identification, expression, and characterization of a range of novel redox partner systems that have "broken the mold" with respect to the well-characterized class

Fig. 1—Cont'd and a substrate radical (viii), which then recombine by "radical rebound" to form the hydroxylated product (ROH) (ix). The images in *gray boxes* adjacent to the compounds I and II show the chemical mechanism by which compound I is converted to compound II, and how the radical rebound step occurs. The image in the *gray box* adjacent to step (ix) shows the dissociation of the hydroxylated product, followed by the binding of a distal water molecule to restore the resting state of the P450. In recent years, a combination of fast reaction (stopped-flow), freeze-quenching, and spectroscopic (UV–visible, Mössbauer and EPR) methods have been used to capture and characterize the transient compounds I and II species (Grant, Hsieh, & Makris, 2015; Grant, Mitchell, & Makris, 2016; Krest et al., 2013; Rittle & Green, 2010).

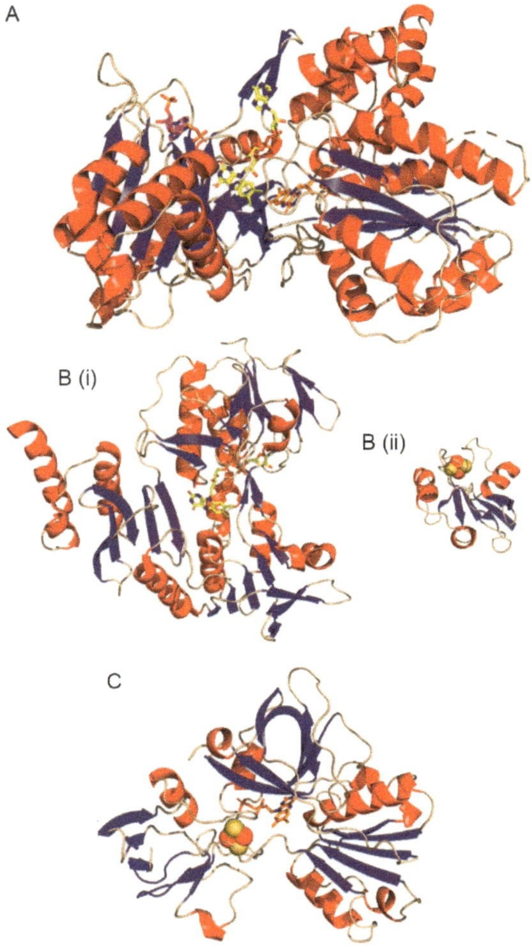

Fig. 2 Structures of native cytochrome P450 redox partners. Panel (A) shows rat CPR (PDB 1AMO) (Wang et al., 1997), the redox partner used by the microsomal P450s. The FAD cofactor is presented as *yellow sticks* and the FMN cofactor as *orange sticks*. An $NADP^+$ cofactor is show as *pink sticks*. Panel (B(i)) shows *P. putida* putidaredoxin reductase with its FAD displayed as *yellow sticks* (PDB 1Q1R) (Sevrioukova, Li, & Poulos, 2004). Panel (B(ii)) shows the *P. putida* 2Fe–2S cluster-binding putidaredoxin (PDB 1PDX) (Pochapsky, Jain, Kuti, Lyons, & Heymont, 1999) which shuttles electrons from putidaredoxin to P450cam at the relevant points in its catalytic cycle. Panel (C) shows *Burkholderia cepacia* phthalate dioxygenase reductase (PDOR) (PDB 2PIA) (Correll, Batie, Ballou, & Ludwig, 1992) with its FMN cofactor shown as *orange sticks* and its 2Fe–2S cluster shown in *spacefill*.

I and class II P450 redox systems. These include (i) the well-characterized *Bacillus megaterium* flavocytochrome P450 BM3 (CYP102A1) enzyme—a natural fusion of a P450 fatty acid hydroxylase to a soluble CPR enzyme (Narhi & Fulco, 1986; Noble et al., 1999); (ii) the CYP116B subfamily of P450-phthalate dioxygenase reductase (PDOR) fusion enzymes (Minerdi et al., 2015; Warman et al., 2012); and (iii) the XplA flavodoxin-P450 fusion enzyme that can reductively degrade the explosive RDX (Royal Demolition eXplosive; 1,3,5-trinitro-1,3,5-triazinane) (Bui et al., 2012; Jackson, Rylott, Fournier, Hawari, & Bruce, 2007).

Through their wide range of catalytic activities, the cytochromes P450 have important applications in biotechnology and synthetic biology. Examples include (i) the stereoselective hydroxylation of the natural compound compactin using an engineered variant of the *Amycolatopsis orientalis* CYP105AS1 (named P450prava) to produce the cholesterol-lowering pravastatin at >6 g/L through fermentation in *Penicillium chrysogenum*; (ii) the generation of an efficient alkane hydroxylase variant of the P450 BM3 enzyme using both random mutagenesis and error-prone mutagenesis to produce the BM3 139-3 variant with 11 amino acid substitutions and substantially improved production levels of propanol (C3) through to octanol (C8) from their respective alkanes (Glieder, Farinas, & Arnold, 2002); and (iii) the development of an in vitro, light-driven redox system in which a plant photosystem I (PSI) and the plant P450 CYP79A1 (that catalyzes conversion of L-tyrosine to [*E*/*Z*]-*p*-hydroxyphenylacetaldoxime in the pathway to the production of the cyanogenic glucoside and insect repellent dhurrin) were reconstituted in nonionic detergents and light-driven activation of PSI was used to reduce a ferredoxin that transferred electrons to the P450 to enable catalytic conversion of L-tyrosine to the oxime product (Jensen, Jensen, & Moller, 2011). In the following sections, we will describe routes to the expression, purification, and characterization of P450 enzymes and their redox partner systems, in addition to examples of their biotechnological applications.

2. EXPRESSION AND PURIFICATION OF MICROBIAL CYTOCHROME P450 ENZYMES

Several different types of expression system have been utilized for the production of P450 enzymes, including baculovirus and yeast expression systems (Hausjell, Halbwirth, & Spadiut, 2018; Kost & Kemp, 2016). However, by far the most commonly used expression system for the production of bacterial and other microbial P450 enzymes is *Escherichia coli*. The advent of quick and affordable gene synthesis offered by several companies (e.g., Geneart, Eurofins, Twist Bioscience) has largely replaced the requirement for gene cloning from the host organisms in cases where the relevant gene (or genome) sequence is known. When preparing a gene for synthesis, it is commonplace to obtain a codon-optimized form, where the desired amino acid sequence is produced using a DNA sequence that more effectively matches codon sequences to the most prevalent tRNAs available in *E. coli*. This strategy can result in higher production levels of the engineered protein, and the approach has been used successfully in, e.g., the production of isopropanol and terpene-based biofuels (Hanai, Atsumi, & Liao, 2007; Peralta-Yahya et al., 2011).

At this stage, it is also easy to add any affinity tags (e.g., His_6-tags) and accompanying protease-cleavage sites to enable, in this case, binding of His-tagged protein to nickel-NTA agarose (Thermo-Fisher Scientific) or by using a similar IMAC (immobilized metal affinity column) approach. This can usually provide an effective first purification step with nonspecific proteins removed from the column by washing with the loading buffer, and then selectively eluting the His-tagged protein by increasing the imidazole concentration of the buffer in order for the imidazole to eventually outcompete the His-tag and displace the target protein from the column. In the case of P450 enzymes, their red color makes it simple to identify eluted fractions containing the P450. If the P450 enzyme is destined for structural studies (e.g., protein crystallization and X-ray diffraction studies), then removal of the affinity tag using a protease (e.g., TEV protease) may be beneficial in facilitating crystallization and may improve the level of the resolution of the structure. Restriction endonuclease sites may be added to the target gene to enable easy cloning of the gene into any desired overexpression vector, although use of methods such as Gibson assembly remove the requirement for complementary restriction endonuclease sites (Gibson et al., 2009). Careful consideration should be given to the choice of the expression vector. Many P450s have been cloned into the pCWori vector first developed in the early 1990s by the Waterman group (Barnes, Arlotto, & Waterman, 1991). Versions of this vector have one or more *tac* promoters under control of a $lacI^q$ repressor, allowing for induction of P450 gene expression using IPTG. It is also common to use commercially available vectors such as the pET vectors from Merck (Poole, UK), which offer a huge variety of different features including N- or C-terminal affinity tags, solubility tags, or the ability to clone multiple genes into the same vector. Many P450 proteins can be readily expressed in *E. coli*, but are then found to have large amounts of P450 in the insoluble protein fraction. Various approaches have been used to overcome this issue including: (i) screening of a number of different *E. coli* host strains (usually including members of the K12 and BL21 families), (ii) induction at low cell optical density, (iii) induction with minimal IPTG concentration, and (iv) addition of the heme precursor delta-aminolevulinic acid (to promote heme production and its insertion into the P450) and growth at low temperatures following induction (e.g., Leys et al., 2003). As described earlier, it is now routine to overexpress P450s (and other proteins) with a His-tag for convenient purification by nickel affinity, and then to cleave the tag using a protease site encoded in the tag sequences (e.g., Belcher et al., 2014). The expression and purification of selected bacterial P450 enzymes are described later. In all cases, the level of purification of the P450 can be monitored spectroscopically by calculation of the RZ (Reinheitszahl) value as the ratio of the absorbance of the P450 Soret peak (typically in the low-spin ferric state) and the protein peak at ~280 nm. Online tools such as ProtParam from the Expasy Bioinformatics Resource Portal can give accurate estimates of the aromatic amino acid contributions to the P450 extinction coefficient at ~280 nm and thus enable the user to establish the level of purification of the P450 by comparison of the observed RZ in the UV–vis spectrum (see Section 5.2 for method) relative to the theoretical RZ based on the ratio of the assumed low-spin Soret heme extinction coefficient (typically ca. 90–100 mM^{-1} cm^{-1}) to that of the calculated 280 nm coefficient. It should be noted that the heme cofactor absorbs in the 280 nm range and therefore the theoretical A_{280} for the P450 will be lower than the spectroscopically determined value. Imidazole also absorbs at 280 nm and so must be removed from P450 samples in order to enable calculation of an accurate RZ value. An alternative route to the determination of the extinction coefficient for the low-spin Soret peak is by using the pyridine hemochromagen method, as described in Section 5.3.2 (Berry & Trumpower, 1987). Examples of purification strategies for two important bacterial P450s are described below.

2.1 Expression and Purification of P450 BM3 (CYP102A1)

Flavocytochrome P450 BM3 (CYP102A1, BM3) from *B. megaterium* is a fatty acid hydroxylase that has been widely studied and mutated to produce variants with activity toward, e.g., pharmaceuticals and alkanes. There are also well-established expression and purification protocols available.

It can be studied in its intact, functional flavocytochrome form, as a 1048 amino acid protein of 119 kDa. Alternatively, BM3 can be dissected to produce its constituent major domains: (i) the N-terminal P450 domain of ~472 amino acids (54 kDa) and (ii) the ~576 amino acid, 65 kDa diflavin (FAD- and FMN-binding) CPR domain. The CPR can be further divided into (iii) a NADPH- and FAD-binding domain of ~394 amino acids (44 kDa, a ferredoxin reductase domain) and (iv) a ~193 amino acid FMN-binding domain (21 kDa, a flavodoxin domain) (Govindaraj & Poulos, 1997; Joyce et al., 2012; Miles et al., 1992). The intact flavocytochrome P450 BM3 and all of its component domains can be expressed in high yields in *E. coli*. Several engineered variants of P450 BM3 have been produced which, e.g., bind and oxidize nonnative substrates (such as omeprazole and related gastric proton pump inhibitor drugs), catalyze α,β-desaturation of alkylbenzenes with much greater efficiency than the wild-type BM3, and hydroxylate testosterone to form the 2β and 15β alcohols with complete diastereoselectivity (Butler et al., 2013, 2014; Kille, Zilly, Acevedo, & Reetz, 2011; Whitehouse, Bell, & Wong, 2008). Various different expression vectors have been used to produce intact P450 BM3 and mutants thereof, including (i) the high copy number pUC118, where transcription of BM3 is regulated by an IPTG-inducible lac promoter (Miles et al., 1992) and (ii) the more tightly regulated pET14b/15b vectors which require to be transformed into DE3 lysogen *E. coli* strains (e.g., BL21 Gold (DE3)) in order for the chromosomally integrated T7 RNA polymerase to be specifically expressed from a lacUV5 promoter when induced by IPTG (Butler et al., 2013). The protocol below describes a method for the production of intact WT P450 BM3 and for its F87V/A82F, F87V, and A82F mutants.

2.1.1 Equipment

- Orbital shaker for growth of *E. coli* transformants on an ~6 L scale in 2 L flasks, e.g., Multitron HT (Infors).
- Centrifuge (e.g., Avanti J-26 XP, Beckman-Coulter) for pelleting *E. coli* transformant cells (using Beckman-Coulter JA-8.1 rotor with 1 L tubes) and for clarification of cell extracts after cell disruption (using Beckman-Coulter JA-25.50 rotor with 50 mL tubes).
- Sonicator for breakage of *E. coli* cells (e.g., Bandelin Sonopuls sonicator).
- AKTA purifier system for purification of P450 BM3 using size exclusion chromatography (SEC) on a Sephacryl S-200 column (GE Healthcare, 26 × 600 mm). Optionally, a second purification step using a hydroxyapatite column (Bio-Rad, 16 × 110 mm) for higher purity.

- Spectrophotometer for analysis of P450 BM3 quantity and purity (e.g., Cary 50 UV–visible spectrometer, Agilent, UK) using 1 mL, 1 cm path length quartz cuvettes.
- SDS-PAGE apparatus for analysis of P450 BM3 enzyme purity (e.g., Mini-PROTEAN system, Bio-Rad).
- Lipidex 1000 (if required) for removal of bound lipids to produce a substrate-free, low-spin form of the enzyme(s) (Perkin Elmer).

2.1.2 Buffers and Reagents

- Buffers should be compatible with stability of the P450 BM3 protein.
- Sodium dithionite reductant and carbon monoxide for spectrophotometric analysis of the Fe^{II}–CO P450 complex, using 1-cm quartz cuvette (Merck, BOC UK and Merck, respectively).
- Dialysis tubing of appropriate molecular weight cut-off (Merck).
- Protein concentrators for decreasing the volume of purified BM3 enzymes, e.g., Amicon Ultra centrifugal filter units (Merck).
- Appropriate solvents for solubilizing lipid (e.g., fatty acid) substrates, such as DMSO and ethanol for delivery of substrates to P450 BM3 enzymes for substrate-binding titrations.

Overexpression of P450 BM3 can be achieved under the control of its own promoter system in *E. coli* without induction by IPTG. However, this may require growth of the transformant cells in rich medium (e.g., terrific broth—TB) for an extended period (ca. 36–48 h) for production of useful amounts of the enzyme (Miles et al., 1992). However, better production can be achieved using induction with IPTG when the gene is expressed from a vector with a more tightly regulated T7 polymerase (Butler et al., 2013). Following overexpression, cell lysis and removal of insoluble proteins and cellular debris by centrifugation, a number of strategies have been employed to purify BM3. In constructs containing a histidine tag, purification begins with addition of ammonium sulfate to 30% saturation. Following centrifugation the supernatant is retained and the pellet discarded, rapidly removing a large number of contaminants and leaving BM3 in the supernatant fraction. The slurry can then be purified by nickel affinity purification without removing the ammonium sulfate. Unbound proteins are removed from the resin with 5 m*M* imidazole prior to eluting the protein with 200 m*M* imidazole in a buffer which does not contain ammonium sulfate, therefore also acting as an efficient desalting step. P450 BM3 can be purified use gel filtration chromatography using an AKTA system (Butler et al., 2013).

In constructs without an affinity tag, purification of clarified cell lysates can begin with binding of protein to a DEAE (diethylaminoethyl) anion exchange column, with BM3 eluted with a linear gradient between 0 and 500 m*M* KCl in a suitable buffer. The protein is then further purified with subsequent chromatography steps on hydroxyapatite and Q-Sepharose anion exchange columns, prior to a final polishing step on a gel filtration column as described earlier (Girvan et al., 2004). The BM3 NADPH-binding site can also be used to purify the protein on, e.g., mimetic yellow or 2′,5′-ADP Sepharose affinity columns, with selective elution of the protein by NADP(H) mimics, with 2′- and 3′-ADP mixed isomers being a cost-effective option. The use of a HPLC system such as the AKTA pure (GE Healthcare) allows monitoring of both the heme and the protein (A_{280}) absorbance as the column is run, allowing for efficient assessment of BM3 fraction purity.

2.1.3 Procedure

1. The WT and mutant forms of P450 BM3 (CYP102A1) are cloned into pET14b/pET15b vectors and transformed into an appropriate *E. coli* strain (e.g., BL21 Gold (DE3)) that supports T7 RNA polymerase-driven expression of the *cyp1021A1* gene.
2. *E. coli* transformant cells (typically ~ 6 L) are grown in autoinduction TB medium (Melford, UK) at ~30–37°C for 24–36 h with shaking in an orbital incubator. Cells are then pelleted and resuspended in a small volume of buffer A (50 m*M* potassium phosphate, pH 7.0, containing 250 m*M* NaCl and 10% (v/v) glycerol) with two cOmplete™, EDTA-free protease inhibitor tablets (Merck). The cells are then disrupted by sonication on ice using a Bandelin Sonopuls instrument with 50 cycles of ~5 s on and 25 s off at 40% amplitude.
3. The cell lysate is centrifuged at 20,000 × *g* for 40 min at 4°C and the P450 BM3-containing supernatant is recovered. Ammonium sulfate is added to 30% saturation on ice with slow stirring for ~4 h. Centrifugation (20,000 × *g*, 40 min, 4°C) is used to separate the BM3 enzymes from precipitated material.
4. The P450 BM3 enzymes are purified using nickel-IDA resin (Qiagen), with the clarified extract mixed overnight in buffer A plus 5 m*M* imidazole. Thereafter, the P450 BM3-bound resin is poured into a column and then selectively eluted using buffer A (containing 200 m*M* imidazole). The collected BM3 enzymes are then dialyzed into buffer

B (50 m*M* Tris, 1 m*M* EDTA, pH 7.2) and then further purified on a Sephacryl S-200 column with an AKTA system.

5. Pure P450 BM3 enzymes can be passed through a Lipidex 1000 column to remove bound lipids. Successful removal of lipids can be established by UV–visible spectroscopy through monitoring the loss of high-spin heme content (peak at ~390 nm) and the accumulation of the low-spin state (peak at ~418 nm). The purity can then be confirmed using indicators such as the A_{418}/A_{280} spectral ratio (RZ value) and by identifying a single major band on an SDS-PAGE gel. Purified protein is then concentrated using ultrafiltration (e.g., with a Vivaspin concentrator) and frozen in buffer B plus 50% sterile glycerol with samples stored at −80°C until required.

2.1.4 Notes

- Use of protease inhibitors is crucial for the production of an intact form of the flavocytochrome P450 BM3 enzyme. An interdomain linker between the P450 and CPR domains of the enzyme is particularly prone to proteolysis. *E. coli* proteases may also cleave at the N-terminal part of the P450 (heme) domain.
- Intact P450 BM3 usually binds tightly to its heme prosthetic group (in the P450 domain) and its FAD cofactor (in the CPR domain). However, affinity is weaker for the FMN cofactor in the CPR domain. Addition of exogenous FMN to the purified P450 BM3 may enhance the flavin content and improve BM3 catalytic properties.
- For mutant forms of P450 BM3 and its P450 domain, structural perturbations may affect the efficiency of heme insertion. In such cases, addition of the heme precursor delta-aminolevulinic acid (ΔALA) may increase the P450 heme content.

2.2 Expression and Purification of OleT (CYP152L1) From *Jeotgalicoccus* sp. ATCC 8456

Recent years have seen growing interest in cytochrome P450 enzymes from the CYP152 family, with several of these P450s now characterized (Girhard, Kunigk, Tihovsky, Shumyantseva, & Urlacher, 2013; Lee et al., 2003; Munro, McLean, Grant, & Makris, 2018). The reason for this interest lies mainly in the fact that various members of the CYP152 family were shown recently to catalyze the production of terminal alkenes from fatty acid substrates through oxidative decarboxylation (in addition to alpha- and beta-hydroxylated fatty acids which were reported much earlier as products)

(Belcher et al., 2014; Grant et al., 2015, 2016; Matthews et al., 2017; Rude et al., 2011). Rude et al. identified *Jeotgalicoccus* spp. ATCC 8456 as a producer of a number of long-chain terminal alkenes, and confirmed alkene formation in other *Jeotgalicoccus* strains. In further work, they identified that other CYP152 enzymes (from *Corynebacterium efficiens*, *Methylobacterium populi*, *Kocuria rhizophila*, and *Bacillus subtilis*) could produce pentadec-1-ene from the C16 saturated fatty acid palmitic acid (in addition to alpha- and/or beta-hydroxylated fatty acids in most cases) (Rude et al., 2011). The *Jeotgalicoccus* bacterium was first isolated from the Korean fermented fish paste Jeotgal, which contains high salt concentrations.

A peroxygenase of particular interest is OleT (also known as $OleT_{JE}$, and with the formal P450 classification number of CYP152L1). OleT is currently the most productive peroxygenase known in terms of its ability to generate terminal alkene products in a substantial excess over alpha- and beta-hydroxylated fatty acids. OleT and related peroxygenases from other bacteria naturally use hydrogen peroxide to drive catalysis through the "peroxide shunt" method (Munro et al., 2018). OleT is active as a peroxygenase on saturated fatty acids in the range from ~C12 to C20, while other studies using redox partner systems have also shown that shorter chain fatty acids are also substrates and can be decarboxylated using NADH with the putidaredoxin reductase/putidaredoxin system, or by using NADPH with the PDOR domain of the CYP116A1-PDOR fusion enzyme from *Rhodococcus erythropolis* (Dennig et al., 2015; Fang et al., 2017). The expression and purification of OleT has been achieved in various *E. coli* expression systems. In early experiments with OleT in our laboratory, we found that high salt stabilized the protein and that dialysis of the partially purified protein against salt-free buffer resulted in the precipitation of OleT. However, by discarding the supernatant it was found that the OleT-containing pellet could be resuspended in buffer containing 750 m*M* NaCl in a functional state and that it retained spectral properties consistent with normal thiolate coordination of the heme iron. In our route to production of OleT, a His_6-tag was used to enable purification by Ni-NTA chromatography, but later in the process the tag was cleaved from OleT using HRV 3C protease, for which a relevant cleavage site is encoded in the expression vector (Belcher et al., 2014; Hsieh et al., 2017). In subsequent work, we changed the HRV 3C site to a TEV (tobacco etch virus) protease-cleavage site, allowing use of the more active and specific TEV protease (Kapust et al., 2001; Matthews et al., 2017). In the Makris lab, *E. coli* expression of the His-tagged OleT was done with the addition of the pG-TF2 plasmid (Takara) which encodes the GroEL, GroES, and Tig chaperones for efficient protein folding, while in our group chaperones were not used, but the protein was stabilized in a high salt concentration (750 m*M* NaCl). In both cases, delta-aminolevulinic acid was added at the same time as the induction of the expression of *CYP152L1*. In both cases, OleT was first purified by nickel affinity chromatography. Subsequently, Makris' group used hydrophobic chromatography by binding the protein to a butyl-*S*-Sepharose column in 30% ammonium sulfate and then eluted OleT with a linear gradient to ammonium sulfate-free buffer (Grant et al., 2015). The method for OleT production described below led to the successful crystallization and structural analysis of OleT.

2.2.1 Procedure

1. The OleT gene is codon optimized for expression in *E. coli* and cloned into expression vector pET47b (by Genscript). The plasmid is transformed into the C41 (DE3) strain in order to improve expression levels by comparison with those obtained using the BL21 (DE3) strain (Dumon-Seignovert, Cariot, & Vuillard, 2004).
2. Transformant cells are grown at 37°C in 500 mL volumes of 2YT medium in 2 L flasks (total volume of 3 L) containing kanamycin (30 μg/mL) and delta-aminolevulinic acid (500 μ*M*, to promote heme incorporation) with shaking at 200 rpm in an orbital incubator. At $OD_{600}=0.5$, OleT expression is induced with 100 μ*M* IPTG (Melford), growth temperature is decreased to 25°C and cell growth is continued for a further 16 h.
3. Cells are harvested by centrifugation (6000 rpm) using a JLA-8.1000 rotor in a Beckman-Coulter Avanti J-26 XP centrifuge. Cell pellets are resuspended in a small volume of 100 m*M* potassium phosphate (KPi, pH 8.0, buffer A), combined and centrifuged as above, and the combined cell pellet frozen at −80°C. When required, the pellet is thawed (4°C) and resuspended in three volumes of buffer A containing 1 *M* NaCl, 20% glycerol and a cOmplete™, EDTA-free protease inhibitor tablet (Roche). 100 μg/mL of both DNase I (bovine pancreas) and lysozyme (hen egg white) are then added (Merck).
4. The cells are disrupted on ice using a Bandelin Sonopuls sonicator at 45% amplitude and using 20 pulses for 20 s at 1 min intervals. The homogenate is centrifuged as above, the supernatant is recovered and the pH adjusted to 8.0. The sample is then mixed overnight with nickel-iminodiacetic acid (Ni-IDA, Generon) on a rolling table at 4°C.

The mixture is then poured into a column and washed with 10 column volumes (CVs) of 100 m*M* KPi (pH 8.0) containing 750 m*M* NaCl and 20% glycerol (buffer B) plus 50 m*M* imidazole. Subsequently, the column is washed with two CVs of buffer B and then five CVs of buffer B plus 150 m*M* imidazole—with the latter wash eluting the majority of the OleT.

5. The eluted sample is dialyzed against ~15 L of buffer A in a cold cabinet, resulting in the selective precipitation of the OleT. The OleT precipitate is collected by centrifugation using an A-4-62 rotor in an Eppendorf 5810 R centrifuge. The pellet is washed gently in 50 mL of buffer A and centrifuged as before. Then the OleT precipitate is resuspended in 5 mL of buffer A plus 1 *M* NaCl and 10% glycerol, resulting in an OleT sample of high purity. For OleT to be used in crystal trials, HRV 3C protease is incubated with the OleT sample for ~15 h at 4°C with a protease-to-protein ratio of 1:50 (units of protease:μg of OleT protein) in order to cleave the N-terminal His-tag. The proteolyzed sample is then passed onto a column containing 5 mL of nickel Sepharose (GE Healthcare) in order to bind the cleaved His-tag to the column and to elute the cleaved OleT in buffer C (100 m*M* KPi, pH 8.0 plus 750 m*M* NaCl, and 10% glycerol). The pure OleT is then flash frozen and stored at −80°C until required.

2.2.2 Notes

- An alternative route to production of OleT that avoids protein precipitation involves removing the dialysis process post the Ni-IDA step. At this stage, the eluate is diluted fivefold in buffer C and then concentrated by ultrafiltration (e.g., with a Vivaspin concentrator), and centrifuged (16,000 rpm at 4°C in a JA-25.50 rotor, Beckman-Coulter). The supernatant is then applied to a 5 mL Ni-IDA column and washed with five CVs of buffer C plus 50 m*M* imidazole, and then 10 CVs of buffer C plus 100 m*M* imidazole. Finally, His-tagged OleT is eluted using buffer C plus 150 m*M* imidazole. Finally, the sample is concentrated by ultrafiltration, snap frozen in liquid nitrogen in small volumes and stored at −80°C until needed.
- If purified samples of OleT (by either method described earlier) show a notable high-spin heme iron component (evident from development of a Soret band feature at ~390 nm), then it is likely that OleT has retained some fatty acids in its active site during the purification process. In this case, passing the sample through a Lipidex-100 column should result in the removal of substantial amounts of the lipids.

3. EXPRESSION AND PURIFICATION OF EUKARYOTIC P450S IN *E. coli*

In vitro studies of eukaryotic P450s must overcome the fact that the vast majority are membrane bound and therefore usually insoluble in their native form when produced in *E. coli* expression systems. In recent years, a number of common strategies have been adopted to allow for the expression of eukaryotic P450s in *E. coli* and for their purification from the bacterial cells. The first structure of a bacterial P450 (the camphor hydroxylase P450cam, CYP101A1) was published in 1987 (Poulos, Finzel, & Howard, 1987), but it was not until 2000 that the first mammalian P450 structure (rabbit CYP2C5) was published (Williams, Cosme, Sridhar, Johnson, & McRee, 2000). As discussed below, and based on the method pioneered by Williams et al., a common strategy has subsequently been used effectively for the expression and purification of several eukaryotic P450s.

3.1 P450 Construct Design

The rabbit CYP2C5, along with the majority of subsequent successful purifications of eukaryotic P450s, have used a construct with the native N-terminal membrane anchor regions removed (e.g., Bart & Scott, 2017; Cosme & Johnson, 2000; DeVore & Scott, 2012; Porubsky, Meneely, & Scott, 2008; Wang, Savas, Hsu, Stout, & Johnson, 2012; Yano et al., 2004). This strategy aids the solubility of the protein, though further modification is typically required to facilitate improved solubility (see below). Tools such as TMpred can be used to predict the length of the transmembrane helix in different cytochrome P450 enzymes (Ikeda, Arai, Okuno, & Shimizu, 2003). It is routine in the overexpression of eukaryotic P450s to further modify the N-terminus, subsequent to its truncation to remove the transmembrane anchor, by incorporating the consensus sequence MAKKTSSKGK at the N-terminus, and by adding a His-tag to the C-terminus (Cosme & Johnson, 2000; Scott, Spatzenegger, & Halpert, 2001; von Wachenfeldt, Richardson, Cosme, & Johnson, 1997). The specific sequence was first designed to solubilize human CYP2C3, and the consensus sequence is that for the truncated form of the wild-type CYP2C3, with the addition of an alanine to facilitate high levels of expression in *E. coli* (von Wachenfeldt et al., 1997). A second strategy employed in the expression of a number of eukaryotic P450s utilizes a modified CYP17A1 truncated N-terminus with a MALLLAVF sequence. As described earlier, an alanine is introduced immediately after the methionine initiator, and a number of other silent mutations were made to minimize mRNA secondary structure formation and to make the mRNA more adenosine and uridine rich, as is common in native *E. coli* proteins. Once again, a C-terminal His-tag is included to facilitate purification (Barnes et al., 1991). Examples utilizing the CYP17A1 consensus sequence include studies of rabbit CYP4B1 (Cheesman, Baer, Zheng, Gillam, & Rettie, 2003) and human CYP2B6 (Hanna, Reed, Guengerich, & Hollenberg, 2000). There are also some notable exceptions in which the N-terminus is not modified. In work on human CYP3A4, it was initially found to be soluble following a partial deletion of the N-terminal membrane anchor (residues 3–14 removed) (Gillam, Baba, Kim, Ohmori, & Guengerich, 1993; Hosea, Miller, & Guengerich, 2000). In subsequent studies toward crystallographic analysis, Guengerich's group further truncated the N-terminus to completely remove the membrane anchor, without the requirement to add the CYP2C3 or CYP17A1 consensus sequences (Yano et al., 2004). It has also been possible to obtain a crystal structure of human CYP1A2 with its transmembrane helix removed, but without the need for the addition of the CYP2C3 or CYP17A1 consensus sequences. Interestingly, crystals were obtained both with and without the consensus sequence in this study, and the crystal without the consensus sequence was the one which diffracted to a higher resolution (Sansen et al., 2007). In these types of studies, the genes are frequently cloned into the pCWori expression vector, which was first developed to enable the overexpression of bovine CYP17A1 (Barnes et al., 1991).

3.2 Routes to Overexpression of Eukaryotic CYPs in *E. coli*

In many cases, the extent of expression of the P450 of interest in a soluble form can be enhanced by the inclusion of secondary plasmids encoding the chaperones GroEL and GroES (Xu & Sigler, 1998), which can be purchased as a pGro7 plasmid from Takara (e.g., Bart & Scott, 2017). The inclusion of delta-aminolevulinic acid prior to the induction of P450 gene expression should also aid incorporation of heme into the P450. Many microsomal P450s are then purified by first isolating spheroplasts (Wester, Stout, & Johnson, 2002). This is done by pelleting the *E. coli* transformant cells prior to their resuspension in a hypertonic solution with sucrose and/or glycerol added, and in the presence of protease inhibitors. If the following step does not use nickel affinity purification, then EDTA can also be added at this stage. However, the presence of the metal chelator EDTA is undesirable in nickel purification. The suspension is then incubated with lysozyme on ice prior to the addition of an equal volume of ice cold water, followed by centrifugation to recover the spheroplasts. The pelleted spheroplasts can then be resuspended in lysis buffer and lysed by sonication, French press or osmotic shock prior to purification (Wester et al., 2002; Yoshimoto, Gonzalez, Auchus, & Guengerich, 2016). In order to efficiently remove the P450 of interest into the soluble fraction during lysis, it is common to add detergents to the lysis buffer. For example, the purification of CYP2A6 and CYP2E1 was carried out in the presence of 4.8 m*M* CYMAL-5, while the lysis buffer used for CYP2D6 and CYP3A4 contained 14 m*M* CHAPS (Bart & Scott, 2017). However, CYP17A1 does not require the addition of detergents for solubilization (Pallan et al., 2015; Yoshimoto et al., 2016). Screening of a number of different detergents is therefore a common strategy to identify conditions conducive to efficient P450 protein recovery and purification. Subsequent purification steps then often utilize Ni-NTA affinity resin and cation exchange chromatography with, e.g., carboxymethyl cellulose resin and SEC (Bart & Scott, 2017; Pallan et al., 2015; Porubsky et al., 2008; Wang et al., 2012; Yano et al., 2004; Yoshimoto et al., 2016). As with the bacterial P450s, eukaryotic P450 purity is readily assessed by SDS-PAGE. UV–vis spectroscopic analysis should also confirm the presence of a cysteine thiolate-coordinated heme iron. The RZ value can also provide a good indication of the extent of P450 purification through the different stages of the process.

3.3 Expression and Purification of Human CYP2D6

The example below is for the production of human CYP2D6 as reported by Wang et al. (2012).

3.3.1 Equipment and Reagents

- Orbital shaker for *E. coli* DH5α transformant growth, e.g., C24KC refrigerated incubator/shaker (New Brunswick Scientific, Edison, NJ, USA).
- pGro7 plasmid for enhanced expression of chaperone proteins GroEL and GroES (Takara Bio Inc., Shiga, Japan) and pCWori vector for CYP2D6 cloning and P450 expression (Barnes et al., 1991).
- IPTG (1 *M*) and arabinose (1 *M*) stock solutions to enable induction of the expression of CYP2D6 and the chaperones GroEL/GroES, respectively.

- Nickel-NTA agarose (Qiagen) and hydroxylapatite-agarose beads (HA Ultrogel, BioSepra Inc.) for protein purification steps.
- Centrifuge and rotor for pelleting of bacterial cells (e.g., Avanti J-26 XP, Beckman-Coulter centrifuge with Beckman-Coulter JA-10 rotor), prior to their resuspension in a hypertonic buffer (with protease inhibitor(s)), leading to treatment with lysozyme and further centrifugation to recover spheroplasts.
- Sonicator (e.g., Bandelin Sonopuls sonicator) or osmotic shock used to lyse spheroplasts.
- Amicon ultracentrifugal filtration device with a 50kDa molecular weight exclusion limit (Millipore) for concentration of CYP2D6 protein purified from spheroplasts.
- Other chemicals: phenylmethylsulfonyl chloride (PMSF) as protease inhibitor, glycerol, potassium phosphate (buffer), ampicillin and chloramphenicol for plasmid selection, and TB medium for bacterial growth.

3.3.2 Procedure

1. The *E. coli* strain DH5α is transformed with both the CYP2D6 expression plasmid (with the expression cassette cloned between NdeI and HindIII sites in the pCWori vector) and the pGro7 plasmid (Takara Bio Inc., Shiga, Japan). Cells are plated on LB agar containing both ampicillin and chloramphenicol antibiotics.
2. 5mL LB medium is prepared containing the above antibiotics. The medium is inoculated with a single transformant colony and grown at 37°C overnight with shaking at 220rpm. Thereafter 500mL of TB containing ampicillin and chloramphenicol are inoculated with 5mL of the overnight culture and cells are grown at 37°C with shaking at 220rpm until an OD_{600} of 0.5 is reached. Thereafter, the temperature is lowered to 30°C and incubation continued with shaking at 190rpm.
3. Once the OD_{600} reaches 0.7–0.8, delta-aminolevulinic acid (5m*M*), IPTG (1m*M*), and arabinose (4g/L) (Merck) are added to induce the expression of P450 2D6 and of the chaperones GroEL and GroES. Cells are harvested after a further 24h growth by centrifugation at 3000 × *g* for 10min at 4°C.
4. Spheroplasts are prepared by resuspending the pelleted cells in 50mL of 20m*M* potassium phosphate buffer (pH 7.4), containing 20% glycerol, 1m*M* PMSF, and 10m*M* β-mercaptoethanol. To produce spheroplasts, the suspension is treated with 0.2mg/mL lysozyme for 30min with

stirring at 4°C. Thereafter an equal volume of cold water is added slowly to the mixture with stirring for 10 min. The resulting spheroplasts are then pelleted by centrifugation at 5000 × *g* for 10 min at 4°C.

5. To lyse the cells, the pellet is frozen in liquid nitrogen, defrosted, and resuspended in 100 mL of 500 m*M* potassium phosphate (pH 7.4) containing 20% glycerol, 10 m*M* β-mercaptoethanol, 14 m*M* CHAPS, and 1 m*M* PMSF. The lysate is then applied to Ni-NTA affinity chromatography resin (Qiagen, Valencia, CA, USA) and washed with several column volumes of the above buffer. The P450 2D6 enzyme is eluted with 10 m*M* potassium phosphate buffer, pH 7.4 containing 30 m*M* histidine, 1 *M* NaCl, 14 m*M* CHAPS, 10 m*M* β-mercaptoethanol, 1 m*M* PMSF, and 20% (v/v) glycerol.
6. The eluted P450 is dialyzed against the same buffer with the NaCl concentration lowered to 150 m*M* and without histidine prior to applying to a column containing hydroxylapatite-agarose beads (HA Ultrogel, BioSepra Inc.) equilibrated with the same buffer.
7. The protein is eluted from the column in 120 m*M* potassium phosphate (pH 7.4), containing 20% (v/v) glycerol, 10 m*M* β-mercaptoethanol, 14 m*M* CHAPS, and 1 m*M* PMSF and stored until required.

3.3.3 Notes

- The plasmid transformation process is best done sequentially by transforming with the pGro7 plasmid first and by selection using chloramphenicol on LB agar plates. Subsequently, a single pGro7 transformant colony is selected and grown in selective culture, prior to transformation with the engineered pCWori vector and colony selection in presence of both chloramphenicol and ampicillin.

3.4 Purification of the Native Form of a *Saccharomyces cerevisiae* CYP51 Enzyme

To date, the only published crystal structure for a eukaryotic P450 that has not required the removal of the transmembrane anchor region from the P450 is that of the CYP51 lanosterol demethylase enzyme from *S. cerevisiae*. In this case, soluble protein was obtained by the expression of the *cyp51* gene with the addition of a C-terminal 6-His-tag in *S. cerevisiae* cells. This CYP51 was isolated from the cells by extraction of the membranes with N-decyl-β-D-maltoside followed by purification on Ni-NTA affinity resin and then SEC (Monk et al., 2014). The strategies have been comprehensively reviewed by Zelasko et al., and we direct the reader to this chapter for additional material on the purification of the *S. cerevisiae* CYP51 enzyme (Zelasko, Palaria, & Das, 2013).

3.5 Examples of Purification of Cytochrome P450 Enzymes From Other Eukaryotes

There are several examples of the expression of P450 enzymes from other nonmammalian eukaryotic organisms. These include P450s from, e.g., insects, plants, and fish. Examples include the expression/purification of *Drosophila melanogaster* CYP6G1, which was shown to catalyze the *O*-dealkylation of *p*-nitroanisole. In this case, the *cyp6a1* gene constructs were expressed from a pCW vector in *E. coli* strain DH5α in different forms—including the native CYP6G1 sequence and a variant in which a MALLLAVF sequence replaced the first eight residues of the native enzyme. Cells were disrupted using a French press and membrane fractions were collected using differential ultracentrifugation. The proteins were then successfully purified from bacterial membranes using detergent treatment in both cases, and the CYP6G1 was further purified using nickel affinity chromatography (Cheesman et al., 2013).

In other work, the taxadiene-5-hydroxylase (CYP725A4) from the Japanese yew tree (*Taxus cuspidata*) was also expressed in *E. coli* (Rouck et al., 2017). This enzyme catalyzes the oxidation of taxadiene to taxadiene-5-ol in the first functionalization step in the synthesis of the anticancer drug taxol. The *cyp725A4* gene was expressed in different N-terminally modified forms to improve expression. These were ones in which (i) the N-terminal 59 amino acids are removed and replaced by a Met-Ala sequence previously shown to increase P450 expression levels (Barnes et al., 1991); (ii) the addition of the 16-residue peptide MAKKTSSKGKLPPGPS, taken from the N-terminus sequence of the rabbit CYP2B1 and shown previously to enhance solubility of an insect P450 (Budriang, Rongnoparut, & Yuvaniyama, 2011), and (iii) addition of an N-terminal MALLLAVF sequence (from the N-terminus of bovine CYP17A1 as described earlier) to the N-terminus of a 24 amino acid N-terminal truncated form of CYP725A4. In the final case, the modification had been used previously by the authors and shown to improve expression of recombinant proteins in *E. coli* (Muller et al., 2008). Transformant *E. coli* cells containing the *cyp725A4* gene constructs in plasmid p5Trc were grown in LB medium containing 0.1 mg/mL ampicillin overnight, and then a sample of the culture was used to inoculate 500 mL of TB containing ampicillin, delta-aminolevulinic acid, and several trace minerals. Following induction of *cyp725A4* gene expression, cells were grown for 44 h at 250 rpm/22°C in an orbital shaker. Cells were then pelleted by centrifugation. The cell pellet was resuspended in buffer containing 0.02 mg/mL lysozyme in 75 m*M* Tris–HCl (pH 8.0), 0.25 *M* sucrose, and 0.25 *M* EDTA at 4°C for 30 min, and then spheroplasts were pelleted by centrifugation at 10,000 × *g* and resuspended in lysis buffer: 100 m*M* potassium phosphate (KPi, pH 7.4) containing 20% glycerol, 6 m*M* $MgCl_2$, 0.1 m*M* DTT, 0.2 m*M* PMSF, and 5 mg of deoxyribonyuclease. Spheroplasts were then lysed by sonication, and the *E. coli* membrane fraction was pelleted by centrifugation and resuspended in extraction buffer: 100 m*M* KPi (pH 7.4) containing 20% glycerol, 200 m*M* NaCl, 0.1 m*M* DTT and 1% cholate, and mixed for 5 h at 4°C. Thereafter, a high speed (140,000 × *g*) centrifugation step separated the CYP725A4-containing supernatant, and the P450 was further purified by Ni-NTA column chromatography (in buffer containing 20% glycerol and 0.1% cholate), with the CYP725A4 proteins eluted from the column with 200 m*M* imidazole and 200 m*M* NaCl. The CYP725A4 proteins were then concentrated by ultrafiltration (Amicon 10,000 MWCO concentrators) and dialyzed into 100 m*M* KPi (pH 7.4) containing 20% glycerol, 0.1% cholate, 1 m*M* DTT, and 200 m*M* NaCl in order to remove imidazole prior to storage.

4. EXPRESSION AND CHARACTERIZATION OF P450 REDOX PARTNER SYSTEMS

4.1 Introduction

A number of different redox partner systems are known to support the catalytic activities of cytochrome P450 enzymes through delivery of electrons required for the P450s to generate reactive heme iron-oxo species that are essential for P450 substrate oxidation (see Fig. 1 in Section 1). Early studies of P450 redox partners suggested that bacterial P450s were reduced by NAD(P)H-dependent ferredoxin reductase and ferredoxin enzymes. The model system here is P450cam along with NADH-putidaredoxin reductase (PDR)/putidaredoxin redox partners (Sevrioukova et al., 2004). In contrast, eukaryotic P450s were reduced by a NADPH-dependent CPR in the microsomes of liver and other tissues, and by a bacterial-type system (adrenodoxin reductase and adrenodoxin) operational in the mitochondria of the adrenal gland and responsible for interactions with P450s involved in vitamin D metabolism, bile acid degradation, and steroid hormone biosynthesis (Ewen, Ringle, & Bernhardt, 2012). However, studies in recent years have identified a number of new types of redox partner systems in both bacterial and eukaryotic systems. These include P450 BM3 and related P450-CPR fusions enzymes; members of the CYP116B family (soluble P450-PDOR fusion enzymes) that can oxidatively modify thiocarbamate herbicides; and the RDX-degrading XplA enzyme, as discussed in Section 1 (Munro et al., 2002; Roberts et al., 2003; Rylott, Lorenz, & Bruce, 2011; Warman et al., 2012). Another example is a sterol demethylase (CYP51 family) P450-ferredoxin fusion enzyme from *Methylococcus capsulatus* (Jackson et al., 2002).

Other relatively uncharacterized enzymes are P450-HCP (hybrid cluster protein) fusion enzymes, where the HCP has two iron–sulfur clusters, one of which is 2Fe–2S cluster while the other is a hybrid [4Fe–4S–2O] cluster (Filenko et al., 2007). In this case, it is not clear whether the HCP has a redox partner role, and instead a peroxidase role has been proposed for the HCP module (Almeida, Romao, Lindley, Teixeira, & Saraiva, 2006). Another important bacterial class of P450s is the peroxygenases, in which hydrogen peroxide can be used efficiently for the oxidation of fatty acids to produce alpha- and beta-hydroxylated fatty acids, as well as terminal alkenes (by oxidative decarboxylation) in certain isoforms such as OleT (CYP152L1) from a *Jeotgalicoccus* species (Grant et al., 2015). Exogenous redox partner systems have also been used effectively to drive catalysis in bacterial P450s, for instance through the use of PDR and putidaredoxin (together with a NADH recycling system) to catalyze fatty acid decarboxylation/hydroxylation activity of OleT (Dennig et al., 2015).

In eukaryotes, various P450 fusion enzymes have also been identified. These include animal heme peroxidase-type and DyP (dye decolorizing)-type peroxidase proteins fused to P450s (Brodhun et al., 2009; Sugano, 2009). Fungal homologs of P450 BM3 also exist, with P450foxy (CYP505A1) being the best characterized of these enzymes, and which hydroxylates fatty acids at the ω-1, ω-2, and ω-3 positions (Nakayama, Takemae, & Shoun, 1996).

For most P450s, separate class I and II redox partner systems are used. The sections below describe examples of the expression and purification of the *E. coli* flavodoxin/ferredoxin reductase and its flavodoxin partner (McIver et al., 1998). Equipment requirements are the same as those described in Section 2.1.1, except that a Lipidex step is not required. Buffers and reagents required are described in Section 4.2.

4.2 Procedure (Flavodoxin/Ferredoxin Oxidoreductase)

1. The flavodoxin reductase (FLDR) gene was originally expressed from the plasmid vector pCL21 (Bianchi et al., 1993), but can also be readily expressed to high levels in more modern vectors—e.g., the pET plasmids (Merck Biosciences, Darmstadt, Germany). The protein can be produced from *E. coli* strains such as HMS174 (DE3) or BL21 Gold (DE3), which have the highly processive T7 RNA polymerase integrated into the genome under regulation of an IPTG-inducible promoter.
2. *E. coli* transformants are grown in volumes of ~5 L in LB medium containing ampicillin or carbenicillin at ~100 μg/mL until an $OD_{600} = 1.0$, and then FLDR gene expression is induced by addition of IPTG (1 m*M*). Growth is continued for a further 3–4 h.
3. The cells are collected by centrifugation (5000 × *g*, 10 min, 4°C), washed by resuspension in ice cold buffer A (10 m*M* potassium phosphate, pH 7.5) and lysed by sonication, e.g., using a Bandelin Sonopuls instrument (50 cycles of ~5 s on and 25 s off at 40% amplitude). Cellular debris is removed by centrifugation (15,000 × *g*, 30 min, 4°C).
4. The resulting supernatant is loaded directly onto a HiLoad 26/10 Q-Sepharose High-Performance column (GE Healthcare) that was prewashed with buffer A. The FLDR protein is eluted from the column in a linear gradient of 0–1 *M* NaCl in buffer A, with the FLDR eluting as yellow fractions at ~150 m*M*.
5. Fractions are pooled and loaded onto a 2′-5′-ADP Sepharose 4B column (GE Healthcare) equilibrated with buffer A plus 150 m*M* NaCl. Purified FLDR is eluted from the column by washing with buffer A plus 500 m*M* NaCl.
6. The pure FLDR sample collected is concentrated by ultrafiltration using, e.g., Amicon™ Ultra-0.5 Centrifugal Filters (10,000 MWCO) to a final concentration of ~5–10 mg/mL and then stored at −80°C with protease inhibitor(s) added.

4.3 Procedure (Flavodoxin)

1. As in the case of the FLDR, the flavodoxin (FLD) gene was originally expressed from the plasmid vector pCL21, although other (e.g., pET) vectors are also appropriate. As for the FLDR, the FLD protein can be expressed to high levels using, e.g., HMS174 (DE3) or BL21 Gold (DE3) *E. coli* strains under control of an IPTG-inducible promoter.
2. *E. coli* transformants are grown in volumes of ~5 L in LB medium with ampicillin or carbenicillin as the selective antibiotic at ~100 μg/mL as

described for FLDR, with the exception that cells are harvested 6 h after FLD gene expression is induced with 100 μ*M* IPTG. Riboflavin (5 mg/L) is also added to the culture at the point of induction.

3. Transformant cells are collected by centrifugation as described for FLDR earlier. The cells are then washed by resuspension in ice cold buffer B (50 m*M* Tris–HCl, pH 7.5). Thereafter, cells are broken by sonication as described for FLDR. Protamine sulfate is added to the extract at a concentration of 0.1% (mass/volume) and the mixture centrifuged (15,000 × *g*, 30 min, 4°C).
4. The supernatant is then loaded onto a HiLoad 26/10 Q-Sepharose High-Performance column which is washed in buffer C (100 m*M* sodium acetate, pH 5.0).
5. The FLD is eluted from the column in a linear gradient of 0–1 *M* NaCl in buffer C, with the orange FLD eluting at ~400 m*M* NaCl. The eluted FLD is then dialyzed against buffer B to raise the pH to 7.5.
6. Further purification of the FLD is done by anion exchange chromatography on a Resource Q column (GE Healthcare). The FLD is loaded in buffer B and then eluted in a gradient of 0–1 *M* NaCl in buffer B, with the FLD coming off the column at ~350–400 m*M* NaCl.
7. The purified FLD is then concentrated by ultrafiltration as described for FLDR (to ~5–10 mg/mL) and stored at −80°C with protease inhibitor(s) added.

4.4 Notes

- *E. coli* FLD and FLDR are both robust proteins with good thermostability that should be stable for extended periods if stored at −80°C. The major issue affecting their stability is thus proteolysis, which should be addressed by addition of a sample of a protease inhibitor cocktail, e.g., a cOmplete™ tablet dissolved in buffer and added to FLD/FLDR samples prior to freezing.

5. DETERMINATION OF P450 PURITY, CONCENTRATION, AND OTHER PROPERTIES USING UV–VISIBLE SPECTROSCOPY

5.1 Cytochrome P450 Spectroscopic and Ligand-Binding Properties

A number of properties of P450 enzymes can be established using UV–vis spectroscopy, including their purity, concentration, and the nature of the ligands that bind to these enzymes. Cytochromes P450 are heme-binding monooxygenases that utilize a heme *b* prosthetic group as the catalytic core of the enzyme (Omura & Sato, 1964a; Poulos & Howard, 1987). The heme *b* (ferriprotoporphyrin IX in the low-spin [LS] ferric form of the P450) has an extensively conjugated double bond system that produces a red color in the substrate-free form of the P450. There are typically four main features in the LS, ferric P450 heme spectrum. The Soret peak, also referred to as the gamma band, is the most intense peak observed in the P450 spectrum, and usually shifts substantially on binding to substrates or inhibitor ligands. For many P450 proteins, the substrate-free form of the enzyme has a Soret peak at around 418 nm. In this LS state, the axial ligands are cysteine thiolate (proximal) and a water molecule (distal). Pyrrole nitrogens interact with the heme iron in the plane of the heme. It is the Soret band that shifts to ~450 nm upon heme iron reduction and carbon monoxide binding. To be in the P450 state, the cysteine thiolate must be retained as the proximal ligand, with protonation of this ligand likely underlying the formation of the P420 state (i.e., with the Fe^{II}–CO complex having a Soret maximum at ~420 nm). The other major absorbance bands in the LS, ferric form of a P450 are located at ~565 (alpha) and ~540 nm (beta). A further band (delta) is usually seen at a shorter wavelength (~360–365 nm), but is obscured somewhat in the high-spin (HS) state of a P450, due to the HS Soret feature undergoing a blue shift to ~390 nm when a substrate or substrate-like molecule binds in the P450 active site and displaces the axial water ligand (Munro et al., 2007). At this stage, it is pertinent to note that some P450s may exist naturally in a HS form (e.g., due to presence of a tight-binding substrate), or be in an equilibrium between the HS and LS states. Accompanying the LS-to-HS conversion is an electronic rearrangement in the 3d orbitals. In the substrate-bound HS state, the axial water is lost and the heme iron becomes pentacoordinate. In the 6-coordinate (LS) state the five electrons in the 3_d orbitals are found exclusively in the t_{2g} orbitals with a total spin of $S=1/2$. However, in the five-coordinate (HS) state the energy gap between the t_{2g} and eg orbitals is decreased and the five electrons occupy each of the $3 \times t_{2g}$ and $2 \times$ eg orbitals singly to give a total spin $=5/2$ (Luthra, Denisov, & Sligar, 2011). The conversion to HS is generally associated with an increased reduction potential of the P450 ferric heme iron, and favors heme iron reduction by a redox partner, enabling the ferrous heme iron to bind dioxygen and for the catalytic cycle to progress. As an example, the binding of the substrate arachidonic acid to the heme (P450) domain of the high activity flavocytochrome P450 BM3 enzyme resulted in an increase of the heme iron potential by ~130 mV from −368 ± 6 mV to −239 ± 6 mV (Daff et al., 1997).

As discussed earlier, binding of a substrate to a LS, ferric P450 typically leads to displacement of the axial H_2O ligand and to a Soret peak shift to a HS state at ~390 nm (a type I spectral shift). This enables spectral titrations to be done to determine the binding constant (K_d value) for the substrate (Driscoll et al., 2010). Typically, small volume samples of the substrate (e.g., 0.1–0.2 μL per addition) are added sequentially to the P450 enzyme (usually in the P450 concentration range between ~2 and 5 μ*M*, corresponding to roughly 0.2–0.5 absorbance units for a LS P450 enzyme) with a spectrum recorded following each addition. Assays are often carried out in 1 mL quartz cuvettes (to provide best quality data) with the spectrophotometer sample chamber maintained at a constant temperature (routinely by use of a Peltier system). Substrate stock solution(s) are prepared in appropriate aqueous buffers or solvents, at appropriate concentrations such that the binding titration can be completed without excessive additions of solvent or significant dilutions of the sample. On completion of the titration (i.e., when there are no further changes in the spectral properties on further addition of substrate) a few more additions should be made as the data fitting is generally more accurate when there is a plateau at the end of the titration. The spectral datasets (typically in the range from ~800 to 300 nm) should then be downloaded into an appropriate data analysis program (e.g., Origin Pro 9.1, OriginLab, Northampton, MA, USA) for processing. For most accurate analysis, the initial (substrate-free) spectrum should be subtracted from each consecutive spectrum obtained with increasing additions of the substrate. This will produce a set of difference spectra that describe the transition from a LS spectrum toward a HS spectrum, and (for a simple LS-to-HS transition) identify the wavelengths at which the absorbance peak (A_{max}) and absorbance trough (A_{min}) occur. Assuming that there is little or no P450 or substrate aggregation that causes significant turbidity, there should also be an isosbestic point observed in both the absolute spectra and the difference spectra. In the latter case, the isosbestic point should be located on the *x*-axis with zero absorbance. To produce data for a substrate-binding curve, the value of (A_{max} minus A_{min}) should be calculated from each difference spectrum (i.e., the peak to trough difference), and each of these values should be plotted against the relevant substrate concentration used. Most frequently, this will produce a hyperbolic curve that can be fitted using a hyperbolic function (e.g., the Michaelis–Menten function) (Eq. 1) (Michaelis & Menten, 1913). However, if the substrate binds very tightly to produce (i) a steep and linear dependence of absorbance change on substrate concentration followed by (ii) a sharp end point with only a slight curvature at the inflection point, then the Morrison (tight-binding or quadratic) equation (Eq. 2) should be the best option, as it will more accurately define the K_d value provided that there is an adequate number of accurate points collected and that there are a group of points clustered around the inflection point (Morrison, 1969).

In some instances, apparent cooperative behavior is observed in the titration, leading to a sigmoidal dependence of absorbance change on substrate concentration. In such cases, the Hill equation (Eq. 3) can be used to determine an apparent K_d (midpoint) value and to establish the Hill coefficient (a measure of the cooperativity) (Hill, 1910). On (rare) occasions on which the substrate (or other ligand) is colored and its spectrum impinges on the Soret band, UV–visible titrations can be done by following substrate-induced changes in the Q-band region (~500–600 nm) where the heme alpha and beta bands also undergo spectral shifts on substrate binding. Owing to the lower intensity of the alpha/beta bands, a higher concentration of the P450 will likely be required for an accurate titration (e.g., 10 μM or more). However, by identifying the Q-band wavelength with the largest change in absorbance (and by baselining spectra carefully at a long wavelength where there is negligible absorbance change observed, e.g., at ~800 nm) and by using a slow scan rate for data collection, smoother spectra should be obtained to enable accurate data fitting at the selected wavelength using the relevant equations described above and presented below. Alternatively, a dual beam spectrophotometer may be used where the spectra from two cuvettes are simultaneously measured. The sample cuvette contains the P450 while the reference cuvette has buffer only. Identical ligand samples are added to both cuvettes and a titration is done as described earlier. The "ligand-only" spectrum is automatically subtracted from the P450 sample spectrum to allow a more accurate UV–visible titration.

Also of note is that, in certain circumstances, ligands can bind to P450s, but do not induce a shift in the heme spectrum. While the presence of such "silent" ligands in the P450 active site is not directly revealed by UV–visible spectroscopy, other methods can reveal their presence—e.g., electron paramagnetic resonance (EPR) spectroscopy. However, the binding and affinity of silent ligands can be determined by UV–visible spectroscopy by using a competitive binding assay. As an example, the *Rhodococcus rhodochrous* P450 enzyme XplA (CYP177A1) binds the explosive RDX (hexahydro-1,3,5-trinitro-1,3,5-triazene) and reductively denitrates the nitro groups. XplA also undergoes a conversion toward the HS state on binding RDX

(Rylott et al., 2011). However, the structurally related trinitrobenzene does not induce a spectral shift in XplA. Further quantitative studies showed that RDX could be used to displace the silent trinitrobenzene ligand (in so doing causing a HS heme shift). Through studies in which RDX was titrated into samples of XplA bound to different (known) concentrations of trinitrobenzene, the apparent K_d values for RDX could be determined in each case. The datasets could then be processed using a competitive binding equation, producing a trinitrobenzene K_d value of $198 \pm 41\,\mu M$ (Bui et al., 2012). Another type of spectral shift occasionally observed in P450s is known as a reverse type I shift. In this case, a ligand that binds to a mixed-spin or HS P450 enzyme form results in a Soret shift toward the LS state (at ~418 nm) (Shimada et al., 2009).

5.1.1 Equations for Spectral Data Fitting

(1) Michaelis–Menten (hyperbolic) equation (adapted for equilibrium ligand-binding studies)

$$A_{obs} = \frac{A_{max} \times [L]}{K_d + [L]}$$

where A_{obs} is the observed absorbance change at ligand (substrate or inhibitor) concentration L. A_{max} is the maximum absorbance change observed at apparent saturation with the ligand. K_d is the dissociation constant for the binding of the ligand.

(2) Morrison (tight-binding) equation

$$A_{obs} = \left(\frac{A_{max}}{(2 \times E_t)}\right)$$

$$\times \left(\left((L + E_t + K_d) - \left((L + E_t + K_d)^2\right) - (4 \times L \times E_t)^{0.5}\right)\right.$$

where A_{obs} is the observed absorbance change at ligand concentration L. A_{max} is the maximum absorbance change observed at apparent saturation with the ligand. E_t is the total concentration of enzyme/protein used and K_d is the ligand dissociation constant.

(3) Hill (sigmoidal) equation

$$A_{obs} = \frac{A_{max} \times L^H}{K^H + L^H}$$

where A_{obs} is the observed absorbance change at ligand (substrate or inhibitor) concentration L. A_{max} is the maximum absorbance change observed at apparent saturation with the ligand. K is the ligand concentration at 50% of the final overall absorbance change, and H is the Hill coefficient.

Several inhibitors are also known for P450s—with most of these binding directly to the P450 heme iron to inactivate its function. The best known are nitrogen-containing inhibitors, such as the azole drugs. These are a group of inhibitors of fungal sterol demethylases (CYP51 family) that bind in the active site and ligate the heme iron, typically through a nitrogen atom of an imidazole or triazole group (Peyton, Gallagher, & Hashemzadeh, 2015). This results in a reinforcement of the low-spin state of the heme iron and a red shift of the heme Soret band. Typically, azole drugs induce a Soret shift to ~423–425 nm when fully bound. In studies of the binding of azoles to the *Mycobacterium tuberculosis* CYP121 P450, bulkier azole drugs such as clotrimazole and econazole (K_d = 73 and 24 nM, respectively) were shown to bind much tighter than smaller imidazole compounds such as 2- and 4-phenylimidazole (K_d = 102 and 32 μM, respectively) (McLean et al., 2008). Other heme-coordinating inhibitors of P450s include cyanide (usually in the mM range) and nitric oxide (NO) that binds to both ferric and ferrous heme iron (Quaroni et al., 2004). The binding of these types of inhibitors can be quantified as described earlier for P450 substrates, but in these cases the difference spectra generated will show an absorbance difference peak at ~433 nm and a trough at ~412 nm (for an azole drug), depending on the particular inhibitor.

5.1.2 Equipment Needed for UV–Visible Spectroscopic Studies

- A UV–vis scanning spectrophotometer, e.g., a Cary 50 or Cary 60 UV–visible spectrometer (Agilent, UK).
- A suitable cuvette such as a 1 mL, 1 cm quartz cuvette. Plastic cuvettes can be used for ligand binding and other assays but may result in lower optical transmission and affect accurate absorbance collection at shorter wavelengths.
- For stabilization of the P450 protein and/or ligand, an appropriate buffering system should be used. A temperature control unit, such as a Peltier device, is recommended for ligand-binding titrations to avoid temperature fluctuations.
- A fume cupboard should be used for spectroscopic studies of P450 binding to carbon monoxide (CO) and nitric oxide (NO), and when performing pyridine hemochromagen assays for determining P450 concentration.

5.1.3 Buffers and Reagents

- The buffer system chosen must be able to stabilize the P450 protein used for the duration of the experiment. This may be up to a maximum of around 2 h depending on the specific experiment. The composition of the buffer should be tailored to the properties of the particular P450 used (e.g., detergents will likely be required for eukaryotic P450s). Care should be taken with choices of additives such as imidazole, solvents (e.g., DMSO), and glycerol, which may bind to the P450 and interfere with results (Kuper, Wong, Roccatano, Wilmanns, & Schwaneberg, 2007).
- Sodium dithionite and carbon monoxide for collection of P450–CO-binding spectra. Nitric oxide for collection of P450–NO-binding spectra. A buffer strong enough to maintain a neutral pH should be used, since both dithionite and NO will otherwise acidify the solution and impact on P450 stability.
- Pyridine stock containing 200 m*M* NaOH, 0.8 m*M* potassium ferricyanide, and 40% pyridine (v/v) for pyridine hemochromogen assays.
- Ligand (substrate and inhibitor) stocks should be in a suitable solvent. If an organic solvent, the ligand concentration should be kept high so that excessive additions of solvent (that may inhibit the enzyme) are avoided. As a rough guide, total solvent volumes added should typically not exceed ~5 μL in a 1 mL sample (i.e., 0.5%, v/v).

5.2 Calculating the Cytochrome P450 Protein Purity Ratio

The purity ratio, or Reinheitszahl (RZ) value, of a P450 can be calculated as the ratio of the absorbance of the heme Soret maximum (at ~418 nm) to that of the absorbance of the ~280 nm peak (originating mainly from absorbance contributions from aromatic amino acids in the P450). This technique was first introduced for the determination of peroxidase purity (Theorell & Maehly, 1950), but can be adapted for other hemoproteins, such as cytochromes P450 (Driscoll et al., 2010). The accuracy of this method for calculating purity will obviously be affected if the P450 has low heme content, or if the P450 is in a HS or mixed-spin state (see, e.g., the arachidonic acid spectrum in Fig. 3) in view of overlapping spectral contributions. In the latter case, extensive dialysis of the P450 can be attempted, or the enzyme could be passed through a Lipidex column (Perkin Elmer) to remove lipids from the P450 prior to the assay. A UV–visible spectrum of the P450 sample in an appropriate buffer should be recorded using a

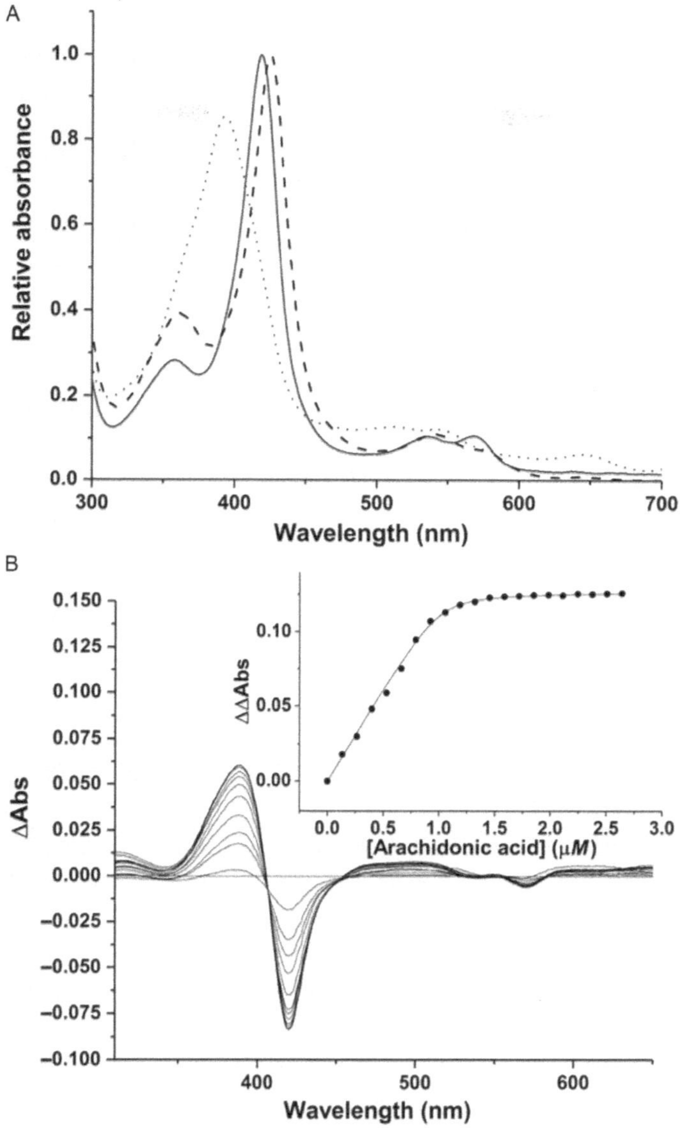

Fig. 3 Binding of inhibitor and substrate compounds to a cytochrome P450 enzyme. The binding of inhibitors and substrates to the cytochromes P450 usually induces characteristic changes in the P450 UV–visible spectra. Panel (A) shows the wild-type P450 BM3 (CYP102A1) heme domain undergoing a spectral shift from a near-complete LS state (peak at 418 nm, *solid line spectrum*) to an extensively HS form (peak at 393 nm, *dotted line spectrum*) on binding of the substrate arachidonic acid. In contrast, the binding of the inhibitor 4-phenylimidazole results in its coordination to the heme iron and in a heme Soret shift to 424 nm (*dashed line spectrum*). Panel (B) shows the absorbance difference spectrum (*main image*) and arachidonic acid-binding curve (*inset*), with the data fitted using the Morrison (tight-binding) equation.

slow-to-medium scan rate in the absorbance range from ~700 to 250 nm to capture both the heme and protein spectra using a high-quality quartz cuvette. Assuming that the P450 heme iron is predominantly in the LS form (i.e., with a sharp absorbance maximum at ~418 nm), the absorbance values at the Soret peak and at the protein peak (i.e., close to 280 nm) should be recorded. By dividing the Soret peak value by the protein peak at ~280 nm value, the RZ value will be obtained. The magnitude of the RZ will depend significantly on the numbers of aromatic amino acids in the P450, with the RZ value of a P450 with relatively few aromatic amino acids being substantially higher than a P450 containing large numbers of these residues. The RZ value can be measured at different stages in the purification of a P450, e.g., following each of the chromatographic steps in the process. The RZ values will provide a guide to the relative efficiencies of P450 purification at each stage in the preparation, and may thus aid in an improved design for a particular P450 purification protocol. The RZ values taken at different stages in the process can also be used to help construct a P450 protein purification table. Imidazole-containing purification buffers will absorb at 280 nm and so caution should be exercised in calculating RZ values after IMAC purification steps with imidazole-containing elution buffers.

5.3 Determining the Concentration of a Cytochrome P450 Enzyme

5.3.1 Calculating P450 Concentration Using Carbon Monoxide Binding

1. The buffer should be degassed by sparging slowly with nitrogen in order to make the buffer anaerobic. The buffer should be suitably strong in order to stabilize the enzyme following its reduction with sodium dithionite (e.g., 100 m*M* potassium phosphate, pH 7). A rubber stopper is used to seal a quartz cuvette. The P450 protein of interest is then delivered into the buffer using a syringe. The quantity of P450 added should be such that the Soret band absorbance is less than 1 unit in the cuvette. The sample is then placed into a spectrophotometer set to a slow scan rate setting to ensure collection of high-resolution data. A suitable speed for scanning is 600 nm/min with data intervals every 1 nm and an average dwell time of 0.1 s to ensure smooth traces for further analysis.
2. A few crystals of sodium dithionite should be added to the cuvette and the cuvette sealed quickly with the rubber stopper. Alternatively, a small aliquot of a concentrated stock of dithionite can be delivered into the sample through a syringe. The resultant spectrum should be altered, displaying a shift of the Soret band to ~407–410 nm for a reduced P450.

Owing to the quite negative heme iron reduction potential of most substrate-free P450s, the process of reduction can be slow. Addition of an excessive amount of dithionite is not recommended in view of potential acidification of the sample. Maintaining a reducing, anaerobic environment is crucial for generating a stable ferrous state of the P450.

3. The final step is to bubble carbon monoxide slowly through the rubber seal using a syringe attached to a carbon monoxide cylinder and with slow introduction of CO bubbles to avoid excessive agitation of the solution that could lead to P450 denaturation. After slow bubbling for 1–2 min, the spectrophotometer should be set to scan for one cycle every minute for 10–30 min (as described in part 1). The CO complex formed should show a Soret band shift to ~450 nm (Fig. 4). The absorbance intensity of the Soret band enables the determination of the P450 concentration. A typical extinction coefficient used is 91 mM^{-1} cm^{-1} at the Soret maximum (Guengerich, Martin, Sohl, & Cheng, 2009). However, various studies have reported a range of extinction coefficients for

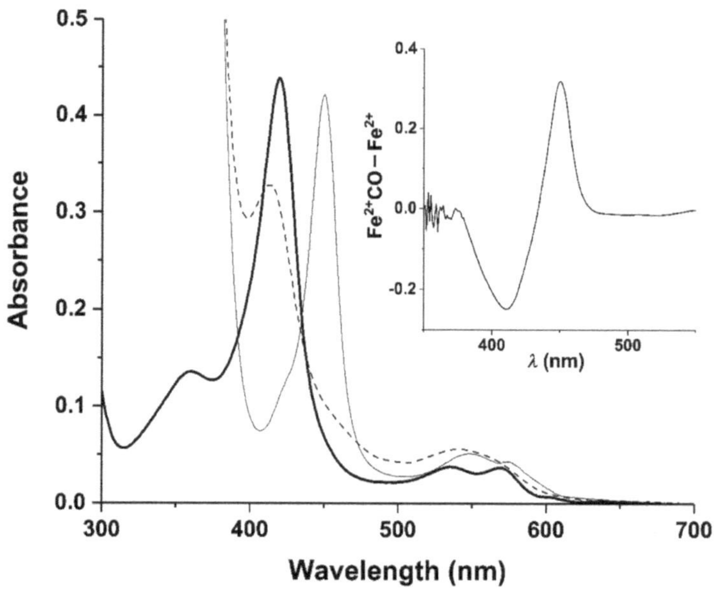

Fig. 4 UV–visible spectral properties of the P450 BM3 heme domain. The *main panel* shows the UV–visible spectrum of the oxidized P450 BM3 heme domain (*heavy, solid line*, ~4.5 μ*M*) with its Soret maximum at 418 nm. The reduction of the heme produces a reduced form with a peak at ~410 nm (*dashed line*), and the subsequent binding of CO produces a near-complete conversion to the P450 form of the enzyme with a Soret absorption maximum at 448 nm (*thin, solid line*). The *inset* shows the absorbance difference spectrum for the ferrous–CO minus ferrous pair with a peak at ~450 nm, as expected for a native P450 enzyme.

different P450s, for instance $\varepsilon_{418} = 95\,\mathrm{m}M^{-1}\,\mathrm{cm}^{-1}$ for the heme domain of P450 BM3.

4. Some P450s are unstable in the P450 (Fe^{II}–CO, cysteinate-coordinated) form and convert toward the P420 (Fe^{II}–CO, cysteine thiol-coordinated) form. In such cases, and assuming a complete conversion to P420 state, it is the best to calculate the heme concentration using a P420-specific coefficient. A completely formed P420 species has a much larger extinction coefficient of $\sim 120\,\mathrm{m}M^{-1}\,\mathrm{cm}^{-1}$ (Dunford et al., 2007).

Early studies of what we now know to be P450s identified these enzymes as a particular class of hemoproteins due to their ability to bind CO in the reduced state, producing a characteristic heme Soret band at ~450 nm (Garfinkel, 1958; Klingenberg, 1958; Omura & Sato, 1964b). The red-shifted 450 nm peak occurs as a result of proximal coordination of the heme iron by a phylogenetically conserved cysteine residue in the thiolate state. The P450 species is considered to be the "active" form of the enzyme (albeit that this Fe^{II}–CO form of the heme is clearly inactive in light of being inhibited by the tight-binding CO distal ligand). The P420 Fe^{II}–CO state of the enzyme is considered an "inactive" species due to its likely proximal coordination by a cysteine thiol. However, two different studies have shown that binding a substrate to the particular P450 enzyme in its P420 (Fe^{II}–CO bound) state can "rescue" the enzymes and return them to their thiolate-coordinated P450 states. Ogura et al. demonstrated that the substrate-free P450epoK enzyme from *Sorangium cellulosum* formed a mixture of P420 and P450 states on reducing the enzymes in the presence of dithionite. Thereafter, the P420 form (with A_{max} at 422 nm) slowly accumulated at the expense of the P450 form. However, with addition of the substrate epothilone D (one of a group of related anticancer drugs) to a preformed P420–CO complex, the enzyme was converted to an extensively P450 form. Similarly, the addition of cholest-4-en-3-one, an excellent substrate for the *M. tuberculosis* CYP142 enzyme, produced a near-complete conversion of the P420 form to the P450 state in this sterol metabolizing enzyme (Driscoll et al., 2010; Ogura et al., 2004). These data clearly indicate that the P420 state of the enzyme can be "rescued," and that substrate binding evidently stabilizes these enzymes and can result in the accumulation of the cysteine thiolate form of the enzymes. In other studies, Dunford et al. demonstrated that the *M. tuberculosis* CYP121 enzyme in its Fe^{II}–CO state could be titrated reversibly in the pH range from 6.1 (mainly P420) to 10.5 (mainly P450), demonstrating clearly the reversibility of the P450/P420 equilibrium in this enzyme (Fig. 5) (Dunford et al., 2007).

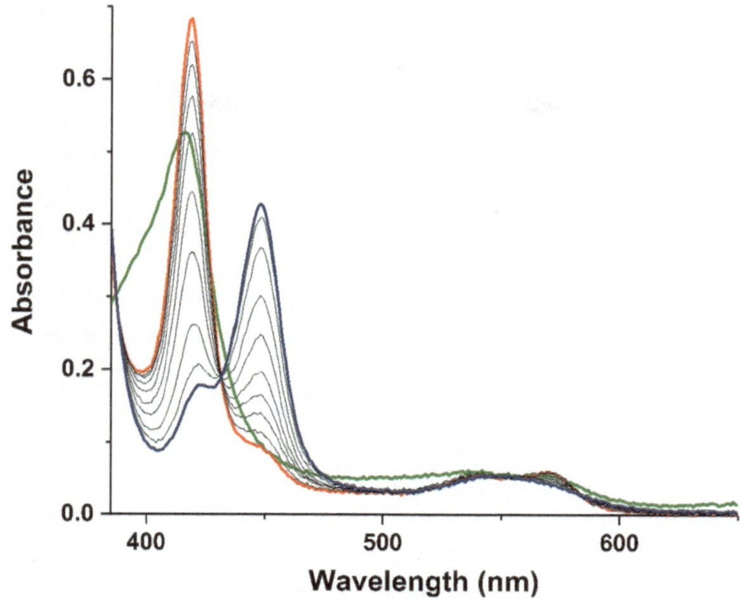

Fig. 5 Substrate-mediated "rescue" of a P450. The *M. tuberculosis* CYP121 enzyme catalyzes C—C bond formation between the aryl groups of its substrate dicyclotyrosine to form the novel product mycocyclosin. The figure shows selected UV–visible spectra for the reduced/CO-bound forms of CYP121 in the pH range from 6.1 (mainly P420, *red spectrum*) to 10.5 (mainly P450, *blue spectrum*). The *green spectrum* shows the reduced CYP121 spectrum collected prior to addition of CO to the sample. The P450/P420 equilibrium is freely reversible in this pH range (Dunford et al., 2007).

5.3.2 Determining a Hemoprotein Extinction Coefficient Using the Pyridine Hemochromogen Method

The pyridine hemochromogen method can be used accurately to determine the extinction coefficient for a P450, or for other heme-containing proteins (Berry & Trumpower, 1987). The method is described below:

1. A stock of P450 enzyme with its Soret maximum at ~1 absorbance unit (~10 μ*M*) should be produced in sufficient quantity for a number of repeats to be carried out. 2 mL should be sufficient to complete the analysis.
2. In a quartz cuvette, the absorbance of 1 mL of the P450 stock is measured with a spectrophotometer set to a slow scan setting to ensure collection of high-resolution data. A suitable speed for scanning is 600 nm/min with data intervals every 1 nm and an average dwell time of 0.1 s to ensure smooth traces for data analysis.

3. 0.5 mL of the P450 stock are then removed and 0.5 mL of a pyridine stock solution added (200 mM NaOH, 0.8 mM potassium ferricyanide, and 40% (v/v) pyridine). The ferricyanide produces a fully oxidized spectrum.
4. Subsequently, several crystals of sodium dithionite are added and a reduced spectrum is produced. This spectrum should produce a distinctive peak between 550 and 570 nm. Pyridine acts as a ligand for the reduced heme cofactor, producing the required spectrum. No stopper is needed for this experiment (Barr & Guo, 2015).
5. This should be repeated three to four times on consecutive samples and the extinction coefficient calculated and averaged. This is done by subtracting the absorbance of the oxidized spectrum from that of the reduced spectrum at the wavelength of maximum absorbance between 550 and 570 nm. Subsequently, this value is divided by the extinction coefficient of pyridine (22.98 mM^{-1} cm^{-1}), taking into account the dilution factor. Assuming there is one heme cofactor per protein, the Beer–Lambert equation is then used to determine the extinction coefficient of the protein using the Soret maximum.

6. USING CYTOCHROMES P450 FOR OXIDATIVE TRANSFORMATIONS

6.1 Assaying P450s for Enzymatic Turnover

Determining the products of P450 reactions is essential for identifying natural enzyme reactions, producing pharmaceutical metabolites for analysis, and for generating high-value compounds for industrial applications. A number of methods can be utilized for metabolite identification, the most common being LC–MS, GC–MS, or NMR. The method should be chosen depending on the chemistry of the ligand. LC–MS may use fewer steps compared to GC–MS (where compounds often require derivatization), but it is not used on ligands lacking ionizable groups such as fatty acids. In this case, GC–MS becomes the technique of choice, although fatty acid samples require derivatization prior to analysis in order to increase the sample volatility and to ensure ionization within the MS. When LC–MS and GC–MS cannot identify the products, NMR can be utilized. Examples include steroidal compounds with multiple conjugated rings and where analysis by MS/MS proves difficult as the energy required to break the molecule causes the production of fragments too small for analysis (Acevedo-Rocha et al., 2018). However, NMR requires much larger sample quantities and the use of deuterated solvents for product identification. Therefore, more reagents and protein are required, increasing the experimental costs. For some P450-mediated reactions, the oxidative turnovers that occur cause another part of the molecule to be lost (e.g., in oxidative deamination or demethylation). This causes a mass change that may be missed in MS reactions where there are several peaks due to the large number of components in the sample. For example, CYP3A4 converts the antidiabetic drug repaglinide to a number of metabolites, one of which results in the loss of two carbon atoms (Bidstrup, Bjornsdottir, Sidelmann, Thomsen, & Hansen, 2003). In these situations, NMR may be useful in molecular analysis.

6.2 Cytochrome P450 Turnover Reactions

Enzymatic turnover reactions can be conducted in vivo or in vitro. If using whole cells as the reaction vessel, appropriate redox partners can be expressed alongside the P450 of interest to enhance activity (e.g., by fusing a redox partner to the P450—for example, a PDOR, as discussed in Section 8.1) (Bleif et al., 2012; Ma et al., 2015). Such in vivo systems can be cost effective, as they do not require the addition of expensive cofactors such as NADPH. However, in vivo experiments routinely require a cell lysis step prior to analysis of metabolites, but it is often possible to use the lysate directly for product analysis without further purification (Sawayama et al., 2009). In vitro experiments usually require NAD(P)H regeneration systems driven by, e.g., glucose-6-phosphate dehydrogenase (NADPH) or glucose dehydrogenase (NADH), as well as an appropriate buffer, the P450 and sufficient substrate concentrations to generate adequate amounts of the product(s) of interest. In vitro reactions offer more control over the system, including avoidance of unwanted substrate or product conversions (which can occur when using in vivo systems) and the ability to regulate the P450 concentration. Solvents used to solubilize the substrate may also inhibit the P450-mediated reaction, and thus should be used at levels that do not affect catalysis significantly.

6.2.1 Reagents Required for P450 Enzyme Assays

- A suitable buffer for the reaction. Factors including pH, ionic strength, and buffer composition are of great importance to optimize, e.g., catalytic rate and stability of the P450. Commercial buffer screens are available that can identify appropriate conditions (e.g., RUBIC buffer screen, Molecular Dimensions, Newmarket, UK).

- Appropriate redox partners to support target P450 enzyme activity (see Section 6.3).
- NADPH or NADH regeneration system (e.g., glucose-6-phosphate dehydrogenase and glucose dehydrogenase, respectively).
- The substrate of interest at an appropriate concentration in aqueous or suitable organic stock solutions.
- Eppendorf thermomixer (Eppendorf, Stevenage, UK) or other shaker with temperature control to perform enzyme assays under optimized conditions.
- As required, a suitable solvent or acid for quenching of the reaction.
- If the product of interest has a chromophore or fluorophore component, a spectrophotometer (e.g., a Cary 50 spectrophotometer) or fluorimeter (e.g., Cary Eclipse instrument, both from Agilent Technologies Ltd., Cheadle, UK) can be used to quantify products.

6.3 Redox Partner Selection for P450 Activity

A number of families of P450 enzymes exist which do not require the addition of exogenous redox partner proteins to shuttle electrons to the heme group for enzymatic turnover. These include P450 BM3 (CYP102A1) and other members of the CYP102 and CYP505 families, which are natural fusions between a P450 domain and a CPR-like diflavin reductase domain, meaning that only NAD(P)H, oxygen, and substrate are required for turnover. There are also other examples of natural P450 redox partner fusions, including *Cupriavidus metallidurans* CYP116B1, which is naturally fused to a PDOR module (Warman et al., 2012). CYP55A1 (P450nor) is capable of utilizing NADH directly in the reduction of nitric oxide without the need for exogenous redox partners (Nakahara, Tanimoto, Hatano, Usuda, & Shoun, 1993). The CYP152 family also does not require redox partners, instead utilizing the peroxide shunt pathway as described in more detail in Section 1 (Matsunaga et al., 1996).

There are two major classifications of exogenous redox partner system employed by the P450s. Class I proteins require two partner enzymes for electron transfer; a ferredoxin/adrenodoxin that contains an iron–sulfur center and a NADPH-dependent ferredoxin/adrenodoxin reductase containing a FAD cofactor. A widely studied example of this system is P450cam (CYP101A1), which uses electrons from NADH delivered through putidaredoxin reductase (PDR) and putidaredoxin (PD) to drive the P450cam enzyme (Katagiri, Ganguli, & Gunsalus, 1968). Class II proteins, found in eukaryotic microsomes in the liver and other tissues, receive their electrons from a NADPH-cytochrome P450 oxidoreductase that contains both an FAD and an FMN cofactor (Lu, Junk, & Coon, 1969; Omura & Sato, 1964a). In the absence of a native redox partner system, exogenous redox partner systems should be investigated. For instance, spinach ferredoxin/ferredoxin reductase, which are commercially available, or, e.g., *E. coli* flavodoxin (FLD)/FLDR are often used as nonnative redox partners (Dong, Yamazaki, Guo, & Guengerich, 1996; Jenkins & Waterman, 1994). It is also routine to test a number of different stoichiometries of redox partners in order to identify the best turnover conditions, and possibly to use both an iron–sulfur cluster-containing ferredoxin and a FMN-containing flavodoxin to establish which is preferred as the redox partner. As a guide, we have routinely used stoichiometries of 1:10:2 P450:FdX:FdR (Driscoll et al., 2010) for *M. tuberculosis* CYP142 or 1:20:5 P450:FLD:FLDR (McLean et al., 2009) for *M. tuberculosis* CYP125.

P450 proteins can be fused to redox partners using a linker region to become catalytically self-sufficient (i.e., requiring only dioxygen [O_2], substrate, and NAD(P)H for activity). These constructs often have higher catalytic activities than their isolated enzyme components, typically due to faster electron transfer between the component domains as a result of a higher rate of productive collisions (Ortega Ugalde et al., 2018). Such proteins have potential uses in production of high-value products for industrial applications, for example in the production of pravastatin (McLean et al., 2015). However, the crafting of these fusion proteins must be undertaken with care to ensure an optimal length (and composition) of the linker in order to enhance catalytic activity and stability (Degregorio et al., 2017). A point to note is that, although the rate of reaction may be faster in such fusion enzymes compared to those of the isolated P450 and partner enzymes, the catalytic rates typically remain much lower than those of naturally evolved fusion enzymes such as P450 BM3 (Noble et al., 1999).

6.4 NAD(P)H Regeneration Systems

For in vitro systems where a source of NAD(P)H is required, it is routine to include a cofactor regeneration system. A commonly used example is glucose-6-phosphate dehydrogenase with its substrate glucose-6-phosphate, which is converted to 6-phosphoglucono-1,5-lactone in a reaction which converts $NADP^+$ to NADPH. It should be noted that, when studying turnover of P450 BM3, it is more effective to add $NADP^+$, rather than NADPH, to the reaction. In this situation, the $NADP^+$ will be converted to NADPH by the glucose-6-phosphate dehydrogenase and thus provide an electron source for P450 BM3, while maintaining a lower NADPH concentration that will prevent the inhibition of P450 BM3 by minimizing its conversion to an inert 3-electron reduced form, i.e., a state in which the FMN cofactor is in the hydroquinone state while the FAD is in the neutral (blue) semiquinone form, with the $FMNH_2$ form being an ineffective electron donor to the P450 heme iron (Murataliev, Klein, Fulco, & Feyereisen, 1997). In systems requiring a source of NADH, glucose dehydrogenase can be used as the recycling enzyme. Alcohol dehydrogenase (ADH) produces organic solvents such as acetone during the regeneration of NADPH, so caution should be applied in its use, as acetone (or other molecules) may accumulate and inhibit turnover or denature the P450.

6.5 In Vitro P450 Reactions

Several methods for in vitro P450 turnovers have been optimized for both the substrate and the enzyme of the reaction (von Buhler & Urlacher, 2014). These protocols can be used as a starting point for new P450 catalytic reactions, but would likely need optimization for each reaction with respect to substrate and enzyme concentrations and stoichiometries, redox partner choice and concentration, optimal pH and temperature conditions, etc. (Funhoff, Bauer, Garcia-Rubio, Witholt, & van Beilen, 2006). In vitro reactions allow the option to add components during the turnover for further catalytic steps, or for optimization of product formation. This can also be accomplished for whole cell reactions, but here there may be other issues, including whether components are able to cross the relevant cell membrane.

6.6 In Vivo P450 Reactions

Many in vivo P450-catalyzed reactions found in the literature focus on using liver hepatocytes with pharmaceutical drugs. Determining the metabolites of several human and animal pharmaceuticals is now essential for FDA approval due to the possibility of side effects arising from toxic metabolites (Gao, Jacobs, White, Booth, & Obach, 2013). Around 70% of the pharmaceuticals that enter the human body are metabolized by the cytochromes P450 (Guengerich, 2003). To determine the structures of these metabolites, in vivo turnovers often make use of harvested hepatocytes which are quenched with a solvent, lysed, and the products analyzed by LC–MS or GC–MS (Bidstrup et al., 2003; Sawayama et al., 2009; Uchiyama et al., 2010). Hepatocytes and microsomes can also be optimized for nondrug compounds, for example fatty acids such as oleic acid (Adas et al., 1998). An advantage of this method is the ability to produce enzyme cascades within the cell, in order to be able to produce high-value compounds from cheap starting chemicals. For example, yeast cells have been used to express a number of cytochromes P450 and other proteins for the conversion of acetyl-CoA to the antimalarial drug artemisinin using a synthetic biology approach (Paddon et al., 2013).

6.7 Preparation of Samples From Turnover Reactions

After the turnover reaction has been completed, products can be separated from contaminants such as the protein itself, components of the NAD(P)H regeneration system, etc., and/or be transferred into a solvent suitable for the analysis method of choice. Where samples can be analyzed in the aqueous phase, the simplest method is to remove the proteins in the sample by protein precipitation. However, further sample preparation can provide much purer samples that provide traces with fewer contaminant peaks, thus allowing easier data analysis. The sections below address routes to improving product recovery.

6.7.1 Protein Precipitation

1. After P450-mediated turnover, the sample is quenched by adding ratios of 1:1, 2:1, or 3:1 (solvent-to-sample) of appropriate solvents, e.g., acetonitrile. Higher solvent-to-sample ratios are recommended for efficient protein precipitation and reaction quenching. However, this will decrease the intensity of the signal observed.
2. Agitation of the solution at high rates and with moderately high temperatures (e.g., 8000 rpm at 50°C for 10 min) should be sufficient to ensure extensive precipitation of proteins while enhancing the solubility of substrates and products [optional]).
3. The resultant solution can then be filtered through a simple filter or protein precipitation columns (for example, Phenomenex Strata Impact protein precipitation plates or columns, Phenomenex, Macclesfield, UK) to obtain a much "cleaner" product sample.

6.7.2 Liquid/Liquid Extraction (LLE)

LLE is routinely used to extract many products (and substrates) from aqueous reaction mixtures. As enzyme turnovers are predominantly conducted in buffer, water will always be one of the liquids utilized. Other appropriate solvents include dichloromethane (DCM), chloroform, and toluene. Acetonitrile and DMSO are not suitable as they will not separate from the water layer effectively, rendering the technique ineffective. The method works on the basis that the products will move from the water layer into the solvent layer due to the polarity of the compounds.

1. After turnover, quench the sample by adding solvent to the sample in a 1:1 ratio.
2. Mix the solutions vigorously and allow the layers to form, or use centrifugation to expedite their formation.
3. Remove the solvent layer, taking care not to disturb the aqueous layer or the interface between the layers.
4. Steps 2 and 3 can be repeated for increased yield, resulting in higher product amounts but more dilute samples.
5. Add solid $MgSO_4$ to remove traces of aqueous material retained during transfer.
6. Centrifuge samples to pellet the $MgSO_4$, then remove the solvent to a clean tube for analysis or derivatization.

An alternative to LLE is SLE (simplified liquid extraction) where the turnover reaction is applied to a column of diatomaceous earth (which is of a larger total volume than the sample itself), and then eluted with the solvent required for downstream analysis. Synthetic alternatives are also available, e.g., Phenomenex Novum (Phenomenex).

6.7.3 Solid-Phase Extraction (SPE)

SPE combines high efficiency of product recovery with the ability to change solvent and to concentrate samples. This technique involves adding the sample to an SPE column (available in either single tube or plate format). The column can then be washed with water or by using a gradient designed for that ligand-column combination to remove any undesired impurities. The product of interest can then be eluted with a solvent, or by applying a change of pH if an ion exchange column is used.

There are many types of SPE columns which separate compounds depending on different properties: reversed-phase (for hydrophobic or nonpolar interactions), normal phase (for hydrophilic or polar interactions), and ion exchange (for ionic interactions with acidic/basic compounds). Within these three modes there are multiple columns that vary in size and in the nature of the adsorbent material within the column itself. Identifying the correct SPE column and protocol for the particular reaction is essential for effective purification of products. A number of online tools are available for this method development, using the structure of the ligand to generate polarity (log *P*) and pK_a values (e.g., www.chemicalize.com).

6.7.4 Derivatization of Samples for GC–MS

Compounds with poor volatility and/or poor thermal stability, or that exhibit column adsorption, will display nonreproducible peak heights and areas when analyzed by GC. This can usually be avoided by derivatization of the sample. It may also be possible to separate chiral compounds by derivatization. Routinely, trimethylsilane (TMS) derivatives are made where the TMS targets –OH, –NH, or –SH groups on the compounds of interest. Examples include studies of fatty acid turnover reactions (Girvan et al., 2011).

6.8 Analyzing P450 Products Using HPLC and LC–MS

It is possible to observe product formation without the need for mass spectroscopy analysis if the retention times of the compounds of interest are known, which is generally done by running suitable standards on HPLC (high-performance liquid chromatography) or UPLC (ultra-performance liquid chromatography). However, these standards may be expensive and require the researcher to know what the expected products are in advance. In many cases, these compounds may not be commercially available. Owing to the promiscuous nature of many P450s, the exact product may not be known in numerous cases, or two or more products may be produced according to the regio- or stereoselectivity of substrate oxidation by the particular P450 used. HPLC and UPLC methods allow product formation to be established, as well as providing an indication of the polarity of the product(s) formed according to their retention time during application of a linear gradient. Many researchers use HPLC or UPLC as a method to ensure correct separation of the parent and product compound(s) before moving on to MS or MS/MS, often within the same instrument setup.

All techniques require a suitable column for analysis. Many researchers use standard reversed-phase C18 columns for separation of products.

Reversed phase refers to the sample having a polar nature, allowing it to bind to the nonpolar column. Compounds are eluted using a gradient from aqueous to organic in order to separate compounds based on their polarity. In normal phase, the opposite situation occurs, i.e., the compounds are nonpolar and the column is polar. Reversed phase is the most common type of HPLC column used as it allows more control of variables such as solvent type, concentration, and pH. Other types of HPLC columns include chiral, size exclusion, and ion exchange columns, available from a number of suppliers (e.g., Merck, Phenomenex and Shimadzu [Milton Keynes, UK]). Extensive purification of samples will allow for longer lifespans for these expensive columns.

For LC–MS, the mass of a product can be determined, but not the position of the chemical modification. MS/MS can give this information by applying collision energy to samples following their passage through the first mass analyzer, causing their fragmentation which is then analyzed within a second analyzer, most commonly a quadrupole or Q-TOF instrument. The fragments can be analyzed using software such as MassHunter (Agilent) or Compound Discoverer (Thermo Scientific), depending on the manufacturer of the MS/MS machine used. In some cases, the fragmentation pattern obtained can still not identify the modification observed, such as the position on an aromatic ring, as the collision energy cannot break it apart without likely destroying the rest of the molecule. In these cases, NMR may be used to identify the product(s) formed. Fig. 6 shows data analysis for the P450 BM3 A82F/F87V mutant-catalyzed hydroxylation of omeprazole.

6.9 Analyzing P450 Products Using GC and GC–MS

GC–MS is also a popular method for organic product analysis and is frequently used to resolve compounds which are more volatile than those studied by LC–MS methods. Samples must be prepared in a volatile solvent and therefore aqueous material must all be removed. Column stationary phases vary in polarity and care should be taken in selecting the correct column. It is important to ensure that compounds of interest have a boiling point low enough to be within the working range of the column, and it should be noted that TMS derivatization agents interact nonreversibly with the PEG column phase and should therefore not be used with wax columns. Also, when using columns with a siloxane containing phase, it is important not to misassign column bleed (i.e., leakage from the column) as TMS in derivatized samples.

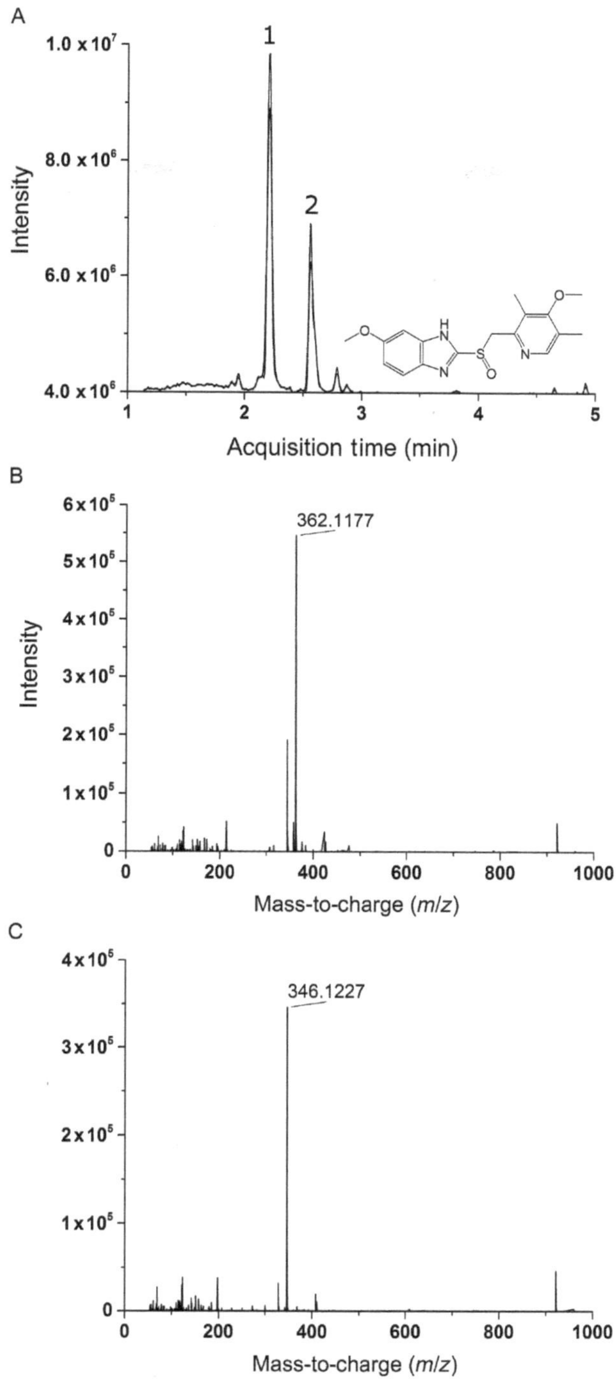

Fig. 6 LC–MS analysis of the oxidation of omeprazole using a BM3 A82F/F87V variant. A reversed-phase C18 column (using an acetonitrile gradient) was used to separate the parent and product compounds. Panel (A) shows the UV peaks from the HPLC. There are two main peaks. Peak 1 is the product (5-hydroxy omeprazole) and peak 2 is the parent compound omeprazole. Panel (B) shows the MS of peak 1 and Panel (C) shows the MS of the parent peak 2.

It is possible to inject gas samples onto the GC instrument by headspace analysis. This technique presents an opportunity to partition the more volatile compound of interest from the liquid matrix into the tube headspace by heating, thereby removing the need to purify the sample. However, if the sample is in an aqueous environment, temperatures must be kept low enough to ensure no aqueous material is introduced onto the column. In order to analyze headspace samples, modifications must be made to the autosampler, the simplest form being a gas-tight autosampler injection system. A number of useful resources are available online from the instrument and column manufacturers (e.g., see Agilent, Waters or Phenomenex websites). Routinely, GC instruments are coupled to an electron ionization source to fragment the column eluate, followed by either a quadrupole or Q-TOF mass spectrometer. NIST libraries can be employed for basic compound assignment. Fig. 7 shows GC–MS analysis of the hydroxylation of dodecanoic (lauric) acid, with the reaction catalyzed by a W1046A BM3 mutant.

6.10 Analyzing P450 Products Using NMR

As stated previously, NMR samples often require large volume turnover reactions to produce the required amount of a compound of interest needed for good NMR signals. These large reaction volumes must then be concentrated and exchanged into a deuterated solvent such as $CDCl_3$. This can be done with SPE columns while purifying the sample from any other contaminants.

Once the samples are prepared and added to the NMR tubes, many modern NMR machines will automatically shim the samples. It should be noted that using mixtures of deuterated solvents can cause misidentification of the lock signal, leading to downstream problems with automation, which can be solved by manually shimming and locking each sample.

Many different types of NMR approaches can be employed to determine product profiles. The most common in the literature is ^{1}H COSY. However, other methods such as HMBC and HSQC are utilized for ^{1}H and ^{13}C. The NMR instrument used must operate at a sufficiently high frequency to allow identification of the relevant compound(s) of interest. A more powerful machine will produce better resolution of crowded spectra and higher sensitivity for lower quantities of samples. Data can be analyzed

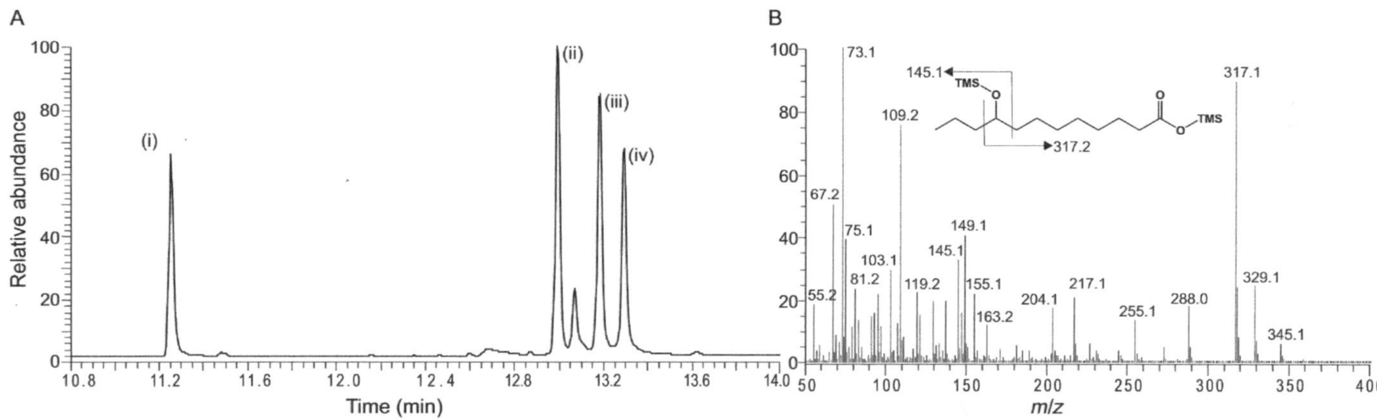

Fig. 7 GC–MS analysis of the hydroxylation of dodecanoic acid by the P450 BM3 W1046A mutant. Panel (A) shows a UV trace of dodecanoic acid (C12:0) and its hydroxylated products. Peak (i) is dodecanoic acid and peaks (ii), (iii), and (iv) are the ω-3, ω-2, and ω-1 hydroxydodecanoic acids, respectively. Panel (B) shows the MS corresponding to the GC peak (ii) and its assignment as ω-3 hydroxydodecanoic acid. A Phenomenex ZB5-MS column was used to resolve these molecules (Girvan et al., 2011).

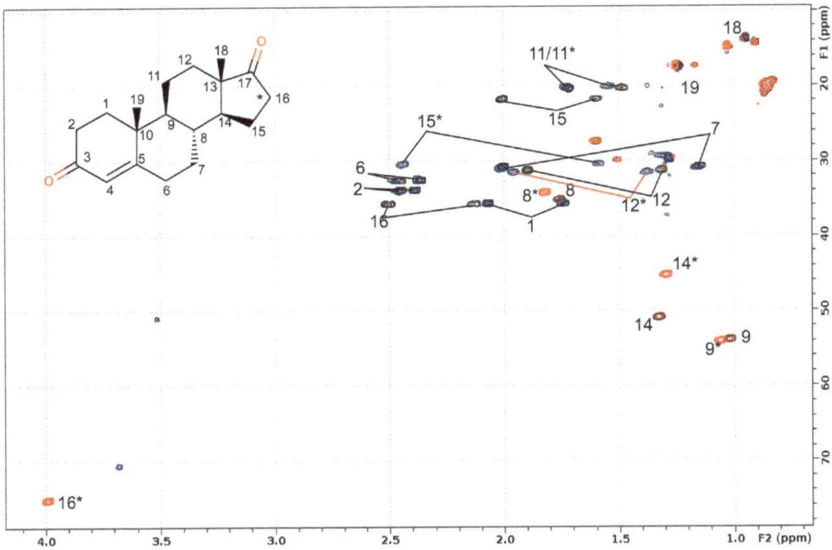

Fig. 8 $^1H–^{13}C$ HSQC of a P450 BM3 variant reacting with androstenedione. Multiplicity edited $^1H–^{13}C$ HSQC at 800 MHz. The reactant (androstenedione) is shown in *black* or *maroon*. Product is shown in *blue* or *red* with any resonances marked with an asterisk indicating significant changes. It was found that a hydroxylation occurred at the C-16 position using a BM3 A82F/F87V/I401P mutant.

manually by determining the splitting of peaks and the δ shift, or by using software such as MestRe Nova (Willcott, 2009). Fig. 8 shows NMR analysis for the conversion of androstenedione to a C16 hydroxylated product.

7. CRYSTALLIZATION AND STRUCTURAL ANALYSIS OF P450S

7.1 Introduction

X-ray crystallography has been utilized from the early 1900s for structure determination, initially for chemical compounds. Myoglobin was the first protein for which a crystal structure was determined in 1958 (Bragg, 1914, 1924; Kendrew et al., 1958). X-rays have a resolution of 0.1–100 Å, allowing researchers to observe even carbon–carbon bonds (1.54 Å). For protein structure elucidation, particularly with respect to ligand binding, resolutions below 2 Å are optimal so that bonds within the crystal can be observed.

7.1.1 Equipment Needed for Protein Crystallization

- Equipment for high-quality protein purification (see, e.g., Section 2.1.1).
- Purchased or self-produced crystallization conditions for sitting drop or hanging drop crystallography methods (e.g., from Molecular Dimensions, Newmarket, UK).
- Liquid handling robot for high-throughput screening of crystallization conditions, such as the Mosquito® from TTP LabTech (Melbourn, UK).
- Crystallization plates (i.e., hanging drop and sitting drop plates for the Mosquito liquid handling robot, such as Swiss SD3 plates [Hampton Research, Aliso Viejo, USA]).
- Light microscope for crystal quality checking.
- Mounting loops (size dependent on crystal size).
- Cryoprotectants.
- Liquid nitrogen storage containers for transport of crystals.

7.2 Protein Construct Engineering and Protein Purification for X-Ray Crystallography

The first cytochrome P450 crystal structure was published in 1985 for the camphor hydroxylase P450cam (CYP101A1) (Poulos, Finzel, Gunsalus, Wagner, & Kraut, 1985). By 2014, the number of P450 crystals structures had grown to 449 on the Protein Data Bank (PDB) (Poulos & Johnson, 2015). Different crystallographic approaches have been utilized to obtain these structures (e.g., different temperatures and drop volumes). However, the P450s present some issues in crystallization due to the inherent flexibility of these highly dynamic proteins. The probability that a P450 or other protein sequence will crystallize can be checked by the online tool Xtal Pred (Slabinski et al., 2007). Low crystallization scores can be improved by, e.g., truncation of helices from the N- or C-termini. The binding of substrates or inhibitors can also stabilize P450s and favor their crystallization (Muller, 2017).

Truncations of eukaryotic P450s that remove the transmembrane helix can help improve their solubility and minimize requirements for detergent addition. Removal of the transmembrane helix will increase the likelihood of crystallization and structural determination (Walsh, Szklarz, & Scott, 2013). However, such engineering may change the structural conformation observed. Other proteins, such as P450 BM3, are natural fusion proteins. These proteins have a labile linker region between two highly dynamic and flexible P450 and CPR domains. P450 BM3 was discovered in the 1970s and its heme domain structure was determined in 1993 (Ravichandran, Boddupalli, Hasermann, Peterson, & Deisenhofer, 1993) (PDB code 2HPD). However, a full-length structure for this protein has not been solved, mainly due to the multidomain protein's conformational flexibility. However, the structures of the component FMN- and FAD-/NADPH-binding domains have been determined individually (PDB codes 4DQK and 1BVY, respectively) (Joyce et al., 2012; Sevrioukova, Li, Zhang, Peterson, & Poulos, 1999).

As for most proteins, the preparation of P450s to a high level of purity and their use soon after production is desirable for successful crystallization and structural analysis. For some P450 proteins, such as the *M. tuberculosis* mycocyclosin synthase CYP121A1, protein frozen immediately after purification can be thawed on ice and used successfully for crystallography, although the final structural resolution may not be as high as for the P450 sample used directly from purification without freezing. For P450s thawed from a frozen stock, it is recommended that the sample is passed through a gel filtration column (such as a GE Healthcare Superdex S200 16/60) to separate the target protein from protein aggregates caused by freezing. For the peroxygenase P450 OleT, purification can be accomplished within 3 days using a Ni-IDA column, TEV protease cleavage to remove the His_6-tag, another Ni-IDA step to remove uncleaved protein, followed by a gel filtration step (Matthews, Belcher, et al., 2017) (see Section 2.2.1). For nontagged proteins, a typical set of purification steps could include ion exchange by Q-Sepharose, followed by further purification on a hydroxyapatite-B column and a final gel filtration step (Butler et al., 2013). The aim is obviously to have the highest P450 purity possible (and at least for the P450 to be >95% pure, as evidenced by SDS-PAGE analysis). We recommend that all purification regimes end with a SEC step to ensure removal of protein aggregates or misfolded protein, thus providing a homogeneous sample before protein crystallization. Sample monodispersity can be confirmed using multiangle laser light scattering (MALLS), which should reveal the presence of monomers, dimers, or higher-order aggregates. Such sample heterogeneity may compromise formation of a regular crystal lattice, and thus further efforts should be made to produce a homogeneous, monodisperse sample. These might include the use of disulfide bond reducing agents such as TCEP (tris(2-carboxyethyl)phosphine hydrochloride) and further P450 protein resolution using SEC. In the preparation of mutant P450 proteins, the purification protocol may require reoptimization in cases where mutations result in structural change and/or protein destabilization.

For some proteins, a His-tag provides a point of flexibility within the protein that may result in improper formation of the crystal lattice and poor crystal diffraction. Therefore, some researchers choose to cleave the His-tag before crystallization using protease sites inserted into the gene (e.g., TEV, HRv3, or thrombin). P450 protein purification can be accomplished without a His-tag present. However, this usually results in a greater number of steps in the purification protocol.

7.3 Crystallography Methods for Proteins Including P450s

Most protein crystal structures deposited on the PDB are crystallized using the sitting drop method due to the development of liquid handling robots such as the Mosquito system from TTP LabTech. This system can also be adapted for the hanging drop method to screen many conditions, although this requires delicately inverting the plate. Hanging drop methods are now more routinely used for proteins or compounds that may degrade plastic, and so a glass slide is favored.

The use of liquid handling systems allows more consistent results (compared to manual pipetting) and the use of smaller amounts of protein enables the screening of a greater number of conditions. For crystallization, protein concentration ranges from 5 to 15 mg/mL have been cited in the literature, depending on the particular P450 used. Higher protein concentrations may lead to more nucleation points, resulting in smaller crystals. However, if there is sufficient protein available in the well, crystallization depends more on the crystal condition than the protein concentration. There are certain conditions which often lead to high-quality crystals (although not specific to cytochromes P450, Molecular Dimensions make the LMB screen, which is a collection of conditions most commonly used at the LMB to obtain high-quality crystals). However, as discussed earlier, mutations and/or the binding of different ligands will often result in the P450 crystallizing under different conditions. Within our lab, we have observed that the P450 BM3 heme domain is notoriously inconsistent with respect to the conditions in which it crystallizes and/or the quality of crystals obtained; whereas CYP121 will always crystallize in a narrow set of conditions producing very large, high-resolution crystals. Many researchers therefore use screening kits, such as those from Molecular Dimensions, to secure a hit. Alternatively, plates can be manually produced using a range of conditions known to favor nucleation and crystallization of the target P450, using multichannel pipettes or liquid handling robots such as the Dragonfly from TTP LabTech.

The protein is added to the plate with the condition of each well (mother liquor) in a 1:1 volume ratio. Drop size can also be optimized for crystal preparation as this will impact on equilibration times and therefore nucleation. For example, within Swiss SD3 plates the drop volume can be varied between 200 and 1200 nL in total. Without access to liquid handling robots, larger drop volumes can be used in 96-well plates or in traditional hanging drop chambers. Both approaches can result in successful crystallization. However, in the absence of robotics, sample preparation is more time consuming, and the process may require a larger amount of protein. As mentioned previously, the Mosquito system can be adapted for hanging drop methods.

Temperature plays an important role in the rate of equilibration between the mother liquor and the protein drop, and therefore impacts on crystal formation. It is routine in our work to incubate all trays at 4°C to slow the equilibration and allow the protein as much time as possible to nucleate into a crystal.

7.4 Improving Crystal Quality for High-Resolution Structures

If poor quality crystals or small crystals are produced, seeding can then be used to improve crystal quality. A "good" crystal is thought to be over 100 μm in length with well-defined edges and uniform color, suggesting a regular lattice within the crystal. A larger crystal will be able to tolerate the X-ray beam for longer periods of time and should result in a more complete dataset.

To seed, a P450 crystal is fragmented by vortexing in a microcentrifuge tube containing the mother liquor of the well from which the crystal originated and a small bead. These seed bead tubes can be bought from Hampton Research. This seed stock is added to the crystallography plate along with the fresh protein and mother liquor. The ratio of seed stock to the fresh protein can be optimized for better quality crystals. However, the ratio of overall protein to the mother liquor should stay constant (1:1). In theory, all wells should then contain a crystal fragment ensuring a nucleation point for a larger crystal. The quality of the seeding crystal should be considered to ensure a good starting point. Multiple rounds of seeding can be done to improve crystal quality systematically (Fig. 9). Of note, many other proteins crystallize without the need of seeding, yet many successful P450 crystals made within our lab are the product of one or multiple seeding rounds.

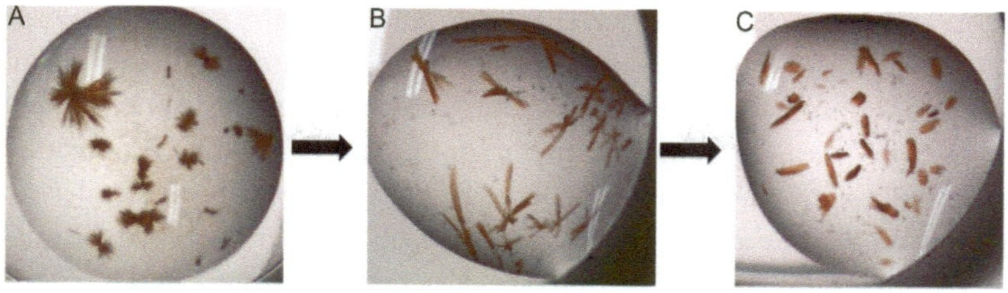

Fig. 9 Systematic improvement of P450 BM3 heme domain crystals. These BM3 heme domain ligand-bound crystals were produced by the sitting drop method using a Mosquito liquid handling system. Each drop contains 200 nL of mother liquor with 150 nL fresh protein and 50 nL seed stock. The crystals become larger and more uniform with seeding going from Panel (A) to Panel (C). Panel (B) contains crystals that, if trimmed, could be used for data collection. Panel (C) contains several good crystals and several irregular crystals.

7.5 Producing Ligand-Bound P450 Crystal Structures

To produce ligand-bound structures, the P450 can be cocrystallized or soaked with the ligand. Cocrystallization involves incubating the protein with the ligand. This can be done immediately prior to setting up crystallography plates, for a few hours, overnight, or during a gel filtration column step in which the P450 is passed through a column using a buffer containing the relevant ligand. The latter approach requires a soluble ligand which can be prepared as a concentrated stock solution. If ligand solubility or excessive solvent is an issue then the P450 can be diluted and placed in a cuvette. The ligand can then be added until a saturation point is reached, assuming that the ligand causes a shift in the UV–visible spectrum of the P450. Thereafter, the protein is concentrated by ultrafiltration and used in crystallographic trials (Haines, Tomchick, Machius, & Peterson, 2001). Soaking provides a relatively straightforward method for obtaining ligand-bound crystals, in which preformed P450 crystals are exposed to the ligand. This method works well for a number of P450s, including CYP121. However, some P450s, once crystallized, do not allow ligands into their active site, and thus cocrystallization is the only available option (McLean et al., 2009). Soaking can be done on an entire crystal or, if the original crystal is large enough, it can be broken into smaller pieces and a range of ligands screened. Soaking can be achieved by directly dissolving solid ligand to saturation with the crystal, or by the addition of an appropriate ligand stock, for example, by adding 2–5 m*M* ligand in an appropriate solvent, such as DMSO, to the mother liquor (Kavanagh et al., 2016).

As of the year 2000, over 75% of the protein structures deposited into the PDB were ligand bound (Berman, Henrick, & Nakamura, 2003). Ligand-bound structures are essential for our knowledge of substrate/inhibitor interactions with the heme prosthetic group, and to identify which amino acid residues are important to facilitate ligand binding. This has real-world applications, for example, by enabling redesign of a pharmaceutical drug to increase the number of interactions the drug makes in the P450 active site to improve drug efficacy. Another benefit in studying ligand-bound P450 structures is the increased homogeneity of the crystal lattice when focusing on very flexible, dynamic proteins such as cytochromes P450. In some proteins, the presence of the ligand may be crucial for stable crystal formation (Muller, 2017).

In order to limit damage by X-rays, crystals are routinely frozen by plunging in liquid nitrogen, forming vitreous ice around the crystal contained within a cryoprotectant. Most cryoprotectants contain polyethylene glycol (PEG) or oils containing Paratone-N or Perfluoropolyether Cryo Oil (PFO), and cryoprotectant choice often relies on trial and error. A number of P450s bind the relatively small PEG molecules, which can be problematic if aiming for a ligand-free crystal or if the PEG displaces the ligand of interest within the active site. However, cryoprotectant oils do not penetrate the solvent channels and so do not offer the same protection as PEG cryoprotectants. Of note is that certain commercial screens contain PEG or other cryoprotectants, such as the Morpheus or BCS screens from Molecular Dimensions, removing the requirement for the addition of cryoprotectant after crystallogenesis.

7.6 Data Collection and Analysis

The instrument used for data collection will dictate the apparatus used for crystal mounting: glass fibers, cryoloops, pins, or in glass capillaries of varying sizes using cryoprotectants. Crystals may also be trimmed of nonuniform sections. This process requires careful and steady handling to ensure no damage occurs to the fragile protein crystals. Once the protein crystal has been mounted onto a loop and flash frozen in liquid nitrogen with the use of cryoprotectants, it can be stored, transported, and analyzed at an X-ray source, such as a synchrotron.

At a synchrotron or other analysis facility, the crystal is mounted on the diffractometer. Rapid screening will allow elimination of poor quality crystals. Similarly, for good quality crystals, rapid screening will inform on diffraction limits and lattice dimensions. At this stage, if the crystal diffracts to a

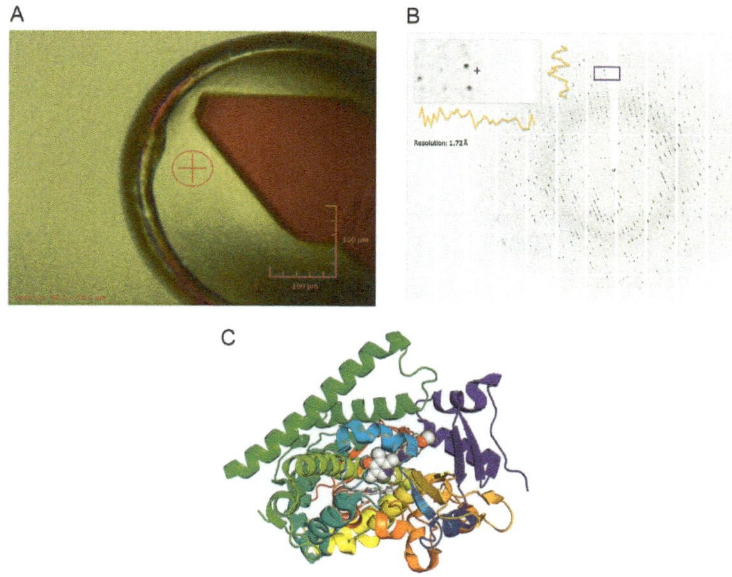

Fig. 10 Stages of crystallographic data collection and analysis. Panel (A) shows a photograph of a BM3 heme domain crystal mounted on a loop at the Diamond Synchrotron directly before analysis. Notice the even color and the clear defined edges. Panel (B) shows the diffraction pattern for a BM3 crystal, produced automatically at the Diamond Synchrotron, Didcot. Panel (C) shows the P450 BM3 A82F/F87V heme domain structure (PDB code 4KEY) with the heme prosthetic group and the ligand (omeprazole) shown in *gray sticks* and *gray spacefill*, respectively.

sufficiently high resolution, a full dataset can be collected. The data are then integrated and scaled, giving result files ready for further processing. For example, molecular replacement (MR), refinement, and rotamer adjustments. MR takes advantage of existing structures to produce accurate models for the new data. Owing to the overwhelming number of P450 structures in the PDB, another P450 structure with a high sequence similarity can often be used to produce a starting model. Refinement is an iterative process where small changes are made to this model in line with the electron density map to produce files with low errors. Programs for such analysis include CCP4 (Winn et al., 2011) or Phenix (Adams et al., 2010). These program suites have a variety of components to fully analyze crystallographic data, produce ligand files, merge data, produce crystallography figures, and more. Rotamers, waters, ligands, and cofactors can be manipulated manually through the use of Coot or other visualization software (Emsley, Lohkamp, Scott, & Cowtan, 2010). Fig. 10 shows key stages from P450 crystals through to structural elucidation. Analyzing X-ray crystallography data is a very lengthy process that requires an in-depth understanding of several programs and so cannot be fully discussed within this chapter. For further reading, please see *Principles of X-ray crystallography* which contains thorough information from sample preparation through to data analysis and structural determination (Ooi, 2010).

8. APPLICATIONS OF P450 ENZYMES IN SYNTHETIC BIOLOGY

8.1 Biotechnological Exploitation of P450s

The P450s catalyze an array of different oxidative and other reactions and are involved in a number of important chemical reactions in eukaryotes and prokaryotes. As examples, P450cam catalyzes the 5-*exo*-hydroxylation of L-camphor to initiate the pathway for its degradation in *Pseudomonas putida* (Katagiri et al., 1968), while P450eryF (CYP107A1) catalyzes the 6*S*-hydroxylation of 6-deoxyerythronolide B (6-DEB) in the first reaction in a pathway that converts 6-DEB to the antibiotic erythromycin (Cupp-Vickery & Poulos, 1995). P450s are also involved in synthesis of several other antibiotics, often by oxidative "tailoring" of a polyketide or other scaffold to produce functional molecules. For instance, CYP107L1 (PikC) from *Streptomyces venezuelae* catalyzes a number of oxidative reactions on both 12- and 14-carbon membered macrolactone rings in order to form different macrolide products. These include the C12 hydroxylation of narbomycin to form the ketolide pikromycin and the C14 hydroxylation of pikromycin to form neopikromycin, as well as dihydroxylation (to a small extent) to form novapikromycin (Lee et al., 2006; Xue, Wilson, Zhao, Liu, & Sherman, 1998). Another example is the antibiotic himastatin, which is a cyclohexadepsipeptide dimer that inhibits growth of methicillin-resistant *Staphylococcus aureus*. The antibiotic is produced by *Streptomyces himastatinicus* ATCC 53653 and involves three different P450-mediated steps. The HmtT P450 catalyzes (i) regio-/stereoselective epoxidation of a C2/C3 double bond of an indole ring (originating from *L*-tryptophan), leading to its cyclization and the formation of a hexahydropyrroloindole moiety. HmtN catalyzes the gamma-hydroxylation of a piperazic acid moiety, while HmtS catalyzes biaryl aromatic coupling of two cyclic depsipeptide monomers, forming the biologically active himastatin dimer (Leet et al., 1996; Ma et al., 2011). *Streptomyces* bacteria contain numerous P450s (e.g., from the CYP105 and CYP107 families) and use these for diverse functions in, e.g., vitamin D hydroxylation (CYP105A1) and in the biosynthesis of biotin

(CYP107H1) and the siderophore coelibactin (CYP105N1) (Cryle & Schlichting, 2008; Lim et al., 2012; Sawada et al., 2004).

Moving away from the "natural" synthetic biology processes in microbes and other organisms, in vitro studies using P450s, as well as expression of P450s in *E. coli* and other bacteria, have enabled production of valuable oxidized products. Studies by Sowden et al. examined the conversion of valencene (found in orange oil) to nootkatone (a valuable fragrance, found in grapefruit juice) using the P450cam and P450 BM3 enzymes (Sowden, Yasmin, Rees, Bell, & Wong, 2005). The P450cam F87A/Y96F double mutant transformed (+)-valencene into (+)-*trans*-nootkatol (70%) and nootkatone (12%) as the major products. A F87L/Y96F mutant increased nootkatone production to 20% of the overall product, while a L244A mutant increased the rate of nootkatone production. The F87V/Y96F/L244A mutant further increased the proportion of nootkatone product to 47%. P450 BM3 mutants R47L/Y51F/F87A and R47L/Y51F/I263A also produced nootkatone, but only at 7.7% and 7.3% of products, indicating lower selectivity and a likelihood of two or more different binding modes in the BM3 mutants. These studies have been extended and work by Luet Wong and others toward enhanced production of nootkatone has continued in the spin-out company Oxford Biotrans in Abingdon, UK.

Recent regulatory advice from the US Food and Drug Administration (FDA) has highlighted the importance of testing of human drug metabolites (particularly those formed in quantities more than 10% of the total steady-state drug exposure) and has led to studies to produce relevant drug metabolites. Examples include the production of various metabolites (including bona fide human metabolites) of gastric proton pump inhibitor drugs (including pravastatin) using the A82F, F87V, and A82F/F87V mutants of P450 BM3 (Butler et al., 2013, 2014). In related work, Kulig et al. fused a PDOR domain to various P450s from *Rhodococcus jostii* RHA1 to create catalytically self-sufficient P450 enzymes that could *N*-demethylate the vasodilatory drug diltiazem and the antidepressant imipramine (Kulig et al., 2015). Kim et al. also reported that wild-type and mutant forms of P450 BM3 could be used to produce the human metabolite piceatannol (3,5,3′,4′-tetrahydroxystilbene) from the anticancer drug *trans*-resveratrol (3,4′,5′-trihydroxystilbene) (Kim, Ahn, Jung, Pan, & Yun, 2009).

An industrial process for the production of the cholesterol-lowering drug pravastatin was developed by Daiichi Sankyo and Bristol-Myers Squibb. Pravastatin (and other statin drugs) inhibit the hydroxymethylglutaryl (HMG) CoA reductase that is the first and rate-limiting step in the cholesterol biosynthesis pathway. A two-step process is used in which compactin is produced by fermentation in *Penicillium citrinum*, and then fed to *Streptomyces carbophilus*, in which P450sca-2 (CYP105A3) catalyzes regio-/stereoselective hydroxylation of compactin to form pravastatin (Sakaki, 2012; Watanabe, Nara, & Serizawa, 1995). Subsequently an alternative, and more productive, process was developed by researchers at DSM (Delft) and the University of Manchester, using *P. chrysogenum* as an efficient production strain. In this process, a P450 enzyme from *A. orientalis* (CYP105AS1) was identified to bind and oxidize compactin. However, the product formed was predominantly the inactive *epi*-pravastatin epimer. The P450 was engineered (using a combination of random and rational evolution) to generate the variant P450prava with five mutations that produced pravastatin in ratios of up to 96:4 pravastatin: *epi*-pravastatin. An esterase activity (that removed the methylbutyrate side chain of compactin) was also identified and the relevant gene deleted from the *P. chrysogenum* genome. At pilot scale production, >6 g/L pravastatin were formed, with less than 0.5 g/L compactin and deacylated statins recovered (McLean, Hans, et al., 2015). In other studies, the *B. megaterium* CYP109E1 was also shown to convert compactin to pravastatin, as well as catalyzing regioselective oxidation of lovastatin (to 6′-β-hydroxy-lovastatin) and simvastatin (to 3′-α-hydroxy-simvastatin) (Putkaradze et al., 2017).

Other important synthetic biology processes involving P450 enzymes include the production of oxygenated taxanes at a level of ~570 mg/L in an engineered *E. coli* strain that also hosts the components of the methylerythritol phosphate pathway that produces the isoprenoid precursors isopentenyl pyrophosphate and dimethylallyl pyrophosphate. These molecules were then polymerized and cyclized by GGPPS (geranylgeranyl pyrophosphate synthase) and taxadiene synthase to form taxa-4(*S*), 11 (12)-diene. At this stage, the P450 (CYP725A4 from *T. cuspidata*, Japanese yew) and its CPR partner (also from *T. cuspidata*) catalyze the conversion of taxadiene to taxadien-5α-ol and other oxygenated taxanes. Both the P450 and the CPR were expressed as N-terminally truncated forms to enhance their solubility (Biggs et al., 2016). A P450 also features prominently in an engineered pathway in *S. cerevisiae* in which the sesquiterpene lactone and antimalarial drug artemisinin is produced by chemical conversion from artemisinic acid. Early steps in the pathway involve the conversion of acetyl-CoA through to amorphadiene. Key steps in the pathway are then performed by the *Artemisia annua* (Sweet wormwood) CYP71AV1, which converts amorphadiene to artemisinic alcohol (by hydroxylation) and then further oxidizes this compound to the aldehyde, and then on to artemisinic acid (Paddon et al., 2013). Finally, the opiate morphine was also synthesized in *S. cerevisiae.* The pathway involved uses the reticuline epimerase from the opium poppy (*Papaver somniferum*), which is a fusion protein of a P450 (CYP82Y2) to an aldo-keto reductase (AKR). The P450 (1,2-dehydroreticuline synthase) catalyzes the transformation of (*S*)-reticuline to 1,2-dehydroreticuline, followed by conversion of this product to (*R*)-reticuline by the AKR domain. Another P450 (CYP719B1) subsequently catalyzes the formation of salutaridine from (*R*)-reticuline, involving a C—C bond coupling reaction step performed by CYP719B1 (Farrow, Hagel, Beaudoin, Burns, & Facchini, 2015; Galanie, Thodey, Trenchard, Filsinger Interrante, & Smolke, 2015; Gesell et al., 2009).

9. SUMMARY AND CONCLUSION

The cytochrome P450 enzymes perform countless different reactions in the vast majority of living organisms. They have crucial roles in organisms including mammals (e.g., steroid biogenesis, lipid transformations and production of eicosanoids, and xenobiotic metabolism), bacteria (e.g., synthesis of antibiotics, antitumor drugs, immunosuppressants, and antifungals), and plants (e.g. production of monoterpenes, diterpenes, triterpenes, and sesquiterpene lactones such as artemensinin; and alkaloids including nicotine, caffeine, and atropine). The advent of high-throughput genome sequencing has led to the identification of huge numbers of cytochrome P450 genes and has enabled research into the structure and function of these P450s, revealing important and/or novel functions such as (i) human host cholesterol degradation [CYP125A1 and CYP142A1] and omega-hydroxylation of menaquinone [CYP128A1] by *M. tuberculosis* P450s; (ii) reductive degradation of the explosive RDX by the XplA P450 from *R. rhodochrous*; and (iii) key roles in the production of plant defense molecules, including CYP79A1 from *Sorghum bicolor* that contributes to the production of the cyanogenic glucoside dhurrin, and the CYP79D subfamily P450s in cassava and in *Lotus japonicus* that participates in the biosynthesis of the cyanogenic glucosides linamarin and lotaustralin, respectively.

Ongoing research has identified novel P450s, their redox partners and biological functions, helping to build a growing database of knowledge on the diverse roles of these enzymes and their potential biotechnological applications. Of particular interest are the catalytically self-sufficient P450 enzymes that are naturally fused to redox partners and require only NAD(P)H, dioxygen and the P450 substrate for activity. The most prominent member of this class is the *B. megaterium* P450 BM3 enzyme, which has been extensively interrogated and engineered to diversify its substrate scope and to enable the enzyme to oxidize nonnatural substrates including steroids, pharmaceuticals, and alkanes. BM3 is recognized as a biotechnologically important enzyme, along with enzymes such as XplA and members of the bacterial CYP152 family (including OleT) that can oxidatively decarboxylate fatty acids to produce terminal alkenes. P450s have already been used effectively in a number of synthetic biology applications, including being key components of pathways that have produced important molecules such as artemisinic acid, morphine, oxygenated taxanes, and pravastatin. The utility of the P450s in such synthetic schemes is now well recognized, particularly in view of the broad range of activities offered by different P450s and since it is now clearly proven that P450s can be engineered efficiently using both rational mutagenesis and directed evolution approaches to produce variants with desired activities. P450s are now accepted as key components in synthetic biology research programs that can perform crucial oxygen insertion (and many other) reactions in pathways to the production of valuable biomolecules.

REFERENCES

Acevedo-Rocha, C. G., Gamble, C. G., Lonsdale, R., Li, A., Nett, N., Hoebenreich, S., et al. (2018). P450-catalyzed regio- and diastereoselective steroid hydroxylation: Efficient directed evolution enabled by mutability landscaping. *ACS Catalysis*, *8*(4), 3395–3410. https://doi.org/10.1021/acscatal.8b00389.

Adams, P. D., Afonine, P. V., Bunkoczi, G., Chen, V. B., Davis, I. W., Echols, N., et al. (2010). PHENIX: A comprehensive python-based system for macromolecular structure solution. *Acta Crystallographica. Section D, Biological Crystallography*, *66*(Pt. 2), 213–221. https://doi.org/10.1107/S0907444909052925.

Adas, F., Berthou, F., Picart, D., Lozac'h, P., Beauge, F., & Amet, Y. (1998). Involvement of cytochrome P450 2E1 in the (omega-1)-hydroxylation of oleic acid in human and rat liver microsomes. *Journal of Lipid Research*, *39*(6), 1210–1219.

Almeida, C. C., Romao, C. V., Lindley, P. F., Teixeira, M., & Saraiva, L. M. (2006). The role of the hybrid cluster protein in oxidative stress defense. *Journal of Biological Chemistry*, *281*(43), 32445–32450. https://doi.org/10.1074/jbc.M605888200.

Barnes, H. J., Arlotto, M. P., & Waterman, M. R. (1991). Expression and enzymatic activity of recombinant cytochrome P450 17 alpha-hydroxylase in *Escherichia coli*. *Proceedings of the National Academy of Sciences of the United States of America*, *88*(13), 5597–5601.

Barr, I., & Guo, F. (2015). Pyridine hemochromagen assay for determining the concentration of heme in purified protein solutions. *Bio-protocol*, *5*(18), e1594.

Bart, A. G., & Scott, E. E. (2017). Structural and functional effects of cytochrome b(5) interactions with human cytochrome P450 enzymes. *Journal of Biological Chemistry*, *292*(51), 20818–20833. https://doi.org/10.1074/jbc.RA117.000220.

Belcher, J., McLean, K. J., Matthews, S., Woodward, L. S., Fisher, K., Rigby, S. E. J., et al. (2014). Structure and biochemical properties of the alkene producing cytochrome P450

OleTJE (CYP152L1) from the Jeotgalicoccus sp 8456 bacterium. *Journal of Biological Chemistry*, *289*(10), 6535–6550. https://doi.org/10.1074/jbc.M113.527325.

Berman, H., Henrick, K., & Nakamura, H. (2003). Announcing the worldwide Protein Data Bank. *Nature Structural Biology*, *10*(12), 980. https://doi.org/10.1038/nsb1203-980.

Berry, E. A., & Trumpower, B. L. (1987). Simultaneous determination of hemes a, b, and c from pyridine hemochrome spectra. *Analytical Biochemistry*, *161*(1), 1–15.

Bianchi, V., Reichard, P., Eliasson, R., Pontis, E., Krook, M., Jornvall, H., et al. (1993). *Escherichia coli* ferredoxin $NADP^+$ reductase: Activation of *E. coli* anaerobic ribonucleotide reduction, cloning of the gene (fpr), and overexpression of the protein. *Journal of Bacteriology*, *175*(6), 1590–1595.

Bidstrup, T. B., Bjornsdottir, I., Sidelmann, U. G., Thomsen, M. S., & Hansen, K. T. (2003). CYP2C8 and CYP3A4 are the principal enzymes involved in the human in vitro biotransformation of the insulin secretagogue repaglinide. *British Journal of Clinical Pharmacology*, *56*(3), 305–314.

Biggs, B. W., Lim, C. G., Sagliani, K., Shankar, S., Stephanopoulos, G., De Mey, M., et al. (2016). Overcoming heterologous protein interdependency to optimize P450-mediated Taxol precursor synthesis in *Escherichia coli*. *Proceedings of the National Academy of Sciences of the United States of America*, *113*(12), 3209–3214. https://doi.org/10.1073/pnas.1515826113.

Bleif, S., Hannemann, F., Zapp, J., Hartmann, D., Jauch, J., & Bernhardt, R. (2012). A new Bacillus megaterium whole-cell catalyst for the hydroxylation of the pentacyclic triterpene 11-keto-beta-boswellic acid (KBA) based on a recombinant cytochrome P450 system. *Applied Microbiology and Biotechnology*, *93*(3), 1135–1146. https://doi.org/10.1007/s00253-011-3467-0.

Bragg, W. H. (1914). X-rays and crystalline structure. *Science*, *40*(1040), 795–802. https://doi.org/10.1126/science.40.1040.795.

Bragg, W. H. (1924). The analysis of crystal structure by X-rays. *Science*, *60*(1546), 139–149. https://doi.org/10.1126/science.60.1546.139.

Brodhun, F., & Feussner, I. (2011). Oxylipins in fungi. *FEBS Journal*, *278*(7), 1047–1063. https://doi.org/10.1111/j.1742-4658.2011.08027.x.

Brodhun, F., Gobel, C., Hornung, E., & Feussner, I. (2009). Identification of PpoA from Aspergillus nidulans as a fusion protein of a fatty acid heme dioxygenase/peroxidase and a cytochrome P450. *Journal of Biological Chemistry*, *284*(18), 11792–11805. https://doi.org/10.1074/jbc.M809152200.

Budriang, C., Rongnoparut, P., & Yuvaniyama, J. (2011). An expression of an insect membrane-bound cytochrome P450 CYP6AA3 in the Escherichia coli in relation to insecticide resistance in a malarial vector. *Pakistan Journal of Biological Sciences*, *14*(8), 466–475.

Bui, S. H., McLean, K. J., Cheesman, M. R., Bradley, J. M., Rigby, S. E., Levy, C. W., et al. (2012). Unusual spectroscopic and ligand binding properties of the cytochrome P450-flavodoxin fusion enzyme XplA. *Journal of Biological Chemistry*, *287*(23), 19699–19714. https://doi.org/10.1074/jbc.M111.319202.

Butler, C. F., Peet, C., Mason, A. E., Voice, M. W., Leys, D., & Munro, A. W. (2013). Key mutations alter the cytochrome P450 BM3 conformational landscape and remove inherent substrate bias. *Journal of Biological Chemistry*, *288*(35), 25387–25399. https://doi.org/10.1074/jbc.M113.479717.

Butler, C. F., Peet, C., McLean, K. J., Baynham, M. T., Blankley, R. T., Fisher, K., et al. (2014). Human P450-like oxidation of diverse proton pump inhibitor drugs by 'gatekeeper' mutants of flavocytochrome P450 BM3. *Biochemical Journal*, *460*(2), 247–259. https://doi.org/10.1042/bj20140030.

Cheesman, M. J., Baer, B. R., Zheng, Y. M., Gillam, E. M., & Rettie, A. E. (2003). Rabbit CYP4B1 engineered for high-level expression in *Escherichia coli*: Ligand stabilization and processing of the N-terminus and heme prosthetic group. *Archives of Biochemistry and Biophysics*, *416*(1), 17–24.

Cheesman, M. J., Traylor, M. J., Hilton, M. E., Richards, K. E., Taylor, M. C., Daborn, P. J., et al. (2013). Soluble and membrane-bound *Drosophila melanogaster* CYP6G1 expressed in *Escherichia coli*: Purification, activity, and binding properties toward multiple pesticides. *Insect Biochemistry and Molecular Biology*, *43*(5), 455–465. https://doi.org/10.1016/j.ibmb.2013.02.003.

Correll, C. C., Batie, C. J., Ballou, D. P., & Ludwig, M. L. (1992). Phthalate dioxygenase reductase: A modular structure for electron transfer from pyridine nucleotides to [2Fe-2S]. *Science*, *258*(5088), 1604–1610.

Cosme, J., & Johnson, E. F. (2000). Engineering microsomal cytochrome P450 2C5 to be a soluble, monomeric enzyme. Mutations that alter aggregation, phospholipid dependence of catalysis, and membrane binding. *Journal of Biological Chemistry*, *275*(4), 2545–2553.

Cryle, M. J., & Schlichting, I. (2008). Structural insights from a P450 carrier protein complex reveal how specificity is achieved in the P450(BioI) ACP complex. *Proceedings of the National Academy of Sciences of the United States of America*, *105*(41), 15696–15701. https://doi.org/10.1073/pnas.0805983105.

Cupp-Vickery, J. R., & Poulos, T. L. (1995). Structure of cytochrome P450eryF involved in erythromycin biosynthesis. *Nature Structural Biology*, *2*(2), 144–153.

Daff, S. N., Chapman, S. K., Turner, K. L., Holt, R. A., Govindaraj, S., Poulos, T. L., et al. (1997). Redox control of the catalytic cycle of flavocytochrome P-450 BM3. *Biochemistry*, *36*(45), 13816–13823. https://doi.org/10.1021/bi971085s.

Degregorio, D., D'Avino, S., Castrignano, S., Di Nardo, G., Sadeghi, S. J., Catucci, G., et al. (2017). Human cytochrome P450 3A4 as a biocatalyst: Effects of the engineered linker in modulation of coupling efficiency in 3A4-BMR chimeras. *Frontiers in Pharmacology*, *8*, 121. https://doi.org/10.3389/fphar.2017.00121.

Denisov, I. G., Makris, T. M., Sligar, S. G., & Schlichting, I. (2005). Structure and chemistry of cytochrome P450. *Chemical Reviews*, *105*(6), 2253–2277. https://doi.org/10.1021/cr0307143.

Dennig, A., Kuhn, M., Tassoti, S., Thiessenhusen, A., Gilch, S., Bulter, T., et al. (2015). Oxidative decarboxylation of short-chain fatty acids to 1-alkenes. *Angewandte Chemie International Edition in English*, *54*(30), 8819–8822. https://doi.org/10.1002/anie.201502925.

DeVore, N. M., & Scott, E. E. (2012). Structures of cytochrome P450 17A1 with prostate cancer drugs abiraterone and TOK-001. *Nature*, *482*, 116. https://doi.org/10.1038/nature10743.

Dong, M. S., Yamazaki, H., Guo, Z., & Guengerich, F. P. (1996). Recombinant human cytochrome P450 1A2 and an N-terminal-truncated form: Construction, purification, aggregation properties, and interactions with flavodoxin, ferredoxin, and NADPH-cytochrome P450 reductase. *Archives of Biochemistry and Biophysics*, *327*(1), 11–19. https://doi.org/10.1006/abbi.1996.0086.

Driscoll, M. D., McLean, K. J., Levy, C., Mast, N., Pikuleva, I. A., Lafite, P., et al. (2010). Structural and biochemical characterization of Mycobacterium tuberculosis CYP142: Evidence for multiple cholesterol 27-hydroxylase activities in a human pathogen. *Journal of Biological Chemistry*, *285*(49), 38270–38282. https://doi.org/10.1074/jbc.M110.164293.

Dumon-Seignovert, L., Cariot, G., & Vuillard, L. (2004). The toxicity of recombinant proteins in *Escherichia coli:* A comparison of overexpression in BL21(DE3), C41(DE3), and C43(DE3). *Protein Expression and Purification*, *37*(1), 203–206. https://doi.org/10.1016/j.pep.2004.04.025.

Dunford, A. J., McLean, K. J., Sabri, M., Seward, H. E., Heyes, D. J., Scrutton, N. S., et al. (2007). Rapid P450 heme iron reduction by laser photoexcitation of Mycobacterium tuberculosis CYP121 and CYP51B1. Analysis of CO complexation reactions and reversibility of the P450/P420 equilibrium. *Journal of Biological Chemistry, 282*(34), 24816–24824. https://doi.org/10.1074/jbc.M702958200.

Emsley, P., Lohkamp, B., Scott, W. G., & Cowtan, K. (2010). Features and development of coot. *Acta Crystallographica. Section D, Biological Crystallography, 66*(Pt. 4), 486–501. https://doi.org/10.1107/S0907444910007493.

Ewen, K. M., Kleser, M., & Bernhardt, R. (2011). Adrenodoxin: The archetype of vertebrate-type [2Fe-2S] cluster ferredoxins. *Biochimica et Biophysica Acta, 1814*(1), 111–125. https://doi.org/10.1016/j.bbapap.2010.06.003.

Ewen, K. M., Ringle, M., & Bernhardt, R. (2012). Adrenodoxin—A versatile ferredoxin. *IUBMB Life, 64*(6), 506–512. https://doi.org/10.1002/iub.1029.

Fang, B., Xu, H., Liu, Y., Qi, F., Zhang, W., Chen, H., et al. (2017). Mutagenesis and redox partners analysis of the P450 fatty acid decarboxylase OleTJE. *Scientific Reports, 7*, 44258. https://doi.org/10.1038/srep44258.

Farrow, S. C., Hagel, J. M., Beaudoin, G. A., Burns, D. C., & Facchini, P. J. (2015). Stereochemical inversion of (S)-reticuline by a cytochrome P450 fusion in opium poppy. *Nature Chemical Biology, 11*(9), 728–732. https://doi.org/10.1038/nchembio.1879.

Filenko, N., Spiro, S., Browning, D. F., Squire, D., Overton, T. W., Cole, J., et al. (2007). The NsrR regulon of *Escherichia coli* K-12 includes genes encoding the hybrid cluster protein and the periplasmic, respiratory nitrite reductase. *Journal of Bacteriology, 189*(12), 4410–4417. https://doi.org/10.1128/jb.00080-07.

Funhoff, E. G., Bauer, U., Garcia-Rubio, I., Witholt, B., & van Beilen, J. B. (2006). CYP153A6, a soluble P450 oxygenase catalyzing terminal-alkane hydroxylation. *Journal of Bacteriology, 188*(14), 5220–5227. https://doi.org/10.1128/JB.00286-06.

Galanie, S., Thodey, K., Trenchard, I. J., Filsinger Interrante, M., & Smolke, C. D. (2015). Complete biosynthesis of opioids in yeast. *Science, 349*(6252), 1095–1100. https://doi.org/10.1126/science.aac9373.

Gao, H., Jacobs, A., White, R. E., Booth, B. P., & Obach, R. S. (2013). Meeting report: Metabolites in safety testing (MIST) symposium-safety assessment of human metabolites: What's REALLY necessary to ascertain exposure coverage in safety tests? *The AAPS Journal, 15*(4), 970–973. https://doi.org/10.1208/s12248-013-9502-6.

Garfinkel, D. (1958). Studies on pig liver microsomes. I. Enzymic and pigment composition of different microsomal fractions. *Archives of Biochemistry and Biophysics*, 77(2), 493–509.

Gesell, A., Rolf, M., Ziegler, J., Diaz Chavez, M. L., Huang, F. C., & Kutchan, T. M. (2009). CYP719B1 is salutaridine synthase, the C–C phenol-coupling enzyme of morphine biosynthesis in opium poppy. *Journal of Biological Chemistry, 284*(36), 24432–24442. https://doi.org/10.1074/jbc.M109.033373.

Gibson, D. G., Young, L., Chuang, R. Y., Venter, J. C., Hutchison, C. A., 3rd, & Smith, H. O. (2009). Enzymatic assembly of DNA molecules up to several hundred kilobases. *Nature Methods, 6*(5), 343–345. https://doi.org/10.1038/nmeth.1318.

Gillam, E. M., Baba, T., Kim, B. R., Ohmori, S., & Guengerich, F. P. (1993). Expression of modified human cytochrome P450 3A4 in *Escherichia coli* and purification and reconstitution of the enzyme. *Archives of Biochemistry and Biophysics, 305*(1), 123–131. https://doi.org/10.1006/abbi.1993.1401.

Girhard, M., Kunigk, E., Tihovsky, S., Shumyantseva, V. V., & Urlacher, V. B. (2013). Light-driven biocatalysis with cytochrome P450 peroxygenases. *Biotechnology and Applied Biochemistry, 60*(1), 111–118. https://doi.org/10.1002/bab.1063.

Girvan, H. M., Dunford, A. J., Neeli, R., Ekanem, I. S., Waltham, T. N., Joyce, M. G., et al. (2011). Flavocytochrome P450 BM3 mutant W1046A is a NADH-dependent fatty acid hydroxylase: Implications for the mechanism of electron transfer in the P450 BM3 dimer. *Archives of Biochemistry and Biophysics, 507*(1), 75–85. https://doi.org/10.1016/j.abb.2010.09.014.

Girvan, H. M., Marshall, K. R., Lawson, R. J., Leys, D., Joyce, M. G., Clarkson, J., et al. (2004). Flavocytochrome P450BM3 mutant A264E undergoes substrate-dependent formation of a novel heme iron ligand set. *Journal of Biological Chemistry*, *279*(22), 23274–23286. https://doi.org/10.1074/jbc.M401716200.
Glieder, A., Farinas, E. T., & Arnold, F. H. (2002). Laboratory evolution of a soluble, self-sufficient, highly active alkane hydroxylase. *Nature Biotechnology*, *20*(11), 1135–1139. https://doi.org/10.1038/nbt744.
Govindaraj, S., & Poulos, T. L. (1997). The domain architecture of cytochrome P450BM-3. *Journal of Biological Chemistry*, *272*(12), 7915–7921. https://doi.org/10.1074/jbc.272.12.7915.
Grant, J. L., Hsieh, C. H., & Makris, T. M. (2015). Decarboxylation of fatty acids to terminal alkenes by cytochrome P450 compound I. *Journal of the American Chemical Society*, *137*(15), 4940–4943. https://doi.org/10.1021/jacs.5b01965.
Grant, J. L., Mitchell, M. E., & Makris, T. M. (2016). Catalytic strategy for carbon–carbon bond scission by the cytochrome P450 OleT. *Proceedings of the National Academy of Sciences of the United States of America*, *113*(36), 10049–10054. https://doi.org/10.1073/pnas.1606294113.
Guengerich, F. P. (2003). Cytochromes P450, drugs, and diseases. *Molecular Interventions*, *3*(4), 194–204. https://doi.org/10.1124/mi.3.4.194.
Guengerich, F. P. (2015). Human cytochrome P450 enzymes. In P. R. Ortiz de Montellano (Ed.), *Cytochrome P450: Structure, mechanism, and biochemistry* (4th ed., pp. 532–785). Springer International Publishing.
Guengerich, F. P., Martin, M. V., Sohl, C. D., & Cheng, Q. (2009). Measurement of cytochrome P450 and NADPH-cytochrome P450 reductase. *Nature Protocols*, *4*(9), 1245–1251. https://doi.org/10.1038/nprot.2009.121.
Haines, D. C., Tomchick, D. R., Machius, M., & Peterson, J. A. (2001). Pivotal role of water in the mechanism of P450BM-3. *Biochemistry*, *40*(45), 13456–13465.
Hanai, T., Atsumi, S., & Liao, J. C. (2007). Engineered synthetic pathway for isopropanol production in Escherichia coli. *Applied and Environmental Microbiology*, *73*(24), 7814–7818. https://doi.org/10.1128/aem.01140-07.
Hanna, I. H., Reed, J. R., Guengerich, F. P., & Hollenberg, P. F. (2000). Expression of human cytochrome P450 2B6 in *Escherichia coli*: Characterization of catalytic activity and expression levels in human liver. *Archives of Biochemistry and Biophysics*, *376*(1), 206–216. https://doi.org/10.1006/abbi.2000.1708.
Hausjell, J., Halbwirth, H., & Spadiut, O. (2018). Recombinant production of eukaryotic cytochrome P450s in microbial cell factories. *Bioscience Reports*, *38*(2). https://doi.org/10.1042/bsr20171290.
Hill, A. V. (1910). The possible effects of the aggregation of the molecules of haemoglobin on its dissociation curves. *Journal of Physiology*, *40*(Suppl), 4–7.
Hosea, N. A., Miller, G. P., & Guengerich, F. P. (2000). Elucidation of distinct ligand binding sites for cytochrome P450 3A4. *Biochemistry*, *39*(20), 5929–5939. https://doi.org/10.1021/bi992765t.
Hsieh, C. H., Huang, X. Y., Amaya, J. A., Rutland, C. D., Keys, C. L., Groves, J. T., et al. (2017). The enigmatic P450 decarboxylase OleT is capable of, but evolved to frustrate, oxygen rebound chemistry. *Biochemistry*, *56*(26), 3347–3357. https://doi.org/10.1021/acs.biochem.7b00338.
Ikeda, M., Arai, M., Okuno, T., & Shimizu, T. (2003). TMPDB: A database of experimentally-characterized transmembrane topologies. *Nucleic Acids Research*, *31*(1), 406–409.
Jackson, C. J., Lamb, D. C., Marczylo, T. H., Warrilow, A. G., Manning, N. J., Lowe, D. J., et al. (2002). A novel sterol 14alpha-demethylase/ferredoxin fusion protein (MCCYP51FX) from Methylococcus capsulatus represents a new class of the

cytochrome P450 superfamily. *Journal of Biological Chemistry, 277*(49), 46959–46965. https://doi.org/10.1074/jbc.M203523200.

Jackson, R. G., Rylott, E. L., Fournier, D., Hawari, J., & Bruce, N. C. (2007). Exploring the biochemical properties and remediation applications of the unusual explosive-degrading P450 system XplA/B. *Proceedings of the National Academy of Sciences of the United States of America, 104*(43), 16822–16827. https://doi.org/10.1073/pnas.0705110104.

Jenkins, C. M., & Waterman, M. R. (1994). Flavodoxin and NADPH-flavodoxin reductase from Escherichia coli support bovine cytochrome P450c17 hydroxylase activities. *Journal of Biological Chemistry, 269*(44), 27401–27408.

Jensen, K., Jensen, P. E., & Moller, B. L. (2011). Light-driven cytochrome p450 hydroxylations. *ACS Chemical Biology, 6*(6), 533–539. https://doi.org/10.1021/cb100393j.

Joyce, M. G., Ekanem, I. S., Roitel, O., Dunford, A. J., Neeli, R., Girvan, H. M., et al. (2012). The crystal structure of the FAD/NADPH-binding domain of flavocytochrome P450 BM3. *FEBS Journal, 279*(9), 1694–1706. https://doi.org/10.1111/j.1742-4658.2012.08544.x.

Kapust, R. B., Tozser, J., Fox, J. D., Anderson, D. E., Cherry, S., Copeland, T. D., et al. (2001). Tobacco etch virus protease: Mechanism of autolysis and rational design of stable mutants with wild-type catalytic proficiency. *Protein Engineering, 14*(12), 993–1000.

Katagiri, M., Ganguli, B. N., & Gunsalus, I. C. (1968). A soluble cytochrome P-450 functional in methylene hydroxylation. *Journal of Biological Chemistry, 243*(12), 3543–3546.

Kavanagh, M. E., Coyne, A. G., McLean, K. J., James, G. G., Levy, C. W., Marino, L. B., et al. (2016). Fragment-based approaches to the development of *Mycobacterium tuberculosis* CYP121 inhibitors. *Journal of Medicinal Chemistry, 59*(7), 3272–3302. https://doi.org/10.1021/acs.jmedchem.6b00007.

Kendrew, J. C., Bodo, G., Dintzis, H. M., Parrish, R. G., Wyckoff, H., & Phillips, D. C. (1958). A three-dimensional model of the myoglobin molecule obtained by x-ray analysis. *Nature, 181*(4610), 662–666.

Kille, S., Zilly, F. E., Acevedo, J. P., & Reetz, M. T. (2011). Regio- and stereoselectivity of P450-catalysed hydroxylation of steroids controlled by laboratory evolution. *Nature Chemistry, 3*(9), 738–743. https://doi.org/10.1038/nchem.1113.

Kim, D. H., Ahn, T., Jung, H. C., Pan, J. G., & Yun, C. H. (2009). Generation of the human metabolite piceatannol from the anticancer-preventive agent resveratrol by bacterial cytochrome P450 BM3. *Drug Metabolism and Disposition, 37*(5), 932–936. https://doi.org/10.1124/dmd.108.026484.

Klingenberg, M. (1958). Pigments of rat liver microsomes. *Archives of Biochemistry and Biophysics, 75*(2), 376–386.

Kost, T. A., & Kemp, C. W. (2016). Fundamentals of baculovirus expression and applications. *Advances in Experimental Medicine and Biology, 896*, 187–197. https://doi.org/10.1007/978-3-319-27216-0_12.

Krest, C. M., Onderko, E. L., Yosca, T. H., Calixto, J. C., Karp, R. F., Livada, J., et al. (2013). Reactive intermediates in cytochrome p450 catalysis. *Journal of Biological Chemistry, 288*(24), 17074–17081. https://doi.org/10.1074/jbc.R113.473108.

Kulig, J. K., Spandolf, C., Hyde, R., Ruzzini, A. C., Eltis, L. D., Gronberg, G., et al. (2015). A P450 fusion library of heme domains from Rhodococcus jostii RHA1 and its evaluation for the biotransformation of drug molecules. *Bioorganic & Medicinal Chemistry, 23*(17), 5603–5609. https://doi.org/10.1016/j.bmc.2015.07.025.

Kuper, J., Wong, T. S., Roccatano, D., Wilmanns, M., & Schwaneberg, U. (2007). Understanding a mechanism of organic cosolvent inactivation in heme monooxygenase P450 BM-3. *Journal of the American Chemical Society, 129*(18), 5786–5787. https://doi.org/10.1021/ja067036x.

Lee, S. K., Park, J. W., Kim, J. W., Jung, W. S., Park, S. R., Choi, C. Y., et al. (2006). Neopikromycin and novapikromycin from the pikromycin biosynthetic pathway of

Streptomyces venezuelae. *Journal of Natural Products*, *69*(5), 847–849. https://doi.org/10.1021/np060026p.

Lee, D. S., Yamada, A., Sugimoto, H., Matsunaga, I., Ogura, H., Ichihara, K., et al. (2003). Substrate recognition and molecular mechanism of fatty acid hydroxylation by cytochrome P450 from Bacillus subtilis. Crystallographic, spectroscopic, and mutational studies. *Journal of Biological Chemistry*, *278*(11), 9761–9767. https://doi.org/10.1074/jbc.M211575200.

Leet, J. E., Schroeder, D. R., Golik, J., Matson, J. A., Doyle, T. W., Lam, K. S., et al. (1996). Himastatin, a new antitumor antibiotic from *Streptomyces hygroscopicus*. III. Structural elucidation. *Journal of Antibiotics (Tokyo)*, *49*(3), 299–311.

Leys, D., Mowat, C. G., McLean, K. J., Richmond, A., Chapman, S. K., Walkinshaw, M. D., et al. (2003). Atomic structure of Mycobacterium tuberculosis CYP121 to 1.06 A reveals novel features of cytochrome P450. *Journal of Biological Chemistry*, *278*(7), 5141–5147. https://doi.org/10.1074/jbc.M209928200.

Lim, Y. R., Hong, M. K., Kim, J. K., Doan, T. T., Kim, D. H., Yun, C. H., et al. (2012). Crystal structure of cytochrome P450 CYP105N1 from *Streptomyces coelicolor*, an oxidase in the coelibactin siderophore biosynthetic pathway. *Archives of Biochemistry and Biophysics*, *528*(2), 111–117. https://doi.org/10.1016/j.abb.2012.09.001.

Lu, A. Y., Junk, K. W., & Coon, M. J. (1969). Resolution of the cytochrome P-450-containing omega-hydroxylation system of liver microsomes into three components. *Journal of Biological Chemistry*, *244*(13), 3714–3721.

Luthra, A., Denisov, I. G., & Sligar, S. G. (2011). Spectroscopic features of cytochrome P450 reaction intermediates. *Archives of Biochemistry and Biophysics*, *507*(1), 26–35. https://doi.org/10.1016/j.abb.2010.12.008.

Ma, L., Du, L., Chen, H., Sun, Y., Huang, S., Zheng, X., et al. (2015). Reconstitution of the in vitro activity of the cyclosporine-specific P450 hydroxylase from Sebekia benihana and development of a heterologous whole-cell biotransformation system. *Applied and Environmental Microbiology*, *81*(18), 6268–6275. https://doi.org/10.1128/AEM.01353-15.

Ma, J., Wang, Z., Huang, H., Luo, M., Zuo, D., Wang, B., et al. (2011). Biosynthesis of himastatin: Assembly line and characterization of three cytochrome P450 enzymes involved in the post-tailoring oxidative steps. *Angewandte Chemie International Edition in English*, *50*(34), 7797–7802. https://doi.org/10.1002/anie.201102305.

Matsunaga, I., Yamada, M., Kusunose, E., Nishiuchi, Y., Yano, I., & Ichihara, K. (1996). Direct involvement of hydrogen peroxide in bacterial alpha-hydroxylation of fatty acid. *FEBS Letters*, *386*(2–3), 252–254.

Matthews, S., Belcher, J. D., Tee, K. L., Girvan, H. M., McLean, K. J., Rigby, S. E., et al. (2017). Catalytic determinants of alkene production by the cytochrome P450 peroxygenase OleTJE. *Journal of Biological Chemistry*, *292*(12), 5128–5143. https://doi.org/10.1074/jbc.M116.762336.

Matthews, S., Tee, K. L., Rattray, N. J., McLean, K. J., Leys, D., Parker, D. A., et al. (2017). Production of alkenes and novel secondary products by P450 OleT(JE) using novel H_2O_2-generating fusion protein systems. *FEBS Letters*, *591*(5), 737–750. https://doi.org/10.1002/1873-3468.12581.

McIver, L., Leadbeater, C., Campopiano, D. J., Baxter, R. L., Daff, S. N., Chapman, S. K., et al. (1998). Characterisation of flavodoxin NADP+ oxidoreductase and flavodoxin; key components of electron transfer in *Escherichia coli*. *European Journal of Biochemistry*, *257*(3), 577–585.

McLean, K. J., Carroll, P., Lewis, D. G., Dunford, A. J., Seward, H. E., Neeli, R., et al. (2008). Characterization of active site structure in CYP121. A cytochrome P450 essential for viability of Mycobacterium tuberculosis H37Rv. *Journal of Biological Chemistry*, *283*(48), 33406–33416. https://doi.org/10.1074/jbc.M802115200.

McLean, K. J., Hans, M., Meijrink, B., van Scheppingen, W. B., Vollebregt, A., Tee, K. L., et al. (2015). Single-step fermentative production of the cholesterol-lowering drug pravastatin via reprogramming of Penicillium chrysogenum. *Proceedings of the National Academy of Sciences of the United States of America, 112*(9), 2847–2852. https://doi.org/10.1073/pnas.1419028112.

McLean, K. J., Lafite, P., Levy, C., Cheesman, M. R., Mast, N., Pikuleva, I. A., et al. (2009). The structure of Mycobacterium tuberculosis CYP125: Molecular basis for cholesterol binding in a P450 needed for host infection. *Journal of Biological Chemistry, 284*(51), 35524–35533. https://doi.org/10.1074/jbc.M109.032706.

McLean, K. J., Luciakova, D., Belcher, J., Tee, K. L., & Munro, A. W. (2015). Biological diversity of cytochrome P450 redox partner systems. *Advances in Experimental Medicine and Biology, 851*, 299–317. https://doi.org/10.1007/978-3-319-16009-2_11.

Michaelis, L., & Menten, M. L. (1913). The kenetics of the inversion effect. *Biochemische Zeitschrift, 49*, 333–369.

Miles, J. S., Munro, A. W., Rospendowski, B. N., Smith, W. E., McKnight, J., & Thomson, A. J. (1992). Domains of the catalytically self-sufficient cytochrome P-450 BM-3. Genetic construction, overexpression, purification and spectroscopic characterization. *Biochemical Journal, 288*(Pt. 2), 503–509.

Minerdi, D., Sadeghi, S. J., Di Nardo, G., Rua, F., Castrignano, S., Allegra, P., et al. (2015). CYP116B5: A new class VII catalytically self-sufficient cytochrome P450 from *Acinetobacter radioresistens* that enables growth on alkanes. *Molecular Microbiology, 95*(3), 539–554. https://doi.org/10.1111/mmi.12883.

Monk, B. C., Tomasiak, T. M., Keniya, M. V., Huschmann, F. U., Tyndall, J. D., O'Connell, J. D., 3rd, et al. (2014). Architecture of a single membrane spanning cytochrome P450 suggests constraints that orient the catalytic domain relative to a bilayer. *Proceedings of the National Academy of Sciences of the United States of America, 111*(10), 3865–3870. https://doi.org/10.1073/pnas.1324245111.

Morrison, J. F. (1969). Kinetics of the reversible inhibition of enzyme-catalysed reactions by tight-binding inhibitors. *Biochimica et Biophysica Acta, 185*(2), 269–286.

Muller, I. (2017). Guidelines for the successful generation of protein–ligand complex crystals. *Acta Crystallographica. Section D, Biological Crystallography, 73*(Pt. 2), 79–92. https://doi.org/10.1107/S2059798316020271.

Muller, P., Warr, E., Stevenson, B. J., Pignatelli, P. M., Morgan, J. C., Steven, A., et al. (2008). Field-caught permethrin-resistant Anopheles gambiae overexpress CYP6P3, a P450 that metabolises pyrethroids. *PLoS Genetics, 4*(11), e1000286. https://doi.org/10.1371/journal.pgen.1000286.

Munro, A. W., Girvan, H. M., Mason, A. E., Dunford, A. J., & McLean, K. J. (2013). What makes a P450 tick? *Trends in Biochemical Sciences, 38*(3), 140–150. https://doi.org/10.1016/j.tibs.2012.11.006.

Munro, A. W., Girvan, H. M., & McLean, K. J. (2007). Variations on a (t)heme—Novel mechanisms, redox partners and catalytic functions in the cytochrome P450 superfamily. *Natural Product Reports, 24*(3), 585–609. https://doi.org/10.1039/b604190f.

Munro, A. W., Leys, D. G., McLean, K. J., Marshall, K. R., Ost, T. W., Daff, S., et al. (2002). P450 BM3: The very model of a modern flavocytochrome. *Trends in Biochemical Sciences, 27*(5), 250–257.

Munro, A. W., McLean, K. J., Grant, J. L., & Makris, T. M. (2018). Structure and function of the cytochrome P450 peroxygenase enzymes. *Biochemical Society Transactions, 46*(1), 183–196. https://doi.org/10.1042/bst20170218.

Murataliev, M. B., Klein, M., Fulco, A., & Feyereisen, R. (1997). Functional interactions in cytochrome P450BM3: Flavin semiquinone intermediates, role of NADP(H), and mechanism of electron transfer by the flavoprotein domain. *Biochemistry, 36*(27), 8401–8412. https://doi.org/10.1021/bi970026b.

Nakahara, K., Tanimoto, T., Hatano, K., Usuda, K., & Shoun, H. (1993). Cytochrome P-450 55A1 (P-450dNIR) acts as nitric oxide reductase employing NADH as the direct electron donor. *Journal of Biological Chemistry, 268*(11), 8350–8355.

Nakayama, N., Takemae, A., & Shoun, H. (1996). Cytochrome P450foxy, a catalytically self-sufficient fatty acid hydroxylase of the fungus *Fusarium oxysporum*. *Journal of Biochemistry, 119*(3), 435–440.

Narhi, L. O., & Fulco, A. J. (1986). Characterization of a catalytically self-sufficient 119,000-Dalton cytochrome P-450 monooxygenase induced by barbiturates in *Bacillus megaterium*. *Journal of Biological Chemistry, 261*(16), 7160–7169.

Noble, M. A., Miles, C. S., Chapman, S. K., Lysek, D. A., MacKay, A. C., Reid, G. A., et al. (1999). Roles of key active-site residues in flavocytochrome P450 BM3. *Biochemical Journal, 339*(Pt. 2), 371–379.

Ogura, H., Nishida, C. R., Hoch, U. R., Perera, R., Dawson, J. H., & Ortiz de Montellano, P. R. (2004). EpoK, a cytochrome P450 involved in biosynthesis of the anticancer agents epothilones A and B. Substrate-mediated rescue of a P450 enzyme. *Biochemistry, 43*(46), 14712–14721. https://doi.org/10.1021/bi048980d.

Omura, T., & Sato, R. (1964a). The carbon monoxide-binding pigment of liver microsomes. I. Evidence for its hemoprotein nature. *Journal of Biological Chemistry, 239*, 2370–2378.

Omura, T., & Sato, R. (1964b). The carbon monoxide-binding pigment of liver microsomes. II. Solubilization, purification, and properties. *Journal of Biological Chemistry, 239*, 2379–2385.

Ooi, L.-L. (2010). *Principals of X-ray crystallography*. United States: Oxford University Press.

Ortega Ugalde, S., Luirink, R. A., Geerke, D. P., Vermeulen, N. P. E., Bitter, W., & Commandeur, J. N. M. (2018). Engineering a self-sufficient *Mycobacterium tuberculosis* CYP130 by gene fusion with the reductase-domain of CYP102A1 from *Bacillus megaterium*. *Journal of Inorganic Biochemistry, 180*, 47–53. https://doi.org/10.1016/j.jinorgbio.2017.12.003.

Paddon, C. J., Westfall, P. J., Pitera, D. J., Benjamin, K., Fisher, K., McPhee, D., et al. (2013). High-level semi-synthetic production of the potent antimalarial artemisinin. *Nature, 496*(7446), 528–532. https://doi.org/10.1038/nature12051.

Pallan, P. S., Nagy, L. D., Lei, L., Gonzalez, E., Kramlinger, V. M., Azumaya, C. M., et al. (2015). Structural and kinetic basis of steroid 17α,20-lyase activity in teleost fish cytochrome P450 17A1 and its absence in cytochrome P450 17A2. *Journal of Biological Chemistry, 290*(6), 3248–3268. https://doi.org/10.1074/jbc.M114.627265.

Peralta-Yahya, P. P., Ouellet, M., Chan, R., Mukhopadhyay, A., Keasling, J. D., & Lee, T. S. (2011). Identification and microbial production of a terpene-based advanced biofuel. *Nature Communications, 2*, 483. https://doi.org/10.1038/ncomms1494.

Peyton, L. R., Gallagher, S., & Hashemzadeh, M. (2015). Triazole antifungals: A review. *Drugs of Today (Barcelona, Spain: 1998), 51*(12), 705–718. https://doi.org/10.1358/dot.2015.51.12.2421058.

Pochapsky, T. C., Jain, N. U., Kuti, M., Lyons, T. A., & Heymont, J. (1999). A refined model for the solution structure of oxidized putidaredoxin. *Biochemistry, 38*(15), 4681–4690. https://doi.org/10.1021/bi983030b.

Porubsky, P. R., Meneely, K. M., & Scott, E. E. (2008). Structures of human cytochrome P-450 2E1. Insights into the binding of inhibitors and both small molecular weight and fatty acid substrates. *Journal of Biological Chemistry, 283*(48), 33698–33707. https://doi.org/10.1074/jbc.M805999200.

Poulos, T. L., Finzel, B. C., Gunsalus, I. C., Wagner, G. C., & Kraut, J. (1985). The 2.6-A crystal structure of *Pseudomonas putida* cytochrome P-450. *Journal of Biological Chemistry, 260*(30), 16122–16130.

Poulos, T. L., Finzel, B. C., & Howard, A. J. (1987). High resolution crystal structure of cytochrome P450 cam. *Journal of Molecular Biology, 195*(3), 687–700. https://doi.org/10.1016/0022-2836(87)90190-2.

Poulos, T. L., & Howard, A. J. (1987). Crystal structures of metyrapone- and phenylimidazole-inhibited complexes of cytochrome P-450cam. *Biochemistry, 26*(25), 8165–8174.

Poulos, T. L., & Johnson, E. F. (2015). In P. R. Ortiz de Montellano (Ed.), *Structures of cytochrome P450 enzymes* (4th ed.). Springer, Vol. 1.

Putkaradze, N., Litzenburger, M., Abdulmughni, A., Milhim, M., Brill, E., Hannemann, F., et al. (2017). CYP109E1 is a novel versatile statin and terpene oxidase from *Bacillus megaterium. Applied Microbiology and Biotechnology, 101*(23–24), 8379–8393. https://doi.org/10.1007/s00253-017-8552-6.

Quaroni, L. G., Seward, H. E., McLean, K. J., Girvan, H. M., Ost, T. W., Noble, M. A., et al. (2004). Interaction of nitric oxide with cytochrome P450 BM3. *Biochemistry, 43*(51), 16416–16431. https://doi.org/10.1021/bi049163g.

Ravichandran, K. G., Boddupalli, S. S., Hasermann, C. A., Peterson, J. A., & Deisenhofer, J. (1993). Crystal structure of hemoprotein domain of P450BM-3, a prototype for microsomal P450's. *Science, 261*(5122), 731–736.

Rittle, J., & Green, M. T. (2010). Cytochrome P450 compound I: Capture, characterization, and C–H bond activation kinetics. *Science, 330*(6006), 933–937. https://doi.org/10.1126/science.1193478.

Roberts, G. A., Celik, A., Hunter, D. J., Ost, T. W., White, J. H., Chapman, S. K., et al. (2003). A self-sufficient cytochrome p450 with a primary structural organization that includes a flavin domain and a [2Fe-2S] redox center. *Journal of Biological Chemistry, 278*(49), 48914–48920. https://doi.org/10.1074/jbc.M309630200.

Rouck, J. E., Biggs, B. W., Kambalyal, A., Arnold, W. R., De Mey, M., Ajikumar, P. K., et al. (2017). Heterologous expression and characterization of plant taxadiene-5alpha-hydroxylase (CYP725A4) in *Escherichia coli. Protein Expression and Purification, 132*, 60–67. https://doi.org/10.1016/j.pep.2017.01.008.

Rude, M. A., Baron, T. S., Brubaker, S., Alibhai, M., Del Cardayre, S. B., & Schirmer, A. (2011). Terminal olefin (1-alkene) biosynthesis by a novel P450 fatty acid decarboxylase from Jeotgalicoccus species. *Applied and Environmental Microbiology*, 77(5), 1718–1727. https://doi.org/10.1128/aem.02580-10.

Rylott, E. L., Lorenz, A., & Bruce, N. C. (2011). Biodegradation and biotransformation of explosives. *Current Opinion in Biotechnology, 22*(3), 434–440. https://doi.org/10.1016/j.copbio.2010.10.014.

Sakaki, T. (2012). Practical application of cytochrome P450. *Biological and Pharmaceutical Bulletin, 35*(6), 844–849.

Sansen, S., Yano, J. K., Reynald, R. L., Schoch, G. A., Griffin, K. J., Stout, C. D., et al. (2007). Adaptations for the oxidation of polycyclic aromatic hydrocarbons exhibited by the structure of human P450 1A2. *Journal of Biological Chemistry, 282*(19), 14348–14355. https://doi.org/10.1074/jbc.M611692200.

Sawada, N., Sakaki, T., Yoneda, S., Kusudo, T., Shinkyo, R., Ohta, M., et al. (2004). Conversion of vitamin D3 to 1alpha,25-dihydroxyvitamin D3 by *Streptomyces griseolus* cytochrome P450SU-1. *Biochemical and Biophysical Research Communications, 320*(1), 156–164. https://doi.org/10.1016/j.bbrc.2004.05.140.

Sawayama, A. M., Chen, M. M., Kulanthaivel, P., Kuo, M. S., Hemmerle, H., & Arnold, F. H. (2009). A panel of cytochrome P450 BM3 variants to produce drug metabolites and diversify lead compounds. *Chemistry, 15*(43), 11723–11729. https://doi.org/10.1002/chem.200900643.

Scott, E. E., Spatzenegger, M., & Halpert, J. R. (2001). A truncation of 2B subfamily cytochromes P450 yields increased expression levels, increased solubility, and decreased aggregation while retaining function. *Archives of Biochemistry and Biophysics*, *395*(1), 57–68. https://doi.org/10.1006/abbi.2001.2574.

Sevrioukova, I. F., Li, H., & Poulos, T. L. (2004). Crystal structure of putidaredoxin reductase from *Pseudomonas putida*, the final structural component of the cytochrome P450cam monooxygenase. *Journal of Molecular Biology*, *336*(4), 889–902. https://doi.org/10.1016/j.jmb.2003.12.067.

Sevrioukova, I. F., Li, H., Zhang, H., Peterson, J. A., & Poulos, T. L. (1999). Structure of a cytochrome P450-redox partner electron-transfer complex. *Proceedings of the National Academy of Sciences of the United States of America*, *96*(5), 1863–1868.

Shimada, T., Tanaka, K., Takenaka, S., Foroozesh, M. K., Murayama, N., Yamazaki, H., et al. (2009). Reverse type I binding spectra of human cytochrome P450 1B1 induced by flavonoid, stilbene, pyrene, naphthalene, phenanthrene, and biphenyl derivatives that inhibit catalytic activity: A structure–function relationship study. *Chemical Research in Toxicology*, *22*(7), 1325–1333. https://doi.org/10.1021/tx900127s.

Slabinski, L., Jaroszewski, L., Rychlewski, L., Wilson, I. A., Lesley, S. A., & Godzik, A. (2007). XtalPred: A web server for prediction of protein crystallizability. *Bioinformatics*, *23*(24), 3403–3405. https://doi.org/10.1093/bioinformatics/btm477.

Sowden, R. J., Yasmin, S., Rees, N. H., Bell, S. G., & Wong, L. L. (2005). Biotransformation of the sesquiterpene (+)-valencene by cytochrome P450cam and P450BM-3. *Organic and Biomolecular Chemistry*, *3*(1), 57–64. https://doi.org/10.1039/b413068e.

Sugano, Y. (2009). DyP-type peroxidases comprise a novel heme peroxidase family. *Cellular and Molecular Life Sciences*, *66*(8), 1387–1403. https://doi.org/10.1007/s00018-008-8651-8.

Theorell, H., & Maehly, A. C. (1950). Untersuchungen an Kunstlichen Peroxydasen. *Acta Chemica Scandinavica*, *4*(3), 422–434. https://doi.org/10.3891/acta.chem.scand.04-0422.

Uchiyama, M., Fischer, T., Mueller, J., Oguchi, M., Yamamura, N., Koda, H., et al. (2010). Identification of novel metabolic pathways of pioglitazone in hepatocytes: N-glucuronidation of thiazolidinedione ring and sequential ring-opening pathway. *Drug Metabolism and Disposition*, *38*(6), 946–956. https://doi.org/10.1124/dmd.109.031583.

von Buhler, C. J., & Urlacher, V. B. (2014). A novel P450-based biocatalyst for the selective production of chiral 2-alkanols. *Chemical Communications*, *50*(31), 4089–4091. https://doi.org/10.1039/c4cc00647j.

von Wachenfeldt, C., Richardson, T. H., Cosme, J., & Johnson, E. F. (1997). Microsomal P450 2C3 is expressed as a soluble dimer in *Escherichia coli* following modifications of its N-terminus. *Archives of Biochemistry and Biophysics*, *339*(1), 107–114. https://doi.org/10.1006/abbi.1996.9859.

Walsh, A. A., Szklarz, G. D., & Scott, E. E. (2013). Human cytochrome P450 1A1 structure and utility in understanding drug and xenobiotic metabolism. *Journal of Biological Chemistry*, *288*(18), 12932–12943. https://doi.org/10.1074/jbc.M113.452953.

Wang, M., Roberts, D. L., Paschke, R., Shea, T. M., Masters, B. S., & Kim, J. J. (1997). Three-dimensional structure of NADPH-cytochrome P450 reductase: Prototype for FMN- and FAD-containing enzymes. *Proceedings of the National Academy of Sciences of the United States of America*, *94*(16), 8411–8416.

Wang, A., Savas, U., Hsu, M.-H., Stout, C. D., & Johnson, E. F. (2012). Crystal structure of human cytochrome P450 2D6 with prinomastat bound. *Journal of Biological Chemistry*, *287*(14), 10834–10843. https://doi.org/10.1074/jbc.M111.307918.

Warman, A. J., Robinson, J. W., Luciakova, D., Lawrence, A. D., Marshall, K. R., Warren, M. J., et al. (2012). Characterization of *Cupriavidus metallidurans* CYP116B1-A thiocarbamate herbicide oxygenating P450-phthalate dioxygenase reductase fusion protein. *FEBS Journal*, *279*(9), 1675–1693. https://doi.org/10.1111/j.1742-4658.2012.08543.x.

Watanabe, I., Nara, F., & Serizawa, N. (1995). Cloning, characterization and expression of the gene encoding cytochrome P-450sca-2 from *Streptomyces carbophilus* involved in production of pravastatin, a specific HMG-CoA reductase inhibitor. *Gene*, *163*(1), 81–85.

Wester, M. R., Stout, C. D., & Johnson, E. F. (2002). Purification and crystallization of N-terminally truncated forms of microsomal cytochrome P450 2C5. *Methods in Enzymology*, *357*, 73–79.

Whitehouse, C. J., Bell, S. G., & Wong, L. L. (2008). Desaturation of alkylbenzenes by cytochrome P450(BM3) (CYP102A1). *Chemistry*, *14*(35), 10905–10908. https://doi.org/10.1002/chem.200801927.

Willcott, M. R. (2009). MestRe Nova. *Journal of the American Chemical Society*, *131*(36), 13180. https://doi.org/10.1021/ja906709t.

Williams, P. A., Cosme, J., Sridhar, V., Johnson, E. F., & McRee, D. E. (2000). Mammalian microsomal cytochrome P450 monooxygenase. *Molecular Cell*, *5*(1), 121–131. https://doi.org/10.1016/S1097-2765(00)80408-6.

Winn, M. D., Ballard, C. C., Cowtan, K. D., Dodson, E. J., Emsley, P., Evans, P. R., et al. (2011). Overview of the CCP4 suite and current developments. *Acta Crystallographica. Section D, Biological Crystallography*, *67*(Pt. 4), 235–242. https://doi.org/10.1107/S0907444910045749.

Xu, Z., & Sigler, P. B. (1998). GroEL/GroES: Structure and function of a two-stroke folding machine. *Journal of Structural Biology*, *124*(2–3), 129–141. https://doi.org/10.1006/jsbi.1998.4060.

Xue, Y., Wilson, D., Zhao, L., Liu, H., & Sherman, D. H. (1998). Hydroxylation of macrolactones YC-17 and narbomycin is mediated by the pikC-encoded cytochrome P450 in *Streptomyces venezuelae*. *Chemistry and Biology*, *5*(11), 661–667.

Yano, J. K., Wester, M. R., Schoch, G. A., Griffin, K. J., Stout, C. D., & Johnson, E. F. (2004). The structure of human microsomal cytochrome P450 3A4 determined by X-ray crystallography to 2.05-A resolution. *Journal of Biological Chemistry*, *279*(37), 38091–38094. https://doi.org/10.1074/jbc.C400293200.

Yoshimoto, F. K., Gonzalez, E., Auchus, R. J., & Guengerich, F. P. (2016). Mechanism of 17alpha,20-lyase and new hydroxylation reactions of human cytochrome P450 17A1: 18O labelling and oxygen surrogate evidence for a role of a perferryl oxygen. *Journal of Biological Chemistry*, *291*(33), 17143–17164. https://doi.org/10.1074/jbc.M116.732966.

Zelasko, S., Palaria, A., & Das, A. (2013). Optimizations to achieve high-level expression of cytochrome P450 proteins using *Escherichia coli* expression systems. *Protein Expression and Purification*, *92*(1), 77–87. https://doi.org/10.1016/j.pep.2013.07.017.

SECTION IV

Genome Engineering for Enzyme Pathways

CHAPTER NINE

Metabolic Engineering of *Saccharomyces cerevisiae* Using a Trifunctional CRISPR/Cas System for Simultaneous Gene Activation, Interference, and Deletion

Carl Schultz*,†, Jiazhang Lian*,‡, Huimin Zhao*,†,1

*Department of Chemical and Biomolecular Engineering, University of Illinois at Urbana-Champaign, Urbana, IL, United States
†Carl R. Woese Institute for Genomic Biology, University of Illinois at Urbana-Champaign, Urbana, IL, United States
‡Key Laboratory of Biomass Chemical Engineering of Ministry of Education, College of Chemical and Biological Engineering, Zhejiang University, Hangzhou, China
[1]Corresponding author: e-mail address: zhao5@illinois.edu

Contents

Abstract

Design and construction of an optimal microbial cell factory typically requires overexpression, knockdown, and knockout of multiple gene targets. In this chapter, we describe a combinatorial metabolic engineering strategy utilizing an orthogonal trifunctional CRISPR system that combines transcriptional activation, transcriptional interference, and gene deletion (CRISPR-AID) in the yeast *Saccharomyces cerevisiae*. This strategy enables multiplexed perturbation of the metabolic and regulatory networks in a modular, parallel, and high-throughput manner. To implement this system, three orthogonal Cas proteins were utilized: dLbCpf1 fused to a transcriptional activator, dSpCas9 fused to a transcriptional repressor, and SaCas9 for gene deletion. Deletion was accomplished by the introduction of a 28 bp frame-shift mutation using a homology donor on the guide RNA expression vector. This approach enables the application

Methods in Enzymology, Volume 608
ISSN 0076-6879
https://doi.org/10.1016/bs.mie.2018.04.010

of metabolic engineering to systematically optimize phenotypes of interest through a combination of gain-, reduction-, and loss-of-function mutations. Finally, we describe the construction of the CRISPR-AID system and its application toward engineering an example phenotype, surface display of recombinant *Trichoderma reesei* endoglucanase II.

1. INTRODUCTION

Microbial cell factories, such as the platform organism *Saccharomyces cerevisiae*, have been increasingly engineered to produce fuels, chemicals, and pharmaceuticals using various renewable feedstocks (Du, Shao, & Zhao, 2011; Nielsen & Keasling, 2016). However, microorganisms possess robust metabolic and regulatory networks optimized for survival in nature, rather than for the synthesis of products of industrial interest. Therefore, the metabolisms of these microbial cell factories must be reengineered to maximize product titer, yield, and productivity. Due to the complex and highly interconnected nature of the metabolic networks, it is often necessary to subject multiple metabolic engineering target genes to different types of modulation, for example, the upregulation of genes encoding rate-limiting enzymes, downregulation of commercially detrimental but biologically essential genes, and the complete removal of competing, nonessential pathways (Nielsen & Keasling, 2016). Furthermore, synergistic interactions may exist between various metabolic engineering targets, but their occurrence may be difficult to predict (Lian, Hamedirad, Hu, & Zhao, 2017). Therefore, to perform the most comprehensive optimization of a phenotype of interest, it is necessary to have access to a complete range of expression profiles for multiple genes of interest simultaneously.

Traditionally, genomic engineering can be accomplished in yeast using homologous recombination-based gene replacement, in which a modified gene flanked by homology arms is used to replace part of the host genome using the cell's native DNA repair machinery. However, this method suffers from low efficiency and the necessity of using a selection marker which must be recovered to make successive modifications, making it nonideal for the modification of many targets (Hegemann, Guldener, & Kohler, 2006).

The clustered regularly interspaced short palindromic repeats (CRISPR) system provides a highly efficient and modular alternative method for multiplex genome engineering. In this system, a 20 bp guide RNA (gRNA) recruits a CRISPR nuclease to the complementary region of the genome to create a double-strand break (DSB). This can be readily exploited to achieve the disruption of a targeted gene by simultaneously providing a homology donor containing a frame-shift mutation, which the cell will sometimes use to repair the DSB by undergoing homology-directed repair. Moreover, transcriptional activation or interference can be achieved using the CRISPR system by fusing an activation or repression domain to the nuclease-deficient CRISPR protein (Gilbert et al., 2014; La Russa & Qi, 2015; Maeder et al., 2013; Qi et al., 2013; Zalatan et al., 2015).

In this chapter, we describe a CRISPR system using three orthogonal CRISPR proteins: one for activation, one for interference, and one for deletion to simultaneously achieve all three modes of genetic modification (CRISPR-AID) (Fig. 1). The use of this system can allow for combinatorial optimization of various metabolic engineering targets using these three modes of modification, and exploration of the synergistic interactions between these targets, in *S. cerevisiae*. For example, by activating, repressing,

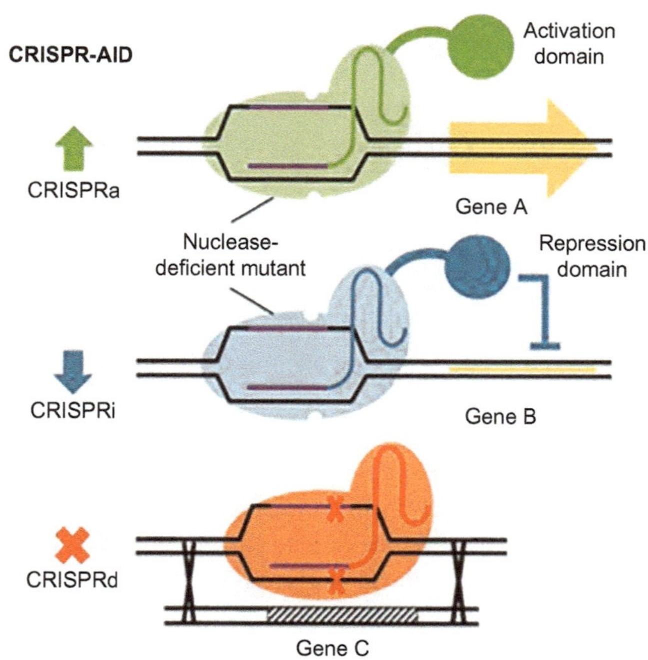

Fig. 1 Design of the CRISPR-AID system with one nuclease-deficient CRISPR protein fused with an activation domain for CRISPRa, another nuclease-deficient mutant fused with a repression domain for CRISPRi, and a third catalytically active CRISPR protein for CRISPRd. *Reproduced with permission from Lian, J., Hamedirad, M., Hu, S., & Zhao, H. (2017). Combinatorial metabolic engineering using an orthogonal tri-functional CRISPR system.* Nature Communications, 8, *1–9.*

and deleting three genes reported to affect β-carotene production, we were able to increase production of this compound in *S. cerevisiae* by more than threefold. We were additionally able to increase surface display of recombinant endoglucanase II (EGII) by 2.5-fold by screening of a library of 1620 different combinations of activation, interference, and deletion targets (Lian et al., 2017). In this chapter, we describe the development of this system, and the procedure by which it was assembled. We provide detailed protocols for the application of the CRISPR-AID system to optimize an example phenotype, surface display of EGII in *S. cerevisiae*, including guide RNA expression vector assembly, yeast transformation, and high- and low-throughput screening methods.

2. DESIGN OF CRISPR-AID SYSTEM FOR METABOLIC ENGINEERING IN *S. cerevisiae*

Three orthogonal Cas proteins were used for the implementation of the trifunctional CRISPR-AID system to prevent crosstalk between the gRNAs. After screening various Cas9 orthologs and *trans*-effector domains, catalytically inactivated *Lachnospiraceae bacterium* Cpf1 (dLbCpf1) fused to the activation domains VP64 and p65AD was chosen for gene activation (dLbCpf1-VP) (Chavez et al., 2015). Catalytically inactivated *Streptococcus pyogenes* Cas9 (dSpCas9) fused to the *S. cerevisiae* repression domains RD11, RD5, and RD2 (derived from various *S. cerevisiae* transcription factors) was chosen for gene interference (dSpCas9-RD1152) (Lian et al., 2017). *Staphylococcus aureus* Cas9 (SaCas9) was chosen for gene deletion, utilizing the HI-CRISPR gene deletion method (Bao et al., 2015). It was found that using an SV40 nuclear localization sequence on both the N- and C-termini was optimal for nuclease activity in yeast. The endoribonuclease Csy4 additionally must be expressed for gRNA processing if multiplexed guide RNA expression is driven by a single type II promoter. In this case, the gRNAs will be expressed in one cassette but separated by Csy4 recognition sites, where the single transcript will be cleaved into the individual gRNAs for each Cas protein (Fig. 2).

dLbCpf1-VP, Csy4, dSpCas9-RD1152, and SaCas9, expressed from the constitutive *TDH3*, *ENO2*, *TPI1*, and *TEF1* promoters, respectively, were cloned into the plasmid pRS41K-CEN-Delta using the DNA assembler method to create the pAID6 plasmid (Shao, Zhao, & Zhao, 2009). This plasmid was linearized with *PmeI* and transformed into *Saccharomyces* strain CEN.PK2 to integrate it into a delta site in the genome, followed by

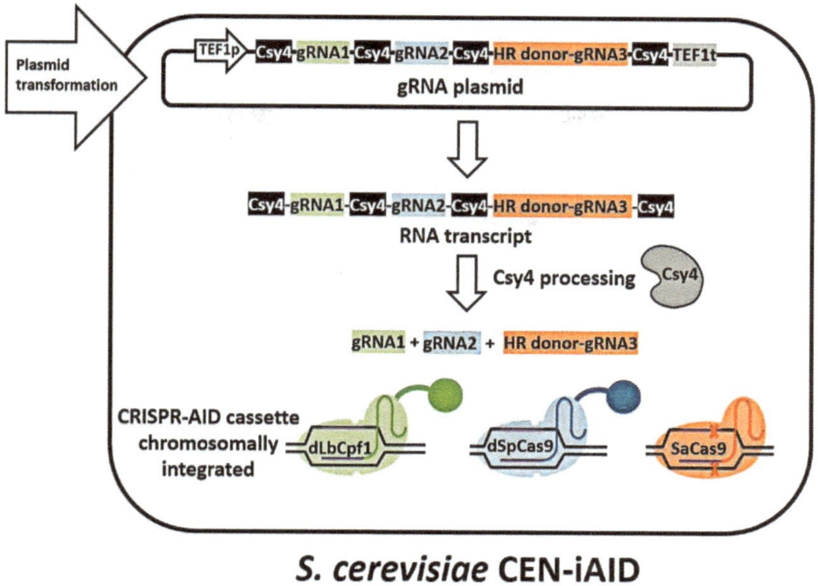

Fig. 2 Process of targeting the CRISPR-AID system to genes of interest. A gRNA expression plasmid is transformed into a strain expressing the CRISPR-AID system, where the gRNA cassette is transcribed and processed to yield the final individual gRNAs. The plasmid also serves as the homology donor for gene deletion.

selection for G418 resistance to create the CRISPR-AID-expressing strain used in this study, CEN-iAID.

Guide RNA expression for individual gRNAs was assisted through the creation of helper plasmids, containing the type III promoter *SNR52p*, two *Bsa*I sites, appropriate gRNA scaffold sequences, and *SUP4t* based on a modified *Bsa*I-free pRS423 vector. The target sequences were then ordered as short oligos which were annealed and phosphorylated, then ligated into the *Bsa*I-digested helper plasmid. For expression of multiple gRNAs, individual *SNR52p*-gRNA-*SUP4t* cassettes were pieced together using Golden Gate assembly (Engler, Kandzia, & Marillonnet, 2008) or arrays of gRNAs driven by the type II promoter *TEF1p* and separated by *Csy4* recognition sites were ordered as gBLOCKs and cloned into a pRS423 vector using restriction digestion–ligation.

Gene deletion was performed through the introduction of a 28-nt frame-shift deletion to the beginning of the target gene. A gRNA targeting Cas9 to the deletion site as well as a homology donor containing the frame-shift mutation (which includes the PAM sequence of the Cas protein, ending cleavage of the site once the mutation is introduced) between two 50 bp homology arms are transformed into the yeast. It was previously shown that the homology donor can be simply included at the 5' end of the gRNA in the expression plasmid using the HI-CRISPR method, and this design was also used for gene deletion in the CRISPR-AID method (Bao et al., 2015).

3. CRISPR-AID FOR COMBINATORIAL METABOLIC ENGINEERING

The CRISPR-AID system can be used for combinatorial optimization of the expression of several genes to achieve a desired phenotype. Here, we describe the optimization of surface display of *Trichoderma reesei* endoglucanase II (EGII) in *S. cerevisiae*. Endoglucanase surface display, in conjunction with display of other proteins, confers to the yeast the ability to directly digest cellulose into monosaccharides for use as a carbon source (Wen, Sun, & Zhao, 2009). Fourteen targets were chosen for activation, 17 for interference, and 5 for deletion, including genes involved in protein folding, secretion, super-glycosylation, and degradation (Lian et al., 2017). A plasmid library of all possible combinations of an activation, interference, and deletion target expressed in multiplex ($15 \times 18 \times 6 = 1620$) was generated and transformed to *S. cerevisiae*.

3.1 gRNA Expression Vector Assembly

The guide RNAs targeting each of these genes were designed using the Benchling CRISPR tool for optimal on-target and off-target score. On-target and off-target scores were not available for LbCpf1, so several design rules were followed: 35%–65% GC content, no polyT, no secondary structure, and <12 bp match by BLAST to the yeast genome. Note that each Cas ortholog has a different PAM sequence: 5' TTTV for LbCpf1 (activation), 3' NGG for SpCas9 (interference), and 3' NNGRRT for SaCas9 (deletion). It was found that in previous studies as well as this one that targeting ~200 bp upstream of the transcription start site (TSS) was optimal for CRISPRa and ~50–100 bp upstream of the TSS for CRISPRi (Lian et al., 2017).

For the expression of individual guide RNAs, a helper plasmid was designed containing two outward-facing *Bsa*I sites such that *Bsa*I digestion would linearize the plasmid with a 3' CTAG overhang ((−) strand, the end of the *SNR52* promoter) and a 5' GTTT overhang ((+) strand, the start of SaCas9 and SpCas9 gRNA scaffold) or a 3' TCTA overhang ((−) strand, the end of the LbCpf1 scaffold) and a 5' TTTT overhang ((+) strand, the start of

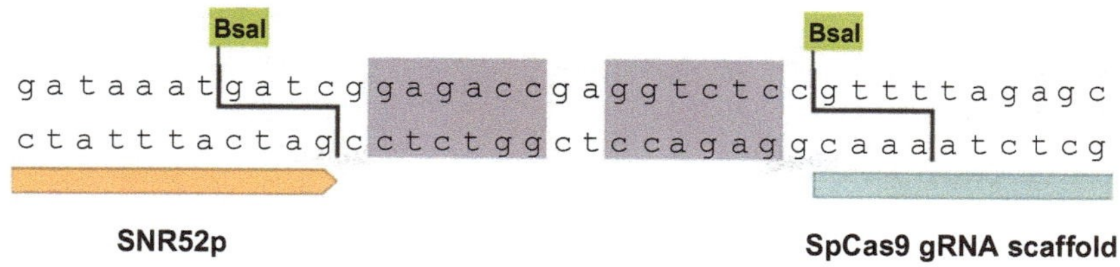

Fig. 3 Helper plasmid design for SpCas9 gRNA expression.

the *SUP4* terminator) (example for SpCas9 shown in Fig. 3). The guide RNAs can then be ordered as two ssDNA oligos, each with an appropriate complementary overhang to the plasmid, which are annealed, phosphorylated, and ligated into the vector backbone. For example, if the targeting sequence is CTCACGGACGGAGACCAGGA, then the two oligos for SpCas9 gRNA expression would be:

a. Forward: 5′-**GATC** CTCACGGACGGAGACCAGGA-3′
b. Reverse: 5′-**AAAC** TCCTGGTCTCCGTCCGTGAG-3′

To anneal and phosphorylate the oligos

1. Suspend each oligo to 100 μ*M* in double-distilled water (ddH_2O).
2. Combine 6.5 μL ddH_2O, 1 μL of each oligo, 1 μL of T4 DNA Ligase buffer (New England Biolabs (NEB)), and 0.5 μL of T4 Polynucleotide Kinase (NEB).
3. Using a thermocycler, incubate the mixture at 37°C for 30 min, heat to 95°C for 5 min, then cool to 4°C at 0.1°C/s.

The final DNA concentration is calculated to be approximately 150 ng/μL for 24 bp oligos. A single pair of annealed oligos can be inserted to the helper plasmid by *Bsa*I digestion–ligation.

Restriction digestion–ligation

1. Combine ~2.5 μg of the helper plasmid, 2 μL of *Bsa*I at 10 U/μL (NEB), and 10 μL CutSmart buffer (NEB), add ddH_2O to 100 μL total volume.
2. Incubate at 37°C for 4 h.
3. Isolate the linearized vector backbone using gel electrophoresis and gel extraction (Qiagen QIAquick Gel Extraction Kit).
4. Combine 25–50 ng digested backbone, 0.5 μL annealed oligo, 1 μL T4 DNA Ligase buffer (NEB), 0.5 μL T4 DNA Ligase (NEB), and add ddH_2O up to 10 μL total volume.
5. Incubate at 16°C overnight or at room temperature for 1 h.

To construct plasmids containing multiple gRNA cassettes, gRNA arrays under a type II promoter were synthesized as gBLOCKs and cloned into a receiving plasmid with restriction digestion/ligation using the above protocol. Alternatively, individual expression cassettes were given compatible overhangs using PCR and assembled into a backbone vector using Golden Gate assembly.

Golden Gate assembly

1. Combine 0.5 μL annealed oligo, 25–50 ng helper plasmid, 1 μL T4 DNA Ligase buffer (NEB), 0.8 μL *Bsa*I (NEB), 0.2 μL T4 DNA Ligase (NEB), and add ddH_2O up to 10 μL total volume.
2. Perform the Golden Gate assembly on a thermocycler: 5 min at 37°C, then 20 cycles of 5 min at 37°C followed by 10 min at 25°C; after the 20 cycles then 20 min at 25°C, then 30 min at 37°C, then 6 min at 75°C (total time: approximately 6 h 30 min).

Once the restriction digestion/ligation or Golden Gate assembly is complete

1. Transform 1 μL of the reaction to NEB 5-alpha *z*-competent *E. coli* or NEB 5-alpha Electrocompetent *E. coli* following the manufacturer's suggested protocol.
2. Spread the transformed cells to LB/Amp agar plates and incubate overnight at 37°C.
3. Pick several colonies and verify that they have the correct insert using colony PCR.
   a. Use a forward primer targeting the *SNR52* promoter and for the reverse primer use the reverse gRNA oligo.
4. Recover the plasmid from a verified colony and further confirm the insert by sequencing (ensure the sequencing primer is >50 nt upstream of the insert).

3.2 Yeast Transformation

After assembly, the vectors are transformed to yeast using the LiAc/PEG/ssDNA heat shock method (Gietz & Schiestl, 2007). Initially, a pooled library of plasmids was transformed using a high transformation efficiency protocol. After the top hits were identified, individual plasmids were retransformed for verification.

To transform the gRNA expression vectors into *S. cerevisiae*

1. Inoculate a colony from an AID-integrated yeast strain to 4 mL YPD, incubate at 30°C and 250 rpm for 24 h
2. Inoculate yeast to OD of 0.2 (individual transformation) or 0.5 (library transformation) in 50 mL YPD (provides enough cells for 10 transformations)
3. Incubate at 30°C and 250 rpm for 4 h

4. Collect yeast by centrifuging at 4000 rpm for 5 min, resuspend in 50 mL sterile water
5. Collect yeast by centrifuging at 4000 rpm for 5 min, resuspend in 1 mL 0.1 *M* LiAc
6. Transfer 100 μL of cells to microfuge tubes, centrifuge at 4000 rpm for 5 min and remove supernatant
7. Resuspend yeast in 240 μL of 50% w/v PEG-3350(aq), 50 μL salmon sperm DNA, 36 μL 1 *M* LiAc, X μL plasmid, 34-X μL sterile water (500 ng DNA is sufficient for single transformation, for library-scale use 10 μg)
8. Heat shock at 42°C for 40 min (single transformation) or 60 min (library transformation). A library transformation will have higher cell density, so periodically invert the tubes to prevent cell settling
9. Collect yeast by centrifuging at 4000 rpm for 5 min
10. (Library transformation only) Recover yeast for 4 h in 4 mL YPD at 30°C and 250 rpm. Longer incubation will result in loss of plasmid
11. (Library transformation only) Remove YPD by centrifuging at 4000 rpm for 5 min
12. (Library transformation only) Spread a 10^3-, 10^4-, and 10^5-fold dilution of cells to selective plates for quantification of library coverage (>100 transformants per plasmid in library indicates a representative library)
13. Spread the transformed yeast to selective plates and grow until colonies are visible (individual transformation) or to selective liquid media for 36 h (library transformation) at 30°C

3.3 Phenotype Screening

The EGII protein recombinant protein was tagged with a 6 × His tag to allow high-throughput immunofluorescence-based screening of the protein display level using fluorescence-activated cell sorting (FACS).

To sort cells using FACS based on immunofluorescence

1. Spin down 1 mL yeast cells and wash in 1 mL PBSA (PBS with 0.5% BSA) buffer.
2. Resuspend in 200 μL PBSA containing monoclonal mouse anti-histidine tag antibody (1:100 dilution, Bio-Rad, Raleigh, NC).
3. Incubate on ice for ~1 h. Resuspend the yeast cells every 20 min and try to avoid light.
4. Spin down 5 min at 4000 rpm and wash in 1 mL PBSA (PBS with 0.5% BSA) buffer.

5. Resuspend in 200 μL PBSA containing goat antimouse IgG (H + L) secondary antibody, Biotin-XX conjugate (1:100 dilution, ThermoFisher Scientific, Rockford, IL).
6. Incubate on ice for ~1 h. Resuspend the yeast cells every 20 min and try to avoid light.
7. Spin down and wash in 1 mL PBSA (PBS with 0.5% BSA) buffer.
8. Resuspend in 200 μL PBSA containing Streptavidin, R-phycoerythrin conjugate (1:100 dilution, ThermoFisher Scientific).
9. Incubate on ice for ~1 h. Resuspend the yeast cells every 20 min and try to avoid light.
10. Spin down and wash three times in 1 mL PBSA (PBS with 0.5% BSA) buffer.
11. FACS was performed on a BD FACS Aria III cell sorting system (BD Biosciences, San Jose, CA).
12. In the first round of sorting, collect around 30,000 cells representing the top 0.5% highest fluorescence into 1 mL selective medium.
13. Transfer to a culture tube with additional 2 mL selective medium.
14. Culture at 30°C for ~3 days
15. Repeat steps 1–12 for the second round of sorting, collect 96 × 4 individual yeast cells with the highest fluorescence (top 0.1%) into a 96-well microplate, 200 μL selective medium in each well.
16. Incubate the microplates at 30°C for ~3 days.
17. Transfer 20 μL of each colony that grows into 2 mL fresh selective medium in culture tubes and culture at 30°C for ~3 days until saturation.
18. Extract plasmids using Zymoprep Yeast Plasmid Miniprep kit (Zymo Research), retransform to *E. coli*, miniprep and sequence to identify gRNA.
19. Retransform the plasmids individually into yeast for verification.

The EGII display level of each top hit was then directly quantified using an assay to measure cellulose digestion. The 3,5-dinitrosalicylic acid (DNS) quantifies the amount of reducing sugars present in solution by reacting to form 3-amino-5-nitrosalicylic acid, which absorbs light at 540 nm.

1. Transfer 40 μL into 2 mL fresh selective medium in culture tubes and culture at 30°C overnight.
2. Measure the cell density.
3. 400 μL yeast cells from culture were washed twice with ddH_2O.
4. Centrifuge at 4000 rpm for 5 min.

5. Resuspend in the same volume of 1% (w/v) carboxymethyl cellulose (CMC) solution (0.1 *M* sodium acetate, pH 5)
6. Incubate at 30°C for 16 h with vigorous shaking.
7. Mix 50 μL supernatant with 50 μL DNS reagent (0.044 *M* 3,5-dinitrosalicylic acid, 0.035 *M* sodium potassium tartrate tetrahydrate, 0.50 *M* NaOH, store away from light).
8. Incubate at 99°C for 5 min in thermocycler.
9. Add 40 μL into 160 μL ddH$_2$O (fivefold dilution).
10. Measure the absorbance at 540 nm.
11. Normalize to cell density.

4. CONCLUSION

The CRISPR-AID method has been developed to combine transcriptional activation, transcriptional interference, and gene deletion in *S. cerevisiae* using the CRISPR/Cas9 system. To implement this method, the CRISPR-AID expression cassette is integrated into a host cell genome and the host is transformed with gRNA expression plasmids to probe the effect of modulating a target gene or genes. This tool enables metabolic engineering to systematically optimize phenotypes of interest through a combination of gain, decrease, and complete loss of function mutations in a parallel and high-throughput manner.

ACKNOWLEDGMENTS

This work was supported by the U.S. Department of Energy (DE-SC0018260).

REFERENCES

Bao, Z., et al. (2015). Homology-integrated CRISPR-Cas (HI-CRISPR) system for one-step multigene disruption in *Saccharomyces cerevisiae*. *ACS Synthetic Biology*, *4*, 585–594.

Chavez, A., et al. (2015). Highly efficient Cas9-mediated transcriptional programming. *Nature Methods*, *12*, 326–328.

Du, J., Shao, Z., & Zhao, H. (2011). Engineering microbial factories for synthesis of value-added products. *Journal of Industrial Microbiology & Biotechnology*, *38*, 873–890.

Engler, C., Kandzia, R., & Marillonnet, S. (2008). A one pot, one step, precision cloning method with high throughput capability. *PLoS One*, *3*, 3647.

Gietz, R. D., & Schiestl, R. H. (2007). High-efficiency yeast transformation using the LiAc/SS carrier DNA/PEG method. *Nature Protocols*, *2*, 31–34.

Gilbert, L. A., et al. (2014). Genome-scale CRISPR-mediated control of gene repression and activation. *Cell*, *159*, 647–661.

Hegemann, J. H., Guldener, U., & Kohler, G. J. (2006). Gene disruption in the budding yeast *Saccharomyces cerevisiae*. *Methods in Molecular Biology*, *313*, 129–144.

La Russa, M. F., & Qi, L. S. (2015). The new state of the art: Cas9 for gene activation and repression. *Molecular and Cellular Biology*, *35*, 3800–3809.
Lian, J., Hamedirad, M., Hu, S., & Zhao, H. (2017). Combinatorial metabolic engineering using an orthogonal tri-functional CRISPR system. *Nature Communications*, *8*, 1–9.
Maeder, M. L., et al. (2013). CRISPR RNA-guided activation of endogenous human genes. *Nature Methods*, *10*, 977–979.
Nielsen, J., & Keasling, J. D. (2016). Engineering cellular metabolism. *Cell*, *164*, 1185–1197.
Qi, L. S., et al. (2013). Repurposing CRISPR as an RNA-guided platform for sequence-specific control of gene expression. *Cell*, *152*, 1173–1183.
Shao, Z., Zhao, H., & Zhao, H. (2009). DNA assembler, an *in vivo* genetic method for rapid construction of biochemical pathways. *Nucleic Acids Research*, *37*, e16.
Wen, F., Sun, J., & Zhao, H. (2009). Yeast surface display of trifunctional Minicellulosomes for simultaneous Saccharification and fermentation of cellulose to ethanol. *Applied and Environmental Microbiology*, *76*, 1251–1260.
Zalatan, J. G., et al. (2015). Engineering complex synthetic transcriptional programs with CRISPR RNA scaffolds. *Cell*, *160*, 339–350.

CHAPTER TEN

YeastFab: High-Throughput Genetic Parts Construction, Measurement, and Pathway Engineering in Yeast

Eva Garcia-Ruiz*, Jamie Auxillos*, Tianyi Li†, Junbiao Dai†, Yizhi Cai*,[1]

*Manchester Institute of Biotechnology (MIB), School of Chemistry, The University of Manchester, Manchester, United Kingdom

†Centre for Synthetic Genomics, Institute of Synthetic Biology, Shenzhen Institutes of Advanced Technology, Chinese Academy of Sciences, Shenzhen, China

[1]Corresponding author: e-mail address: yizhi.cai@manchester.ac.uk

Contents

Methods in Enzymology, Volume 608
ISSN 0076-6879
https://doi.org/10.1016/bs.mie.2018.05.003

Abstract

For many years, researchers have devised elegant techniques to assemble genetic parts into larger constructs. Recently, increasing needs for complex DNA constructs has driven countless attempts to optimize DNA assembly methods for improved efficiency, fidelity, and modularity. These efforts have resulted in simple, robust, standardized, and fast protocols that enable the implementation of high-throughput DNA assembly projects for the fabrication of large synthetic genetic constructs. Recently our groups have developed the YeastFab assembly, a highly efficient method for the design and construction of DNA-building blocks based on the native elements from *Saccharomyces cerevisiae*. Furthermore, these standardized DNA parts can be readily characterized and assembled into transcriptional units and pathways. In this chapter, we describe the protocols to assemble pathways from characterized standardized yeast parts using YeastFab.

1. INTRODUCTION

Many endeavors have been taken to develop efficient and reliable DNA assembly methods to create complex synthetic metabolic pathways for biobased production of natural products, commodity chemicals, and biofuels (Chen & Nielsen, 2013; Liu, Shin, Li, & Liu, 2015). Most metabolic pathways involve multiple genes that should function in coordination, under tight regulatory control. However, engineering microorganisms to produce exogenous chemicals is usually laborious and time consuming. Limited successful examples such the manufacture of the antimalarial drug artemisinin (Paddon et al., 2013) or painkillers such as hydrocodone (Galanie, Thodey, Trenchard, Filsinger Interrante, & Smolke, 2015) in yeast illustrate the difficulties and bottlenecks to overcome in order to produce the target product. It is well described that the introduction of functional DNA in a host microorganism usually generates a burden for the cells (Da Silva & Srikrishnan, 2012; Gorgens, van Zyl, Knoetze, & Hahn-Hagerdal, 2001; Karim, Curran, & Alper, 2013); therefore, fine regulation is needed in order to increase the yield of the desired product. Combinatorial pathway library construction is an effective approach for pathway optimization. In this case, basic modular genetic parts are randomly combined to search for combinations that improve the overall performance of the pathway. Modular and highly efficient DNA assembly techniques are desirable for the rapid generation of the large number of parts and combinations. In fact, the generation and screening of large DNA libraries should preferably be performed by high-throughput (HTP) workflows. To this objective, highly efficient, robust, and standardized protocols are essential for HTP DNA assembly.

In the last decade, a number of DNA assembly methods to assemble multiple DNA parts in a single step have been developed, including the Golden Gate method (Engler, Kandzia, & Marillonnet, 2008), Gibson assembly (Gibson et al., 2009), and in vivo recombination (Shao, Zhao, & Zhao, 2009) among others. These methods have been optimized to increase efficiency, fidelity, and modularity resulting in several synthetic biology toolkits that simplify design and handling (Andreou & Nakayama, 2018; Guo et al., 2015; Lampropoulos et al., 2013; Lee, DeLoache, Cervantes, & Dueber, 2015; Martella, Matjusaitis, Auxillos, Pollard, & Cai, 2017; Sarrion-Perdigones et al., 2013; Weber, Engler, Gruetzner, Werner, & Marillonnet, 2011).

Here we present YeastFab (Guo et al., 2015; Yuan et al., 2017; Zhou et al., 2018), a hierarchical DNA assembly toolkit for metabolic engineering in yeast *Saccharomyces cerevisiae*. YeastFab enables the rapid construction of libraries from well-characterized modular and reusable DNA parts and their assembly into larger DNA constructs with biological functions according to standard modular design. In addition, HTP and automated workflows can be readily implemented.

YeastFab utilizes with defined biological parts for its basic modular building blocks: promoters, open reading frames (ORFs), and transcriptional terminators. These parts can be cloned and subsequently released by alternating two Type IIS restriction endonucleases, *Bsa*I and *Bsm*BI, producing complementary sticky overhangs. Harnessing the high fidelity and efficiency of golden gate assembly (Engler et al., 2008), libraries of modularized parts can be generated within a week with nearly 100% success rate. These parts can be easily characterized and assembled into transcriptional units (TUs) and pathways using the same golden gate approach by alternating *Bsm*BI and *Bsa*I enzymes. Due to its modular design, combinatorial assembly of building blocks can be readily accomplished within a week.

In this chapter we describe a step-by-step protocol for: (1) generating libraries of standard biological parts, (2) characterization of these modular parts, (3) construction of TUs from characterized modular parts, and (4) (combinatorial) assembly of pathways.

2. WORKFLOW

The YeastFab system hierarchically assembles standardized biological parts into pathways using golden gate assembly (Fig. 1). The workflow followed to generate and utilize these biological parts for the construction and engineering of a given pathway can be divided into four Steps:

Step 1. Libraries of biological parts (namely, Promoters, ORFs, and Transcriptional terminators) can be easily generated by PCR amplification from the yeast genome with primers containing standardized overhangs and subsequently cloned into the HCKan_P/O/T standard vectors by golden gate assembly using *Bsa*I.

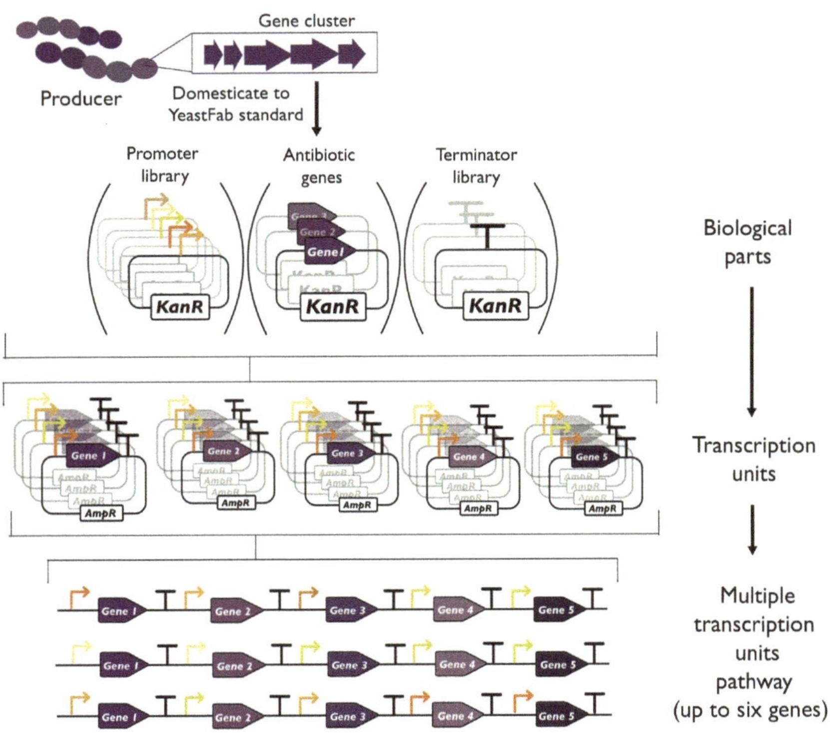

Fig. 1 Scheme of the workflow to construct standard biological parts, transcription units, and pathways. Standard biological parts are generated by PCR amplification from native sequences and cloned into the HCKan standard vectors. Each part from each library can be assembled with other parts to form a TU. TUs can be assembled to form pathways that can be integrated into the yeast genome or into a plasmid.

Step 2. Regulation, expression, and function of these standardized parts can be tested in yeast using a reporter vector. For instance, a promoter library could be characterized under different growth and stress conditions after cloning the parts into the reporter vector Re2.8 by golden gate assembly using *Bsm*BI (or its isoschizomer *Esp*3I that has optimal activity at 37°C).

Step 3. Each individual part from the libraries created in Step 1 can be cherry picked and used to generate TUs containing a promoter, an ORF, and a terminator. These TUs are assembled into the destination POT standard vectors by golden gate using *Bsm*BI (*Esp*3I). Each TU can be readily screened after transforming the POT vector into yeast.

Step 4. Multiple TUs are assembled together into a multi-TU pathway using a *Bsa*I-driven digestion reaction. Finally, the assembled multi-TU can be directly integrated into the yeast genome or kept in a self-replicating vector.

3. DNA PART LIBRARY CONSTRUCTION

3.1 Design of Standard Biological Parts

Three biological parts in yeast were defined: promoter (P), 500 bp upstream of the gene's start codon or up to the last gene boundary; open reading frame (O), the coding region from the start to the stop codon; and transcriptional terminator (T), 200 bp downstream of the gene's stop codon or up to the next gene boundary. The YeastFab system also required the design of standard cloning vectors, HCKan_P, HCKan_O, and HCKan_T, to host each type of parts, respectively (Fig. 2A).

Standardization of biological parts in the YeastFab system exploits the ability of Type IIS restriction enzymes (for this example, *Bsa*I and *Bsm*BI or *Esp*3I) to cleave DNA outside of the recognition sequence leaving 4-bp overhangs (Engler et al., 2008). Biological parts and YeastFab HCKan_P/O/T cloning vectors are designed to place the *Bsa*I site in such a way that after the assembly the *Bsa*I recognition site is removed from the final construct and the DNA-building block is ordered and assembled into the appropriate HCKan_P/O/T vector.

In its simplest form, this process requires the *Bsa*I recognition site, and the designed 4-base overhangs that direct the assembly added to both ends of the DNA of interest by PCR. For this example, when designing the PCR primers, the prefixes and suffixes shown in Table 1 should be added to the

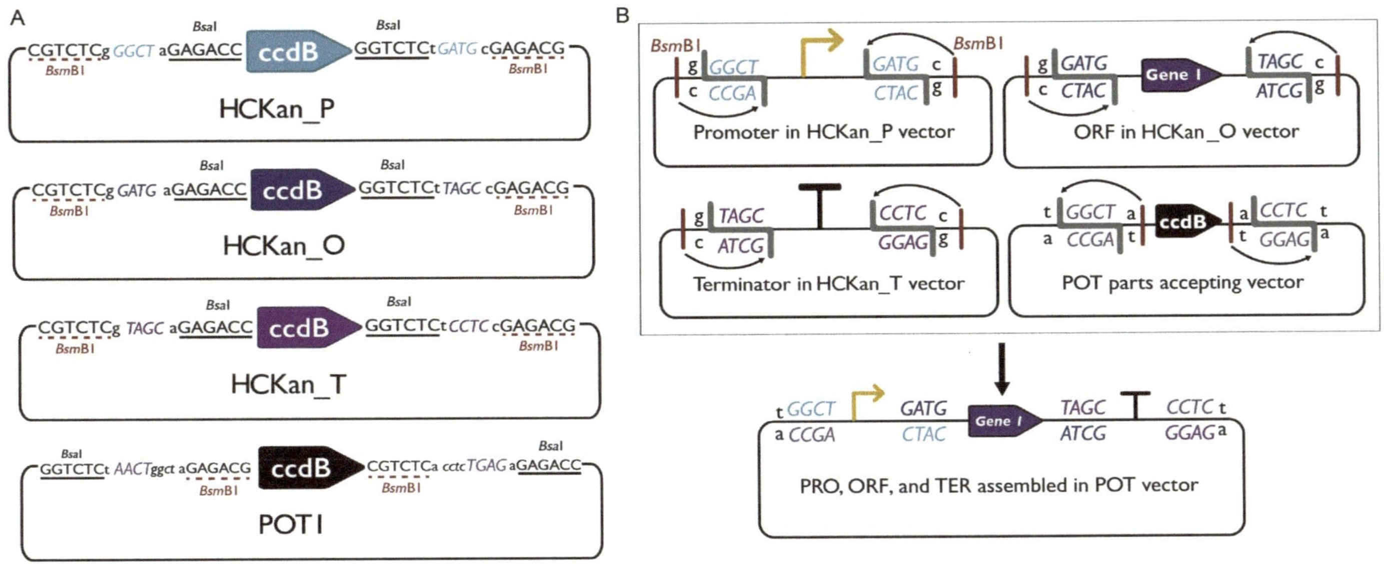

Fig. 2 Detailed representation of the overhangs of YeastFab acceptor vectors and TU assembly. (A) HCKan acceptor vectors and POT vector. In case of HCKan vectors, they contain two different type IIs restriction enzyme recognition sites that produce the same 4-bp overhangs. *Bsa*I releases the ccdB marker. *Bsm*BI is used to put different parts together to construct TUs. In case of POT, *Bsm*BI is used to release the *ccd*B gene, producing the 4-bp overhangs in lowercase italics. *Bsa*I is used to assemble various TUs together, producing the 4-bp overhangs in uppercase italics. (B) Detailed representation of TUs assembly.

Table 1 Prefix Sequences for the Amplification of Standard Biological Parts

Prefix Name	Sequence[a]	Biological Part to be Amplified	Destination Cloning Vector
HCKan_P_Forward	5′ agcgtgGGTCTCa*GGCT*	Promoter	HCKan_P
HCKan_P_Reverse	5′ agcgtgGGTCTCg*CATC*	Promoter	HCKan_P
HCKan_O_Forward	5′ agcgtgGGTCTCa*GATG*	ORF	HCKan_O
HCKan_O_Reverse	5′ agcgtgGGTCTCg*GCTA*	ORF	HCKan_O
HCKan_T_Forward	5′ agcgtgGGTCTCa*TAGC*	Transcriptional terminator	HCKan_T
HCKan_T_Reverse	5′ agcgtgGGTCTCa*GAGG*	Transcriptional terminator	HCKan_T

[a]*Bsa*I recognition site is underlined; four-base overhangs compatible with the destination cloning vector released after digestion are shown in italics.

Table 2 Prefix and Suffix Sequences in HCKan Standard Vectors

Vector Name	Prefix Sequence	Suffix Sequence	Biological Part to be Cloned
HCKan_P	5′ CGTCTCg*GGCT*aGAGACC	5′ GGTCTCt*GATG*cGAGACG	Promoter
HCKan_O	5′ CGTCTCg*GATG*aGAGACC	5′ GGTCTCt*TAGC*cGAGACG	ORF
HCKan_T	5′ CGTCTCg*TAGC*aGAGACC	5′ GGTCTCt*CCTC*cGAGACG	Transcriptional terminator

*Bsa*I recognition site is underlined; *Bsm*BI is double underlined; four-base overhangs compatible with the inserted part are shown in italics.

corresponding primer that binds to the sequence to be amplified. This will allow the user to generate a version of chosen part flanked by *Bsa*I restriction sites compatible with the destination HCKan vector.

It is important to pay special attention when designing the primers to amplify ORFs. Notice that the 4-bp overhangs left after cleaving with *Bsa*I include the start codon (GATG) and the stop codon (TAGC).

Another feature worth mentioning is that once a biological part is cloned into the corresponding HCKan_P/O/T standard vector, it can be released with *Bsm*BI (*Esp*3I) (producing the exact same 4-bp overhangs as *Bsa*I) and assembled in TU in the next level of assembly (Table 2; Fig. 2A).

3.1.1 Note

It is highly recommended that *Bsa*I/*Bsm*BI sites are not present in the sequence of your parts, as efficiency of golden gate assembly will drop dramatically. Therefore, to detect correct assemblies you may need to analyze increased number of colonies, which will boost the labor and cost.

3.2 Construction of P/O/T Libraries

Here we describe the protocol to clone a library of promoters from yeast genomic DNA into the YeastFab destination vector HCKan_P. This protocol can also be followed for generating a library of ORFs and/or transcriptional terminators with only the modification of using the appropriate YeastFab destination vector (HCKan_O or HCKan_T) and consequently, the proper primer design.

3.2.1 Polymerase Chain Reaction

Run PCR to amplify your promoter library as explained later. For this example, PrimeStar HS Premix is used because it is formulated containing a low error rate Taq polymerase (that minimizes mutations), dNTPs, and buffer in one master mix that reduce the risk of pipetting mistakes and contamination. Additionally, its hot-start feature prevents false initiation events during the reaction setup due to mispriming and primer digestion.

3.2.1.1 Buffers and Reagents

- PrimeStar HS Premix (Takara Clontech)
- Genomic DNA template from BY4741 (*MATa*, *his3Δ1*, *ura3Δ0*, *leu2Δ0*, *met15Δ0*) yeast strain (ATCC 201388) (100 ng/μL)
- Forward and reverse DNA primers containing the prefixes described in Table 1 and distributed in 96-well PCR plates (10 μ*M* conc.)

3.2.1.2 Procedure

1. For one 96-well plate, set up on ice a reaction master mix containing 505 μL of 2 × PrimeStar HS master mix, 202 μL of genomic DNA template (100 ng/μL), and 252.5 μL of nuclease-free sterile water. Distribute 9.5 μL of master mix into each well of a 96-well PCR plate.
2. Dispense 0.5 μL of primers (forward and reverse) into the prefilled wells of the plate (0.25 μ*M* final concentration) using a multichannel pipette.
3. Seal and spin down the reaction plate.
4. Run the following thermal cycling protocol:
   Step 1: 98°C for 1 min
   Step 2: 98°C for 10 s
   Step 3: 55°C for 5 s
   Step 4: 72°C for 1 min (go to Step 2 and cycle 30 times)
   Step 5: 72°C for 5 min
   Step 6: 4°C forever

5. Correct PCR amplification is verified by agarose gel electrophoresis. Run 1 μL of each sample on a 1% (w/v) agarose gel. One band of the size of your promoter should be detected.
6. PCR products are cleaned up and diluted to a final concentration of 50 ng/μL.

3.2.1.3 Note

PCR cleanup can be performed with commercial kits like PureLink PCR purification kit (Invitrogen).

The concentration of the purified PCR product was measured using the fluorescent DNA-binding dye QuantiFluor dsDNA system (Promega) following the manufacturer's instructions. This dye binds specifically to dsDNA. Use ≤1 μL of purified PCR reactions to quantify the DNA (you may need to adjust this volume). You can use other methods to quantify DNA concentration.

3.2.2 Golden Gate Assembly Reaction

The digestion and ligation of PCR DNA (promoter) together with HCKan_P vector happens in the same tube during the Golden Gate reaction.

3.2.2.1 Buffers and Reagents

- *Bsa*I (10 U/μL; NEB)
- T4 DNA Ligase (400 U/μL; NEB)
- 10 × T4 DNA Ligase Reaction Buffer (NEB)
- HCKan_P plasmid DNA (100 ng/μL)
- Purified PCR products (50 ng/μL)

3.2.2.2 Procedure

1. On ice, prepare a Golden Gate assembly reaction mix containing 50.5 μL of HCKan_P plasmid DNA (100 ng/μL), 38 μL of *Bsa*I (10 U/μL), 101 μL of 10 × T4 DNA ligase reaction buffer, 12.6 μL of T4 DNA ligase (400 U/μL), and 606 μL of nuclease-free sterile water. Distribute 8 μL of master mix into each well of a 96-well PCR plate.
2. Next, using a multichannel pipette add 2 μL of purified PCR product into each well of the plate.
3. Seal and spin down the reaction plate.
4. Run the following DNA assembly thermal cycling protocol:

   Step 1: 37°C for 5 min

   Step 2: 16°C for 10 min (go to Step 1 and cycle 30 times)

Step 3: 37°C for 30 min
Step 4: 80°C for 20 min
Step 5: 4°C forever

3.2.3 Bacterial Heat Shock Transformation

Transform *E. coli* competent cells with your golden gate reaction. Commercial chemically or electrocompetent cells can be purchased from different providers (follow the manufacturer's instructions for your competent cells).

Additionally, HCKan_P/O/T vectors contain a *ccd*B gene between the two *Bsa*I sites as a cloning marker, so it is important that your *E. coli* strain is sensitive to the toxic effect of the CcdB protein. Therefore, only the *E. coli* colonies containing the vector carrying your insert will survive.

Here, we describe how to transform homemade DH5alpha *E. coli* (calcium chloride) chemically competent cells aliquoted into 96-well PCR plates.

3.2.3.1 Buffers and Reagents

- DH5alpha *E. coli* chemically competent cells prealiquoted into 96-well PCR plates (20 μL/well)
- 1 × Lysogeny broth (LB)
- 8 × LB + kanamycin (50 μg/mL) OmniTray agar plates (Thermo Fisher Scientific)

3.2.3.2 Procedure

1. Thaw the competent cells plate on ice for 5 min.
2. On ice, unseal the plate and add 2.5 μL of the assembly mixture to each well of the competent cell plate.
3. Seal the plate (gas permeable seal) and incubate the mixture on ice for 10 min.
4. Heat shock the plate for 45 s at 42°C in a preheated PCR machine.
5. Incubate the plate on ice for another 5 min.
6. Unseal the plate and add 200 μL of 1 × LB broth to each well of the plate.
7. Reseal the plate with the gas permeable seal and incubate it at 37°C for 1 h.
8. With a multichannel pipette, dispense 20 μL of each culture on a tilted OmniTray agar plate (12 samples/plate; LB + kanamycin (50 μg/mL)); after dispensing the culture let the drop drip down the agar surface (see Fig. 3).

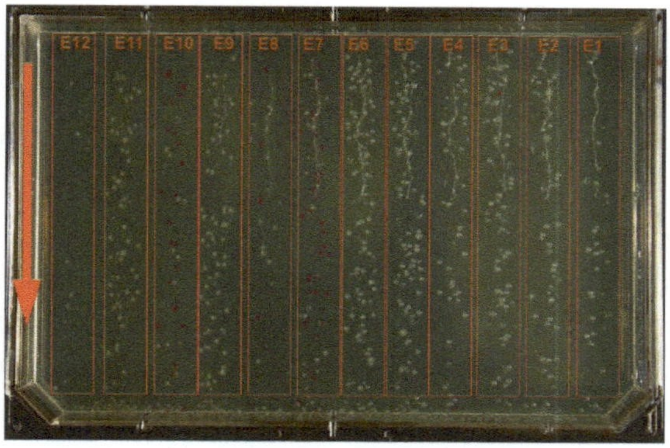

Fig. 3 Example of plating transformed *E. coli* cells in LB OmniTray agar plates. Competent cells transformed with golden gate reaction are dispensed into a tilted LB OmniTray agar plate. This allows isolation of single colonies for 12 transformations (E1–E12) in one agar plate. In this example, the HCKan vector used in the golden gate assembly reaction contained RFP (red fluorescence protein) flanked by *Bsa*I sites, which facilitates the visualization of (in)correct assemblies (white colonies potentially contain correct assemblies).

9. Dry OmniTray agar plates in the tilted position and, once dried, incubate them at 37°C overnight.
10. After 16h, colonies should appear.

3.2.4 Colony PCR

The number of incorrect assemblies is expected to be very low because the HCKan_P (as well as HCKan_O/T) contains the toxic *ccd*B gene between the *Bsa*I cloning sites. As mentioned before, this will dramatically reduce the survival of *E. coli* cells transformed with residual empty vectors, so from two colonies analyzed, at least one should contain the correct assembly.

Pick 2 individual bacterial colonies for each promoter to check for successful assemblies (2 × 96 colonies).

3.2.4.1 Buffers and Reagents

- 2 × GoTaq Green DNA Master Mix (Promega)
- HCKan chc Fwd primer (GATCCTTTGATTTTCTACCG) (10 μ*M*) and HCKan chc Rev primer (CTCGATAACTCAAAAAATACG) (10 μ*M*), these primers bind to all HCKan vectors outside of the inserted DNA
- 10,000 × SYBR Safe DNA Gel Stain (Thermo Fisher Scientific)
- Agarose (Sigma-Aldrich)
- 1 × TAE buffer (40 m*M* Tris pH 7.6, 20 m*M* acetic acid, 1 m*M* EDTA)
- 2-Log DNA ladder (NEB)

3.2.4.2 Procedure

1. Prepare two 96-well PCR plates containing 20 μL of sterile nuclease-free water in each well. Pick one colony for each cloned promoter and inoculate it in the corresponding well of the 96-well PCR plate. Repeat this step with the second 96-well PCR plate.
2. For each 96-well reaction plate (2 plates in total), prepare a reaction master mix on ice containing 525 μL of 2× GoTaq Green DNA Master Mix, 10.5 μL of HCKan chc Fwd primer (10 μ*M*), 10.5 μL of HCKan chc Rev primer (10 μ*M*), and 504 μL of sterile nuclease-free water.
3. On ice, aliquot 9 μL of the master mix into each well of the 96-well PCR plate (×2 plates).
4. Next, add 1 μL of the bacterial solution prepared in step 1 to each well of the plate (×2 plates). Do not dispose of the remaining bacterial solution, as it will be used to further analyze the assemblies.
5. Seal the plate and spin it quickly.
6. Run the following thermal cycling protocol:
   Step 1: 94°C for 10 min
   Step 2: 94°C for 10 s
   Step 3: 50°C for 15 s
   Step 4: 72°C for 1 min (go to Step 2 and cycle 35 times)
   Step 5: 72°C for 7 min
   Step 6: 4°C forever
7. Once finished, run 6 μL of each sample (no loading dye needed) on a 1% w/v agarose gel. One band of the size of your promoter should be detected.

3.2.5 Verification of Successful Assembled Promoters

From the two colonies analyzed in Section 3.2.4, only one will be further analyzed by restriction digest and, finally, sequencing.

3.2.5.1 Buffers and Reagents

- LB + kanamycin (50 μg/mL) selective medium
- Wizard® MagneSil® Plasmid DNA Purification System (Promega)
- *Bsm*BI (10 U/μL; NEB)
- 10× NEB Buffer 3.1 (NEB)
- Purified plasmid DNA (100 ng/μL)
- 10,000× SYBR Safe DNA Gel Stain (Thermo Fisher Scientific)
- Agarose (Sigma-Aldrich)
- 1× TAE buffer (40 m*M* Tris pH 7.6, 20 m*M* acetic acid, 1 m*M* EDTA)
- 2-Log DNA ladder (NEB)

3.2.5.2 Procedure

1. Using a multichannel pipette, aliquot 1 mL of LB + kanamycin (50 μg/mL) media into each well of the 2 mL 96-deep well plate.
2. Inoculate 10 μL from the bacterial solution prepared in step 1 of the procedure described in Section 3.2.4 into each well of the 96-deep well plate.
3. Seal the plate with a gas permeable seal and incubate the culture overnight at 37°C in a shaking incubator.
4. After 18 h, centrifuge the 96-deep well plate and remove the supernatant.
5. Purify the DNA with Wizard MagneSil Plasmid DNA Purification System (Promega), following the manufacturer's instructions. The isolation of plasmid DNA is based on a magnetic resin, which does not require the use of centrifugation or vacuum manifolds. Other purification methods are also available on the market. Follow the manufacturer's instructions.
6. On ice, prepare a restriction reaction mix containing 25.25 μL of *Bsm*BI restriction enzyme, 101 μL of 10 × NEB buffer 3.1 and 379 μL of sterile nuclease-free water. Distribute 5 μL of master mix into each well of a 96-well PCR plate.
7. Next, using a multichannel pipette add 5 μL of purified plasmid DNA (from step 5) into each well of the plate.
8. Seal and spin down the reaction plate.
9. Run the following restriction digest thermal cycling protocol:
   Step 1: 55°C for 2 h
   Step 2: 80°C for 20 min
   Step 3: 4°C forever
10. Once finished, run 6 μL of each sample (add loading dye) on a 1% (w/v) agarose gel. One band of the size of the HCKan_P vector and one band of the size of the promoter should be detected.
11. Cloned promoters that show the correct restriction pattern should be verified by sequencing.
12. Finally, once the sequence is verified, make stocks of the correct clones in 20% glycerol and store them at −80°C.

Now, your biological parts are ready for the next steps.

3.3 Characterization of a Promoter Library

Characterization of promoters is performed in a reporter vector, Re2.8. This vector contains the toxic *ccd*B gene flanked by a pair of *Bsm*BI (*Esp*3I) sites just upstream of the YFP gene that after cleavage will leave 4-bp overhangs compatible with promoter parts. Measurement of YFP fluorescence will provide information about the promoter activity. The reporter vector also contains an internal control, an mCherry gene under the control of the TEF2 promoter and ADH1 terminator. The activity of each promoter is calculated as the ratio of YFP to mCherry fluorescence intensities.

Here, we describe the protocol to clone a collection of promoters into the reporter vector for their characterization.

3.3.1 Golden Gate Assembly Reaction

3.3.1.1 Buffers and Reagents

- *Esp*3I (10 U/μL; Thermo Scientific)
- T4 DNA Ligase (400 U/μL; NEB)
- 10 × T4 DNA Ligase Reaction Buffer (NEB)
- Purified DNA of promoters cloned into HCKan_P vector distributed in 96-well plates (20 ng/μL)
- Re2.8 plasmid DNA (100 ng/μL)

3.3.1.2 Procedure

1. On ice, prepare a Golden Gate assembly reaction mix containing 10.1 μL of Re2.8 plasmid DNA (100 ng/μL), 30.3 μL of *Esp*3I (10 U/μL), 101 μL of 10 × T4 DNA ligase reaction buffer, 12.6 μL of T4 DNA ligase (400 U/μL), and 805.5 μL of sterile nuclease-free water. Distribute 9.5 μL of master mix into each well of a 96-well PCR plate.
2. Next, using a multichannel pipette add 0.5 μL of each promoter cloned in HCKan_P into each well of the plate.
3. Seal and spin down the reaction plate.
4. Run the following DNA assembly thermal cycling protocol:
   Step 1: 37°C for 5 min
   Step 2: 16°C for 10 min (go to Step 1 and cycle 30 times)
   Step 3: 37°C for 30 min
   Step 4: 80°C for 20 min
   Step 5: 4°C forever
5. The verification of correct assemblies by colony PCR and sequencing follows the protocol described in Section 3.2 with the only exception that after transformation of *E. coli* competent cells, ampicillin (100 μg/mL) should be used as selection marker for Re2.8 instead of

kanamycin. For colony PCR and sequencing use the following primers: Forward: GCGTATATATACCAATCTAAGTCT and Reverse: GTCAATTTACCGTAAGTAGCATC.

6. Correct reporter assemblies are transformed into yeast following the standard LiAc/ssDNA/PEG protocol (Gietz & Schiestl, 2007) and selected on synthetic complete medium lacking leucine (SC-Leu). Independent clones can be tested in different growth conditions. Once the behavior of promoters in these conditions is profiled, the information can be used to in silico model multigene synthetic pathways before their assembly.

3.3.1.3 Note

Carbenicillin (100 μg/mL) can be used in place of ampicillin. Carbenicillin is more stable and will prevent the emergence of satellite colonies.

4. ASSEMBLY OF SYNTHETIC PATHWAYS

Once the biological parts are in the YeastFab standard, they can be easily assembled together to generate functional units (TUs) and pathways.

First, we describe the protocol to assemble a promoter, ORF, and transcriptional terminator into a POT standard vector.

4.1 Assembly of TUs

In this assembly the three donor vectors (HCKan_P/O/T) containing the promoter, ORF, and terminator of interest are cloned into a receiver POT vector by golden gate using *Bsm*BI (*Esp*3I) (Fig. 2B). To obtain high assembly efficiencies it is recommended that each donor and destination vector are added in equimolar amounts.

4.1.1 Buffers and Reagents

- *Esp*3I (10 U/μL; Thermo Fisher)
- T4 DNA Ligase (400 U/μL; NEB)
- 10 × T4 DNA Ligase Reaction Buffer (NEB)
- Purified DNA of cloned promoter, ORF, and terminator (HCKan_P, HCKan_O, and HCKan_T)
- POT vector (100 ng/μL)

4.1.2 Procedure

1. On ice, prepare a reaction mix containing 3.5 U of *Esp*3I, 1 × T4 DNA ligase reaction buffer, 100 U of T4 DNA ligase, 50 ng of POT destination vector, and equimolar amounts of HCKan_P, HCKan_O, and HCKan_T and bring the volume to 10 μL with sterile nuclease-free water.
2. Run the following DNA assembly thermal cycling protocol:
   Step 1: 37°C for 5 min
   Step 2: 16°C for 10 min (go to Step 1 and cycle 30 times)
   Step 3: 37°C for 30 min
   Step 4: 80°C for 20 min
   Step 5: 4°C forever
3. After the reaction, use 5 μL to transform *E. coli* competent cells following the standard protocol described before. Plate transformed cells onto LB ampicillin (100 μg/mL) agar plates.
4. After 16 h, colonies should appear. Pick two colonies and verify the assemblies by colony PCR and restriction digestion.

The TUs assembled in POT vectors can not only be directly used to transform yeast competent cells to test functional expression but also can be assembled to generate a multigene pathway.

4.2 Assembly of Pathways

Metabolic pathways usually include more than one gene. The YeastFab system comes with 11 POT overhang-compatible vectors that enable the rapid combinatorial assembly of up to six genes in one step by golden gate using *Bsa*I. If more than six genes need to be assembled, the number of compatible POT vectors can be extended. The combination of these POT vectors is based on the number of TUs to be assembled as shown in Fig. 4. Table 3 shows the 4-bp overhangs of each POT vector. The TUs should be assembled in the correct POT vector prior the construction of the pathway. Then,

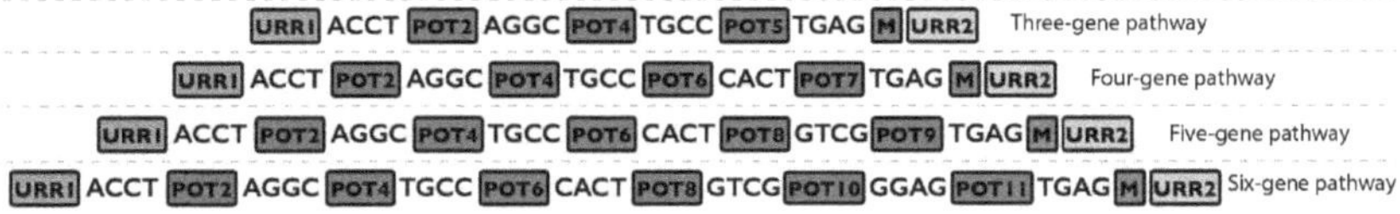

Fig. 4 Representation of assembling a pathway with up to six genes. Different POT vectors should be used based on the number of genes to be assembled.

Table 3 Prefix and Suffix Sequences in POT Standard Vectors

Vector Name	Prefix Sequence	Suffix Sequence	Total TUs[a]
POT1	5′ GGTCTCt*ACCT*ggctaGAGACG	5′ CGTCTCacctc*TGAG*aGAGACC	1
POT2	5′ GGTCTCt*ACCT*ggctaGAGACG	5′ CGTCTCacctc*AGGC*aGAGACC	2,3,4,5,6
POT3	5′ GGTCTCt*AGGC*ggctaGAGACG	5′ CGTCTCacctc*TGAG*aGAGACC	2
POT4	5′ GGTCTCt*AGGC*ggctaGAGACG	5′ CGTCTCacctc*TGCC*aGAGACC	3,4,5,6
POT5	5′ GGTCTCt*TGCC*ggctaGAGACG	5′ CGTCTCacctc*TGAG*aGAGACC	3
POT6	5′ GGTCTCt*TGCC*ggctaGAGACG	5′ CGTCTCacctc*CACT*aGAGACC	4,5,6
POT7	5′ GGTCTCt*CACT*ggctaGAGACG	5′ CGTCTCacctc*TGAG*aGAGACC	4
POT8	5′ GGTCTCt*CACT*ggctaGAGACG	5′ CGTCTCacctc*GTCG*aGAGACC	5,6
POT9	5′ GGTCTCt*GTCG*ggctaGAGACG	5′ CGTCTCacctc*TGAG*aGAGACC	5
POT10	5′ GGTCTCt*GTCG*ggctaGAGACG	5′ CGTCTCacctc*GGAG*aGAGACC	6
POT11	5′ GGTCTCt*GGAG*ggctaGAGACG	5′ CGTCTCacctc*TGAG*aGAGACC	6

[a]Number of TUs that can be assembled with each POT.
*Bsa*I recognition site is underlined; *Bsm*BI is double underlined; four-base overhangs released after *Bsa*I cleavage are shown in italics; four-base overhangs released after *Bsm*BI (*Esp*I) cleavage are shown in uppercase italics.

for a two-gene pathway use POT2 and POT3; for a three-gene pathway use POT2, POT4, and POT5; for a four-gene pathway use POT2, POT4, POT6, and POT7; for a five-gene pathway use POT2, POT4, POT6, POT8, and POT9; finally for a six-gene pathway use POT2, POT4, POT6, POT8, POT10, and POT11.

At this point, two strategies can be used to introduce a synthetic pathway in yeast: (1) integration of the exogenous pathway into the yeast genome and (2) insertion of the exogenous pathway into a self-replicating vector.

For integration into the yeast genome two unique recombination regions (URR1 and URR2) of 500 bp (Guo et al., 2015) were designed and cloned into the HO locus on chromosome IV (flanking a selection marker *HIS3*), generating a specific target to accommodate the pathway. Therefore, the assembled multi-TU flanked by the selection marker *LEU2* and URR1 and URR2 will be used to transform yeast competent cells, and by homologous recombination through URR1 and URR2 the multi-TU will replace the selection marker *HIS3* at the HO locus. Then, the yeast colonies carrying the successful assemblies inserted into the HO locus will be able to grow in selective media lacking leucine but not histidine.

Additionally, a yeast shuttle vector derived from pRS4XX series was modified to contain the two URRs. In this case, assembled multi-TU flanked by the selection marker *LEU2* and URR1 and URR2 will be used to cotransform yeast competent cells together with linearized pRS4XX. Then, yeast homologous recombination machinery will recombine the multi-TU and the pRS4XX through URR1 and URR2 generating a circular self-replicating plasmid.

For this example, we will describe the integration of a three-gene pathway into the HO locus. The production of β-carotene in yeast can be achieved by introducing three genes from the β-carotene pathway of *Xanthophyllomyces dendrorhous*: *crtE* (GGPP synthase), *crtI* (phytoene desaturase), and *crtYB* (bifunctional phytoene synthase and lycopene cyclase) (Guo et al., 2015; Verwaal et al., 2007). Each gene was cloned into the HCKan_O vector and assembled into POT2, POT4, and POT5 vectors, respectively, with three different promoters (pTDH3, pTEF2, and pCYC1, representing strong, medium, and weak promoters) and the ADH1 terminator. A total of 27 possible combinations were assembled and tested in parallel.

This protocol requires three steps: (1) releasing URR1, URR2, and *LEU2* from pMV vector by *Bsm*BI; (2) releasing *crtE*, *crtI*, and *crtYB* from POT vector by *Bsa*I; and (3) ligation of URR1, URR2, *LEU2*, *crtE*, *crtI*, and *crtYB*.

4.2.1 Buffers and Reagents

- *Bsa*I (10 U/μL; NEB)
- 10 × NEB CutSmart Buffer (NEB)
- T4 DNA Ligase (400 U/μL; NEB)
- 10 × T4 DNA Ligase Reaction Buffer (NEB)
- Purified DNA of POT2-*crtE*, POT4-*crtI*, and POT5-*crtYB*
- Purified pMV-URR1, pMV-URR2, and pMV-*LEU2*
- *Bsm*BI (10 U/μL; NEB)
- 10 × NEB Buffer 3.1 (NEB)
- Blocking oligos for *Bsm*BI system LRsm (5′-aAGAGACgCAAGACA-CTGCGGATa and 5′-aAGAGACgGAGTCACTGCCAACa) (10 μ*M*)
- Blocking oligos for *Bsa*I system LRcom (5′-aAGAGACCCAAGACAC-TGCGGATa and 5′-aAGAGACCGAGTCACTGCCAACa) (10 μ*M*)

4.2.2 Procedure

4.2.2.1 Releasing URRs and Selectable Marker

URR1, URR2, and the selectable marker *LEU2* are in pMV plasmids flanked by *Bsm*BI sites.

1. On ice, prepare a reaction mix containing 3 U of *Bsm*BI, 1 × buffer 3.1, 250 ng of each plasmid pMV and bring the volume to 8 μL with sterile nuclease-free water.
2. Run the following DNA assembly thermal cycling protocol:
   Step 1: 55°C for 3 h
   Step 2: 4°C forever
3. Add 0.9 μL of 10 μ*M* LRsm to the restriction reaction.
4. Run the following protocol in a thermocycler:
   Step 1: 83°C for 6 min
   Step 2: 80°C for 3 min
   Step 3: 75°C for 90 s
   Step 4: 70°C for 90 s
   Step 5: 65°C for 90 s
   Step 6: 60°C for 90 s
   Step 7: 55°C for 90 s
   Step 8: 50°C for 90 s
   Step 9: 4°C forever

4.2.2.2 Releasing TUs

The TUs are in POT vectors flanked by *Bsa*I sites.

1. On ice, prepare a reaction mix containing 3 U of *Bsa*I, 1 × CutSmart buffer, 250 ng of each plasmid POT containing *crtE*, *crtI*, and *crtYB* and bring the volume to 8 μL with sterile nuclease-free water.
2. Run the following DNA assembly thermal cycling protocol:
   Step 1: 37°C for 3 h
   Step 2: 4°C forever
3. Add 0.9 μL of 10 μ*M* LRcom to the restriction reaction.
4. Run the following protocol in a thermocycler:
   Step 1: 83°C for 6 min
   Step 2: 80°C for 3 min
   Step 3: 75°C for 90 s
   Step 4: 70°C for 90 s
   Step 5: 65°C for 90 s
   Step 6: 60°C for 90 s
   Step 7: 55°C for 90 s
   Step 8: 50°C for 90 s
   Step 9: 4°C forever

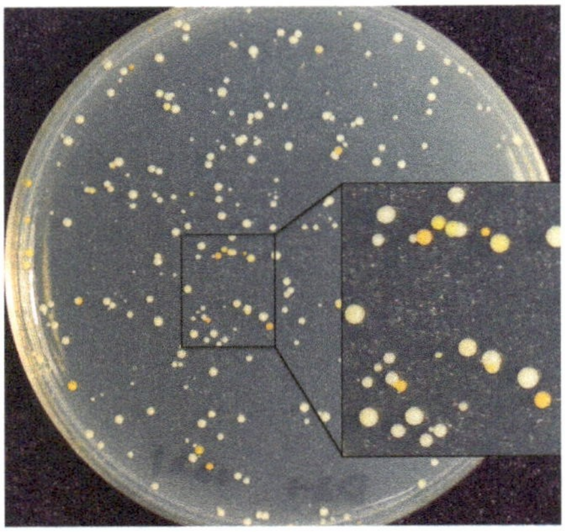

Fig. 5 *S. cerevisiae* cells transformed with a pool of β-carotene pathways growing on a selective medium agar plate. The presence of yellow-orange colonies indicates the cells could produce β-carotene successfully. The plate was photographed after 60h of incubation at 30°C.

4.2.2.3 Ligation

1. On ice, prepare a ligation reaction containing 4μL of URRs-marker digestion and 4μL of TUs digestion reaction, 1 × T4 DNA ligase reaction buffer, 100U of T4 DNA ligase and bring the volume to 10μL with sterile nuclease-free water.
2. Run the following protocol:
   Step 1: 16°C for 16h
   Step 2: 65°C for 20min
   Step 3: 4°C forever
3. Run 1μL of ligation reaction (add loading dye) on a 0.7% w/v agarose gel. A major band of the size of the construct URR1-multi-TU-*LEU2*-URR2 should be visualized.
4. Transform yeast competent cells (with URR1-*HIS3*-URR2 at HO locus) with the remaining 9μL of ligation reaction following the standard LiAc/ssDNA/PEG protocol. Plate transformed cells onto SC-LEU agar plates.
5. After 3 days, colonies with different yellow-orange intensity expressing β-carotene should appear (Fig. 5).

5. COMBINATORIAL ASSEMBLY OF TUs

The modularity of the YeastFab system makes it possible to fine tune gene expression of a multi-TU pathway. As we showed earlier, the combination of three promoters enables different β-carotene production levels (Fig. 5). In this example, gene expression is balanced by combinatorial assembly of a library of 150 promoters with *crtE*, *crtI*, and *crtYB* to optimize the production of β-carotene by yeast.

5.1 POT Pool Generation

5.1.1 Buffers and Reagents

- *Esp*3I (10U/μL; Thermo Fisher)
- T4 DNA Ligase (400U/μL; NEB)
- 10× T4 DNA Ligase Reaction Buffer (NEB)
- Pool of 150 promoters clone into HCKan_P
- Purified DNA of HCKan_O-*crtE*, HCKan_O-*crtI*, and HCKan_O-*crtYB*
- Purified DNA of HCKan_T-ADH1
- POT2, POT4, and POT5 vector (100ng/μL)
- Plasmid safe ATP-dependent DNase (10U/μL; Epicentre)
- 25m*M* ATP solution

5.1.2 Procedure

1. Generate a pool of 150 promoters (previously cloned into HCKan_P) in a tube by adding the same mole amount of each promoter. For example, add 2μL of 1μ*M* of each HCKan_P to the same tube.
2. On ice, prepare a reaction mix to construct a pool of POT vectors containing *crtE* under the control of different promoters. Add 3.5U of *Esp*3I, 1× T4 DNA ligase reaction buffer, 100U of T4 DNA ligase, 50ng of POT2 vector, and equimolar amounts of the promoter pool, HCKan_O-*crtE* and HCKan_T-ADH1, and bring the volume to 10μL with sterile nuclease-free water.

   In parallel, prepare two reaction mixes. One should contain 3.5U of *Esp*3I, 1× T4 DNA ligase reaction buffer, 100U of T4 DNA ligase, 50ng of POT4 vector, and equimolar amounts of the promoter pool, HCKan_O-*crtI* and HCKan_T-ADH1. The other one should contain 3.5U of *Esp*3I, 1× T4 DNA ligase reaction buffer, 100U of T4 DNA ligase, 50ng of POT5 vector, and equimolar amounts of the promoter pool, HCKan_O-*crtYB* and HCKan_T-ADH1, and bring the volume to 10μL with sterile nuclease-free water.
3. Run the following DNA assembly thermal cycling protocol:

   Step 1: 37°C for 5min
   Step 2: 16°C for 10min (go to Step 1 and cycle 30 times)
   Step 3: 37°C for 30min
   Step 4: 80°C for 20min
   Step 5: 4°C forever

4. After the reaction, add 0.25 μL of Plasmid safe ATP-dependent DNase and 0.8 μL of ATP solution to each tube.
5. Run the following digestion protocol in a thermocycler:
   Step 1: 37°C for 30 min
   Step 2: 70°C for 20 min
   Step 3: 4°C forever
6. Use 5 μL of each digestion reaction to transform *E. coli* competent cells following the standard protocol described before. Plate transformed cells onto LB Carbenicillin (100 μg/mL) agar plates. After 16 h, colonies should appear.
7. Add 1 mL of LB liquid medium on the agar plate and scrape all the colonies for each plate (each pool). Extract the plasmid pool from the cells using a commercial miniprep kit like Qiagen plasmid miniprep kit.

Now, you have three pools of vectors, POT2-*crtE*, POT4-*crtI*, and POT5-*crtYB*, with different promoters.

5.1.3 Note

Addition of Plasmid ATP-dependent DNase in step 4 will reduce background and false positives. During step 3 of the DNA assembly protocol, remaining undigested HCKan and POT vectors will be cleaved and linearized. The ATP-dependent DNase will digest linear dsDNA selectively, while closed-circular DNA vectors (correctly assembled POT vectors) will stay undigested. You can also add 3 U of *Esp*3I together with plasmid safe ATP-dependent DNase to ensure that after the golden gate reaction all HCKan and POT vectors that did not participate in the assembly are linearized and hence, digested by the DNase.

5.2 Combinatorial Assembly

5.2.1 Buffers and Reagents

- *Bsa*I (10 U/μL; NEB)
- 10 × NEB CutSmart Buffer (NEB)
- T4 DNA Ligase (400 U/μL; NEB)
- 10 × T4 DNA Ligase Reaction Buffer (NEB)
- Pools of POT2-*crtE*, POT4-*crtI*, and POT5-*crtYB*
- Purified pMV-URR1, pMV-URR2, and pMV-*LEU2*
- *Bsm*BI (10 U/μL; NEB)
- 10 × NEB Buffer 3.1 (NEB)

- Blocking oligos for *Bsm*BI system LRsm (5′-aAGAGACgCAAGACA-CTGCGGATa and 5′-aAGAGACgGAGTCACTGCCAACa) (10 μ*M*)
- Blocking oligos for *Bsa*I system LRcom (5′-aAGAGACCCAAGACA-CTGCGGATa and 5′-aAGAGACCGAGTCACTGCCAACa) (10 μ*M*)

5.2.2 Procedure

5.2.2.1 Releasing URRs and Selectable Marker

Follow the protocol described in the procedure for "releasing URRs and selectable marker" in Section 4.2.2.1.

5.2.2.2 Releasing Pool of TUs

1. On ice, prepare a reaction mix containing 3 U of *Bsa*I, 1 × CutSmart buffer, 250 ng of each pool of POTs containing *crtE*, *crtI*, and *crtYB* and bring the volume to 8 μL with sterile nuclease-free water.
2. Continue the protocol from step 2 of the procedure for "Releasing TUs" in Section 4.2.2.2.

5.2.2.3 Ligation

1. On ice, prepare a ligation reaction containing 4 μL of URRs-marker digestion and 4 μL of pool of TUs digestion reaction, 1 × T4 DNA ligase reaction buffer, 100 U of T4 DNA ligase and bring the volume to 10 μL with sterile nuclease-free water.
2. Continue the protocol from step 2 of the procedure for "Ligation" in Section 4.2.2.3.

6. VERIFICATION OF CORRECT ASSEMBLIES

Despite extensive optimization of protocols to assemble DNA parts into pathways, verification of the assembly is still a necessity. Prior to confirming the constructs by Sanger sequencing, multiple clones are screened by PCR over junctions between individual parts and restriction pattern analysis. These approaches help to narrow the number of samples to sequence and hence, cut down the total cost considerably.

6.1 PCR Over Junctions

Junction PCR analysis checks for correctly ordered assembly. In our particular case, PCR is performed using one primer specific to one of the parts (e.g., POT2-*crtE*) and the other specific to the next expected part (e.g., POT4-*crtI*). Junction PCR also checks for correct integration of multi-TUs into the proper yeast genome locus. For this, one of the primers is designed to be specific to the genomic sequence outside of the homology URR and the other is specific to the insert. You need to design as many pairs of primers as junctions.

The correct assembly of our β-carotene pathway integrated into the HO locus will require four pairs of primers to amplify by PCR the junctions between: (1) HO locus and *crtE*; (2) *crtE* and *crtI*; (3) *crtI* and *crtYB*; and (4) *crtYB* and HO locus.

In addition, to simplify the process, the PCR over junctions can be performed directly on yeast colonies which avoids the need for prior extraction of genomic or plasmid DNA.

6.1.1 Buffers and Reagents

- 2 × GoTaq Green DNA Master Mix (Promega)
- Forward and reverse primers (10 μ*M*)
- 10,000 × SYBR Safe DNA Gel Stain (Thermo Fisher Scientific)
- Agarose (Sigma-Aldrich)
- 1 × TAE buffer (40 m*M* Tris pH 7.6, 20 m*M* acetic acid, 1 m*M* EDTA)
- 2-Log DNA ladder (NEB)
- 20 m*M* NaOH

6.1.2 Procedure

1. Pick yeast colonies and inoculate into a tube or 96-well PCR plate containing 20 μL of sterile nuclease-free water.
2. In a new tube or 96-well PCR plate, aliquot 10 μL of 20 m*M* NaOH.
3. Transfer 10 μL of resuspended cells to the wells containing 20 m*M* NaOH to obtain a final concentration of 10 m*M* NaOH. Do not dispose of the remaining non-NaOH-treated solution prepared in step 1 as it will be used to further analyze the assemblies.
4. Incubate the NaOH-treated cell suspension for 10 min at 95°C, then cool to 4°C.
5. For each PCR reaction prepare a reaction master mix on ice containing 5 μL of 2 × GoTaq Green DNA Master Mix, 0.5 μL of Forward primer (10 μ*M*), 0.5 μL of Reverse primer (10 μ*M*), and 2 μL of sterile nuclease-free water.
6. Next, add 2 μL of NaOH-treated cell suspension prepared in step 4 to each well, for yeast colony PCR.

7. Run the following thermal cycling protocol:
   Step 1: 94°C for 10 min
   Step 2: 94°C for 10 s
   Step 3: 50°C for 1 min 30 s
   Step 4: 72°C for 1 min/kb (go to Step 2 and cycle 35 times)
   Step 5: 72°C for 7 min
   Step 6: 4°C forever
8. Once finished, run 6 μL of each sample (no loading dye needed) on a 1% w/v agarose gel. One band of the predicted size should be detected.
9. Choose two or three colonies that are PCR positive for all the junctions analyzed and recover them from the cell suspension (step 1).

6.2 Restriction Pattern Analysis

In the case that the multi-TUs are assembled into a self-replicating vector, analysis of the restriction pattern will facilitate the detection of the correct assemblies before Sanger sequencing.

Selection of the right restriction enzyme (RE) is essential to obtain a restriction pattern providing relevant information. The RE or combination of REs should not generate extremely small fragments (that cannot be detected in an agarose gel) and should not generate extremely large fragments (where small missing regions cannot be observed). The size of the fragments should lie between 8 and 0.5 kb, preferably. Ideally the RE should cleave in the sequence of each assembled part and release fragments of different size that can be separated and clearly visualized as individual bands in an agarose gel.

6.2.1 Buffers and Reagents

- Zymoprep yeast plasmid miniprep II kit (ZymoResearch)
- Qiagen plasmid miniprep kit (Qiagen)
- Restriction enzyme and reaction buffer recommended by the provider
- 10,000 × SYBR Safe DNA Gel Stain (Thermo Fisher Scientific)
- Agarose (Sigma-Aldrich)
- 1 × TAE buffer (40 m*M* Tris pH 7.6, 20 m*M* acetic acid, 1 m*M* EDTA)
- 2-Log DNA ladder (NEB)

6.2.2 Procedure

1. After recovering the yeast colonies, selected from the PCR junction screen step, in selective media, the vector is recovered using the Zymoprep yeast plasmid miniprep II kit following the manufacturer's instructions.

2. *E. coli* competent cells are transformed with 2 μL of miniprepped DNA following the standard protocol and plated onto selective LB ampicillin (100 μg/mL) agar plates.
3. After 16 h, colonies should appear. Pick one colony and inoculate into 5 mL of selective LB ampicillin (100 μg/μL) liquid medium. Incubate at 37°C for 16 h in a shaking incubator. Extract the vector with Qiagen plasmid miniprep kit.
4. For each restriction reaction, prepare a master mix containing 10 U of RE, 1 μg of DNA, 1 × restriction enzyme buffer and bring the volume to 15 μL with sterile nuclease-free water.
5. Run the following restriction digest thermal cycling protocol:
   Step 1: 37°C for 3 h (use the enzyme's optimum incubation temperature)
   Step 2: 80°C for 20 min
   Step 3: 4°C forever
6. Once finished, run 10 μL of the restriction reaction (add loading dye) on a 0.7% (w/v) agarose gel. The electrophoresis data of the restriction digest should match the expected digest pattern.

6.3 Sequencing

The last step is to sequence verify the assembled multi-TUs. You must design enough primers to sequence the entire multi-TUs region (primer walking).

The vector containing the multi-TUs can be directly used in the sequencing reaction as a template. However, to sequence the integrated multi-TUs, you would need to amplify the inserted DNA by PCR and sequence the PCR product. Follow the instructions of your Sanger sequencing facility.

7. TROUBLESHOOTING

7.1 Failure of PCR

Achieving specific amplification with high yield of your DNA parts requires careful primer design. Melting temperature (T_m), GC content, secondary structures, cross homology, and dimerization are some of the parameters that can cause PCR failure. Thus, it is essential to use a primer design software which assists in PCR primer design.

In addition to a thorough design, keeping the reactions cold and using hot-start polymerases may possibly reduce nonspecific products due to nonspecific priming events.

If nonspecific products are still observed, you may try "touchdown PCR" (TD-PCR) (Korbie & Mattick, 2008). Briefly, TD-PCR is a traditional PCR where the initial annealing temperature is higher than the optimal melting temperature of the primers (10°C above the T_m) and is gradually reduced over subsequent cycles until the calculated T_m temperature is reached.

7.2 Golden Gate Reaction Failure

High-efficiency golden gate assembly can be hampered by many parameters. Some of these have been already optimized in the protocols explained earlier. Here we will mention those that require special attention.

The amount of enzyme added should not exceed 10% of the total reaction volume. This ensures that the total glycerol concentration does not exceed 5% (v/v), which may promote star activity. Bear in mind that restriction enzyme and T4 ligase are added into the same reaction, so you may prefer to use concentrated T4 ligase to reduce the total amount of glycerol.

T4 DNA ligase requires ATP to function. ATP will degrade after multiple freeze-thaws so make aliquots of the reaction buffer and use fresh buffer for every reaction.

Accurate equimolar ratios of the amount of the DNA will benefit the performance of the golden gate reaction. We recommend quantifying the DNA as accurately as possible. There are commercial kits based on fluorescent dyes that specifically bind to dsDNA that will avoid underestimation of the DNA amount due to other contaminants in the tube. Additionally, as this method is based on fluorescence measurements it can be easily adopted to HTP workflows.

Quality of DNA also affects the final performance of the golden gate reaction. We recommend DNA purification whenever possible.

The presence of internal recognition sites in your DNA parts will dramatically reduce fidelity. Removing these internal sites is imperative to improve the overall performance of the DNA assembly. If the internal site is in the ORF part it can be easily removed without affecting the function. However, when the site appears in the promoter or terminator part, modification of the sequence will modify function and you may need to characterize the performance of that part which may be different from the original one. Another option is to remove those promoters and terminators from the library.

7.3 High Colony Background

After transformation of *E. coli* with the golden gate reaction mix, screening a couple of colonies is necessary to find the correct one. As explained in the text introduction of *ccd*B gene in our plasmids dramatically reduces the number of colonies to be analyzed. At the same time it is important to use an *E. coli* strain sensitive to the toxic effect of *ccd*B gene. Although DH5alpha is usually the lab workhorse in many labs for standard cloning, sometimes this strain "reverts" and it is able to survive the toxic effect of the *ccd*B gene resulting in false positives. You may need to check different *E. coli* strains to find the one with least background. This is especially important in the case of combinatorial assemblies as pools of plasmids are generated from pool of colonies that have not been subjected to assembly verification.

Another step mentioned in Section 5.2 is the addition of plasmid safe ATP-dependent DNase together with the restriction enzyme after golden gate reaction and before *E. coli* transformation. Although this approach can be used for every golden gate reaction reducing the background it will increase the cost especially in HTP workflows.

7.4 Recombination in Yeast

Yeast is well known for its high homologous recombination efficiency. If the assembled multi-TUs contain the same sequence repeated, e.g., the same promoter controls the expression of more than one gene, then an event of recombination could happen and a piece of DNA could be deleted, duplicated, or translocated. A careful design avoiding repeating the same sequence may help to reduce homologous recombination.

8. SUMMARY AND CONCLUSIONS

We present here a hierarchical assembly system that allows us to expedite the construction and optimization of a complete pathway in yeast. First, we systematically designed and constructed standard biological parts based on the native genes and regulatory elements in *S. cerevisiae*. The assembly routine exploits the existing Golden Gate method by alternating between two Type IIS enzymes in each level of the assembly scheme. Optimization of protocols is a critical step for the success of the assembly. We have demonstrated the high efficiency of YeastFab by generating libraries of hundreds of biological parts, e.g., promoters with a success rate nearly 100%.

Further, standardized parts are easily characterized under various culture conditions, which assists in the rational design of metabolic pathways. Finally, these well-characterized elements can be assembled into functional pathways in few simple steps. Furthermore, the performance of the metabolic pathway can be finely optimized through combinatorial assembly using hundreds of regulatory parts.

REFERENCES

Andreou, A. I., & Nakayama, N. (2018). Mobius assembly: A versatile golden-gate framework towards universal DNA assembly. *PLoS One*, *13*(1), e0189892. https://doi.org/10.1371/journal.pone.0189892.

Chen, Y., & Nielsen, J. (2013). Advances in metabolic pathway and strain engineering paving the way for sustainable production of chemical building blocks. *Current Opinion in Biotechnology*, *24*(6), 965–972. https://doi.org/10.1016/j.copbio.2013.03.008.

Da Silva, N. A., & Srikrishnan, S. (2012). Introduction and expression of genes for metabolic engineering applications in *Saccharomyces cerevisiae*. *FEMS Yeast Research*, *12*(2), 197–214. https://doi.org/10.1111/j.1567-1364.2011.00769.x.

Engler, C., Kandzia, R., & Marillonnet, S. (2008). A one pot, one step, precision cloning method with high throughput capability. *PLoS One*, *3*(11), e3647. https://doi.org/10.1371/journal.pone.0003647.

Galanie, S., Thodey, K., Trenchard, I. J., Filsinger Interrante, M., & Smolke, C. D. (2015). Complete biosynthesis of opioids in yeast. *Science*, *349*(6252), 1095–1100. https://doi.org/10.1126/science.aac9373.

Gibson, D. G., Young, L., Chuang, R.-Y., Venter, J. C., Hutchison Iii, C. A., & Smith, H. O. (2009). Enzymatic assembly of DNA molecules up to several hundred kilobases. *Nature Methods*, *6*, 343. https://doi.org/10.1038/nmeth.1318.

Gietz, R. D., & Schiestl, R. H. (2007). High-efficiency yeast transformation using the LiAc/SS carrier DNA/PEG method. *Nature Protocols*, *2*, 31. https://doi.org/10.1038/nprot.2007.13.

Gorgens, J. F., van Zyl, W. H., Knoetze, J. H., & Hahn-Hagerdal, B. (2001). The metabolic burden of the PGK1 and ADH2 promoter systems for heterologous xylanase production by *Saccharomyces cerevisiae* in defined medium. *Biotechnology and Bioengineering*, *73*(3), 238–245.

Guo, Y., Dong, J., Zhou, T., Auxillos, J., Li, T., Zhang, W., et al. (2015). YeastFab: The design and construction of standard biological parts for metabolic engineering in *Saccharomyces cerevisiae*. *Nucleic Acids Research*, *43*(13), e88. https://doi.org/10.1093/nar/gkv464.

Karim, A. S., Curran, K. A., & Alper, H. S. (2013). Characterization of plasmid burden and copy number in Saccharomyces cerevisiae for optimization of metabolic engineering applications. *FEMS Yeast Research*, *13*(1), 107–116. https://doi.org/10.1111/1567-1364.12016.

Korbie, D. J., & Mattick, J. S. (2008). Touchdown PCR for increased specificity and sensitivity in PCR amplification. *Nature Protocols*, *3*, 1452. https://doi.org/10.1038/nprot.2008.133.

Lampropoulos, A., Sutikovic, Z., Wenzl, C., Maegele, I., Lohmann, J. U., & Forner, J. (2013). GreenGate—A novel, versatile, and efficient cloning system for plant transgenesis. *PLoS One*, *8*(12), e83043. https://doi.org/10.1371/journal.pone.0083043.

Lee, M. E., DeLoache, W. C., Cervantes, B., & Dueber, J. E. (2015). A highly characterized yeast toolkit for modular, multipart assembly. *ACS Synthetic Biology*, *4*(9), 975–986. https://doi.org/10.1021/sb500366v.
Liu, Y., Shin, H.-d., Li, J., & Liu, L. (2015). Toward metabolic engineering in the context of system biology and synthetic biology: Advances and prospects. *Applied Microbiology and Biotechnology*, *99*(3), 1109–1118. https://doi.org/10.1007/s00253-014-6298-y.
Martella, A., Matjusaitis, M., Auxillos, J., Pollard, S. M., & Cai, Y. (2017). EMMA: An extensible mammalian modular assembly toolkit for the rapid design and production of diverse expression vectors. *ACS Synthetic Biology*, *6*(7), 1380–1392. https://doi.org/10.1021/acssynbio.7b00016.
Paddon, C. J., Westfall, P. J., Pitera, D. J., Benjamin, K., Fisher, K., McPhee, D., et al. (2013). High-level semi-synthetic production of the potent antimalarial artemisinin. *Nature*, *496*, 528. https://doi.org/10.1038/nature12051.
Sarrion-Perdigones, A., Vazquez-Vilar, M., Palaci, J., Castelijns, B., Forment, J., Ziarsolo, P., et al. (2013). GoldenBraid 2.0: A comprehensive DNA assembly framework for plant synthetic biology. *Plant Physiology*, *162*(3), 1618–1631. https://doi.org/10.1104/pp.113.217661.
Shao, Z., Zhao, H., & Zhao, H. (2009). DNA assembler, an *in vivo* genetic method for rapid construction of biochemical pathways. *Nucleic Acids Research*, *37*(2), e16. https://doi.org/10.1093/nar/gkn991.
Verwaal, R., Wang, J., Meijnen, J. P., Visser, H., Sandmann, G., van den Berg, J. A., et al. (2007). High-level production of beta-carotene in *Saccharomyces cerevisiae* by successive transformation with carotenogenic genes from *Xanthophyllomyces dendrorhous*. *Applied and Environmental Microbiology*, *73*(13), 4342–4350. https://doi.org/10.1128/AEM.02759-06.
Weber, E., Engler, C., Gruetzner, R., Werner, S., & Marillonnet, S. (2011). A modular cloning system for standardized assembly of multigene constructs. *PLoS One*, *6*(2), e16765. https://doi.org/10.1371/journal.pone.0016765.
Yuan, T., Guo, Y., Dong, J., Li, T., Zhou, T., Sun, K., et al. (2017). Construction, characterization and application of a genome-wide promoter library in Saccharomyces cerevisiae. *Frontiers of Chemical Science and Engineering*, *11*(1), 107–116. https://doi.org/10.1007/s11705-017-1621-7.
Zhou, Y., Li, G., Dong, J., Xing, X. H., Dai, J., & Zhang, C. (2018). MiYA, an efficient machine-learning workflow in conjunction with the YeastFab assembly strategy for combinatorial optimization of heterologous metabolic pathways in *Saccharomyces cerevisiae*. *Metabolic Engineering*, *47*, 294–302. https://doi.org/10.1016/j.ymben.2018.03.020.

SECTION V

Bioengineered Strains with Designed Enzyme Pathways

CHAPTER ELEVEN

Halomonas and Pathway Engineering for Bioplastics Production

Jiang Xiao-Ran[*,†,‡], **Yin Jin**[*,†,‡], **Chen Xiangbin**[*,†,‡], **Chen Guo-Qiang**[*,†,‡,§,1]

[*]MOE Lab of Bioinformatics, School of Life Sciences, Tsinghua University, Beijing, China
[†]Center for Synthetic and Systems Biology, Tsinghua University, Beijing, China
[‡]Tsinghua-Peking Center for Life Sciences, Tsinghua University, Beijing, China
[§]Manchester Institute of Biotechnology, Centre for Synthetic Biology, The University of Manchester, Manchester, United Kingdom
[1]Corresponding author: e-mail address: chengq@mail.tsinghua.edu.cn

Contents

Abstract

Traditional microbial chassis, including *Escherichia coli*, *Bacillus subtilis*, *Ralstonia eutropha*, and *Pseudomonas putida*, are grown under neutral pH and mild osmotic pressure for production of chemicals and materials. They tend to be contaminated easily by many microorganisms. To address this issue, next-generation industrial biotechnology employing halophilic *Halomonas* spp. has been developed for production of bioplastics polyhydroxyalkanoates (PHAs) and other chemicals. *Halomonas* spp. that can be grown contamination free under open and unsterile condition at alkali pH and high NaCl have been

ISSN 0076-6879
https://doi.org/10.1016/bs.mie.2018.04.008

engineered to produce several PHA polymers in elongated or enlarged cells. New pathways can also be constructed both in plasmids and on chromosomes for *Halomonas* spp. Synthetic biology approaches and parts have been developed for *Halomonas* spp., allowing better control of their growth and product formation as well as morphology adjustment. *Halomonas* spp. and their synthetic biology will play an increasingly important role for industrial production of large volume chemicals.

1. INTRODUCTION

Industrial biotechnology aims to produce chemicals, materials, and biofuels to partially replace petrochemical industry for a sustainable world. However, industrial biotechnology based on microbial transformation suffers from several disadvantages including being prone to contamination, necessitating complicated processing and production procedures, expensive equipment, and complex downstream processing. To address this challenge, next-generation industrial biotechnology based on extremophile microorganisms has been developed (Chen & Jiang, 2018). In this approach, traditional microbial chassis, such as *Escherichia coli*, *Bacillus subtilis*, *Ralstonia eutropha*, and *Pseudomonas putida*, are replaced by halophilic *Halomonas* spp. that are able to grow under alkaline conditions at high NaCl levels in cultures in mineral media. The highly alkaline NaCl-rich medium allows *Halomonas* spp. to grow under open and continuous conditions without danger of contamination, reducing process complexity (Li, Guo, Qiao, & Chen, 2016; Yin, Chen, Wu, & Chen, 2015). Metabolic engineering and synthetic biology approaches have been developed for engineering *Halomonas* spp., not only for producing a variety of products but also for changing their morphology for easier separation (Jiang & Chen, 2016). In this chapter, we will introduce how metabolic engineering and synthetic biology on *Halomonas* spp. have been applied to diversify products and to improve separation efficiency. It is expected that *Halomonas* spp. will become increasingly important as a chassis for industrial biotechnology.

2. CONTROLLABLE MORPHOLOGY ENGINEERING OF *HALOMONAS* sp.

The halophile *Halomonas campaniensis* strain LS21 was reported to allow a seawater-based open and continuous process for polyhydroxyalkanoate (PHA) production utilizing a low-cost carbon source. *H. campaniensis* LS21 was continuously grown contamination free in a bioreactor for 65 days without sterilization, reducing energy consumption for sterilization. *H. campaniensis* LS21 is expected to become a platform strain for PHA industrial production (Yue et al., 2014). However, PHA contents accumulated by wild-type *H. campaniensis* LS21 range from 20% to 60% of cell dry weight (CDW) even though the entire cellular spaces were filled with PHA granules (Fig. 1A). Most bacterial cells have small sizes ranging from 0.5 to 2 μm; this not only increases the difficulty in separating the bacterial biomass from the fermentation broth but also limits the amount of PHA granules to be stored in each cell. It is therefore important to investigate what to do to engineer the bacterial morphology for convenient separation of cells from broth and the creation of a larger space for inclusion body storage (Jiang, Yao, & Chen, 2017).

2.1 Cell Morphology Study

2.1.1 Bacteria Cell Walls Maintain the Cell Morphology

Most bacteria have cell walls that maintain cell morphology and protect the cells against osmotic lysis. The main stress-bearing component of bacterial cell walls is peptidoglycan, which provides the mechanical strength to resist osmotic and mechanical challenges (Vollmer, Blanot, & de Pedro, 2008). The peptidoglycan of the cell wall is a macromolecule consisting of glycan chains cross-linked by short peptides (Wang, Furchtgott, Huang, & Shaevitz, 2012). Penicillin-binding proteins (PBPs) are membrane-associated proteins involved in polymerizing and modifying peptidoglycan. According to their molecular weights, sequences, enzymatic, and cellular functions, PBPs are categorized into five high-molecular-weight PBPs (PBPs 1a, 1b, 1c, 2, and 3) and seven low-molecular-weight PBPs (PBPs 4, 5, 6, 7, AmpC, DacD, and AmpH) (Spratt, 1975).

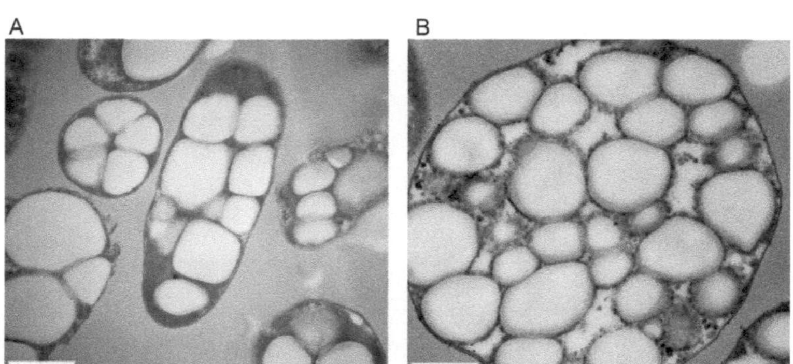

Fig. 1 Morphology and PHB accumulation by *H. campaniensis* LS21 (A) and engineered *H. campaniensis* LS21 with a large size (B). Scale bar: 0.5 μm.

PBP3 is an essential protein for cell wall synthesis at the cell division site. Inhibition of PBP3 blocks division and leads to bacterial elongation (Denome, Elf, Henderson, Nelson, & Young, 1999). PBP2 is involved in cell wall extension during *E. coli* cell growth (Vollmer & Bertsche, 2008). Inactivation of PBP2 alters the cells from a rod shape to a spherical one (de Pedro, Donachie, Holtje, & Schwarz, 2001). The cells are grown as spheres with their volume continue expanding, eventually broken up when the PBP2 and PBP3 were inactivated at the same time (de Pedro et al., 2001). Of the nonessential penicillin-binding proteins, PBP5 seems to have a particularly important role, as PBP5 inactivation mutants have an aberrant cellular morphology such as irregular branching (Fig. 2) (Nelson & Young, 2000).

2.1.2 Membrane-Bound and Membrane-Associated Proteins Decide the Cell Morphology

Membrane-bound and membrane-associated proteins can transmit morphology-determining information across the cytoplasmic membrane by linking the peptidoglycan and the cytoskeleton. This group of shape determinants includes cytoskeletal protein complex proteins and cell divisome proteins, such as MreB, MreC, MreD, RodZ, RodA, FtsZ, and FtsA (Cabeen & Jacobs-Wagner, 2005).

The bacterial cell elongation during fission is dependent on the insertion of the peptidoglycan precursor into the sidewall of the bacterial cylindrical structure (Gitai, 2005). PBP2 is required for cell wall elongation in *E. coli* and its activity is similar, in whole or in part, to the cytoskeleton protein MreB. The cytoskeletal protein MreB exerts spatial and temporal control over peptidoglycan synthesis, plays a critical role in maintaining the rod shape of many species, and is often required for viability of bacteria such as

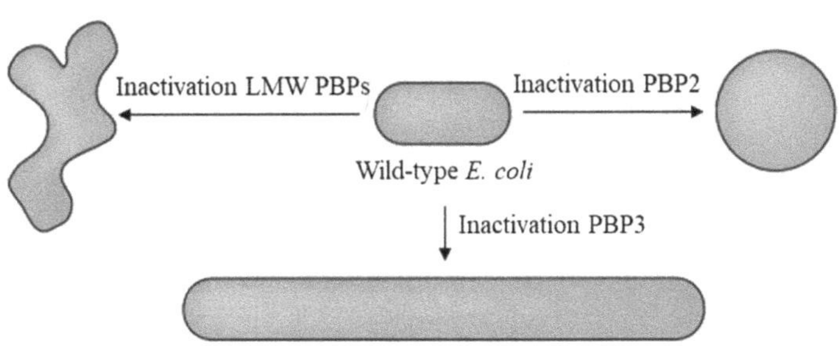

Fig. 2 Phenotypes of *E. coli* inactivated with its different PBP proteins. The width increased and the morphology changed from a rod to a sphere after inactivating PBP2 protein. The length increased after inactivating PBP3 protein. When low-molecular-weight PBP proteins were inactivated, cells became irregular in shape.

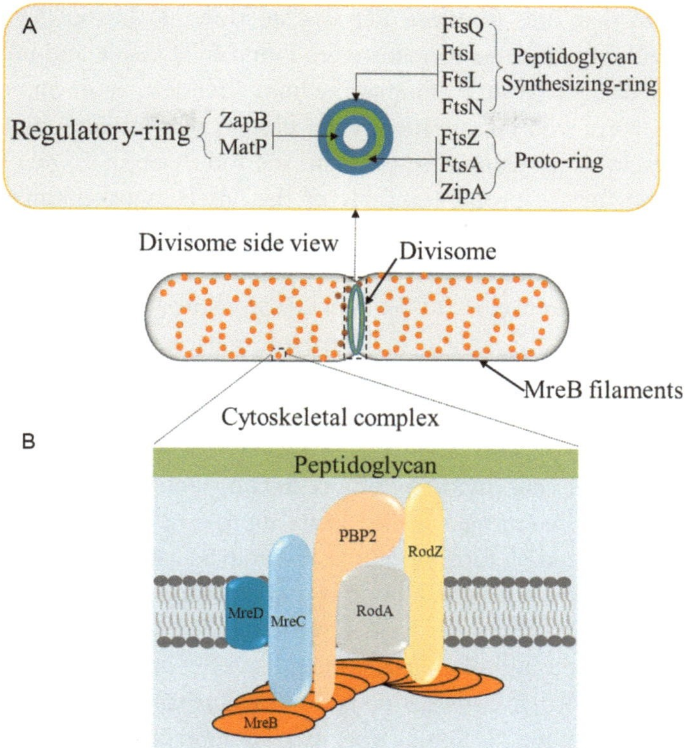

Fig. 3 Composition of main membrane-bound shape determinants involved in maintaining the morphology of rod-shaped and forming divisome during cell elongations. Pairwise images represent *Z*-rings containing different sets of proteins (A) and cytoskeletal protein complex containing different sets of proteins (B). Note that only divisome proteins that have assigned to a particular ring and cytoskeletal complex proteins that interacted to MreB are shown.

E. coli and *B. subtilis* during elongation (Gitai, 2005; Wang et al., 2012). MreB forms a cytoskeletal complex with different sets of proteins (Fig. 3B). MreB protein forms helical filamentous structures, serving as a scaffold for the peptidoglycan synthesis machinery (Carballido-López & Errington, 2003; Figge, Divakaruni, & Gober, 2004). The MreB cytoskeleton determines cellular stiffness and cell width of a cell wall (Wang, Arellano-Santoyo, Combs, & Shaevitz, 2010). MreB tethers to the cell membrane by interacting with the conserved innermembrane proteins MreC, MreD, and RodZ (Shih, Le, & Rothfield, 2003). Bitopic membrane protein MreC and integral membrane protein MreD, both transcribed as an operon and function in the same morphogenetic pathway, were also reported to be important for cell shape control (van den Ent, Johnson, Persons, de Boer, & Löwe, 2010). Cells mutated with any of these proteins were found fully viable and propagated with a spherical shape in a minimal medium (Bendezú & de Boer, 2008). RodZ is a conserved bitopic membrane protein that interacts with MreB, and both factors are required to maintain the rod shape of *E. coli*. RodZ is also required for the proper assembly of the MreB cytoskeleton (van den Ent et al., 2010). RodA protein belongs to a family of polytopic membrane proteins, is required for normal cell shape formation, and is not strictly essential for cell viability (Shiomi et al., 2013). In *B. subtilis*, disruption of the *mbl* encoding MreB-like proteins that play a cytoskeletal role in bacterial cell morphogenesis leads to slow growth and highly abnormal cell shapes (Abhayawardhane & Stewart, 1995), such as irregular bends, twists, and bulges (Cabeen & Jacobs-Wagner, 2005).

Most bacterial cells divide by binary fission, which is the splitting of a bacterial cell into two progeny cells that are approximately equal in size. This process is conducted by forming a protein complex called the divisome, the assembly of which is directed by a conserved tubulin homologue FtsZ. In several model bacteria, the formation of a divisome is divided into two stages. First, microtubule-like FtsZ undergoes GTP-dependent polymerization into filaments that form *Z*-ring at the site of cytokinesis. Z-ring tethers to the cell membrane via interacting with membrane-associated proteins, such as FtsA and ZipA in *E. coli*, which forms initial complex called the proto-ring (Fig. 3A). FtsZ is universally required for cell division in nearly all bacteria (Bi & Lutkenhaus, 1991; Erickson, Anderson, & Osawa, 2010). Then, the proto-ring recruits a second group of divisome proteins that are involved in cell wall (septum) synthesis, such as FtsI, FtsN, and FtsBLQ in *E. coli*, thereby forming the mature divisome (Fig. 3A) (Erickson et al., 2010). These also include the PBPs, such as PBP3 (also known as FtsI) in *E. coli*, which synthesize the peptidoglycan required for the formation of the cell division septum. The divisome then constricts the cytoplasmic membrane, and at the same time, the synthesis of new peptidoglycan at the division septum and invagination of the outer membrane achieves cytokinesis and cell separation (Cabeen & Jacobs-Wagner, 2010).

When FtsZ is overproduced at least sevenfolds above wild-type levels, cell division is also inhibited (Ma, Ehrhardt, & Margolin, 1996; Ward & Lutkenhaus, 1985), and cells are grown into long filaments, possibly due to distortion of the normal peptidoglycan synthesis in the septum (Wang, Henk, & Gayda, 1993). On the other hand, inactivation of FtsZ induces an immediate cessation of cell division followed by filamentary growth.

SulA is a cell division inhibitor produced in response to DNA damage as a component of the SOS response in *E. coli* (Huisman & D'Ari, 1981; Huisman, D'Ari, & Gottesman, 1984). When interacting with cell division FtsZ ring, SulA overexpression blocks binary division, turning the normally rod-shaped *E. coli* into filamentary cells (Bi & Lutkenhaus, 1993). Filamentation increases the cell volume as it extends the cell length (Wang, Wu, Jiang, & Chen, 2014, Wang, Yin, & Chen, 2014). Induction of the cell division inhibitor MinCD, a part of the *min* system, blocks formation of the FtsZ ring leading also to filamentation (Bi & Lutkenhaus, 1993). The negative effector MinC prevents assembly of the *Z*-ring by interacting directly with FtsZ.

2.2 Engineering Bacterial Morphology for Biomass Separation

The in-frame deletion of *mreB* gene in *E. coli* changes the rod shape to a sphere (Jiang, Chen, et al., 2015; Jiang, Wang, Shen, & Chen, 2015). After complementary expression *mreB* using a lower-copy number plasmid, morphology engineered *E. coli* can be restored to its original rod shape with a lower cellular stiffness than that of its wild type. Morphology engineered *E. coli* expanded to larger spherical cells together with a highly enhanced PHA production when grown in the glucose-containing medium (Jiang, Chen, et al., 2015; Jiang, Wang, et al., 2015). The PHA production of morphology engineered *E. coli* has been increased from 5.72 to 9.29 g/L in shake flasks (Jiang, Chen, et al., 2015; Jiang, Wang, et al., 2015).

A certain degree of the filamentary *E. coli* resulted from *sulA* overexpression could create more space for intracellular PHA accumulation (Jiang, Chen, et al., 2015; Jiang, Wang, et al., 2015). Filamentary cells formed by SulA inhibition on the formation of FtsZ ring lead to more PHB storages compared to that of the wild type, 27% and 13% more, respectively, due to an enlarged space for more PHA granules (Wang, Wu, et al., 2014; Wang, Yin, & Chen, 2014). Similar to the SulA-induced formation of *E. coli* filamentary cells, the inducible overexpression of the inhibitor MinCD during the growth stationary phase in *Halomonas* TD08 generated filamentous cells, resulting in enhanced PHA content from 69% to 82% (Tan, Wu, Chen, & Chen, 2014). Inducible overexpression of the FtsZ inhibitor MinCD, *Halomonas bluephagenesis* TD08 growth as filamentous cells that have a larger cell volume and much more PHA granule accumulation was observed in each cell (Tan et al., 2014).

In *H. campaniensis* LS21, essential morphology-related genes *mreB* and *ftsZ* were deleted or weakened in a controllable way by manipulating the

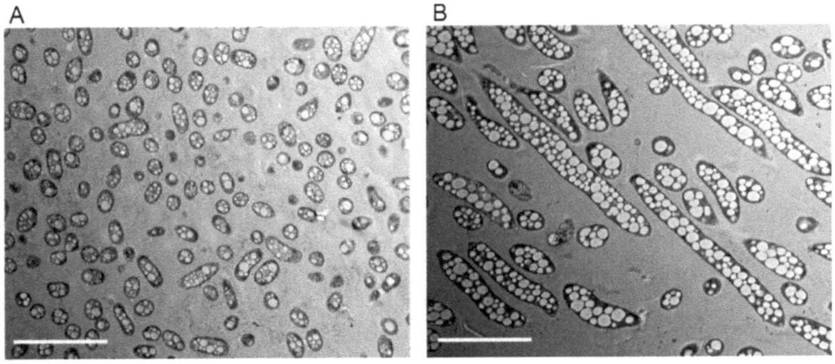

Fig. 4 The morphology and PHB accumulation by *H. campaniensis* LS21 (A) and longer filamentous *H. campaniensis* LS21 (B). Scale bar: 5 μm.

expressions of *mreB* or *ftsZ* on a temperature-sensitive plasmid expression system after deleting the relevant gene on the genome. Larger spherical or longer filamentous cells were achieved via disruption actin-like protein MreB or *Z*-ring formation protein FtsZ. Subsequently, recombinants larger spherical *H. campaniensis* LS21 and longer filamentous *H. campaniensis* LS21 grown to 16.6 g/L containing 75.9% PHB and 16.5 g/L containing 78.7% PHB, respectively, after totally 48 h of controllable morphology changes (Jiang et al., 2017), compared to wild-type *H. campaniensis* LS21 only grown to 12.7 g/L containing 56.8% PHB under the same culture condition. In this study, the controllable morphology engineering of *H. campaniensis* LS21 was constructed for both sufficient cell growth and more PHB accumulation (Figs. 1B and 4B). Morphology engineered *H. campaniensis* LS21 has a larger cell volume which promoted cell separation from medium under gravity; this significantly simplifies the downstream processing of biomass.

3. CHROMOSOME ENGINEERING OF *HALOMONAS* sp. FOR BIOPLASTIC PHA PRODUCTION

Since the first bacterial plasmid-based vector was constructed in 1972 (Cohen, Chang, & Hsu, 1972), this breakthrough has opened the door of molecular biology and plasmid-based vectors that have been widely used today for genetic circuit construction, gene expression, and protein production. Although plasmids are easy and quick to manipulate, they are normally unstable (Friehs, 2004), requiring the presence of antibiotics to maintain the stability. This instability of plasmids and the high cost of antibiotics prohibit the application of plasmid-based vectors in large-scale industrial production.

Chromosomal engineering has attracted attentions since heterologous genes can be expressed stably (Tyo, Ajikumar, & Stephanopoulos, 2009). *Halomonas* sp., being a potential platform strain for industrial applications, strongly requires chromosome engineering for gene expression and metabolic engineering. This section reviews the methods and applications of chromosome engineering in *Halomonas* sp.

3.1 Chromosome Engineering Methods in *Halomonas* sp.

Similar to other nonmodel microorganisms, homologous recombination is the basis of chromosome engineering in *Halomonas* sp. The first case of gene knock-out in *H. bluephagenesis* was reported in 2014 (Fig. 5) (Fu et al., 2014). Upstream and downstream DNA of the target gene (also termed homologous arms) are joined and cloned into a suicide vector. The suicide vector carries a replicon unable to replicate in *H. bluephagenesis*. Thus, under the selective pressure of an antibiotic, the suicide vector is integrated into the genome via homologous recombination. This step is called single crossover. Double crossover occurs when stimulated by a helper plasmid. The helper plasmid harbors the gene of an endonuclease I-*Sce*I. When the endonuclease gene is expressed, the I-*Sce*I recognizes its corresponding site on the suicide vector (now integrated into the genome) and creates double-strand break in the genome. Consequently, homologous recombination happens again to

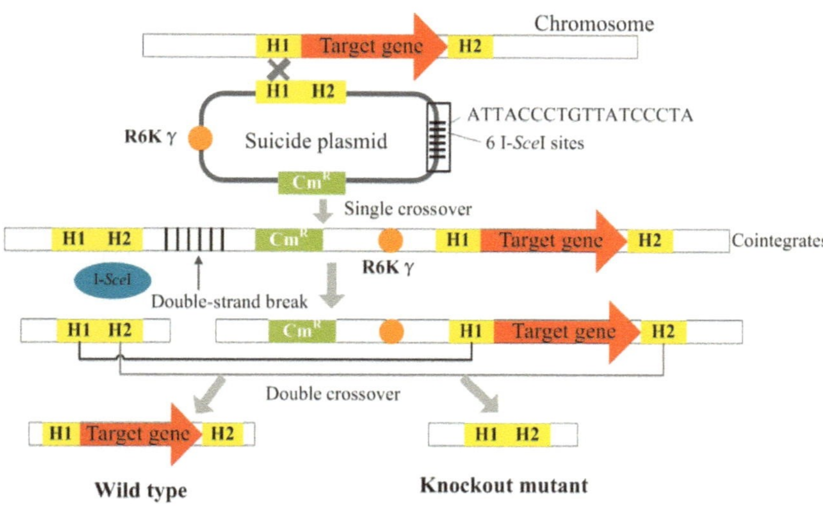

Fig. 5 Schematic description of homologous recombination-based chromosome engineering (Fu et al., 2014). The whole process includes two rounds of homologous recombination, single and double crossover, respectively.

eliminate the double-strand break from the genome, resulting to either gene knock-out mutant or wild-type strain. Finally, the gene knock-out mutant is verified by PCR and sequencing. Using this method, the gene encoding 2-methylcitrate synthase was knocked out for poly(3-hydroxybutyrate-*co*-3-hydroxyvalerate) (PHBV) synthesis by *H. bluephagenesis* grown in the presence of propionic acid, which is a precursor for 3-hydroxyvalerate (3HV). The conversion efficiency of propionic acid to 3HV monomer fraction of PHBV is increased from 10% to almost 100% (Fu et al., 2014).

This homologous recombination-based double crossover method can also be used for genomic integration of heterologous gene(s). The procedure of gene knock-in is almost the same as gene knock-out. The only difference is that the suicide vector carries the gene to-be-integrated, flanked with homologous arms upstream and downstream of integration site. In *H. bluephagenesis*, the strongly expressed *porin* operon was developed as an integration site (Yin, Fu, Wu, Chen, & Chen, 2014). The heterologous gene, without promoter, was integrated downstream *porin* gene, forming an artificial gene cluster; thus the inserted gene was transcribed together with *porin* and strongly expressed (Yin et al., 2014). In other cases, heterologous genes under a promoter, forming a separate operon, were integrated into the chromosome of *H. bluephagenesis* (Chen et al., 2017; Zhao, Wei, et al., 2016; Zhao, Zhang, et al., 2016). This enables metabolic pathway regulation using different promoters to tune the expression level of inserted genes.

A *pyrF*-based method was developed for efficient gene knock-out in *H. campaniensis* (Jiang et al., 2017). First, the *pyrF* gene encoding orotic acid 5′-phosphate decarboxylase was deleted from the chromosome of *H. campaniensis*. Then the I-*Sce*I gene in the above-mentioned suicide vector was replaced with *pyrF*. After a single crossover, the *pyrF* gene together with the whole suicide vector was integrated into the chromosome. In the presence of 5-fluoroacid acid (5-FOA), PyrF converts 5-FOA into a toxic metabolite 5-fluorouracil. Therefore, *pyrF* can act as a negative selection marker, which stimulates double crossover to eliminate the *pyrF* and suicide vector from the chromosome. In *H. campaniensis*, this *pyrF*-based gene knock-out method showed higher efficiency than I-*Sce*I-based method.

Recently, a rapid, efficient, and scarless method via CRISPR/Cas9 system was developed for chromosome engineering in both *H. bluephagenesis* and *H. campaniensis* (unpublished by this lab). The mechanism of CRISPR/Cas9-based genome editing tool is similar with the ones used in other bacteria (Jiang, Chen, et al., 2015, Jiang, Wang, et al., 2015). The Cas9 protein, guided by sgRNA, cuts the chromosome at the corresponding site and thus makes the double strand break. In the presence of homologous arms, recombination happens to generate mutant strains. Since the Cas9-mediated genome restriction is lethal to wild-type strain and acting as a selection marker, the efficiency of genome editing is 100% theoretically. This chromosome editing strategy speeds up the engineering studies in *Halomonas* sp.

3.2 Applications of Chromosome Engineering for PHA Production in *Halomonas* sp.

Wild-type *H. bluephagenesis* is a strong PHB producer (Tan, Xue, Aibaidula, & Chen, 2011). However, it has poor ability to produce other PHA, which limits the industrial application for this halophilic bacterium. The variety and productivity of PHA have been expanded via chromosome engineering.

H. bluephagenesis is able to produce PHBV [poly(3-hydroxybutyrate-*co*-3-hydroxyvalerate)] when glucose and propionic acid are used as carbon sources. Yet 3HV (3-hydroxyvalerate) precursor, namely, propionyl-CoA converted from propionic acid, is consumed rapidly by the methylcitric acid cycle (MCC). As mentioned in Section 3.1, when *prpC* gene encoding 2-methylcitrate synthase was deleted from the chromosome, the mutant strains *H. bluephagenesis* TD04 (*prpC* knock-out) and TD08 (*prpC* and 3 *phaZ* knock-out) converted almost 100% of propionic acid into 3HV as monomers for PHBV (Fu et al., 2014). *H. bluephagenesis* TD08 was further engineered for PHBV production from glucose as the sole carbon source. The *scpAB* gene cluster, encoding methylmalonyl-CoA mutase and methylmalonyl-CoA decarboxylase, respectively, was inserted into the site downstream *porin* gene. The mutant *H. bluephagenesis* TD08AB produced PHBV with 5–12 mol% 3HV fraction (Yin, Chen, et al., 2015; Yin, Wang, et al., 2015).

Wild-type *H. bluephagenesis* TD01 is not able to synthesize 4-hydroxybutyrate (4HB) containing poly(3-hydroxybutyrate-*co*-4-hydroxybutyrate) or P3HB4HB, even in the presence of related carbon sources such as γ-butyrolactone. The gene *orfZ* from *Clostridium kluyveri*, encoding 4-hydroxybutyryl-CoA transferase, was integrated into the genome of *H. bluephagenesis* TD01. The recombinant strain *H. bluephagenesis* TD40 converted 4HB via spontaneous hydrolysis of γ-butyrolactone into 4-hydroxybutyryl-CoA, which was consequently polymerized into the copolymer P3HB4HB (Chen et al., 2017).

The productivity of PHA can be enhanced via overexpression of PHA synthesis pathway in *H. bluephagenesis* (Zhao, Wei, et al., 2016; Zhao, Zhang, et al., 2016). In initial attempts, the well-studied T7 expression system was used to develop high-level expression system in *H. bluephagenesis*, yet inexplicably failed. Alternatively, part mining from a phage genome database was carried out to source novel T7-like expression system. Three novel RNA polymerase–promoter pairs, namely MmP1, K1F, and VP4, were obtained displaying orthogonality, tight regulation, and high efficiency. The MmP1 RNAP module was integrated into the chromosome of *H. bluephagenesis* TD01. The mutant strain termed TD-MmP1, similar to *E. coli* BL21 (DE3), can be used as a chassis for high-level gene expression. The MmP1 promoter module was used to overexpress *phaCAB* operon from *R. eutropha* for enhanced PHA accumulation. When this promoter module driving *phaCAB* operon was integrated into the genome of TD-MmP1, the recombinant strain accumulated PHA up to 92% of cell dry weight in 7 L bioreactor. This novel T7-like expression system displayed strong power for chromosomal metabolic engineering in *Halomonas* sp.

3.3 Chromosome Engineering in *Halomonas* sp. for Other Applications

The T7-like expression system can also be used for protein production in recombinant *Halomonas* sp. PhaP, also known as Phasin, is a PHA granule binding protein (Lan, Zhao, Chen, & Chen, 2016). Due to its amphiphilic and emulsion properties, PhaP holds promise to be developed as biosurfactant (Lan et al., 2016). A cassette containing T7-like promoter and the *phaP* gene from *Aeromonas hydrophila* 4AK4 was integrated into the chromosome of *H. bluephagenesis* TD-MmP1, generating the mutant strain TD-CG-4P (Lan et al., 2016). The PHA synthase gene *phaC* was deleted from *H. bluephagenesis* TD-CG-4P and the mutant was named *H. bluephagenesis* TD-ΔCG-4P. Interestingly, *H. bluephagenesis* TD-CG-4P produced more PhaP in the insoluble fraction than that of *H. bluephagenesis* TD-ΔCG-4P. This was attributed to PhaP bound to the surface of PHA granules and coprecipitated. In a fed-batch fermentation process, *H. bluephagenesis* TD-ΔCG-4P produced the highest PhaP in soluble form with 19% of total cellular soluble proteins and with a yield of 1.86 g/L. This study demonstrated that T7-like expression system-based chromosome engineering can be used for high-yield production of protein in *H. bluephagenesis*.

During fermentation processes, oxygen availability during late phase is the key factor limiting cell density. Expression of bacterial *Vitreoscilla* hemoglobin (VHb) can enhance cellular oxygen availability for high cell density growth. In *H. bluephagenesis* TD01, the VHb-encoding gene *vgb* was integrated downstream *porin* gene for strong and stable expression

(Ouyang et al., 2017). Moreover, the twin-arginine translocase (Tat) pathway was used by adding Tat signal peptide in front of VHb to export the protein into the periplasm for the intermittent storage of hemoglobin-bound oxygen. The recombinant strain *H. bluephagenesis* TD-tv, harboring Tat-*vgb* in the chromosome, performed better than its wild-type *H. bluephagenesis* TD01 as a control, reaching 55.7 g/L and 27.6 g/L CDW in 1.5 L computer-controlled fermentor, respectively.

3.4 Future Prospects

Halomonas spp. are suitable chassis for development of the next-generation industrial biotechnology, due to their ability to grow under unsterile and continuous conditions and prevent microbial contamination (Chen & Jiang, 2018). Chromosome engineering speeds up strain development and opens up a wide area for developing next-generation industrial biotechnology. To expand the variety of products by *Halomonas* spp., metabolic engineering and synthetic biology should be employed to widen the diversity to medium-chain-length and long-chain-length PHA as well as block copolymer PHA. Chemicals other than PHA, such as amino acids, organic acids, and proteins, should be produced in PHA-synthesis-deficient *Halomonas* mutants. Improved strain properties, including controllable morphology changes, wider feedstock uptake, and high conversion efficiencies, are required to systematically reduce the production cost. In summary, chromosome engineering associated with next-generation industrial biotechnology should be able to make biochemical economics competitive to petrochemical economics.

4. INDUSTRIAL PRODUCTIONS OF PHA BY *HALOMONAS*

4.1 Challenges Facing Traditional PHA Industrialization

In the past decades, global efforts have been made to industrialize PHA. It can go back to the 1970s and 1980s when the large-scale production of PHB and PHBV was achieved by Chemie Linz AG, Austria, and Imperial Chemical Industries, UK, respectively. In the early 1990s, Metabolix, USA, successfully commercialized poly(3-hydroxybutyrate-*co*-4-hydroxybutyrate) (P3HB4HB), and in the late 1990s, with the joint efforts of Tsinghua University, KAIST, and P&G, poly(3-hydroxybutyrate-*co*-3-hydroxyhexanoate) (PHBHHx) became available in large quantity (Chen, 2009). Companies worldwide joined this effort, attempting to make PHA more cost competitive in comparison with traditional petroleum-based plastics.

However, the cost of PHA remained high due to many reasons. On one hand, the PHA industry is essentially a fermentation-based one, which heavily consumed freshwater and energy (for sterilization, oxygen supply, and agitation as well as downstream processing) during fermentation process, and was relatively complicated in downstream product separation and purification. Glucose is usually served as a carbon source for cell growth. Raw materials usually account for as high as 50% of the total cost (Koller et al., 2005). The production cost is difficult to significantly reduce unless robust strains are developed that can grow rapidly in low cost feedstocks under simple conditions. On the other hand, the expectation that petroleum prices would increase significantly to a level to offset the high cost of PHA has not happened yet. Instead, low petroleum price combined with new energy sources including solar power, shale gas, and wind power allows products based on petroleum even more competitive in cost. Very limited success has been actually achieved for PHA industrial production as witnessed by only 1.4% share of PHA in the bioplastic packaging market.

Microorganisms are urgently required that are grown robustly under extreme conditions such as acidic, basic, high temperature, or high osmotic pressure can reduce the complexity of bioprocesses and avoid contamination. Efforts should be made to develop this biotechnology based on strains including acidophiles, alkaliphiles, psychrophiles, thermophiles, xerophiles, methanotrophs, and halophiles (Chen & Jiang, 2017). Important progress of PHA production has been achieved based on the halophiles, with success in pilot-scale production of several commercial PHA materials including PHB, PHBV, and P3HB4HB. Industrialization of PHA production based on *Halomonas* is being demonstrated to be of great potential to significantly lower the production costs and make the final products much more competitive with petroleum-based plastics.

4.2 Progress in Industrial Production of PHA Based on *Halomonas*

Halophilic bacteria are suitable for developing open and unsterile fermentation process as they can be grown under high salt concentration to prevent contamination from nonhalophilic microorganisms. Wild-type *H. bluephagenesis* has been demonstrated to produce around 80% PHB in cell dry weight. When cultivated under open and continuous fermentation conditions for 14 days, over 40 g/L cell dry weight containing 60 wt% PHA was achieved (Tan et al., 2011). This first trial of PHA fermentation based on *H. bluephagenesis* indicated the feasibility of open and unsterile process.

To produce PHA with better properties such as PHBV and P3HB4HB, *H. bluephagenesis* should be metabolically engineered. Native *H. bluephagenesis* showed poor ability to produce PHBV as it rapidly consumed 3HV precursor propionyl-CoA for methylcitric acid cycle. Deletion on *prpC* gene encoding 2-methylcitrate synthase responsible for pushing propionyl-CoA to methylcitric acid cycle (Fig. 6) allows mutant *H. bluephagenesis* TD04 to accumulate

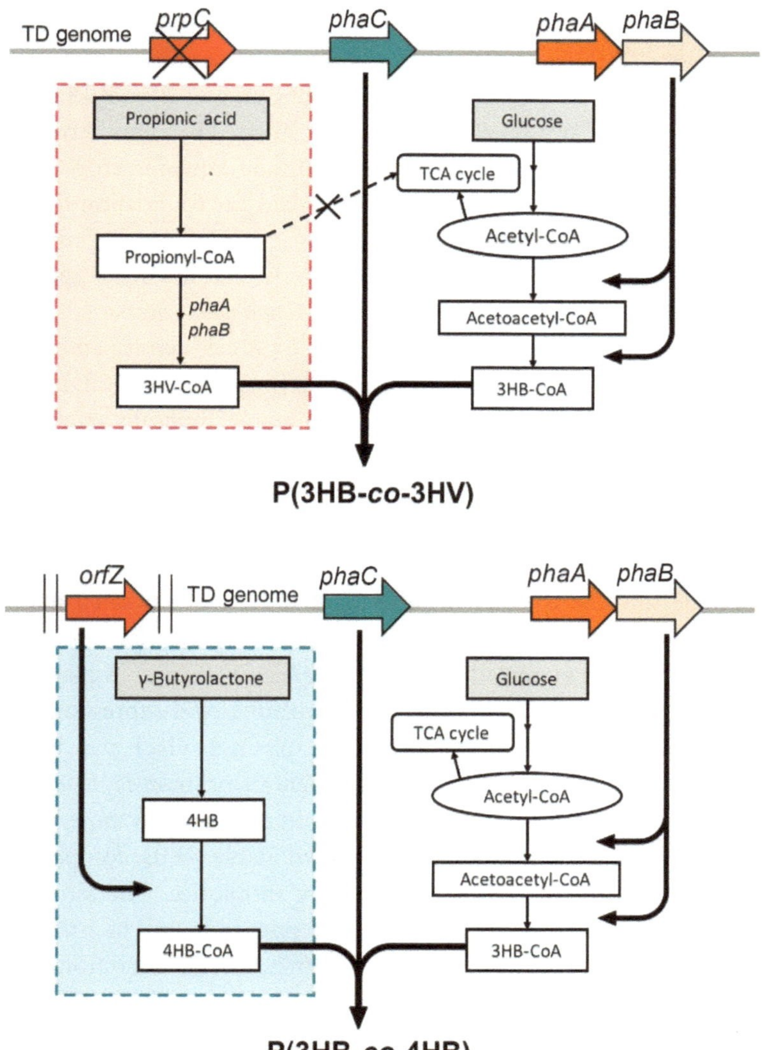

Fig. 6 Genetic manipulation of *H. bluephagenesis* for production of P P(3HB-*co*-3HV) and P(3HB-*co*-4HB). Genes: *orfZ*, encoding 2-methylcitrate synthase; *phaA*, encoding β-ketothiolase; *phaB*, encoding NADPH-dependent acetoacetyl-CoA reductase; *phaC*, encoding PHA synthase; *prpC*, encoding 2-methylcitrate synthase.

higher 34 mol% 3HV in the presence of 1.5 g/L propionic acid, while the wild-type produced only 18 mol% 3HV in the presence of 3 g/L propionic acid (Fu et al., 2014). Moreover, the conversion efficiency of propionic acid to 3HV was significantly higher in *H. bluephagenesis* TD04 (96%) than that in *H. bluephagenesis* TD01 (10%). The mutant strain *H. bluephagenesis* TD08 with further gene deletion on three PHA depolymerase *phaZ* genes was employed in a 500-L fed-batch fermentation process. In the optimized fermentation process, the strain reached 112 g/L cell dry weight consisting of approximately 70 wt% PHB after 70 h growth. If 0.5 g/L propionic acid was constantly fed into the culture, 80 g/L dry cell biomass consisting of 70 wt% PHBV containing 8 mol% 3HV was produced (Fu et al., 2014). The entire fermentation was conducted under unsterile conditions without contamination. The successful pilot-scale fermentation of PHBV has laid the foundation for further industrial production of PHA.

P3HB4HB is another attractive PHA material that has many applications in different fields. Most studies are based on *Cupriavidus necator* (also known as *R. eutropha*) as it can easily be grown to a high cell density and accumulate P3HB4HB from related precursors like γ-butyrolactone (GBL). To use *H. bluephagenesis* as a platform strain, it has to be metabolically engineered as it cannot synthesize P3HB4HB from precursors like GBL due to the absence of the 4HB-CoA transferase that converts 4HB to 4HB-CoA. Thus the 4HB-CoA transferase *orfZ* was cloned from *C. kluyveri*. It was found that the overexpression of *orfZ* led to random mutations in constitutive P_{porin} promoter, indicating the toxicity of 4HB-CoA transferase encoded in *orfZ* to *H. bluephagenesis*. T7-like expression system termed MmP1 was a novel-inducible expression system developed for *Halomonas* spp., displaying high efficiency and tight regulation. It was used to tune *orfZ* expression and the genetic part with the best performance in a plasmid which was integrated into the genome to form a chromosome recombinant strain *H. bluephagenesis* TD40 (Fig. 6). This strain showed similar results compared to those harboring plasmids in shake flask studies and achieved stable 4HB composition in P3HB4HB during fermentation without any antibiotics. The fermentation was further scaled up from 1-L, 7-L, to 1000-L pilot scale, in which the recombinant was grown to a high density under open nonsterile conditions, to form over 80 g/L CDW containing over 60% P(3HB-*co*-16 mol% 4HB) in 48 h of growth. The P3HB4HB copolymer also shows good mechanical properties, with an elongation at break of around 1000% (Chen et al., 2017). The scale-up study demonstrated that *H. bluephagenesis* is a suitable industrial strain for large-scale production processes under open and unsterile conditions.

4.3 *Halomonas* as a Platform for PHA-Associated Chemical Production

Halomonas spp. can also be used as a platform for other biochemical production. One example was the production of ectoine which was one of the most widely found compatible solutes in halophilic bacteria with applications as bioprotection agents (Ning et al., 2016). *Halomonas elongate* was first patented for ectoine production for its special property of broad salt tolerance, which not only favors a simplified process but also makes it possible to achieve large-scale internal metabolite extraction using a "bacterial milking" procedure (Pastor et al., 2010).

PhaP, known as Phasin, a protein regulating PHA synthesis, is a useful biosurfactant (Zhao, Wei, et al., 2016; Zhao, Zhang, et al., 2016). To produce PhaP in high level, the native PHA synthesis pathway was blocked by deleting *phaC* gene. The *phaP* and T7-like expression system was integrated into *H. bluephagenesis* genome. The chromosome expression system produced 19% PhaP in total cellular soluble proteins with a yield of 1.86 g/L, and the purified PhaP was easily obtained using a simple "heat lysis and salt precipitation" method (Lan et al., 2016). Similarly, another PHA-associated protein PhaR was also produced in a similar way by *phaC* mutant *H. bluephagenesis* (Fu et al., 2014).

There have also been some trials to coproduce intracellular PHA and extracellular soluble and high value-added chemical to improve the economy of PHA production. 5-Aminolevulinic acid (ALA), an important cell metabolic intermediate with potentials for cancer treatment, was produced by recombinant *H. bluephagenesis* expressing codon optimized mitochondrial 5-aminolevulinic acid synthase. 2 g/L ALA and 22% PHB were accumulated in shake flask studies (Li et al., 2016). Coproduction could be competitive under continuous and unsterile conditions. Potentials of *Halomonas* as a platform for industrial production of more chemicals are strong.

5. SUMMARY AND CONCLUSION

Halomonas spp. have demonstrated their advantages as chassis for next-generation industrial biotechnology. Various metabolic engineering and synthetic biology approaches have been successfully developed to widen the product spectrum and other properties of *Halomonas* spp. Engineered *H. bluephagenesis* has been successfully grown in pilot 500–1000 L scale to produce PHBV and P3HB4HB. The protein surfactants PhaP and PhaR, as well as the low-molecular-weight chemical 5-aminolevulinic acid, have also been produced by engineered *H. bluephagenesis*. In addition, synthetic biology allows developing controllable circuits to change the morphology of *Halomonas* spp. from small bar shapes to large spheres or elongated fibers, reducing the difficulty of separating *Halomonas* spp. from the fermentation broth. Further developments in the synthetic biology of *Halomonas* spp. should lead to more products and more interesting properties for enhancing the competitiveness of next-generation industrial biotechnology.

REFERENCES

Abhayawardhane, Y., & Stewart, G. C. (1995). *Bacillus subtilis* possesses a second determinant with extensive sequence similarity to the *Escherichia coli mreB* morphogene. *Journal of Bacteriology, 177*, 766–773.

Bendezú, F. O., & de Boer, P. A. (2008). Conditional lethality, division defects, membrane involution, and endocytosis in *mre* and *mrd* shape mutants of *Escherichia coli*. *Journal of Bacteriology, 190*, 1792–1811.

Bi, E., & Lutkenhaus, J. (1991). FtsZ ring structure associated with division in *Escherichia coli*. *Nature, 354*, 161–164.

Bi, E., & Lutkenhaus, J. (1993). Cell division inhibitors SulA and MinCD prevent formation of the FtsZ ring. *Journal of Bacteriology, 175*, 1118–1125.

Cabeen, M. T., & Jacobs-Wagner, C. (2005). Bacterial cell shape. *Nature Reviews Microbiology, 3*, 601–610.

Cabeen, M. T., & Jacobs-Wagner, C. (2010). The bacterial cytoskeleton. *Annual Review of Genetics, 44*, 365–392.

Carballido-López, R., & Errington, J. (2003). The bacterial cytoskeleton: *In vivo* dynamics of the actin-like protein Mbl of *Bacillus subtilis*. *Developmental Cell, 4*, 19–28.

Chen, G. Q. (2009). A microbial polyhydroxyalkanoates (PHA) based bio- and materials industry. *Chemical Society Reviews, 38*, 2434–2446.

Chen, G. Q., & Jiang, X. R. (2017). Next generation industrial biotechnology based on extremophilic bacteria. *Current Opinion in Biotechnology, 50*, 94–100.

Chen, G. Q., & Jiang, X. R. (2018). Engineering microorganisms for improving polyhydroxyalkanoate biosynthesis. *Current Opinion in Biotechnology, 53*, 20–25.

Chen, X., Yin, J., Ye, J., Zhang, H., Che, X., Ma, Y., et al. (2017). Engineering *Halomonas bluephagenesis* TD01 for non-sterile production of poly(3-hydroxybutyrate-*co*-4-hydroxybutyrate). *Bioresource Technology, 244*, 534–541.

Cohen, S. N., Chang, A. C., & Hsu, L. (1972). Nonchromosomal antibiotic resistance in bacteria: Genetic transformation of *Escherichia coli* by R-factor DNA. *Proceedings of the National Academy of Sciences of the United States of America, 69*, 2110–2114.

de Pedro, M. A., Donachie, W. D., Holtje, J. V., & Schwarz, H. (2001). Constitutive septal murein synthesis in *Escherichia coli* with impaired activity of the morphogenetic proteins RodA and penicillin-binding protein 2. *Journal of Bacteriology, 183*, 4115–4126.

Denome, S. A., Elf, P. K., Henderson, T. A., Nelson, D. E., & Young, K. D. (1999). *Escherichia coli* mutants lacking all possible combinations of eight penicillin binding proteins: Viability, characteristics, and implications for peptidoglycan synthesis. *Journal of Bacteriology, 181*, 3981–3993.

Erickson, H. P., Anderson, D. E., & Osawa, M. (2010). FtsZ in bacterial cytokinesis: Cytoskeleton and force generator all in one. *Microbiology and Molecular Biology Reviews, 74*, 504–528.

Figge, R. M., Divakaruni, A. V., & Gober, J. W. (2004). MreB, the cell shape-determining bacterial actin homologue, co-ordinates cell wall morphogenesis in *Caulobacter crescentus*. *Molecular Microbiology, 51*, 1321–1332.

Friehs, K. (2004). Plasmid copy number and plasmid stability. In T. H. Scheper (Ed.), *New trends and development in biochemical engineering: Vol. 86* (pp. 47–82). Berlin, Heidelberg: Springer.

Fu, X. Z., Tan, D., Aibaidula, G., Wu, Q., Chen, J. C., & Chen, G. Q. (2014). Development of *Halomonas* TD01 as a host for open production of chemicals. *Metabolic Engineering, 23*, 78–91.

Gitai, Z. (2005). The new bacterial cell biology: Moving parts and subcellular architecture. *Cell, 120*, 577–586.

Huisman, O., & D'Ari, R. (1981). An inducible DNA replication–cell division coupling mechanism in *E. coli*. *Nature, 290*, 797–799.

Huisman, O., D'Ari, R., & Gottesman, S. (1984). Cell-division control in *Escherichia coli*: Specific induction of the SOS function SfiA protein is sufficient to block septation. *Proceedings of the National Academy of Sciences of the United States of America, 81*, 4490–4494.

Jiang, X. R., & Chen, G. Q. (2016). Morphology engineering of bacteria for bio-production. *Biotechnology Advances, 34*, 435–440.

Jiang, Y., Chen, B., Duan, C., Sun, B., Yang, J., & Yang, S. (2015). Multigene editing in the *Escherichia coli* genome via the CRISPR-Cas9 system. *Applied and Environmental Microbiology, 81*, 2506–2514.

Jiang, X. R., Wang, H., Shen, R., & Chen, G. Q. (2015). Engineering the bacterial shapes for enhanced inclusion bodies accumulation. *Metabolic Engineering, 29*, 227–237.

Jiang, X. R., Yao, Z. H., & Chen, G. Q. (2017). Controlling cell volume for efficient PHB production by *Halomonas*. *Metabolic Engineering, 44C*, 30–37.

Koller, M., Bona, R., Braunegg, G., Hermann, C., Horvat, P., Kroutil, et al. (2005). Production of polyhydroxyalkanoates from agricultural waste and surplus materials. *Biomacromolecules, 6*, 561–565.

Lan, L. H., Zhao, H., Chen, J. C., & Chen, G. Q. (2016). Engineering *Halomonas* spp. as a low-cost production host for production of bio-surfactant protein PhaP. *Biotechnology Journal, 11*, 1595–1604.

Li, T., Guo, Y. Y., Qiao, G. Q., & Chen, G. Q. (2016). Microbial synthesis of 5-aminolevulinic acid and its coproduction with polyhydroxybutyrate. *ACS Synthetic Biology, 5*, 1595–1604.

Li, T. T., Li, T., Ji, W. Y., Wang, Q. Y., Zhang, H. Q., Chen, G. Q., et al. (2016). Development of constitutive and inducible expression systems enabled by defining core promoter regions in *Halomonas* sp. TD01. *Biotechnology Journal, 11*, 219–227.

Ma, X., Ehrhardt, D. W., & Margolin, W. (1996). Colocalization of cell division proteins FtsZ and FtsA to cytoskeletal structures in living *Escherichia coli* cells by using green fluorescent protein. *Proceedings of the National Academy of Sciences of the United States of America, 93*, 12998–13003.

Nelson, D. E., & Young, K. D. (2000). Penicillin binding protein 5 affects cell diameter, contour, and morphology of *Escherichia coli*. *Journal of Bacteriology, 182*, 1714–1721.

Ning, Y., Wu, X., Zhang, C., Xu, Q., Chen, N., & Xie, X. (2016). Pathway construction and metabolic engineering for fermentative production of ectoine in *Escherichia coli*. *Metabolic Engineering, 36*, 10–18.

Ouyang, P., Wang, H., Hajnal, I., Wu, Q., Guo, Y., & Chen, G. Q. (2017). Increasing oxygen availability for improving poly(3-hydroxybutyrate) production by *Halomonas*. *Metabolic Engineering, 45*, 20–31.

Pastor, J. M., Salvador, M., Argandoña, M., Bernal, V., Reinabueno, M., Csonka, L. N., et al. (2010). Ectoines in cell stress protection: Uses and biotechnological production. *Biotechnology Advances, 28*, 782–801.

Shih, Y. L., Le, T., & Rothfield, L. (2003). Division site selection in *Escherichia coli* involves dynamic redistribution of Min proteins within coiled structures that extend between the two cell poles. *Proceedings of the National Academy of Sciences of the United States of America, 100*, 7865–7870.

Shiomi, D., Toyoda, A., Aizu, T., Ejima, F., Fujiyama, A., Shini, et al. (2013). Mutations in cell elongation genes *mreB*, *mrdA* and *mrdB* suppress the shape defect of RodZ-deficient cells. *Molecular Microbiology, 87*, 1029–1044.

Spratt, B. G. (1975). Distinct penicillin binding proteins involved in the division, elongation, and shape of *Escherichia coli* K12. *Proceedings of the National Academy of Sciences of the United States of America, 72*, 2999–3003.

Tan, D., Wu, Q., Chen, J. C., & Chen, G. Q. (2014). Engineering *Halomonas* TD01 for the low-cost production of polyhydroxyalkanoates. *Metabolic Engineering, 26C*, 34–47.

Tan, D., Xue, Y. S., Aibaidula, G., & Chen, G. Q. (2011). Unsterile and continuous production of polyhydroxybutyrate by *Halomonas* TD01. *Bioresource Technology, 102*, 8130–8136.

Tyo, K. E. J., Ajikumar, P. K., & Stephanopoulos, G. (2009). Stabilized gene duplication enables long-term selection-free heterologous pathway expression. *Nature Biotechnology, 27*, 760–765.

van den Ent, F., Johnson, C. M., Persons, L., de Boer, P., & Löwe, J. (2010). Bacterial actin MreB assembles in complex with cell shape protein RodZ. *EMBO Journal, 29*, 1081–1090.

Vollmer, W., & Bertsche, U. (2008). Murein (peptidoglycan) structure, architecture and biosynthesis in *Escherichia coli*. *Biochimica et Biophysica Acta, 1778*, 1714–1734.

Vollmer, W., Blanot, D., & de Pedro, M. A. (2008). Peptidoglycan structure and architecture. *FEMS Microbiology Reviews, 32*, 149–167.

Wang, S., Arellano-Santoyo, H., Combs, P. A., & Shaevitz, J. W. (2010). Actin-like cytoskeleton filaments contribute to cell mechanics in bacteria. *Proceedings of the National Academy of Sciences of the United States of America, 107*, 9182–9185.

Wang, S., Furchtgott, L., Huang, K. C., & Shaevitz, J. W. (2012). Helical insertion of peptidoglycan produces chiral ordering of the bacterial cell wall. *Proceedings of the National Academy of Sciences of the United States of America, 109*, 595–604.

Wang, H., Henk, M. C., & Gayda, R. C. (1993). Overexpression of *ftsA* induces large bulges at the septal regions in *Escherichia coli*. *Current Microbiology, 26*, 175–181.

Wang, Y., Wu, H., Jiang, X., & Chen, G. Q. (2014). Engineering *Escherichia coli* for enhanced production of poly(3-hydroxybutyrate-*co*-4-hydroxybutyrate) in larger cellular space. *Metabolic Engineering, 25*, 183–193.

Wang, Y., Yin, J., & Chen, G. Q. (2014). Polyhydroxyalkanoates, challenges and opportunities. *Current Opinion in Biotechnology, 30*, 59–65.

Ward, J. E., Jr., & Lutkenhaus, J. (1985). Overproduction of FtsZ induces minicell formation in *E. coli*. *Cell, 42*, 941–949.

Yin, J., Chen, J. C., Wu, Q., & Chen, G. Q. (2015). *Halomonas* spp, a rising star for industrial biotechnology. *Biotechnology Advances, 33*, 1433–1442.

Yin, J., Fu, X. Z., Wu, Q., Chen, J. C., & Chen, G. Q. (2014). Development of an enhanced chromosomal expression system based on porin synthesis operon for halophile *Halomonas* sp. *Applied Microbiology and Biotechnology, 98*, 8987–8997.

Yin, J., Wang, H., Fu, X. Z., Gao, X., Wu, Q., & Chen, G. Q. (2015). Effects of chromosomal gene copy number and locations on polyhydroxyalkanoate synthesis by *Escherichia coli* and *Halomonas* sp. *Applied Microbiology and Biotechnology, 99*, 5523–5534.

Yue, H. T., Ling, C., Yang, T., Chen, X. B., Chen, Y. L., Deng, H. T., et al. (2014). A seawater-based open and continuous process for polyhydroxyalkanoates production by recombinant *Halomonas campaniensis* LS21 grown in mixed substrates. *Biotechnology for Biofuels, 7*, 108.

Zhao, H., Wei, H., Liu, X., Yao, Z., Xu, M., Wei, D., et al. (2016). Structural insights on PHA binding protein PhaP from *Aeromonas hydrophila*. *Scientific Reports, 6*, 39424.

Zhao, H., Zhang, H. M., Chen, X., Li, T., Wu, Q., Ouyang, Q., et al. (2016). Novel T7-like expression systems used for *Halomonas*. *Metabolic Engineering, 39*, 128–140.

CHAPTER TWELVE

An Engineering Approach for Rewiring Microbial Metabolism

Sebastian Wenk, Oren Yishai, Steffen N. Lindner, Arren Bar-Even[1]
Max Planck Institute of Molecular Plant Physiology, Potsdam-Golm, Germany
[1]Corresponding author: e-mail address: bar-even@mpimp-golm.mpg.de

Contents

Abstract

The introduction of synthetic pathways into microbes often requires substantial modifications of the host metabolism. Here, we present and discuss key experimental aspects required for modifying microbial central metabolism. We introduce the concept of dividing pathways into metabolic modules, the activity of which can be selected for and optimized in dedicated gene-deletion strains. We provide a comprehensive methodology for systematic pathway implementation in vivo, ranging from gene-deletion methods for the creation of selection strains to cloning strategies that allow fine-tuned expression of individual pathway enzymes in synthetic operons. We further describe pathway testing and validation via high-throughput growth experiments and ^{13}C-labeling measurements. While we focus on *Escherichia coli* as bacterial host, the holistic approach we present could be easily adapted for the metabolic engineering of other microbes.

Methods in Enzymology, Volume 608
ISSN 0076-6879
https://doi.org/10.1016/bs.mie.2018.04.026

1. INTRODUCTION

As synthetic biology matures as a field, it acquires properties usually attributed to more traditional engineering fields, such as modular design, systematic construction of parts, and rigorous testing and optimization. Here, we demonstrate how these properties can be harnessed to rewire the central metabolism of the model microbe *Escherichia coli*. We specifically elaborate and exemplify four conceptual and methodological aspects: (i) selection for the activity of pathway modules within dedicated selection strains, i.e., gene-deletion strains which are auxotrophic for metabolites produced by the module, (ii) spanning gene expression levels using multiple regulatory elements, (iii) conducting high-throughput growth experiments to identify optimal pathway activity, and (iv) applying ^{13}C labeling to confirm pathway activity (Fig. 1).

To demonstrate our approach to metabolic engineering, we use a concrete project as an example: the establishment of an altered serine pathway in *E. coli*. The serine pathway operates in several methylotrophic bacteria and is responsible for the assimilation of C1 feedstocks, e.g., methanol and formate (Anthony, 2011). Establishing this pathway in *E. coli* could enable this bacterium to grow on formate as sole carbon source. Previously, we suggested several modified structures of the serine pathway that would specifically fit the endogenous metabolism of *E. coli* (Bar-Even, 2016). The PFL-threonine cycle, shown in Fig. 2, seems especially interesting as it presents two complementary entry points for formate into central metabolism. The enzyme formate-tetrahydrofolate ligase (FTL) catalyzes the assimilation of formate into 10-formyltetrahydrofolate, which is then reduced to 5,10-methylenetetrahydrofolate (5,10-CH_2-THF) and condensed with glycine to give serine. Serine is directly deaminated to pyruvate which is further metabolized to oxaloacetate using the endogenous anaplerotic pathway. Oxaloacetate is then converted to glycine and acetyl-CoA via threonine biosynthesis and degradation. Finally, acetyl-CoA is condensed with formate to produce another pyruvate molecule, as catalyzed by the oxygen-sensitive pyruvate-formate lyase (PFL), thus completing the autocatalytic bicycle.

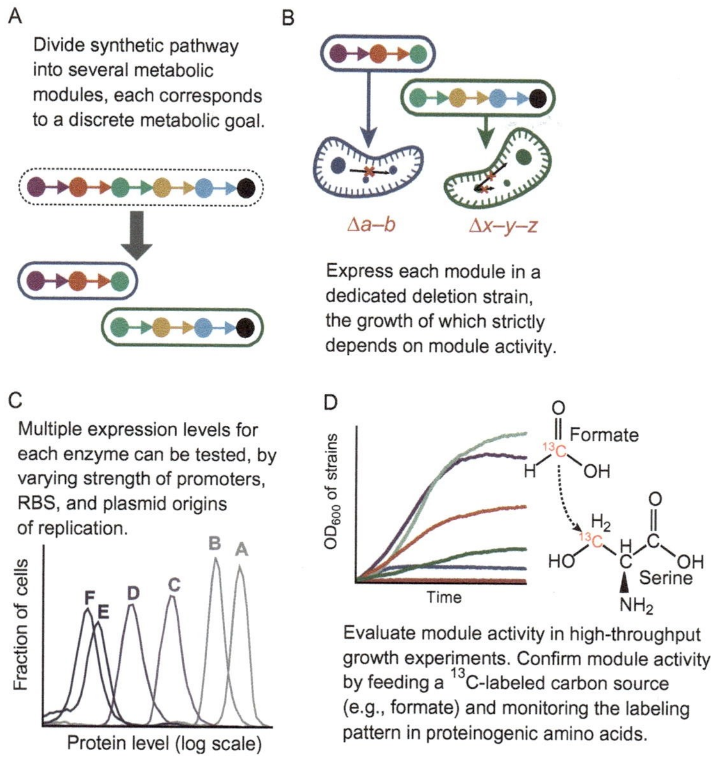

Fig. 1 Schematic representation of the metabolic engineering approach described in this manuscript. (A) To facilitate in vivo implementation, a synthetic pathway is first divided into subpathways (modules) that can be individually tested in dedicated deletion strains, the growth of which (on a given carbon source) is dependent on module activity (B). (C) Module enzymes are expressed in synthetic operons, in which the expression level of individual genes can be manipulated by a combination of ribosome-binding sites (RBSs), origins of replication and promoters. (D) High-throughput growth experiments are conducted to evaluate module activity which is further confirmed via ^{13}C-labeling experiments. *The figure is partly adapted from Yishai, O., Goldbach, L., Tenenboim, H., Lindner, S. N., & Bar-Even, A. (2017). Engineered assimilation of exogenous and endogenous formate in* Escherichia coli. ACS Synthetic Biology, *6, 1722–1731; Zelcbuch, L., Antonovsky, N., Bar-Even, A., Levin-Karp, A., Barenholz, U., Dayagi, M., et al. (2013). Spanning high-dimensional expression space using ribosome-binding site combinatorics.* Nucleic Acids Research, 41. *https://doi.org/10.1093/nar/gkt151.*

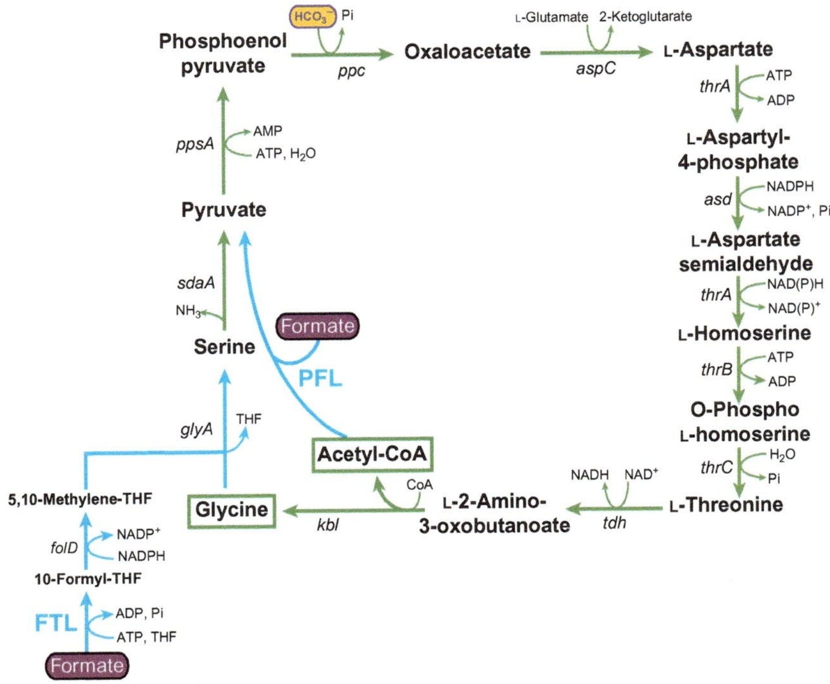

Fig. 2 The PFL-threonine cycle—a synthetic formate assimilation pathway—serves as an example for pathway implementation. The pathway overlaps with *E. coli's* central metabolism and employs only one foreign enzyme, formate-tetrahydrofolate ligase (FTL). This pathway is in fact an autocatalytic bicycle, where its products, acetyl-CoA and glycine, are reassimilated to the pathway via condensation with C1 units.

2. STRAIN ENGINEERING: IN VIVO SELECTION FOR PATHWAY MODULES

Establishing a fully operational synthetic pathway is challenging, especially if it is intended to replace a part of central metabolism (Erb, Jones, & Bar-Even, 2017). In most cases, overexpression of pathway enzymes is not sufficient to establish the required flux, which might be limited by various factors such as suboptimal activity of an enzyme, limited availability of pathway precursors, and competition with endogenous pathways (Keasling, 2010). Therefore, to facilitate the in vivo establishment of a synthetic route, it is advisable to divide it into metabolic modules, i.e., subpathways consisting of short reaction sequences (Yishai, Goldbach, Tenenboim, Lindner, & Bar-Even, 2017).

Each pathway module is expressed in a selection strain—a strain carrying multiple gene deletions such that its growth, under a specific set of conditions (e.g., specific carbon sources in minimal medium), depends on the activity of the module. These selection strains are designed to be "biosensors" for module activity, serving as reporters with the simple readout: growth, no growth, or, occasionally, retarded growth. Multiple such selection strains can be constructed for each module, where the growth of each strain depends to a different extent on module activity. This enables to sequentially increase the selection pressure for module activity, e.g., one strain would use the module to synthesize a single amino acid, another would use it as source for an entire segment of central metabolism, while in yet another strain all biomass would be derived via the module.

If a selection for module activity fails, or results in retarded growth, different solutions can be tested, for example, fine-tuning the expression of module components and replacing module enzymes with orthologues. The selection strains could be reengineered, deleting, for example, endogenous pathways that might deleteriously compete with or disrupt module activity. Taken together, these selection strains provide a testing platform in which the activity of each module can be tinkered with and optimized separately.

Only when all modules are confirmed to be sufficiently active as separate units, should they be integrated to constitute a complete pathway. Toward this aim, further selection strains can be created, the growth of which depends on the combined activity of several modules, thus enabling stepwise implementation of the complete synthetic route.

While pathways can theoretically be broken into many different modules, we suggest several guidelines for their design: (1) module activity must result in a clear growth phenotype in a conceivable gene deletion strain, that is, relief of an auxotrophy; (2) modules should be short, enabling their expression from a single operon and allowing for fast testing of expression levels; (3) together, the modules should cover the entire pathway, where overlaps—enzymes shared between modules—are advantageous as they enable testing the required enzyme activities within different metabolic contexts; (4) preferably, modules should be easy to combine; that is, their integrated activities can be selected for within a dedicated selection strain.

We note that establishing module activity within a selection strain does not necessarily mean that the module, once expressed as part of a whole pathway, will behave in the same manner. Indeed, an enzyme that supports high flux within a particular selection strain might fail to retain its high activity within another metabolic context for many possible reasons, including changes in reactant concentrations, allosteric regulation, or the cellular energetic state. Yet, in our experience, modules tend to behave similarly across many selection regimes, and the information gained in establishing module activity facilitates the engineering of the entire pathway.

We use the PFL-threonine cycle to demonstrate how a pathway can be divided into modules. As shown in Fig. 3, we can divide this synthetic pathway into several modules. One basic module is acetyl-CoA condensation with formate to give pyruvate, as catalyzed by PFL (Fig. 3A). A previous study showed that we can select for the activity of this module in an *E. coli* strain in which the glyoxylate shunt is deleted (*ΔaceAB*) such that acetate can be assimilated only via its condensation with formate (Zelcbuch et al., 2016). Another basic module is the production of glycine and serine via threonine biosynthesis and degradation, which can be selected for in a strain deleted in the endogenous serine biosynthesis route (*ΔserA*) (Yishai et al., 2017). Fig. 3B shows a selection scheme for the activity of both modules, in which acetate and formate serve as carbon sources and glycine and serine are produced via threonine metabolism. An additional module is the assimilation of formate into the THF system, providing 5,10-CH_2-THF which is condensed with glycine to provide serine. The activity of this module was previously tested and confirmed in a strain deleted in the endogenous serine biosynthesis route (*ΔserA*) as well as in the glycine cleavage system (*ΔgcvTHP*) which normally cleaves glycine for the generation of 5,10-CH_2-THF (Yishai et al., 2017). As shown in Fig. 3C, selecting for the activity of the three basic modules together would support the activity of an almost complete pathway, in which formate is required for both pyruvate and serine biosynthesis.

The ultimate step would be the removal of acetate and selection for growth on formate as a sole carbon source where acetyl-CoA is regenerated from threonine cleavage. Clearly, this would be highly challenging. Hence, it is helpful to first engineer a strain which would depend on the activity of the complete autocatalytic pathway for the production of only a fraction of its biomass. As shown in Fig. 4A, it is possible to separate *E. coli's* central metabolism into segments: deletion of enolase (*Δeno*) would metabolically separate "upper" metabolism, i.e., glycolysis, from "middle" metabolism, while deletion of citrate synthase (*ΔgltA ΔprpC*) and 2-ketoglutarate dehydrogenase (*ΔsucAB*) would separate "lower" metabolism, i.e., glutamate-derived metabolism, from "middle" metabolism. In this strain, glycerol and 2-ketoglutarate would be added to the medium to provide biomass precursors for "upper" and "lower" metabolism, while the PFL-threonine cycle would use formate to provide all biomass precursors for "middle" metabolism, e.g., acetyl-CoA, pyruvate, and oxaloacetate. Only once the synthetic cycle is shown to support sustainable growth by feeding "middle" metabolism, could it be tested to provide biomass precursors for "upper" and "lower" metabolism as well, as shown in Fig. 4B.

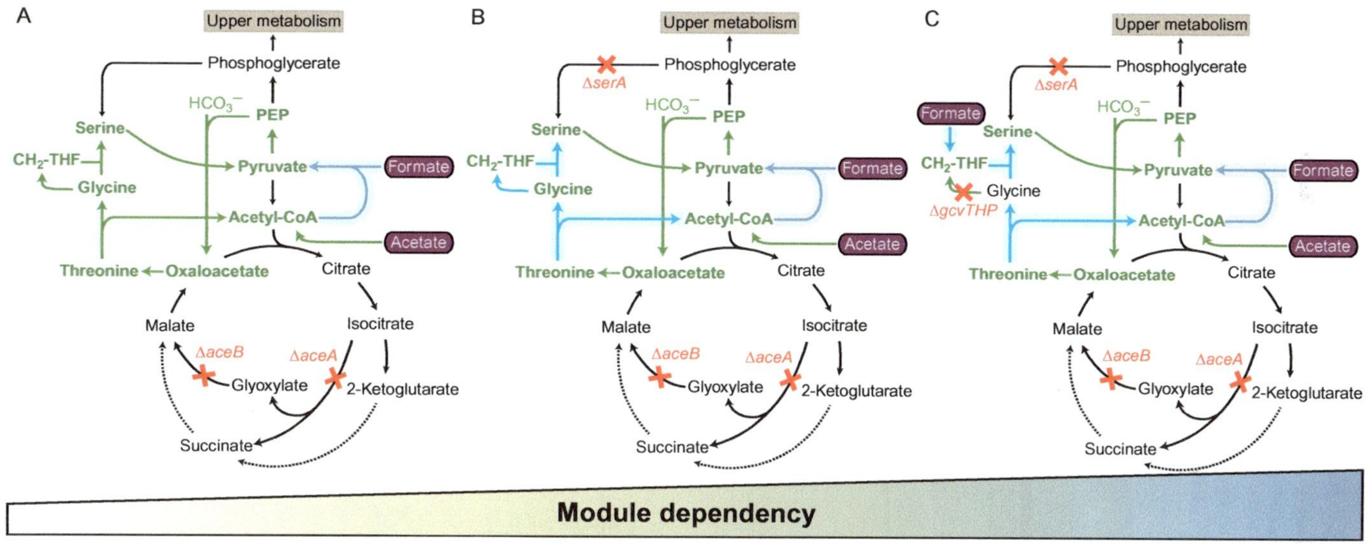

Fig. 3 Selection schemes for the stepwise implementation of the modules of the PFL-threonine cycle in *E. coli*. (A) The first module—assimilatory activity of PFL—is tested in a strain whose growth depends on the condensation of acetyl-CoA and formate as it cannot assimilate acetate via the glyoxylate shunt (*ΔaceAB*). (B) The activity of two pathway modules is tested together: PFL-dependent assimilation of acetate and formate and production of glycine and serine from threonine metabolism. (C) Selection for the activity of three modules: PFL activity, glycine production from threonine metabolism, and serine production via formate assimilation via the tetrahydrofolate system.

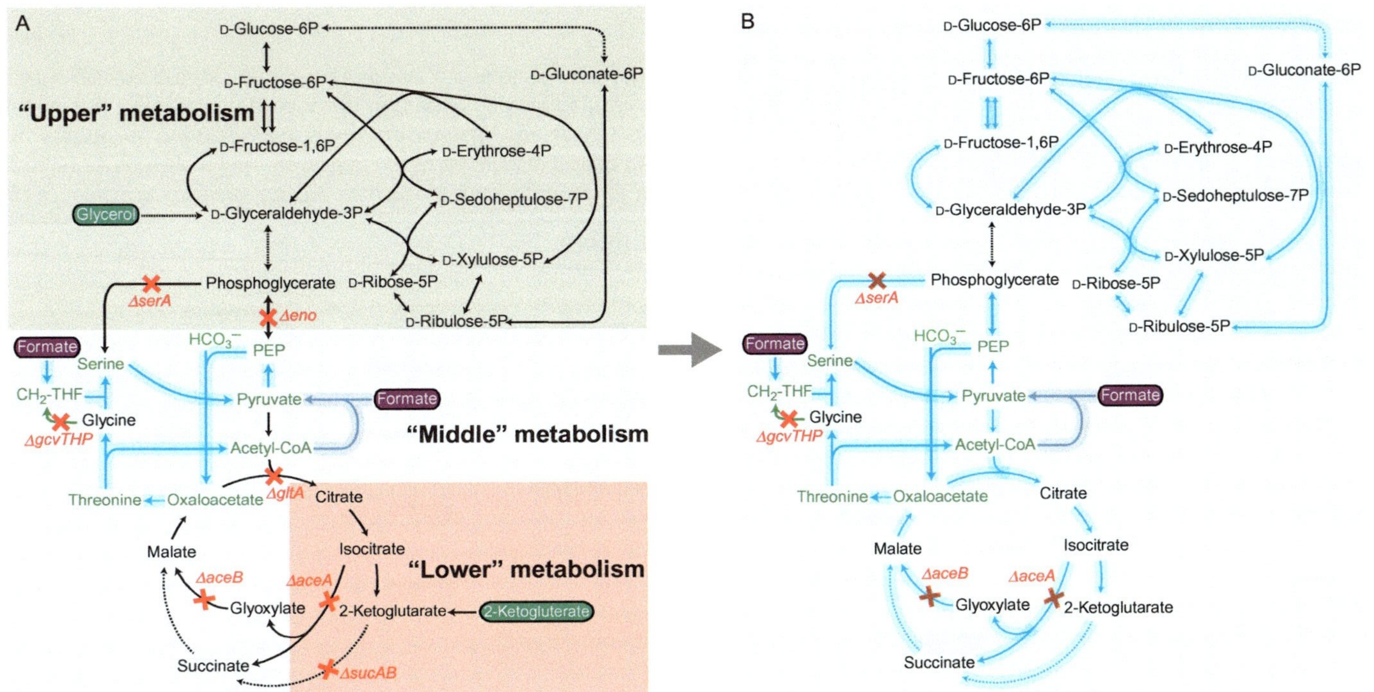

Fig. 4 Stepwise establishment of the complete PFL-threonine cycle in *E. coli*. (A) Deletion of key central metabolism enzymes, enolase (*eno*) and citrate synthase (*gltA*), separate cellular metabolism into several segments. The synthetic pathway provides biomass precursors only to the middle metabolism segment, such that low pathway flux is sufficient. (B) Next, by reconnecting all segments of cellular metabolism, it would be would be possible to select for pathway activity to provide all biomass precursors for cellular growth.

2.1 Gene Deletion Methods

The construction of *E. coli* selection strains requires the deletion of multiple genes. A common gene-deletion approach relies on homologous recombination (HRC) to replace a genomic region with a selection marker, e.g., an antibiotic resistance cassette (Zhang, Buchholz, Muyrers, & Francis Stewart, 1998). Gene-deletion systems based on bacteriophage λ Red recombinase facilitate HRC-based gene deletion using a PCR-amplified dsDNA fragment with relatively short homology arms (Datsenko & Wanner, 2000). To enable HRC, the target strain must be first transformed with a temperature-sensitive λ Red helper plasmid which harbors the recombineering genes (Murphy, 1998). After deletion is completed, cells are plasmid-cured at elevated temperature. As multiple gene deletions are usually required, the selection marker must be recycled, i.e., removed from the genome. To facilitate this, the antibiotic cassette is flanked by FRT (Flippase Recognition Target) sites (Senecoff, Bruckner, & Cox, 1985). The cell is transformed with a flippase (FLP) recombinase helper plasmid, which carries a gene encoding FLP which recombines at the FRT sites and removes the antibiotic cassette (Senecoff et al., 1985). Elevated temperature is subsequently used to cure the cell from the FLP plasmid. Taken together, the deletion of multiple genes using this method requires numerous cycles of plasmid transformation and curing, making the entire process labor intensive and time consuming.

To overcome the need for cotransformation with helper plasmids and speed up the procedure, Nielsen et al. integrated the λ Red recombineering genes and the FLP gene into the genome of *E. coli* K-12 MG1655, where the former are regulated by an L-arabinose-induced promoter and the latter by an L-rhamnose-induced promoter (Jensen, Lennen, Herrgård, & Nielsen, 2015). As the resulting SIJ488 strain supports a considerably faster gene-deletion procedure, we use it as a basis for constructing all selection strains. However, we found that some genes could not be easily deleted using this method. In these cases we commonly get false-positive colonies in which the cassette is integrated in a wrong genomic locus. Furthermore, it is often the case that gene deletion becomes more challenging if a strain already carries multiple deletions, where the genomic FRT "scars" may recombine with the introduced deletion template or react in the flippase reaction to hamper marker recycling (Datsenko & Wanner, 2000).

When gene deletion via λ Red recombination fails, we employ an alternative method which relies on the transfer of a genetic marker from a donor strain to a recipient strain by P1 phage transduction (Lennox, 1955). P1 phage is a commonly used transducing agent because it can package and transfer random sections of the host chromosome. This enables the transfer of a genomic region in which a gene is replaced with an antibiotic marker (Ikeda & Tomizawa, 1965). As a source of donor strains, the Keio collection provides single-gene deletion strains of all nonessential *E. coli* K-12 genes (when grown on LB), where coding regions were replaced with a kanamycin (Kan) cassette flanked by FRT sites (Baba et al., 2006). If a gene deletion of interest is not available in the Keio collection, which means it is essential under LB culturing conditions, we can delete the gene in the SIJ488 strain using dedicated growth conditions and use the resulting strain as donor for P1 phage transduction.

As P1 phage transfers very big genomic segments (up to 100 kb), high efficiency of native homologous recombination in *E. coli* can be achieved and false-positive colonies are very rare (Thomason, Costantino, & Court, 2007). However, the P1 phage also transfers sequences that flank the deleted gene, which can introduce undesired mutations present in the donor genome to the accepting genome. This is especially problematic when trying to delete a gene that is adjacent to the locus of a previously deleted gene: the P1 phage may reintroduce the WT copy of the deleted gene into the genome (Thomason et al., 2007). Consequently, we limit the use the P1 phage method to genes which we fail to delete using the genomic λ Red recombineering enzymes. With the combination of the two gene deletion methods we can delete almost any gene in a fast and accurate manner.

2.2 Protocol for Gene Deletion in the SIJ488 Strain

As deletion of genes is expected to affect the host's physiology, the cultivation medium should be adapted to support the growth of the gene-deleted strain, that is, if special carbon sources or supplements are necessary for its growth, they should be added to the medium. This is especially relevant when deleting central metabolism enzymes like enolase that effectively divide cellular metabolism into disconnected segments, each requiring a dedicated carbon source (as shown in Fig. 4). Another example is the deletion of phosphoglycerate mutase which was shown to require supplementation with a carbon source for "upper" metabolism (e.g., glycerol) and a carbon source for "lower" metabolism (e.g., pyruvate) (Zelcbuch at al., 2015).

Equipment

- Standard PCR equipment
- Standard gel electrophoresis equipment, 1% agarose gels
- Cooling benchtop centrifuge
- Electroporation device for bacterial transformation (e.g., Biorad Gene Pulser)
- Electroporation cuvettes
- Thermoshaker (e.g., Eppendorf Thermomixer)
- 37°C shaking incubator for bacteria cultivation
- 37°C plate incubator for bacteria cultivation

Buffers and reagents

- Standard PCR reagents
- High-fidelity polymerase (e.g., PrimeStar, TaKaRa; Phusion, Thermo Fisher Scientific)
- PCR purification kit
- Gel purification kit
- 1 *M* L-arabinose solution
- 1 *M* L-rhamnose solution
- 10% glycerol solution
- Nonselective and selective medium (e.g., LB liquid medium and LB agar plates) supplemented with antibiotics, depending on the introduced antibiotic cassette: 50 μg/mL kanamycin; 30 μg/mL chloramphenicol (Cam)

Strains

- *E. coli* SIJ488

Preparations

- Design primers to amplify an FRT-flanked antibiotic cassette and add homology arms—corresponding to the sequences flanking the desired deletion region—on each side of the cassette. Use the same procedure for the homology arms as described in Baba et al. (2006).
- Amplify FRT-flanked antibiotic cassette from a plasmid using the above primers. Use a high-fidelity polymerase to avoid introduction of point mutations.
- Apply 5 μL of the PCR on a 1% agarose gel and verify the size of the PCR product. If the correct band size is obtained, purify the PCR products. If unspecific bands are visible, purify the deletion cassette by gel purification.

- Design gene-specific primers for the verification of the desired deletion, binding upstream and downstream of the gene of interest (approximately within 200 bp on either side of the deleted region).
- Design forward and reverse primers that can bind to the introduced antibiotic cassette. This is especially important when resistance marker and target gene have the same size.

Day 1: Delete the gene of interest

1. Inoculate SIJ488 in 4 mL LB medium and culture at 37°C, shaking the culture tubes at 240 rpm in a shaking incubator until OD_{600} has reached ~0.4.
2. Induce expression of λ Red recombineering genes with 15 m*M* L-arabinose and culture for 1 h at 37°C, shaking at 240 rpm.
3. Centrifuge 2 mL of culture for 30 s at 11,000 rcf in a cooled benchtop centrifuge (at 2°C). Discard the supernatant and place the pellet on ice.
4. Resuspend the pellet in 1 mL chilled 10% glycerol and mix the suspension. Centrifuge for 30 s at 11,000 rcf in a cooled benchtop centrifuge (at 2°C).
5. Repeat the washing step (discard supernatant, resuspension in glycerol, and centrifugation).
6. Discard the supernatant, leaving 100–150 μL in the tube with the pellet.
7. Add 3–5 μL (~250 ng DNA) of the deletion cassette (purified PCR product) to the cell suspension and mix by pipetting.
8. Transfer the cell suspension from the tube to a chilled electroporation cuvette. If the volume does not reach the top of the electrode, fill up with 10% glycerol.
9. Electroporate cells in an electroporation device (e.g., Biorad Gene Pulser) with the following settings:
   Voltage: 1780 V for 1 mm cuvettes and 2500 V for 2 mm cuvettes
   Capacitance: 25 μF
   Resistance: 200 Ω
10. Resuspend the electroporated cells in 1 mL medium (e.g., LB) without antibiotics and transfer them to a new microfuge tube.
11. Incubate at 37°C in a thermoshaker, shaking at 1000 rpm for 1–2 h.
12. Plate the cells on agar plates (e.g., LB) supplemented with the introduced antibiotic and incubate overnight at 37°C in a plate incubator.

Day 2: Reduce WT background

1. Pick and restreak single colonies (~10–50) to a new selective plate (e.g., LB) supplemented with the introduced antibiotic to get rid of WT background.
2. Incubate at 37°C in a plate incubator overnight.

Day 3: Verify gene deletion

1. Do a colony PCR to confirm the desired gene deletion. Include a WT control for comparison.
2. Analyze the PCR products on an agarose gel. Positive knockouts carrying the antibiotic marker should lead to PCR products that differ in size from the WT band.
3. If the PCR shows ambiguous results, i.e., both WT and deletion band are present for the same sample, streak out several colonies on an new agar plate (e.g., LB) supplemented with antibiotics to remove the WT. Verify removal of WT background with another colony PCR on the following day.

Please note: If the antibiotic marker is of similar length to the deleted gene, design a primer which binds to a sequence within the marker and combine it with one of the flanking verification primers (e.g., primers k1 and k2 from the Keio collection that bind within the kanamycin cassette).

Day 4: Removal of antibiotic resistance cassette

1. Inoculate a colony that carries the desired gene deletion in 4 mL non-selective medium (e.g., LB) and culture at 30°C, shaking the culture tubes at 240 rpm in a shaking incubator until OD_{600} has reached ~0.4.
2. Induce FLP expression with 50 m*M* L-rhamnose and incubate at 30°C, shaking at 240 rpm.
3. After 4–12 h streak out a small amount of culture (<5 μL) with an inoculation loop on a nonselective plate (e.g., LB) to obtain single colonies.

Day 5: Verify removal of antibiotic resistance cassette

1. Pick single colonies and restreak them on two different plates in parallel, one supplemented with the introduced antibiotic, and the other one without addition of antibiotics.
2. Incubate at 37°C in a plate incubator overnight.

Day 6: Verify removal of antibiotic resistance cassette

1. Clones growing on the nonselective plate, but not on the plate supplemented with antibiotics, have successfully removed the selection marker by FLP–FRT recombination.

2. To verify the removal of the antibiotic cassette, perform a colony PCR on the candidate colonies, including a WT as control.

When the antibiotic marker is successfully removed, another gene can be deleted by restarting the protocol. It is also possible to delete two or three genes one after another without removing the antibiotic cassette when working with different antibiotics as selection markers.

2.3 Protocol for Gene Deletion Using P1 Phage Transduction

As P1 phages can infect all *E. coli* strains, it is advisable to use a separate bench and equipment for all work related to phages. The virulent P1 phage mutant, P1vir (Ikeda & Tomizawa, 1965), should be acquired from a supplier (e.g., CGSC, The Coli Genetic Stock Center). Before the actual P1 transduction of a selectable marker can be conducted, a fresh P1 stock lysate should be prepared from a WT *E. coli* strain to ensure a standard marker-free lysate. This lysate is required to generate donor lysates from strains containing desired mutations. Thereafter, the P1 stock lysate can be stored at 4°C for several months and can be used for the preparation of P1 donor lysates from strains in which a gene is replaced with an antibiotic marker.

Equipment

- Sterile disposable culture tubes (e.g., falcon tubes, microfuge tubes)
- Benchtop centrifuge
- Thermoshaker (e.g., Eppendorf Thermomixer)
- Standard PCR equipment
- Standard gel electrophoresis equipment, 1% agarose gels
- Syringe and 0.22-μm filter
- Laminar flow cabinet/Bunsen burner
- 37°C shaking incubator for bacteria cultivation
- 37°C plate incubator for bacteria cultivation

Buffers and reagents

- P1 salt solution: LB containing 10 m*M* $CaCl_2$ and 5 m*M* $MgSO_4$
- 1 *M* $CaCl_2$ solution
- 1 *M* Sodium citrate
- Nonselective and selective medium (e.g., LB liquid medium and LB agar plates) supplemented with antibiotics, depending on the introduced antibiotic cassette: 50 μg/mL kanamycin, 30 μg/mL chloramphenicol with and without 5 m*M* sodium citrate

Strains

- WT *E. coli* strain for the preparation of P1 stock lysate
- *E. coli* nonlysogenic donor strain (containing the genetic marker to be transduced)
- *E. coli* recipient strain (to be transduced with the genetic marker)
- P1 stock lysate (10^9–10^{10} pfu/mL)

Donor lysate preparation

1. Inoculate an overnight culture of your donor strain in 4 mL medium (e.g., LB) and culture at 37°C, shaking the culture tubes at 240 rpm in a shaking incubator.
2. Subculture your overnight strain 1:100 in 3 mL medium (e.g., LB) + 10 μL 1 *M* $CaCl_2$ (final concentration 5 m*M*) in a 50-mL flacon tube. Do not use reusable tubes to eliminate the chance of future contamination. The P1 phage requires $CaCl_2$ for activity.
3. Grow your subculture in a shaking incubator at 37°C, shaking at 240 rpm for 1 h (or until OD_{600} ~0.3, note: overgrown cultures will not lyse).
4. Add 50–100 μL of the P1 stock lysate to the *E. coli* subculture.
5. Continue incubation for 2–4 h until the culture is lysed (solution becomes clear with only cell debris visible). If no lysis occurs, stop the procedure and troubleshoot the P1 stock lysate. The first time a lysate is prepared, it is useful to run a parallel culture without the addition of P1 stock lysate, to estimate usual growth in comparison to lysis.
6. Transfer the lysed culture to 2-mL microfuge tubes and pellet the cell debris at 11,000 rcf for 1 min at room temperature in a benchtop centrifuge.
7. Remove the supernatant and filter it through a 0.22-μm filter into a new 2-mL tube.
8. Store the P1 donor lysate at 4°C (for up to 3 months).

Transduction

1. Inoculate an overnight culture of recipient strain in 4 mL medium (e.g., LB) and culture at 37°C, shaking the culture tubes at 240 rpm in a shaking incubator.
2. The next day, gently pellet 1 mL of the recipient culture at 4000 rcf for 5 min at room temperature in a benchtop centrifuge.
3. Discard the supernatant and resuspend the pellet in 0.5 mL P1 salt solution.

4. Add 100 μL of the recipient cell suspension to a sterile centrifuge tube.
5. Add 40–80 μL of P1 donor lysate containing the desired deletion to the suspension. If the transduction fails, try varying the amount of lysate.
6. Incubate for 30 min at 30°C to allow phage adsorption to the cells.
7. Add 1 mL medium (e.g., LB) + 100 μL of 1 *M* sodium citrate. Citrate chelates the calcium necessary for P1 phage activity and minimizes secondary infection.
8. Incubate the culture for 1 h at 37°C in a thermoshaker, shaking at 1000 rpm.
9. Centrifuge the culture at 11,000 rcf for 30 s at room temperature. Discard the supernatant and wash with 1 mL medium (e.g., LB). Repeat the washing process again.
10. Resuspend the pellet in 100 μL medium and plate the whole amount on selective plates containing 5 m*M* sodium citrate and the introduced antibiotic. Incubate overnight at 37°C in a plate incubator.
11. The next day, analyze colonies via PCR. Streak out the cells two times (or more) on plates (e.g., LB) supplemented with 5 m*M* sodium citrate and appropriate antibiotics to obtain a clean culture void of P1 phages or WT *E. coli*.

As the antibiotic cassette is flanked by FRT sites, the selection marker can be removed by expression of FLP (Senecoff et al., 1985). Assuming that the SIJ488 strain is used, please see the protocol above for the removal of the marker.

3. PATHWAY ENGINEERING: FINE-TUNE EXPRESSION OF ENZYMES

Establishing the activity of a synthetic pathway, or one of its modules, requires the appropriate expression of the genes encoding for its enzymes. Codon adaptation of enzymes from other organisms is highly recommended in order to avoid translation bottlenecks (Gustafsson, Govindarajan, & Minshull, 2004; Sørensen, Kurland, & Pedersen, 1989). For native enzymes that participate in the synthetic pathway, overexpression is only required when the endogenous activity fails to support pathway flux.

Gene expression can be established either on a plasmid or via integration into the host genome. The former provides a more flexible, easily tunable option (Sørensen & Mortensen, 2005), while the latter enables stable and robust expression and propagation (Bassalo et al., 2016; Srinivasan, Barnard, & Gerngross, 2003). In *E. coli*, gene expression from plasmids usually provides a convenient way to test and optimize pathway activity. This is especially useful when the enzymatic components that need to be expressed are not fully predetermined: different orthologues may need to be tested or endogenous genes may require overexpression or modification. Genome integration, which is less flexible, should be used only once pathway enzymes and their expression parameters are determined.

The expression level of key enzymes can have a dramatic effect on pathway activity (Pitera, Paddon, Newman, & Keasling, 2007). This is especially true for enzymes that catalyze thermodynamically or kinetically limiting reactions, enzymes that generate reactive or toxic intermediates, and enzymes that function at a metabolic crossroads between different metabolic processes. We use three types of regulatory elements to span the expression space of pathway enzymes over several orders of magnitude: plasmid origins of replication, promoters, and ribosome-binding sites (RBSs). First, we integrate the required genes into synthetic operons using the method described by Zelcbuch et al. (2013). Each gene is cloned into a high copy cloning plasmid pNiv, where it is attached to one of six well-characterized RBS sequences that together span three orders of magnitude of expression (Fig. 5A). Using an augmented BioBrick cloning strategy (Shetty, Endy, & Knight, 2008), the RBS-modulated genes are combined into one plasmid to generate a synthetic operon (Fig. 5B). This method allows the cloning of any number of genes into a synthetic operon. Instead of cloning each gene to a single pNiv vector, the genes can be cloned into a library of pNiv vectors (i.e., a mixture of vectors), each carrying a different RBS. In this case, integrating the genes results in a library of operons, each characterized by a different pairing of genes and RBSs (Zelcbuch et al., 2013).

The synthetic operon (or the library of which) is subsequently transferred to a pZ expression vector. The pZ vector family comprises three genetic modules: origin of replication, antibiotic resistance marker, and promoter. These modules are integrated in a combinatorial manner to provide a variety of expression plasmids (Lutz & Bujard, 1997). We replaced the inducible promoters of the pZ vectors with constitutive ones based on *E. coli*'s endogenous *pgi* promoter (Froman, Tait, & Gottlieb, 1989), such that three origins of replication with different copy numbers are combined with three promoters of different strength. Fig. 5C shows that these combinations span almost three orders of magnitude of expression. Furthermore, each origin-promoter combination is available with three different antibiotic markers—kanamycin, streptomycin, and tetracyclin—allowing cotransformation with different expression vectors and enhancing the flexibility of the system.

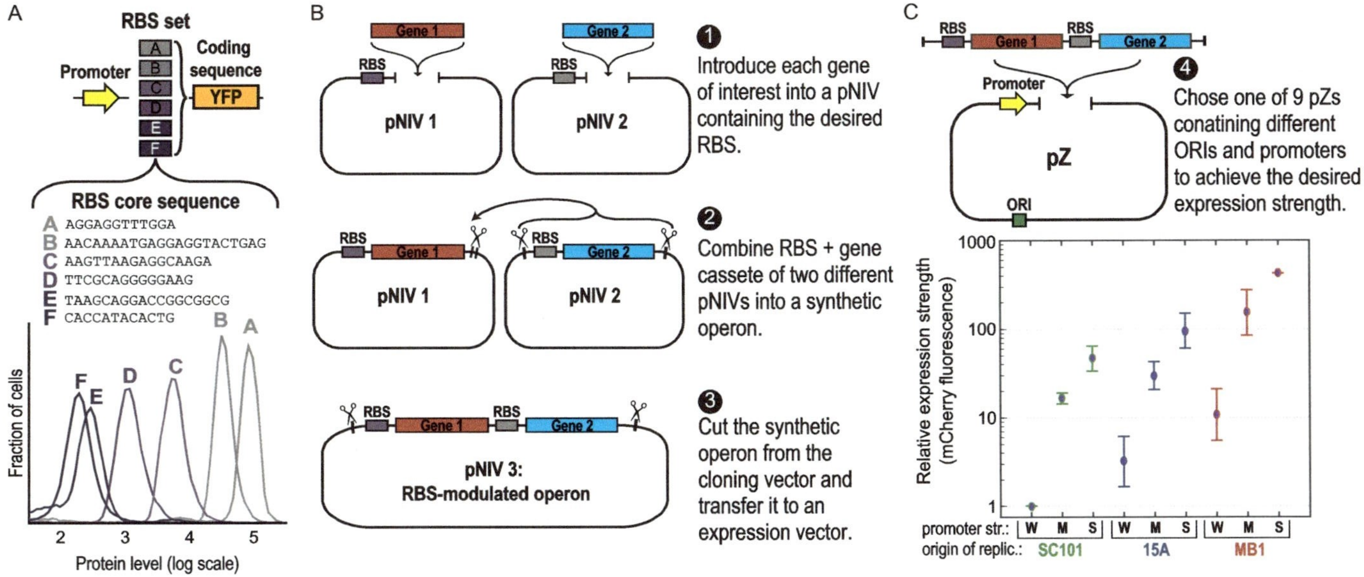

Fig. 5 Schematic representation of the expression strategy used to modify expression of pathway genes in a synthetic operon. (A) Six well-characterized RBS sequences span the expression over three orders of magnitude as shown by Zelcbuch et al. in a flow cytometry experiment. (B) Individual pathway genes are attached to an RBS of a desired strength by cloning into a corresponding pNIV vector. A BioBrick-based cloning approach allows combining of any number of RBS-gene cassettes into a synthetic operon. (C) The operon is transferred to a pZ vector that provides fine-tuning of expression level by combining origins of replication (ORI) of different copy numbers with promoters of different strength. mCherry fluorescence of all the different combinations of ORI and promoter was measured and shown to span almost three orders of magnitude. *Panel (A): Figure adapted from Zelcbuch, L., Antonovsky, N., Bar-Even, A., Levin-Karp, A., Barenholz, U., Dayagi, M., et al. (2013). Spanning high-dimensional expression space using ribosome-binding site combinatorics.* Nucleic Acids Research, 41. *https://doi.org/10.1093/nar/gkt151.*

By combining different RBS sequences with the pZ library, our system can span up to six orders of magnitude of expression. However, we usually find it unnecessary to explore the full expression space when testing a synthetic pathway. In most cases, a combination of a medium-copy-number origin (15A), with a medium or strong promoter, where all genes are regulated by a medium-strength RBS (RBS-C), suffices to establish satisfactory pathway activity. Only when this "default" combination fails or achieves suboptimal results, we explore regulatory combinations that span a larger expression space. For big operons, changing the order of the genes provides another way to modulate their expression, as the first genes tend to be expressed to a higher degree than those at the end of the operon (Zelcbuch et al., 2013).

3.1 Protocol for Integrating Genes in a Synthetic Operon

Native genes can be cloned directly from the genome. To enable cloning of a gene into a pNiv vector, the forward primer needs to include, upstream of the gene, the sequence ATGCATCATCACCATCACCAC which introduces an ATG + HIS-tag and an *Nsi*I restriction site (the primer-binding site needs to start downstream of the endogenous ATG). Equally, the reverse primer should include an *Xba*I restriction site downstream of the gene. If one of the following restriction sites is present in the gene sequence, it should be removed by the introduction of a synonymous mutation, e.g., by an overlap extension PCR: *Eco*RI, *Nhe*I, *Nsi*I, *Pst*I, *Sal*I, *Bcu*I, *Xba*I, *Xho*I, and *Vsp*I. It is recommended to codon optimize foreign genes for expression in *E. coli* (e.g., Jcat website (Grote et al., 2005)). The same restriction sites should be avoided in the codon-optimized sequence and the modifications described above should be included in the sequence.

We note that after each transformation, as described below, a colony PCR should be performed to confirm successful cloning events. To verify the construct, we usually extract the plasmid from two colonies that gave a positive PCR result and send them for sequencing. When assembling a synthetic operon, only the 5′ and 3′ ends of the operon are sequenced using vector-specific primers, as PCR-based mutations do not occur in the BioBrick-like assembly process.

Equipment

- DNA analysis software (e.g., Geneious®)
- Standard restriction enzyme digest cloning equipment
- Standard PCR equipment

- Standard gel electrophoresis equipment, 1% agarose gels
- Benchtop centrifuge
- Thermoshaker (e.g., Eppendorf Thermomixer)
- 37°C shaking incubator for bacteria cultivation
- 37°C plate incubator for bacteria cultivation

Buffers and reagents

- Standard PCR reagents
- High-fidelity polymerase (e.g., PrimeStar, TaKaRa; Phusion, Thermo Fisher Scientific)
- Restriction enzymes: *Nsi*I, *Xba*I, *Xho*I, *Bcu*I, *Sal*I, *Nhe*I, *Vsp*I, *Eco*RI, *Pst*I (e.g., Thermo Fisher Scientific)
- T4 ligase (e.g., Thermo Fisher Scientific),
- Blunt-end cloning kit (e.g., CloneJET PCR Cloning Kit, Thermo Fisher Scientific)
- PCR purification kit
- Gel purification kit
- Nonselective and selective medium LB liquid medium and LB agar plates supplemented with antibiotics, depending on the antibiotic cassette of the vector: 100 μg/mL ampicillin (amp), 50 μg/mL kanamycin, 100 μg/mL streptomycin, or 10 μg/mL tetracyclin

Strains

- Chemically competent *E. coli* DH5α

Cloning procedure

1. PCR amplify the gene of interest from the genome using primers that introduce an ATG + HIS-tag and an *Nsi*I restriction site upstream of the gene and a *Xba*I restriction site downstream of the gene.
2. Ligate the gene product into a linearized blunt vector (e.g., pJET) and transform the ligation product into chemically competent *E. coli* DH5α cells. Plate the transformants on selective LB plates (e.g., Amp for pJET).
3. After confirmation of the gene integrity (colony PCR and sequencing of the whole gene) cut out the gene from the vector using *Nsi*I + *Xba*I.

   Please note: Steps 2 and 3 are essential when preparing a library of operons. If only one defined operon is cloned, the PCR product from step 1 can be digested with *Nsi*I + *Xba*I and directly ligated into a pNiv vector (step 4).

4. Ligate the gene into an *Nsi*I/*Xba*I linearized pNiv vector such that an RBS is added upstream of the gene (it is also possible to use a mixture of pNiv vectors to obtain a library of RBS sequences regulating the gene, please see below).
5. Transform the ligation product into chemically competent *E. coli* DH5α cells and plate the transformants on selective LB plates (e.g., Amp for pNiv).
6. Confirm cloning of the gene into pNiv by colony PCR and sequencing.
7. If this is the only gene in the synthetic operon, jump to step 13.
8. If you are building an operon of two or more genes, clone all genes into pNiv vectors using the RBS sequences of your choice (or as a library of RBS sequences).
9. Incorporate the next gene into the operon as followed:
   i. Digest the acceptor vector—the vector that contains the initial part of the operon downstream of which the next gene shall be integrated—with *Nhe*I and *Xho*I.
   ii. Digest the donor vector—the vector that contains the gene to be incorporated into the operon—with *Bcu*I and *Sal*I to cut out the RBS:gene cassette.
   iii. Ligate the RBS:gene cassette from the donor vector into the acceptor vector.

      NheI and BcuI generate complementary sequences but once ligated form a scar sequence that is not recognized by any of the former restriction enzymes. The same is true for XhoI and SalI.
10. Transform the ligation product into chemically competent *E. coli* DH5α cells and plate the transformants on selective LB plates.
11. Confirm successful cloning by colony PCR and sequencing 5′ and 3′ ends of the operon.
12. Continue to add genes as described above until the synthetic operon is complete.
13. Digest the final pNiv vector—carrying all the required genes—with *Eco*RI/*Pst*I to obtain the entire operon cassette. Ligate the operon into an *Eco*RI/*Pst*I linearized pZ expression vector of choice (27 vectors available: 3 different ORIs, 3 promoters, and 3 antibiotic resistance markers).
14. Transform the ligation product into chemically competent *E. coli* DH5α cells and plate the transformants on selective LB plates (supplemented with antibiotics according to the pZ vector resistance).

15. After verification of the expression vector via colony PCR and sequencing, transform the appropriate selection strain with the pZ plasmid via electroporation (as described in the protocol for gene deletion in the SIJ488 strain).

Please note: For the preparation of a library of synthetic operons, each gene should be cloned into a mixture of pNivs. Therefore the *Nsi*I/*Xho*I-digested gene should be ligated into an equimolar mixture of *Nsi*I/*Xho*I linearized pNiv plasmids and transformed into competent cells. The next day, all resulting colonies are collected by adding 1–2 mL water and scraping gently into solution. Plasmid preparation should then proceed as if it were a dense overnight culture (a very high cloning efficiency that results in many colonies is vital to guarantee sufficient diversity of the library). The final pNiv library can be subsequently transferred into a pZ vector, thus constituting a library of expression vectors ready to be transformed into the appropriate strain.

3.2 Protocol for Genome Integration of Synthetic Operons

Chromosomal integration of a synthetic operon could eliminate plasmid-associated problems, such as genetic instability (plasmid loss) and the emergence of diverse population of plasmids—with slightly or significantly different sequences—within a single cell (Friehs, 2004; Silva, Queiroz, & Domingues, 2012). As gene copy number is substantially reduced when an operon is transferred from a plasmid to the genome, it is important to compensate by using stronger promoters or, as we found effective in multiple studies, replacing the default RBS-C with the stronger RBS-B.

Several genomic loci were previously found to be safe spots, i.e., facilitating both high-genome integration efficiency and high expression levels of genes without interfering with *E. coli*'s genetic elements (Bassalo et al., 2016). The synthetic operons are transferred into these spots using *E. coli's* endogenous homologous recombination mechanism. The availability of several safe spots is highly useful if multiple synthetic operons need to be overexpressed. However, if some of the genes within a synthetic operon also occur in the genome, the homologous recombination can fail to insert the operon into the right genomic locus. Deleting the genomic copy of the gene or shuffling the codon usage of the reintroduced copy —such that the genomic and plasmid sequences lose their homology—could relieve this problem.

We adapted the pDM4 vector (Milton, O'Toole, Hörstedt, & Wolf-Watz, 1996) to our BioBrick-based system, generating the genome integration vector pKI, and enabling subcloning of an operon from pNiv into pKI by restriction enzyme-based cloning. The pKI plasmid carries a chloramphenicol resistance gene (*Cam*^{R}), a levansucrase gene (*sacB*) which converts sucrose into the toxic compound levan (Gay, Le Coq, Steinmetz, Ferrari, & Hoch, 1983), and the conjugation gene *traJI* for the transfer of the plasmid. The plasmid replicates under the control of oriR6K, which does not replicate in WT *E. coli* but only in *E. coli* λpir strains (Kvitko et al., 2012). Thus, the acceptor strain can grow only under chloramphenicol selection if the plasmid has been successfully integrated into the genome by homologous recombination (Fig. 6A). For different integration loci, we constructed different pKI versions that contain two homology regions (HRs) of 600 bp length up- and downstream of the target safe spot. Between the HRs, the plasmid contains a strong promoter, a multiple cloning site for operon integration, and an FRT-flanked kanamycin cassette. After integration of the plasmid into the genome, a second homologous recombination between the integrated homology regions and their genomic homologues can occur. This removes from the genome either the complete plasmid or only the backbone genes (Fig. 6B). To select for the latter event, sucrose and kanamycin are added to the medium. Only cells in which the sacB gene was removed via the second homologous recombination (but that kept the kanamycin resistance gene) survive.

pKI plasmids can be transferred to an *E. coli* acceptor strain either by electroporation (see protocol for gene deletions) or by conjugation, which can be more successful when the recipient strain grows poorly, e.g., when several central metabolism genes are deleted. For conjugation, we use the *E. coli* ST18 strain which can only grow when 5-aminolevulinic acid (ALA) is added to the medium (Thoma & Schobert, 2009). Thus, after conjugation, culturing on a medium that does not contain ALA effectively eliminates the *E. coli* ST18 strain.

Equipment

- Standard restriction enzyme digest cloning equipment
- Standard PCR equipment
- Standard gel electrophoresis equipment, 1% agarose gels
- Benchtop centrifuge
- Thermoshaker (e.g., Eppendorf Thermomixer)

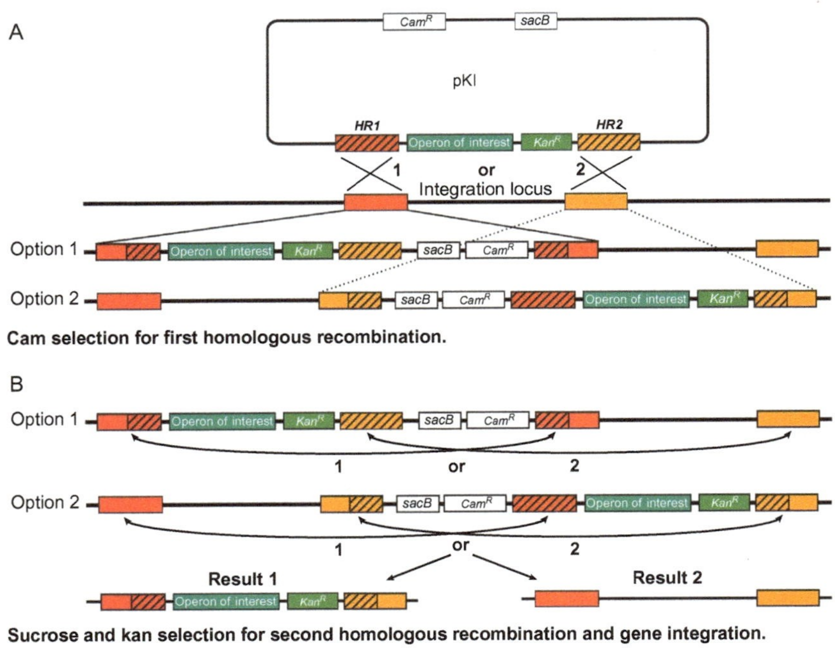

Fig. 6 Schematic representation of the pKI-mediated genome integration of a synthetic operon. (A) The operon of interest is cloned between two homology regions (HR1 and HR2) that are homologous to the desired integration locus, i.e., one of several characterized safe spots. A first homologous recombination occurs either between HR1 and its genomic homologue sequence (option 1) or between HR2 and its genomic homologue sequence (option 2), leading to the integration of the complete plasmid into the genome. As pKI carries a chloramphenicol (Cam) resistance gene, the first HRC can be selected for via cultivation on Cam. (B) Since the HR1 and HR2 and their genomic homologues remain within the genome, a second recombination within the integration region can occur. This can result either in the removal of the complete plasmid, leading to a WT genotype, or in the removal of unrequired backbone genes leading to the desired genotype. To select for the second homologous recombination sucrose is added to the medium. Only cells that have undergone second homologous recombination, such that the sacB gene—encoding for levansucrase an enzyme that converts sucrose into toxic levan—is removed, would survive. To select for the recombination events that resulted in the correct integration of the operon kanamycin (Kan) is added to the medium.

- 37°C shaking incubator for bacteria cultivation
- 37°C plate incubator for bacteria cultivation

Buffers and reagents

- Standard PCR reagents
- Restriction enzymes: *Bcu*I, *Nhe*I (e.g., Thermo Fisher Scientific)

- T4 ligase (e.g., Thermo Fisher Scientific)
- PCR purification kit
- Gel purification kit
- Nonselective and selective LB liquid medium and LB agar plates supplemented with antibiotics (depending on the introduced antibiotic cassette): 100 μg/mL ampicillin (amp), 50 μg/mL kanamycin, 30 μg/mL chloramphenicol
- Supplements for media and plates: ALA (final concentration 50 mg/mL), sucrose (final concentration 10%)

Strains

- Chemically competent *E. coli* DH5α λpir cells. This strain is used for cloning of the pKI vector as it does not require ALA for growth.
- Chemically competent *E. coli* ST18 for conjugation. An improved version of the S17-1λpir strain in which *hemA* is deleted. The strain grows only when ALA is added to the medium, which facilitates selection of positive acceptor strains after conjugation

Cloning the genome integration vector

1. Clone the operon of interest into a pNiv as described above. To achieve sufficient gene expression after genome integration, it is recommended to use RBS-B.
2. Cut out the operon from pNiv by digesting it with *Bcu*I/*Nhe*I.
3. Ligate the operon into a *Bcu*I/*Nhe*I linearized pKI genome integration vector containing homologous regions for the target locus and transform the ligation product into chemically competent *E. coli* DH5α λpir cells by a 42°C heat shock for 1 min. Recover the cells in LB medium for 1 h at 37°C, shaking at 1000 rpm in a thermoshaker.
4. Plate cells on selective LB plates supplemented with chloramphenicol and/or kanamycin (both resistance markers of pKI) and grow at 37°C in a plate incubator overnight.
5. Screen colonies by PCR for the correct insert.
6. Extract the plasmids from two positive colonies and validate the plasmid sequence.

Conjugation and genome integration (first homologous recombination)

1. Transform *E. coli* ST18 cells with the pKI genome integration vector by heat shock as described above (grow *E. coli* ST18 on LB plates supplemented with ALA and chloramphenicol and/or kanamycin).

2. Inoculate the ST18 donor strain and the recipient strain in 5 mL of appropriate medium (e.g. LB-ALA) in the morning and incubate at 37°C and shaking at 240 rpm in a shaking incubator until OD_{600} of 0.5.
3. Mix donor and recipient strain in a ~1:1 OD ratio by combining 500 μL of each culture in a microfuge tube. If the OD differs, harvest higher volume of the strain with the lower OD and resuspend in 500 μL. Mix donor and recipient strain by inverting the tube.
4. Centrifuged the mixture for 3 min at 4000 rcf in a benchtop centrifuge.
5. Discard the supernatant and resuspend the cell pellet in the remaining ~100 μL of the supernatant.
6. Pipette the entire cell suspension as one drop in the middle of an LB-ALA plate and incubate for 4–5 h upside up at room temperature under sterile conditions to allow for mating. Keep the lid of the plate open for the first 30 min to allow the spot to dry.
7. After 4–5 h scoop the cells from the mating spot using an inoculation loop and resuspend 100 μL LB medium. Vortex to stop the mating.
8. Plate 2 μL (20 μL of a 1:10 dilution), 20 μL and the rest of the cell suspension on plates optimized for the recipient strain (e.g., LB) containing just chloramphenicol and no ALA and incubate the plates at 37°C overnight in a plate incubator.
9. Pick single colonies and analyze plasmid integration by PCR.
10. Although *E. coli* ST18 cannot grow without the addition of ALA, plasmid integration strains should be transferred to a new plate at least two times by picking single colonies.

Plasmid removal by second homologous recombination

The second homologous recombination takes place between the integrated homologous sequences leading to either (i) removal of the plasmid backbone such that the synthetic operon is integrated into the genome, or (ii) removal of the entire plasmid resulting in a WT sequence that does not harbor the synthetic operon (Fig. 6B). We select for the former case by adding kanamycin to the medium:

1. Inoculate one or two transconjugants in 5 mL of appropriate medium (e.g., LB) in the morning and incubate at 37°C and 240 rpm in a shaking incubator for 4–5 h.
2. Plate 100 μL of a 10^{0}–10^{-4} dilution on plates containing 10% sucrose and kanamycin and grow for 24 h at 37°C.

Levansucrase, encoded by *sacB* which is encoded on the pKI backbone, converts sucrose into toxic levan; hence sucrose-tolerating cells are likely to have undergone the second homologous recombination.

3. Pick single colonies and streak them on two different LB plates in parallel, one supplemented with Cam and Kan, the other one with Kan only. Incubate overnight at 37°C in a plate incubator.
4. Test colonies growing on Kan but not on Cam and Kan by colony PCR.
5. If desired, the FRT/flanked kanamycin cassette can be removed by FLP as described in a previous section.

4. GROWTH EXPERIMENTS: IN VIVO SELECTION FOR PATHWAY/MODULE ACTIVITY

In most metabolic engineering projects, the synthetic pathway does not contribute to microbial growth; in fact, in most cases, it leads to reduced fitness. This complicates pathway testing and optimization and leads to genetic instability, that is, the activity of synthetic pathway is commonly lost after several generations as mutations suppressing it arise naturally and take over the microbial population (Koffas, Jung, & Stephanopoulos, 2003; Stephanopoulos & Vallino, 2009). In contrast, our metabolic engineering strategy is based on constructing gene-deletion strains in which the activity of a pathway, or a module of which, is coupled to microbial growth. This offers us a clear and simple readout for pathway activity and further ensures that the metabolic phenotype will not be lost during long-term cultivation.

As discussed above, several selection strains can be constructed for the testing of a metabolic route (pathway or module). Furthermore, many expression vectors—each combining different genes at different expression strengths—can be constructed. Altogether, testing pathway activity requires multiple parallel growth experiments which cannot be easily handled manually in flasks or cultivation tubes. For this reason, we perform growth experiments in 96-well plates, which are incubated within microplate readers at defined temperature, shaking, and gas composition. This method enables us to directly compare the growth characteristics of different strains, thereby identifying the most suitable genetic constructs to support a required activity.

There are two ways to monitor microbial growth in the microplate reader. The most common one is the change in absorbance at 600 nm.

A complementary approach is to measure the fluorescence of NADH (excitation at 340 nm, emission at 460 nm). We found that the absorbance and fluorescence curves usually correlate nicely. Deviations of the fluorescence curve from the absorbance curve can usually be interpreted as a major physiological shift and thus provide useful information on the metabolic state of the cell. The microplate growth experiments can be conducted both aerobically and anaerobically, where in the latter case the microplate reader is placed within an anaerobic chamber.

Growth experiments are performed in duplicates or triplicates. This enables us to decipher whether an observed growth is the result of a mutation: if all technical replicates are characterized by identical curves, it is reasonable to assume that the observed growth is not the result a mutation emerging in the population. On the other hand, if one replicate starts to grow and the others do not, it is clear that a mutant has taken over the population. In this latter case, we cannot be sure that the metabolic pathway being tested is indeed active or whether a metabolic bypass has emerged to enable growth under the selective conditions.

To exemplify pathway testing via growth experiments, we focus on the combined activity of the two modules of the PFL-threonine cycle, as shown in Fig. 7A: PFL-dependent condensation of acetate and formate to give pyruvate and threonine metabolism to produce glycine and serine. As all enzymes of these modules are endogenous to *E. coli*, it is unclear which enzyme might limit module activity and hence require overexpression. Clearly, we prefer to overexpress only genes that are strictly limiting rather than all pathway enzymes. As shown in Fig. 7B, we found that only a few enzymes require overexpression to support the combined activity of the modules in a minimal medium supplemented with acetate and formate: ThrA*, the first enzyme of threonine biosynthesis carrying a point mutation that makes it insensitive to product inhibition (Wang, Liu, & Qi, 2014), and Tdh and Kbl, which together catalyze threonine cleavage to acetyl-CoA and glycine (all genes were regulated by RBS-C and the expression vector had the medium copy number 15A origin and a strong promoter). Overexpression of Asd, ThrB, and ThrC—of the threonine biosynthesis pathway—leads to decreased growth which might be related to increased protein burden. Furthermore, the overexpression of PFL, its downstream enzymes leading to threonine biosynthesis, and the enzymes converting glycine to serine, are completely unnecessary as the endogenous expression levels suffice to support growth under the selective conditions.

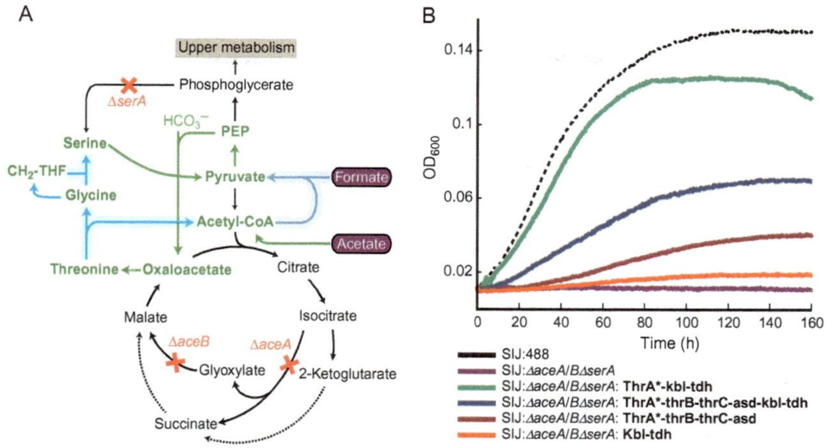

Fig. 7 Testing module activity, as exemplified by the PFL-threonine cycle. (A) The two metabolic modules selected for: PFL-dependent assimilation of acetate and formate together with glycine and serine biosynthesis from threonine metabolism. (B) A high-throughput growth experiment testing which endogenous genes must be overexpressed in order to achieve pathway activity. Strains were grown in a 96-well plate in minimal medium supplemented with 30 m*M* acetate, 30 m*M* formate, and 50 m*M* nitrate under anaerobic conditions. Growth was monitored in a Tecan Infinite® 200 PRO plate reader. As shown, overexpression of ThrA*—the first enzyme of threonine biosynthesis that is insensitive to product inhibition—as well as Tdh and Kbl—which cleaves threonine to glycine and acetyl-CoA—suffices to establish pathway activity and thus support growth. The endogenous activity of other enzymes seems to be sufficiently high as to support the desired flux.

4.1 Protocol for Growth Measurements With a Microplate Reader

Equipment

- Culture tubes for precultures
- Spectrophotometer (e.g., OD600 DiluPhotometer™, IMPLEN)
- Transparent 96-well plate (e.g., Nunclon™ Delta Surface, Cat. N° 167008, Thermo Scientific)
- Plate reader (e.g., Tecan Infinite® 200 PRO)
- Laminar flow cabinet
- Benchtop centrifuge
- Multichannel pipette
- Anaerobic chamber, if experiments are to be conducted anaerobically (e.g., COY Lab Products)
- Gas control unit (e.g., COY Lab Products)
- 37°C shaking incubator for bacteria cultivation

Buffers and reagents

5 × M9 stock solution

- Suspend 52.5 g M9 Medium Broth powder (Amresco) in deionized water
  - o 33.9 g/L Na_2HPO_4
  - o 15 g/L KH_2PO_4
  - o 5 g/L NH_4Cl
  - o 2.5 g/L NaCl
- Autoclave or filter sterilize

Trace elements stock solution ×100

- 5 g EDTA in 800 mL water, pH 7.5 (adjusted with NaOH)
- 498 mg $FeCl_3$ (anhydrous)
- 84 mg $ZnCl_2$
- 765 μL of 0.1 *M* $CuCl_2 \cdot 2H_2O$
- 210 μL of 0.2 *M* $CoCl_2 \cdot 6H_2O$
- 1.6 mL of 0.1 *M* H_3BO_3
- 8.1 μL of 1 *M* $MnCl_2 \cdot 4H_2O$
- Add water to a final volume of 1 L
- Autoclave or filter sterilize

M9 minimal medium (1 L)

- 200 mL of 5 × M9 stock solution (autoclaved)
- 10 mL of 100 × trace element mix (autoclaved)
- Add sterile ddH_2O to 1 L
- 2 mL of 1 *M* $MgSO_4$ (sterile filtered: 0.22 μm)
- 100 μL of 1 *M* $CaCl_2$ (sterile filtered: 0.22 μm)
- Mix well until the medium is clear

Strains

- *E. coli* WT strain (e.g., MG1655) for growth comparison
- *E. coli* selection strain—negative control
- *E. coli* selection strains transformed with different combinations of pathway enzymes

Day 1: Preculture

1. Inoculate strains in 5 mL relaxing medium: a nonselective minimal medium on which all strains should grow—that is, having all necessary carbon sources and supplements—even if the pathway/module is not active. Incubate the test tubes overnight in a shaking incubator at 37°C, shaking at 240 rpm.

Please note: Depending on the genetic background of the selection strain, precultures might take longer to grow to a dense culture.

2. Prepare all required selective media by supplementing M9 minimal medium with the appropriate concentration of the carbon source to be tested.

   *For anaerobic experiments: transfer media to the anaerobic chamber for overnight degassing. Vortex and open the lid to allow oxygen diffusion out of the media.

Day 2: Setup of the growth experiment

1. Harvest ~1 mL of densely grown precultures into 1.5 mL microfuge tubes and centrifuge for 3 min at 4000 rcf in a benchtop centrifuge.
2. Carefully discard the supernatant, using a pipette.
3. Add 1 mL of fresh M9 minimal medium, resuspend the pellet, and centrifuge for 3 min at 4000 rcf.
4. Repeat washing of cells (steps 2 and 3) two more times.
5. Resuspend the pellet in 1 mL M9 minimal medium and determine the OD_{600} of the cell suspension (e.g., by measuring the OD_{600} of a 1:10 dilution in a spectrophotometer).
6. Determine the volume of cell suspension that needs to be transferred to 1 mL of selective medium to receive the desired starting OD_{600}. (A starting OD_{600} of 0.005 is recommended.)
7. Transfer the determined volume of selection strains to microfuge tubes containing 1 mL of selective medium.

   *For anaerobic experiments: transfer the resuspended cells to the anaerobic chamber. Open the lids for degassing before transferring the appropriate volume of cells to the selective media.
8. Mix cells and selective medium by inverting the tubes.
9. Transfer 150 μL of the cells in selective medium to each well of the 96-well plate (using a multichannel pipette if possible). Technical duplicates or triplicates are highly recommended. Starting OD_{600} in each well is 0.005.
10. Cover the 150 μL of cell suspension in the well with 50 μL mineral oil to avoid evaporation during the experiment. Mineral oil enables free diffusion of gas.
11. Transfer the 96-well plate to the plate reader. Take off the lid and start the experiment by starting the program (parameters given below) on the computer.

Set up the following (or similar) program to run long-term growth measurements (app. 9 days) on a Tecan Infinite® 200 plate reader (Table 1).

Table 1 Parameters for Setting up a Long-term Growth Measurements on a Tecan Infinite® 200 Plate Reader

Incubation Parameters for an Aerobic Growth Experiment		**Incubation Parameters for an Anaerobic Growth Experiment**
1000 kinetic cycle of the following steps		1000 kinetic cycle of the following steps
1. Absorbance measurement:		*1. Absorbance measurement:*
Wavelength: 600 nm N° of flashes: 3 Settle time: 50 ms		Wavelength: 600 nm N° of flashes: 3 Settle time: 50 ms
2. Shaking/incubation (12 min)		*2. Shaking/Incubation (12 min)*
Linear shaking for 60 s Amplitude: 3 mm Frequency: 452.1 rpm	3 times	Wait: 10 min 30 s
Orbital shaking for 60 s Amplitude: 3 mm Frequency: 218.3 rpm		
Linear shaking for 60 s Amplitude: 2 mm Frequency: 579.8 rpm		Orbital shaking for 60 s Amplitude: 3 mm Frequency: 218.3 rpm
Orbital shaking for 60 s Amplitude: 2 mm Frequency: 280.8 rpm		Wait: 30 s
3. Return to step 1		*3. Return to step 1*

5. PATHWAY CONFIRMATION: ^{13}C LABELING OF PROTEINOGENIC AMINO ACIDS

Growth experiments provide a clear readout for the activity of a pathway or a module. However, even if all proper control and repetition experiments have been carefully conducted, it could still be the case that the observed growth does not correspond to the desired activity but rather stems from the emergence of an unrelated metabolic bypass. We use steady-state ^{13}C-labeling experiments to follow the metabolic conversion of a substrate and thus unequivocally confirm or disprove pathway activity (Szyperski, 1995). In these experiments, one of the carbon sources in the selective medium is replaced with an analog in which one or more carbon atoms are substituted with the stable isotope ^{13}C. The cells are cultivated in this medium until reaching late exponential stage, where they are harvested and boiled in concentrated acid. This procedure extracts the proteinogenic amino acids, whose labeling can then be measured using liquid chromatography–mass spectrometry (LC–MS) (Opiteck, Lewis, Jorgenson, & Anderegg, 1997). The labeling patterns of these amino acids confirm or disprove pathway activity.

A comprehensive metabolomics analysis of small molecules could potentially give a more accurate picture of cellular metabolism (Bennett, Kimball, & Gao, 2009). However, in almost all cases, this level of resolution is not required, and the more accessible proteinogenic amino acids, whose extraction is simple and straightforward, provide a clear view of the key metabolic fluxes. Hence, the rather complex quenching and extraction procedures associated with quantifying the levels and labeling of small molecules can usually be avoided.

Several LC–MS/MS procedures can further fragment the amino acids and test the mass associated with each fragment. This can reveal which of the carbons in the amino acid is labeled, and thus allows a more precise view of the metabolic flow of carbon. In some cases, the identity of the labeled carbon could be inferred even if further breakdown of the amino acids is not possible. For example, if alanine was found to be labeled once, pyruvate could be labeled on each of its three carbons. However, if valine is also labeled once, we can deduce that pyruvate is labeled on its carboxylic carbon. Conversely, if valine is labeled twice, the labeled carbon of pyruvate cannot be the carboxyl. We can deduce this as the biosynthesis of valine involves the condensation of two pyruvate molecules, in which one carboxylic group is lost; therefore, valine contains only one carboxylic carbon from pyruvate but two carbons that originate both from the second and third positions of pyruvate.

To demonstrate the use of ^{13}C-labeling experiments, we focus on formate assimilation into the tetrahydrofolate system—an essential part in the PFL-threonine cycle (Fig. 2). As shown in Fig. 8, the C1 moiety of 10-formyltetrahydrofolate contributes one carbon to histidine (carbon colored in yellow), and the C1 moiety of 5-methyltetrahydrofolate provides a carbon for methionine (carbon colored in pink). Hence, the assimilation of labeled ^{13}C-formate via the tetrahydrofolate system is expected to result in singly labeled histidine and singly labeled methionine. Indeed, and as we elaborate in a previous study (Yishai et al., 2017), the expression of a formate-tetrahydrofolate ligase within a Δ*glyA* Δ*gcvTHP* strain—which cannot produce the essential C1-activated THF molecules from serine

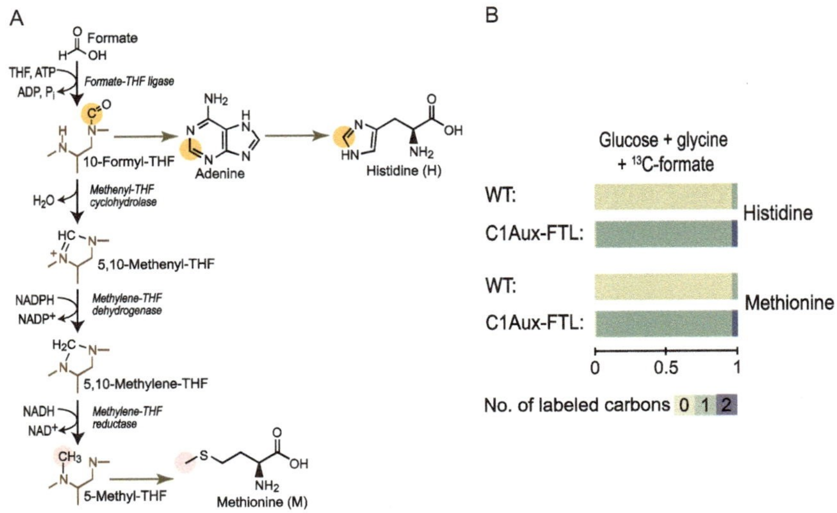

Fig. 8 A demonstration for the use of ^{13}C labeling to confirm the activity of a metabolic module. (A) Assimilation of labeled formate into the tetrahydrofolate (THF) system can be detected by monitoring the labeling of histidine and methionine, where one of the carbons of the former amino acid is derived from 10-formyltetrahydrofolate (marked in *yellow*) and one of the carbons of the latter amino acid originates from 5-methyltetrahydrofolate (marked in *pink*). (B) The results of the labeling experiment: histidine and methionine are singly labeled in a Δ*glyA* Δ*gcvTHP* strain in which formate serves as the only source of C1-activated tetrahydrofolate; these amino acids are unlabeled in a WT strain in which C1-activated tetrahydrofolate molecules are derived from serine and glycine.

and glycine—resulted in fully labeled histidine and methionine once ^{13}C-formate was added to the medium (Fig. 8B). As expected, these amino acids were unlabeled when ^{13}C-formate was added in the cultivation of a WT strain.

5.1 Protocol for ^{13}C-Labeling Analysis of Proteinogenic Amino Acids

Equipment

- Benchtop centrifuge
- Thermoshaker (e.g., Eppendorf Thermomixer)
- Cap locks for sealing microfuge tubes
- Air stream for drying the samples
- Standard equipment for bacteria culture
- 37°C shaking incubator for bacteria cultivation

- LC–MS device: Liquid chromatography system (e.g., Waters Acquity UPLC System)/Mass spectrometer (e.g., Exactive mass spectrometer, Thermo Scientific)
- LC–MS data analysis software (e.g., Xcalibur, Thermo Scientific)

Buffers and reagents
- M9 minimal medium
- ^{13}C-Labeled carbon source
- 6 *M* hydrochloric acid (HCl)
- Amino acid standards (Sigma-Aldrich)

Strains
- Transformed selection strain that showed clear pathway activity in growth experiments
- WT *E. coli* strain (used for the creation of the selection strain) as control

Growth on labeled carbon sources

Inoculate strains into 2mL selective medium where one of the carbon sources is replaced with an analogue in which one or more carbon atoms are substituted with ^{13}C. Incubate the test tubes in a shaking incubator at 37°C and 240rpm.

Amino acid extraction
1. At the late exponential stage (OD_{600} ~0.8), harvest the equivalent of 1mL of culture at OD_{600} =1 by spinning down the cells in a benchtop centrifuge for 1min at 18,500rcf.
2. Discard supernatant and resuspend the pellet in 1mL of water.
3. Spin down cells in a benchtop centrifuge for 1min at 18,500rcf and discard the supernatant.
4. Add 1mL of 6 *M* HCl to the cell pellet, seal the tubes with a cap lock, and boil at 95°C in a thermoshaker overnight to hydrolyze the cells.
5. Open the lid of the tubes and dry the samples at 95°C under a constant stream of air (4–8h).
6. When the samples have completely dried, dissolve the pellet in 1mL water.
7. Centrifuge the solution in a benchtop centrifuge at 18,500rcf for 5min and transfer supernatant to a new tube.
8. Submit the samples for analysis by LC–MS.

Mass analysis by LC–MS
1. Separate hydrolyzed amino acids with an LC system (e.g., Waters Acquity UPLC system), using a reversed phase column (e.g., HSS T3

C_{18}, 100 mm × 2.1 mm, 1.8 μm; Waters) at a flow rate of 0.4 mL/min and elute off the column using a hydrophobicity gradient.

2. Use the buffers (A) 0.1% formic acid in water and (B) 0.1% formic acid in acetonitrile with the following gradient for the mobile phase:
   - 0–3 min—100% of A;
   - 3–9 min—100% of A to 100% of B;
   - 9–13 min—100% of B;
   - 13–14 min—100% of B to 100% of A;
   - 14–20 min—100% of A;
3. Use a mass spectrometer coupled to the LC device for the acquisition of mass spectra of the proteinogenic amino acids (e.g., Exactive mass spectrometer, Thermo Scientific).
4. Use the following parameters for the acquisition of amino acid spectra:
   a. Use argon as the collision gas with a flow rate of 0.22 mL/min
   b. Cone voltage 25 V
   c. Capillary set to 3 kV
   d. Source temperature 150°C
   e. Desolvation temperature 500°C
   f. Desolvation gas flow 700 L/min
   g. Source offset 50
   h. Cone gas flow 250 L/min
   i. Collision energy 14 eV
5. Positive ionization mode is sufficient for identification of most of the desired amino acids.
6. The mass/scan range should be 50.0–300.0 m/z to detect all amino acids from glycine (the lightest amino acid) to tryptophan (the heaviest amino acid).
7. Record mass spectra for 20 min.
8. Analyze MS data using an appropriate analysis software (e.g., Xcalibur, Thermo Scientific).

When analyzing MS data for the first time, the accurate retention times and ionization parameters of the amino acids need to be determined and optimized by including and analyzing amino acid standards.

6. CONCLUDING REMARKS

In this chapter, we presented our systematic approach for the rewiring of microbial central metabolism. The key aspects of our approach involve the division of a synthetic pathway into small modules whose activity can be selected for in dedicated deletion strains, construction of expression vectors that can span the expression space of the pathway enzymes, high-throughput growth experiments to validate pathway activity, and ^{13}C labeling of proteinogenic amino acids to confirm the desired metabolic flux. While the protocols provided here relate to the metabolic engineering of *E. coli*, they can be easily adapted for the rewiring of central metabolism of other bacteria and yeast hosts.

One aspect we have not covered relates to question: what can we do when selection for pathway activity fails after testing multiple expression strategies? In this case, the best approach is to apply adaptive laboratory evolution (Dragosits & Mattanovich, 2013), in which the strain, expressing the necessary genes (preferably after genome integration), is continuously cultivated under selective conditions that promote the emergence of mutations that facilitate pathway activity. For example, to establish the activity of the full PFL-threonine cycle, long-term cultivation at saturating amounts of formate but limiting amounts of acetate is expected to slowly evolve the strain toward reduced utilization of the latter feedstock, until growth on formate as sole carbon source is reached. The combination of a direct selection approach as described in this chapter with adaptive laboratory evolution has the potential to dramatically reshape microbial metabolism into novel and biotechnologically productive architectures.

ACKNOWLEDGMENTS

This work was funded by the Max Planck Society. The authors thank Charlie Cotton, Hai He, and Hezi Tenenboim for discussions and critique reading of the manuscript.

REFERENCES

Anthony, C. (2011). How half a century of research was required to understand bacterial growth on C1 and C2 compounds; the story of the serine cycle and the ethylmalonyl-CoA pathway. *Science Progress*, *94*, 109–137.

Baba, T., Ara, T., Hasegawa, M., Takai, Y., Okumura, Y., Baba, M., et al. (2006). Construction of *Escherichia coli* K-12 in-frame, single-gene knockout mutants: The Keio collection. *Molecular Systems Biology*. *2*, https://doi.org/10.1038/msb4100050.

Bar-Even, A. (2016). Formate assimilation: The metabolic architecture of natural and synthetic pathways. *Biochemistry*, *55*, 3851–3863.

Bassalo, M. C., Garst, A. D., Halweg-Edwards, A. L., Grau, W. C., Domaille, D. W., Mutalik, V. K., et al. (2016). Rapid and efficient one-step metabolic pathway integration in *E. coli*. *ACS Synthetic Biology*, *5*, 561–568.

Bennett, B., Kimball, E., & Gao, M. (2009). Absolute metabolite concentrations and implied enzyme active site occupancy in *Escherichia coli*. *Nature Chemical Biology*, *5*, 593–599.

Datsenko, K. A., & Wanner, B. L. (2000). One-step inactivation of chromosomal genes in *Escherichia coli* K-12 using PCR products. *Proceedings of the National Academy of Sciences of the United States of America*, *97*, 6640–6645.

Dragosits, M., & Mattanovich, D. (2013). Adaptive laboratory evolution—Principles and applications for biotechnology. *Microbial Cell Factories*, *12*, 64. https://doi.org/10.1186/1475-2859-12-64.

Erb, T. J., Jones, P. R., & Bar-Even, A. (2017). Synthetic metabolism: Metabolic engineering meets enzyme design. *Current Opinion in Chemical Biology*, *37*, 56–62.

Friehs, K. (2004). Plasmid copy number and plasmid stability. *Advances in Biochemical Engineering/Biotechnology*, *86*, 47–82.

Froman, B., Tait, R., & Gottlieb, L. (1989). Isolation and characterization of the phosphoglucose isomerase gene from *Escherichia coli*. *Molecular and General Genetics MGG*, *217*, 126–131.

Gay, P., Le Coq, D., Steinmetz, M., Ferrari, E., & Hoch, J. A. (1983). Cloning structural gene sacB, which codes for exoenzyme levansucrase of Bacillus subtilis: Expression of the gene in *Escherichia coli*. *Journal of Bacteriology*, *153*, 1424–1431.

Grote, A., Hiller, K., Scheer, M., Münch, R., Nörtemann, B., Hempel, D. C., et al. (2005). JCat: A novel tool to adapt codon usage of a target gene to its potential expression host. *Nucleic Acids Research*, *33*, W526–W531. https://doi.org/10.1093/nar/gki376.

Gustafsson, C., Govindarajan, S., & Minshull, J. (2004). Codon bias and heterologous protein expression. *Trends in Biotechnology*, *22*, 346–353.

Ikeda, H., & Tomizawa, J. I. (1965). Transducing fragments in generalized transduction by phage P1: III. Studies with small phage particles. *Journal of Molecular Biology*, *14*, 120–129.

Jensen, S. I., Lennen, R. M., Herrgård, M. J., & Nielsen, A. T. (2015). Seven gene deletions in seven days: Fast generation of *Escherichia coli* strains tolerant to acetate and osmotic stress. *Scientific Reports*, *5*, 17874. https://doi.org/10.1038/srep17874.

Keasling, J. D. (2010). Manufacturing molecules through metabolic engineering. *Science*, *330*, 1355–1358.

Koffas, M. A. G., Jung, G. Y., & Stephanopoulos, G. (2003). Engineering metabolism and product formation in Corynebacterium glutamicum by coordinated gene overexpression. *Metabolic Engineering*, *5*, 32–41.

Kvitko, B. H., Bruckbauer, S., Prucha, J., McMillan, I., Breland, E. J., Lehman, S., et al. (2012). A simple method for construction of pir+ Enterobacterial hosts for maintenance of R6K replicon plasmids. *BMC Research Notes*, *5*, 157. https://doi.org/10.1186/1756-0500-5-157.

Lennox, E. S. (1955). Transduction of linked genetic characters of the host by bacteriophage P1. *Virology*, *1*, 190–206.

Lutz, R., & Bujard, H. (1997). Independent and tight regulation of transcriptional units in *Escherichia coli* via the LacR/O, the TetR/O and AraC/I1-I2 regulatory elements. *Nucleic Acids Research*, *25*, 1203–1210.

Milton, D. L., O'Toole, R., Hörstedt, P., & Wolf-Watz, H. (1996). Flagellin A is essential for the virulence of Vibrio anguillarum. *Journal of Bacteriology*, *178*, 1310–1319.

Murphy, K. C. (1998). Use of bacteriophage λ recombination functions to promote gene replacement in *Escherichia coli*. *Journal of Bacteriology*, *180*, 2063–2071.

Opiteck, G. J., Lewis, K. C., Jorgenson, J. W., & Anderegg, R. J. (1997). Comprehensive on-line LC/LC/MS of proteins. *Analytical Chemistry*, *69*, 1518–1524.

Pitera, D. J., Paddon, C. J., Newman, J. D., & Keasling, J. D. (2007). Balancing a heterologous mevalonate pathway for improved isoprenoid production in *Escherichia coli*. *Metabolic Engineering*, *9*, 193–207.

Senecoff, J. F., Bruckner, R. C., & Cox, M. M. (1985). The FLP recombinase of the yeast 2-micron plasmid: Characterization of its recombination site. *Proceedings of the National Academy of Sciences of the United States of America*, *82*, 7270–7274.

Shetty, R. P., Endy, D., & Knight, T. F. (2008). Engineering BioBrick vectors from BioBrick parts. *Journal of Biological Engineering*, *2*, 5. https://doi.org/10.1186/1754-1611-2-5.

Silva, F., Queiroz, J. A., & Domingues, F. C. (2012). Evaluating metabolic stress and plasmid stability in plasmid DNA production by *Escherichia coli*. *Biotechnology Advances*, *30*, 691–708.

Sørensen, M. A., Kurland, C. G., & Pedersen, S. (1989). Codon usage determines translation rate in *Escherichia coli*. *Journal of Molecular Biology*, *207*, 365–377.

Sørensen, H. P., & Mortensen, K. K. (2005). Advanced genetic strategies for recombinant protein expression in *Escherichia coli*. *Journal of Biotechnology*, *115*, 113–128.

Srinivasan, S., Barnard, G. C., & Gerngross, T. U. (2003). Production of recombinant proteins using multiple-copy gene integration in high-cell-density fermentations of Ralstonia eutropha. *Biotechnology and Bioengineering*, *84*, 114–120.

Stephanopoulos, G., & Vallino, J. J. (2009). Network rigidity and metabolite engineering in metabolic overproduction. *Science*, *252*, 1675–1681.

Szyperski, T. (1995). Biosynthetically directed fractional ^{13}C-labeling of proteinogenic amino acids. An efficient analytical tool to investigate intermediary metabolism. *European Journal of Biochemistry*, *232*, 433–448.

Thoma, S., & Schobert, M. (2009). An improved Escherichia coli donor strain for diparental mating. *FEMS Microbiology Letters*, *294*, 127–132.

Thomason, L. C., Costantino, N., & Court, D. L. (2007). *E. coli* genome manipulation by P1 transduction. *Current Protocols in Molecular Biology*. Chapter 1, 1.17.1–1.17.8.

Wang, Q., Liu, X., & Qi, Q. (2014). Biosynthesis of poly(3-hydroxybutyrate-co-3-hydroxyvalerate) from glucose with elevated 3-hydroxyvalerate fraction via combined citramalate and threonine pathway in Escherichia coli. *Applied Microbiology and Biotechnology*, *98*, 3923–3931.

Yishai, O., Goldbach, L., Tenenboim, H., Lindner, S. N., & Bar-Even, A. (2017). Engineered assimilation of exogenous and endogenous formate in *Escherichia coli*. *ACS Synthetic Biology*, *6*, 1722–1731.

Zelcbuch, L., Antonovsky, N., Bar-Even, A., Levin-Karp, A., Barenholz, U., Dayagi, M., et al. (2013). Spanning high-dimensional expression space using ribosome-binding site combinatorics. *Nucleic Acids Research*, *41*, e98. https://doi.org/10.1093/nar/gkt151.

Zelcbuch, L., Lindner, S. N., Zegman, Y., Vainberg Slutskin, I., Antonovsky, N., Gleizer, S., et al. (2016). Pyruvate formate-lyase enables efficient growth of Escherichia coli on acetate and formate. *Biochemistry*, *55*(17), 2423–2426.

Zelcbuch, L., Razo-Mejia, M., Herz, E., Yahav, S., Antonovsky, N., Kroytoro, H., et al. (2015). An in vivo metabolic approach for deciphering the product specificity of glycerate kinase proves that both E. coli's glycerate kinases generate 2-phosphoglycerate. *PLoS ONE*, *10*(3), e0122957.

Zhang, Y., Buchholz, F., Muyrers, J. P. P., & Francis Stewart, A. (1998). A new logic for DNA engineering using recombination in *Escherichia coli*. *Nature Genetics*, *20*, 123–128.

CHAPTER THIRTEEN

Multifragment DNA Assembly of Biochemical Pathways via Automated Ligase Cycling Reaction

Christopher J. Robinson*,†,2, Mark S. Dunstan*,†,2, Neil Swainston‡, James Titchmarsh§, Eriko Takano*,†, Nigel S. Scrutton*,†, Adrian J. Jervis*,†,1

*Manchester Centre for Synthetic Biology of Fine and Speciality Chemicals (SYNBIOCHEM), Manchester Institute of Biotechnology, The University of Manchester, Manchester, United Kingdom
†School of Chemistry, The University of Manchester, Manchester, United Kingdom
‡School of Chemistry, Faculty of Science and Engineering, Manchester Synthetic Biology Research Centre for Fine and Speciality Chemicals (SYNBIOCHEM), Manchester Institute of Biotechnology, University of Manchester, Manchester, United Kingdom
§Hamilton Sales & Service UK Limited, Birmingham, United Kingdom
[1]Corresponding author: e-mail address: adrian.jervis@manchester.ac.uk

Contents

Abstract

The microbial production of commodity, fine, and specialty chemicals is a driving force in biotechnology. An essential requirement is to introduce biosynthetic pathways to

[2] Authors contributed equally to this work.

Methods in Enzymology, Volume 608
ISSN 0076-6879
https://doi.org/10.1016/bs.mie.2018.04.011

the target compound(s) into chassis organisms. First suitable enzymes must be selected and characterized, and then genetic pathways must be designed and assembled into suitable expression vectors. The design of these pathways is crucial for balancing the pathway for efficient in vivo activity. This can be achieved through optimization of the pathway regulation by altering transcription and translation rates. The possible permutations of a multigene pathway create a vast design space which is intractable to explore using traditional time-consuming and laborious pathway assembly methods. The advent of multifragment DNA assembly technologies has enabled simultaneous, multiplexed pathway construction allowing an increased capability to sample the design space. Furthermore, the implementation of laboratory automation allows error-reduced, high-throughput (HTP) construction of pathways. In this chapter, we present a workflow that combines automated in silico design of DNA parts followed by pathway assembly using the ligase cycling reaction on robotics platforms, to allow multiplexed assembly of plasmid-borne gene pathways with high efficiency. Details and considerations in designing DNA parts for expression bacterial chassis are discussed followed by laboratory protocols for HTP pathway assembly and screening using robotics platforms. This workflow is employed in the SYNBIOCHEM Synthetic Biology Research Center, providing the capability to assemble over 96 plasmids simultaneously, with over 40% of clones from each assembly harboring the correctly assembled plasmids. This workflow is easy to modify for use in other laboratories and will help to accelerate synthetic biology projects with diverse applications.

1. INTRODUCTION

One core area of synthetic biology is to develop microorganisms capable of producing commodity, fine, and specialty chemicals via cheap and reliable fermentation technology. This technology aims to supplant traditional petrochemical synthesis to provide greener more sustainable routes to existing chemicals (e.g., biofuels, pharmaceuticals, flavorings, fragrances) and to exploit the regio- and stereospecificity of enzyme catalysis for the discovery and production of novel compounds with desirable new properties.

There are many examples of researchers developing chemical production chassis by methods which can broadly be grouped under pathway discovery and chassis engineering, often a combination of the two. Pathway engineering refers to the identification and optimization of an enzymatic route from a substrate, often a naturally occurring metabolite, to a target compound. This may be a natural biosynthetic pathway, or a novel recombinant route combining enzymes from multiple organisms (Carbonell, Parutto, Baudier, Junot, & Faulon, 2013; Delépine, Duigou, Carbonell, & Faulon, 2018; Galanie, Thodey, Trenchard, Interrante, & Smolke, 2015;

Leonard, Chemler, Lim, & Koffas, 2016; Martin, Pitera, Withers, Newman, & Keasling, 2003; Yim et al., 2011), exploiting enzyme promiscuity or employing engineered enzymes (Currin, Swainston, Day, & Kell, 2015; McLean et al., 2015; Menon et al., 2015). Once individual enzymes have been selected, the creation of a production chassis necessitates that the genes for these enzymes are built into a functional genetic pathway for in vivo expression. In addition to the gene parts pathway regulation must also be designed considering elements such as promoter location and strength, constitutive vs inducible control, translation initiation rates (TIRs), and plasmid copy numbers (Jeschek, Gerngross, & Panke, 2016; Smanski et al., 2014). Traditionally, these pathways have been designed and built in an informed but limited manner, by the iterative cloning of genes into standard expression vectors. This approach severely constrains the construction of many pathway variants for optimization of performance. The advent of simultaneous multifragment DNA assembly methods, such as Gibson (Gibson et al., 2009) or Golden Gate assembly (Engler, Kandzia, & Marillonnet, 2008), has made this process more efficient with the ability to produce a multigene pathway in a single reaction allowing the rapid production of pathway variant libraries.

There are a multitude of complementary assembly methods reported which have been previously reviewed (Casini, Storch, Baldwin, & Ellis, 2015). One attractive method is the ligase cycling reaction (LCR), which has been demonstrated for the high-efficiency construction of 12kb plasmids from up to 20 DNA parts in *Escherichia coli* (Chandran, 2017; de Kok et al., 2014). LCR works by joining phosphorylated DNA parts through the action of single-stranded "bridging oligonucleotides" which act like dominoes, each half recognizing a different DNA part and temporarily joining them for ligation during a thermal cycle utilizing a thermostable ligase. Among the key benefits of this approach are that it is scarless (and therefore the final sequence can be fully defined), individual DNA parts can be reused and rearrangement of novel DNA part combinations can be created by simply changing the pool of bridging oligonucleotides added, without redesigning the parts themselves.

This chapter presents the workflow used in SYNBIOCHEM (Carbonell et al., 2016; Le Feuvre et al., 2016) for the in silico design of DNA parts compatible with LCR, and their subsequent high-throughput (HTP) assembly into biosynthetic pathways using laboratory-based robotics platforms (Fig. 1).

Fig. 1 Schematic view of the laboratory-based, automated multifragment DNA assembly workflow using the ligase cycling reaction and robotic platforms. First DNA part and bridging oligonucleotide sequences are designed in silico using PartsGenie and PlasmidGenie software, respectively, and synthesized by commercial vendors. Subsequent steps are carried out using laboratory automation. Parts are PCR amplified and processed into a final Part Plate and quality controlled using capillary electrophoresis to check size and integrity of each part. Bridging oligonucleotides are combined into unique, assembly-specific Bridging Oligo Pools. The Part and Bridging Oligo Pools plates are used as source plate to set up LCRs which are subsequently transformed into *E. coli* competent cells. Four colonies from each plate are then picked and grown for storage and plasmid extraction. Plasmids are screened for correct assembly by restriction digest followed by capillary electrophoresis. Plasmids displaying the correct restriction mapping are sequence verified.

2. DESIGN OF DNA PARTS AND ASSEMBLY RECIPES

The design of DNA parts is influenced by the requirements of the assembly method chosen, biological activity (e.g., enzymatic or regulatory), and limitations in DNA synthesis technology. Here we describe how we can design the full pathway assembly workflow from considerations and methods for designing DNA parts (gene parts, regulatory parts, and bridging oligonucleotides) to recipes/worklists compatible for automated assembly in the laboratory using custom in silico tools.

2.1 Design of DNA Parts

2.1.1 Assembly and Synthesis Design Considerations

A key design requirement for parts suitable for assembly by LCR is that they have unique 5′ and 3′ termini to which complementary bridging oligonucleotide sequences can be designed. The length of these sequences varies, but should have a predicted annealing temperature of 70°C (de Kok et al., 2014). The efficiency of assembly is also influenced by the size of the DNA parts in two ways: (i) smaller parts are assembled more efficiently than larger ones (although a smaller number of large parts can facilitate larger final constructs) (de Kok et al., 2014); and (ii) heterogeneity in part sizes for an assembly can have a negative impact. Therefore, where possible, it is recommended to perform LCR with small, approximately equal-sized parts (within ~0.5 kb of each other). Fulfilling this recommendation is often difficult when considering gene-based parts (typically 0.5–3 kb), regulatory parts (e.g., promoters and ribosome-binding sites (RBS), which are typically ~20 bp), and vectors (typically 2–6 kb). This said, there is some tolerance in part size homogeneity while maintaining good assembly efficiency, and assembly of up to 10 parts which range between 0.5 and 3 kb in size is routinely performed with good efficiency. If an assembly includes very large parts, it is recommended to redesign these as multiple smaller fragments. For example, a useful strategy is to break vectors into several modules such as the origin of replication, antibiotic resistance marker, and promoter-repressors. This also has the advantage of allowing combinatorial assembly of vector backbone components, in a similar manner to that used to build the BglBrick family of vectors (Lee et al., 2011).

The SYNBIOCHEM approach involves designing parts for assembly by LCR, while also making them compatible with other popular methods including Golden Gate (Engler et al., 2008) and BglBrick cloning (Anderson et al., 2010) toward community parts sharing. To this end, parts are designed to be free of the corresponding restriction sites used in these methods (specifically *Bsa*I, *Eco*RI, *Bgl*II, *Bam*HI, and *Hin*dIII). We routinely use BglBrick vectors for pathway assembly, resulting in unique restriction sites flanking the pathway inserts. This facilitates diagnostic restriction digest analysis of clonal plasmids for simple quality control (QC).

There are also a number of constraints which need to be considered for tractable DNA synthesis, including the removal of repetitive regions, avoiding multiple consecutive bases, and controlling GC content.

SYNBIOCHEM has developed an open-access online DNA parts designer, PartsGenie (Swainston et al., 2018), to facilitate part design while simultaneously optimizing these multiple requirements and constraints. The PartsGenie design tool is available online (https://parts.synbiochem.co.uk/), where a detailed user guide can be found. PartsGenie also interfaces with the inventory of composable elements (ICE) database (Ham et al., 2012) allowing for storage and curation of designed parts.

2.1.2 Biological Considerations

Two main types of parts are used in a pathway: gene parts, which contain a single open reading frame (ORF), and regulatory parts which contain control elements such as transcriptional promoters and terminators.

ORFs require a 5′ UTR containing an RBS, which are typically short (~60 bp) and would reduce the efficiency of assembly. Therefore, bespoke RBS sequences are included in our gene part design, reducing the number of parts and producing unique 5′ sequences in the process. PartsGenie uses algorithms based on the RBS calculator (Salis, Mirsky, & Voigt, 2009), allowing for the design of RBS sequences with predictable TIRs. This design process also adds short insulating sequences 5′ of the RBS, to reduce the impact of upstream sequences on translation levels. PartsGenie will also detect and remove any aberrant RBS sequences within an ORF to prevent unproductive translation events. The 3′ ends of genes are usually unique, but if this is not the case, then short random sequences can be added after the stop codon to provide unique sequences for bridging oligonucleotide hybridization. This can also be a useful strategy if several parts are to be treated as interchangeable (e.g., multiple homologues of a single enzyme), in which case consistent sequences must be added to the 5′ and 3′ ends of each interchangeable part. ORFs are also codon-optimized for the organism of choice. All of the operations (codon optimization, RBS generation, 5′ and 3′ sequence generation, protection of restriction sites, maintenance of desirable GC content, etc.) are applied simultaneously. Details for how to design a gene part are in the procedure below (Section 2.1.3).

Regulatory parts can be designed according to the user's specifications. For an example promoter part, a unique 5′ assembly sequence is designed, followed by a transcriptional terminator, a random spacer sequence (200–300 bp), a promoter sequence, and a 3′ assembly sequence. These parts are typically 0.5 kb in size to facilitate assembly efficiency.

2.1.3 Procedure

1. In a web browser navigate to https://parts.synbiochem.co.uk/.
2. Provide a name for the part you are designing in the text box below the feature icons.
3. Drag-and-drop required elements from the feature icons into the design window in the desired order. Parameters for each feature can be selected by clicking the feature and editing in the right-hand feature panels. For LCR-compatible gene parts, the following parameters are used:
   a. **Sequence of defined melting temperature** (set at 70°C).
   b. Set **Ribosome binding site** with the recommended TIR of 25,000 (can be varied from 100 up to 200,000).
   c. For **Coding sequence** specify a UniProt ID or amino acid sequence.
   d. *Optional.* **Sequence of defined melting temperature** (set at 70°C). If required this 3′ flanking sequence (along with that at the 5′ end of the design) can be defined for subsequent parts if you wish to create interchangeable parts that can be assembled using the same bridging oligonucleotides.
4. In the **Organism** text box select the organism of choice (e.g., type Escherichia, then select "Escherichia coli str. K-12 substr. MG1655" from the options).
5. In **Filters** select **Excluded restriction sites** such as *Bam*HI, *Bgl*II, *Bsa*I, *Eco*RI, and *Xho*I.
6. Click the **Submit** button at the top of the **Parts** window. After the design has completed, information is displayed in a **Results** window to the right (including the DNA sequence of the designed part).
7. Save the results to an ICE repository. Use the **ICE** button to input the **URL**, **Username**, and **Password** of the desired ICE repository, and **Login**. Doing so will enable the **Save** button in the **Results** panel. Alternatively, if not using an ICE database, sequences can be viewed and copied for storage.
8. Order the designed part(s) from a commercial vendor.

2.2 Design of Bridging Oligonucleotides

Bridging oligonucleotide design is simple once parts are designed because the sequences have already been designed at the parts design stage (Section 2.1). However, this can become time-consuming and error-prone when attempting to multiplex pathway assembly or when producing large combinatorial libraries of pathways. To automate this process the online tool PlasmidGenie has been developed (https://parts.synbiochem.co.uk/plasmidGenie). The PlasmidGenie software is used to design bridging oligos, according to requirements previously described (de Kok et al., 2014), and once designed the SYNBIOCHEM LCR software can be used to output a list of all the bridging oligonucleotides (and parts) that require ordering. Oligonucleotides should be ordered in defined positions in multiwell plates to facilitate the generation of robotic work-lists and oligonucleotide tracking.

2.2.1 Procedure

1. Design plasmids in a software package capable of exporting data in a tab-delimited format (e.g., Excel). The format allows for the design of multiple plasmids, which are arranged on a line-per-line basis. The first column represents the plasmid name, with subsequent columns representing the individual parts (in order) from which the plasmid will be assembled. The identifiers of the parts must match IDs of entries in the ICE database (see file plasmids.txt for an example of the data format).
2. Navigate to https://parts.synbiochem.co.uk/plasmidGenie and log into ICE as described earlier.
3. Choose the plasmid design file (from step 1) in the **Design file** option.
4. Specify the default **Melting temperature** of 70°C.
5. In **Digest Reagents**.
6. **Submit** the task, and on completion, save the results to ICE, noting the ICE IDs of each plasmid after they have been saved. (These will be visible under **Links** in the **Results** panel.)
7. Use the SYNBIOCHEM LCR software to generate list of parts and bridging oligonucleotides required for purchase.
   a. Install Python if not already present (https://www.python.org/downloads/).
   b. Install the Python tool, pip, if not already present (https://packaging.python.org/tutorials/installing-packages/).
   c. Install the SYNBIOCHEM generic libraries through the following command: `pip install synbiochem-py`

d. Download and unzip the SYNBIOCHEM LCR software from https://github.com/synbiochem/lcr/archive/master.zip.
e. In a Terminal or Command Prompt, navigate to the directory in which the LCR software was unzipped, and type: export PYTHONPATH=.:$PYTHONPATH.
f. Using the ICE IDs from step 6, above, generate the list of parts by typing: `python lcr_utils/order.py [ICE_URL] [ICE_USERNAME] [ICE_PASSWORD] [ICE_PLASMID_ID_1] [ICE_PLASMID_ID_2] [ICE_PLASMID_ID_N]`.
g. A full list of required parts for all plasmids will be generated in the output file *out.csv*. Use this file (ideally open with Excel) to generate an order of the oligonucleotides required from a vendor in step 8.

8. Order basic salt-free oligonucleotides (standard purification is adequate) presuspended at 100 μ*M* in H_2O, in defined wells in 96-well plate format.

2.3 Production of Bridging Oligonucleotide Pools and Assembly Recipes

Two key recipes are needed to provide the correct assembly instructions for the automated robotics platform: the unique bridging oligonucleotide pool for each individual assembly (i.e., all the bridging oligonucleotides required for assembly of a complete pathway design into a host compatible vector), and the LCR recipe containing all the individual components for any given assembly (i.e., parts, reagents, bridging oligonucleotide pool). Automated production of these recipes as spreadsheets is an important step toward the generation of laboratory automation worklists discussed later and should be considered where possible. Manual annotation of these recipes can often incorporate errors, particularly for more complex assemblies and when throughput is high. To generate automated worklists, follow the procedure below.

2.3.1 Procedure

1. Assuming that the SYNBIOCHEM software has already been installed (see Step 7 of Procedure in Section 2.2.1), navigate to the lcr software directory and type: `python lcr_utils/lcr.py [ICE_URL] [ICE_USERNAME] [ICE_PASSWORD] [ICE_PLASMID_ID_1] [ICE_PLASMID_ID_2] [ICE_PLASMID_ID_N]`
2. Upon completion, an *assembly* directory will be generated, containing five text files in comma-separated format. Three of these files represent plates

(*components.csv*, *domino_pools.csv*, and *MastermixTrough.csv*), and two contain worklists (*domino_pools_worklist.csv* and *lcr_worklist.csv*). (*components.csv* and *MastermixTrough.csv* specify well positions for synthetic DNA (parts and bridging oligos) and reagents required for LCR, respectively. These must be adhered to in the following automated assembly steps.)

3. Edit the two worklist files (ideally, use Excel as an editing tool) to ensure that the column names *DestinationPlateBarcode* and *SourcePlateBarcode* correspond with the actual barcodes of the plates being used in the automated assembly.

2.4 Production of Robotic Worklists and Programs

Below are details of how we program and create worklists to drive robotics platforms to create bridging oligonucleotide pools, perform LCR assemblies, and run QC checks on our assembled pathways. Each of these operations requires either predefined programs which can be run routinely or worklist-driven programs to carry out specific liquid handling steps for each sample.

The worklists are defined by both the specific recipe and the automated procedure required, and prepared in a spreadsheet format (comma-separated values, CSV) which can be produced by hand, but automated generation allows multiplexing and reduces error. They define a list of pipetting instructions to take individual components from barcoded "Source Plates" and a "Master Mix Trough" (containing premixed reagents) and dispense them into defined wells in "Destination Plates." The order in which components are added is defined by source plate type (Master Mix Trough is always first) and the order each pipetting action appears in the worklist by row.

Worklists are organized to minimize the number of plates required typically a maximum of three Source Plates and a single Destination Plate and define barcodes to all Destination and Source plates.

2.4.1 Bridging Oligonucleotide Pool Worklist

Each pathway assembly reaction requires a unique set of bridging oligonucleotides which are mixed in a pool, and the bridging oligonucleotides pool worklist defines the automated pipetting steps needed to combine these oligonucleotides for each individual LCR. This worklist draws information from the bridging oligonucleotide recipes (which oligos are required for each assembly; Section 2.3.1) and the bridging oligonucleotide order spreadsheet (Section 2.2.1) which defines the well locations of individual oligonucleotides in the Bridging Oligos plate. A Destination Plate (Bridging Oligo Pools) is created such that each pool is defined in a single well and the mapping information and plate can then be used to create a subsequent LCR worklist and reaction (Section 2.4.2). An example of the bridging oligonucleotide pool worklist can be seen in Fig. 2A.

2.4.2 LCR Worklist

Each LCR requires the same standard reagents but also unique combinations of DNA parts and oligonucleotide pools. To remove error during HTP assembly of unique pathways an LCR worklist is produced which draws from the LCR recipes (which parts are required for each assembly) and the Part Plate and Bridging Oligo Pools Plate mapping definitions from the previous steps. A destination plate (LCR) is created such that each assembly is defined in a single well (Fig. 2B).

2.4.3 Plate-Based Restriction Analysis Setup

We use a standard liquid handling protocol (not worklist driven) to set up restriction digests of our plasmids typically across a full 96-well plate. To this end we have developed a program which uses the 8-channel head to transfer a digestion master mix from a single source column to all wells in a Destination plate followed by the transfer of plasmid samples into the same destination plate using the 96-channel head. The plate is then sealed manually, centrifuged, and incubated as detailed in Section 3.5.3.

2.4.4 Setup of Fragment Analyzer Capillary Electrophoresis Plates

Similar to the restriction analysis, we use a fixed program to dilute PCR products or restriction digest reactions into a 96-well PCR plate for analysis on a capillary electrophoresis machine as detailed in Section 3.5.3.

3. PATHWAY ASSEMBLY VIA AUTOMATED LCR

3.1 Robot Specifications

Here we present some general guides to the specifications of robotic hardware and operational considerations associated with our DNA assembly workflow. These are guidelines and would need alteration for use on other robotic platforms but will serve as a guide and to highlight important considerations.

A

DestinationPlateBarcode	DestinationPlateWell	SourcePlateBarcode	SourcePlateWell	Volume	ComponentName	plasmid_id
BRIDGING_OLIGO_POOLS_1	A1	BRIDGING_OLIGOS_1	G2	1.5	PART1_PART2	p001
BRIDGING_OLIGO_POOLS_1	A1	BRIDGING_OLIGOS_1	H2	1.5	PART2_PART3	p001
BRIDGING_OLIGO_POOLS_1	A1	BRIDGING_OLIGOS_1	A3	1.5	PART4_PART5	p001
BRIDGING_OLIGO_POOLS_1	A1	BRIDGING_OLIGOS_1	B3	1.5	PART5_PART1	p001
BRIDGING_OLIGO_POOLS_1	A1	BRIDGING_OLIGOS_1	F3	1.5	PART6_PART7	p001
BRIDGING_OLIGO_POOLS_1	A1	BRIDGING_OLIGOS_1	G3	1.5	PART7_PART8	p001
BRIDGING_OLIGO_POOLS_1	A1	BRIDGING_OLIGOS_1	H3	1.5	PART3_PART9	p001
BRIDGING_OLIGO_POOLS_1	A1	BRIDGING_OLIGOS_1	A4	1.5	PART9_PART6	p001
BRIDGING_OLIGO_POOLS_1	A1	MastermixTrough	A1	482	H2O	

B

DestinationPlateBarcode	DestinationPlateWell	SourcePlateBarcode	SourcePlateWell	Volume	ComponentName	Description	plasmid_id
LCR_REACTION_1	A1	MastermixTrough	E1	8.5	H2O	H2O	p001
LCR_REACTION_1	A1	PART_PLATE_1	A9	1	lacI-PT7	Backbone promoter	p001
LCR_REACTION_1	A1	PART_PLATE_1	C9	1	ColE1	Origin	p001
LCR_REACTION_1	A1	PART_PLATE_1	D9	1	AmpR	Resistance cassette	p001
LCR_REACTION_1	A1	PART_PLATE_1	E8	1	GENE#1	ORF	p001
LCR_REACTION_1	A1	PART_PLATE_1	G8	1	GENE#2	ORF	p001
LCR_REACTION_1	A1	PART_PLATE_1	F8	1	GENE#3	ORF	p001
LCR_REACTION_1	A1	PART_PLATE_1	H8	1	GENE#4	ORF	p001
LCR_REACTION_1	A1	BRIDGING_OLIGO_POOLS_1	D3	1	domino pool	domino pool	p001
LCR_REACTION_1	A1	MastermixTrough	D1	7	mastermix	mastermix	p001
LCR_REACTION_1	A1	MastermixTrough	G12	1.5	Ampligase	Ampligase	p001

Fig. 2 Examples of liquid handling worklists for Hamilton Microlab STAR Robotics platforms. (A) Bridging oligonucleotide pool mixing in a "Bridging Oligo Pools" plate from individual bridging oligonucleotides in a "Bridging Oligo" source plate and water from a Mastermix plate (MastermixTrough). (B) LCR setup in an "LCR reaction" plate from two source plates (Bridging Oligos and Part Plate) and a Mastermix plate (MastermixTrough).

3.1.1 Hardware

Hamilton Microlab STAR robotic platforms have modular deck and a versatile selection of add-on equipment (incubators, centrifuges, etc.) allowing a tailored setup directed by applications. Pertinent features of our customization included 96-channel (96-Probe Head) and 8-channel (1-mL Channel) pipetting heads, integrated Biometra TRobot thermocycler (Analytik Jena), iSWAP and CO-RE Grips (for plate transfers), an integrated barcode reader for sample tracking, an integrated light deck, and an integrated CLARIOstar plate reader (BMG Lab Tech). A typical deck layout can be seen in Fig. 3.

3.1.2 Pipetting

Careful consideration should be paid to liquid class determination at low volumes (0.5–2 μL). Significant errors in liquid transfer are common in poorly optimized liquid classes and can reduce assembly efficiency (see notes in Section 3.3.4). Pipetting accuracy can be maintained by transferring volumes of >1 μL (75 nm) using 50 μL tips (Hamilton Nestable NTR CO-RE 50 μL Tips) and reagent concentrations should be adjusted to work in this range.

Every source well will require excess of its reagent to allow accurate pipetting and this must be taken into consideration when calculating reagent volume requirements. This excess volume is termed the "dead volume." Using conductive tips, it is recommended to always use "liquid detection" during pipetting for confidence that the correct transfers have occurred. If sample volumes are below the required dead volumes, the robotic error warnings can be used to pause and flag insufficient liquid errors allowing manual intervention if required. It is essential that any liquid detection errors are checked at the end of every run to ensure there was sufficient volume available in the source well prior to liquid transfer to ensure the component was correctly transferred. Recommended dead volumes are 15 μL for 96-well PCR plates and 40 μL for 96-deepwell plates.

For accurate pipetting it is also crucial to optimize the "liquid class" particularly when low volume liquid transfer using the 8-channel pipetting head, as in our protocol. A recommended approach is to optimize liquid transfer using the fluorescent compound fluorescein and check relative fluorescence using a CLARIOstar plate reader (or equivalent). Fluorescein can be used to track the relative consistency of liquid transfer across a 96-well plate or as a direct comparison to manual pipetting. All individual component parts of the LCR should be checked and optimized where needed.

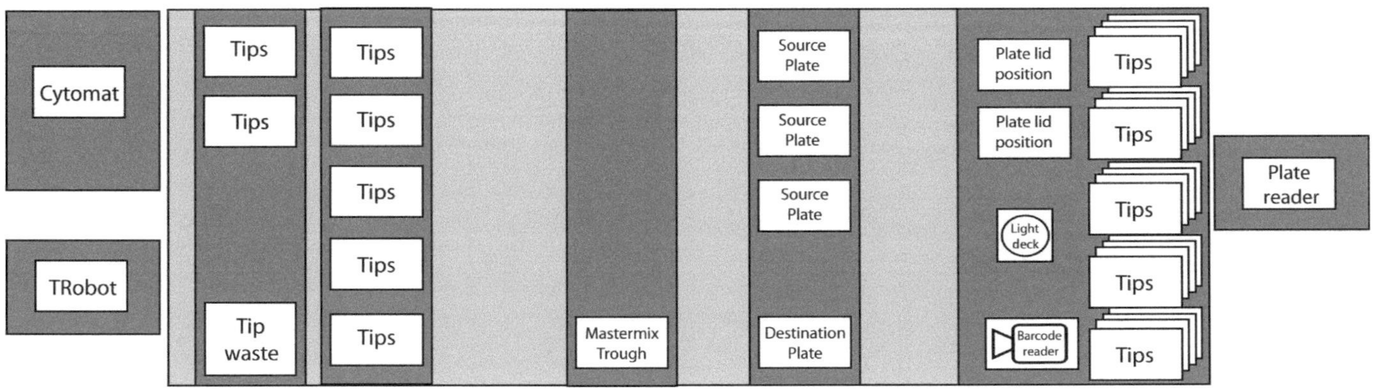

Fig. 3 Hamilton Microlab STAR deck layout for DNA preparation and assembly. Platform includes integrated Trobot (thermocycler), barcode reader, light deck for colony picking, and temperature-controlled cooling unit (Cytomat). Labware positions include 3 × Source Plates, 1 × Destination Plate, 1 × MastermixTrough, and 50 μL NTR nested tips (Max 1920).

Factors found to improve the pipetting accuracy at low volumes are: (1) increased the dispense speed, (2) increased the blowout speed, (3) increased the over-aspirate volume, and (4) increased settling time.

3.1.3 Colony Picking

Colony picking can be performed on the Hamilton Microlab STAR if equipped with an integrated light deck and the software easyPickII. Cultures grown on barcoded Petri dishes must be diluted to achieve distributed single colonies to facilitate colony recognition and single colony selection for inoculation in destination 96-deepwell blocks containing media. Output CSV files are produced postcolony pick and facilitate sample tracking of each clone.

3.1.4 Sample Tracking

The inclusion of a barcode reader is essential for sample tracking and confidence that the correct plates are placed on deck in the correct location. On-screen instructions direct the user where to place barcoded plates on deck which are then subsequently read prior to any pipetting step to ensure correct setup. Placing Source Plates in the wrong position will result in a barcode mismatch, triggering on-screen guides that allow the user to replace in the correct position. Only when plate barcodes match the requested barcodes (i.e., match those in the worklist) and are placed in the correct position on deck will pipetting begin. This information can be held in worklists and outputs from robotics platforms allowing both tracking of plates and the definition of wells. It is also recommended to use colored plates where possible to allow easy identification of plate types.

3.2 Preparation of DNA Parts

DNA parts are prepared by standard PCR from commercially synthesized templates using a high-fidelity DNA polymerase. PCR primers are purchased with 5′ phosphorylation and are used to amplify blunt-ended parts, to allow ligation during the LCR. Nonphosphorylated primers can be used, provided an additional polynucleotide kinase reaction step is added to the workflow to phosphorylate parts (de Kok et al., 2014).

3.2.1 Equipment

- Standard thermocycler (96-well plate format preferable for automation)
- DNA electrophoresis equipment (standard agarose gel electrophoresis or capillary electrophoresis system)

- Standard microcentrifuge
- Nanodrop spectrophotometer or equivalent
- 96-Well PCR Part Plate
- Adhesive PCR seal

3.2.2 Buffers and Reagents

- High-fidelity DNA polymerase Master Mix, such as Phusion Hot Start Flex (New England Biolabs (NEB), M0536) or CloneAmp HiFi (Clontech, #639298)
- *DpnI* with 10 × CutSmart buffer (NEB, R0176)
- Spin column DNA cleanup kit (e.g., Macherey-Nagel NucleoSpin kit, #740609)
- Nuclease-free, deionized water

3.2.3 Procedure

1. PCR amplify part from 10 ng of plasmid template with 25 pmol each of forward and reverse primers, in a 50 μL final volume using standard conditions.
2. Analyze 1 μL of PCR product via gel electrophoresis for correct size and integrity.
3. Add 122.5 μL of nuclease-free deionized water, 25 μL of 10 × CutSmart buffer and 2.5 μL (50 U) of *Dpn*I enzyme.
4. Mix gently and briefly centrifuge.
5. Incubate for 2 h in 37°C water bath.
6. Denature at 90°C for 30 min.
7. Clean reaction using "NucleoSpin Gel and PCR Cleanup kit" (Macherey-Nagel) or equivalent, eluting in 25 μL.
8. Check concentration and purity using a Nanodrop or equivalent (target 260/280 ratio of ~1.8, 260/230 ratio of 2.0–2.2).
9. Dilute parts to 75 n*M* by adding nuclease-free deionized water.
10. Transfer parts to defined wells in a barcoded, 96-well PCR Part Plate and seal with removable seal.
11. Use immediately or store at 4°C.

3.2.4 Notes

Parts should be made fresh each time for assembly and stored at 4°C for <1 week. Freezing is not recommended as freeze–thaw cycles will reduce assembly efficiency.

3.3 Preparation of Bridging Oligonucleotide Pools

The procedure for preparing bridging oligonucleotide pools is straightforward, but when performed in HTP, errors are easily introduced when carried out manually. It is therefore strongly recommended to perform this step using a worklist-driven robotics platform.

3.3.1 Equipment

- Hamilton Microlab STAR robotics platform or equivalent
- "Bridging Oligo Pools" destination 96-well format deepwell plate
- "Mastermix Trough" 96-deepwell plate

3.3.2 Buffers and Reagents

- "Bridging Oligos" plate, commercially synthesized bridging oligonucleotides (100 μ*M*)
- Nuclease-free, deionized water

3.3.3 Procedure

1. Place a barcoded Bridging Oligo Pools (Destination Plate), Bridging Oligo deepwell plate (Source Plate) and a Mastermix Trough containing nuclease-free deionized water in position on deck.
2. Run the bridging oligonucleotide pool worklist, to add nuclease-free deionized water and 3.75 μL of each oligonucleotide to defined target wells in the Destination Plate, to produce 500 μL pools with a final concentration of 0.75 μ*M* for each oligonucleotide.

3.3.4 Notes

Bridging oligonucleotide pools can be stored at 4°C for <2 weeks or −20°C for many weeks.

3.4 Automated LCR

LCRs are driven by a previously written worklist (Section 2.4.2) and follow the standard reaction setup devised by de Kok et al. (2014). Parts are diluted to 75 n*M* to ensure accurate pipetting of 1 μL allowing a total of 16 parts per assembly in a total reaction volume of 25 μL.

3.4.1 Equipment

- Hamilton Microlab STAR robotics platform or equivalent
- Standard 96-well plate format thermocycler (integrated/separated)
- "LCR Reaction" Destination Plate (96-well PCR plate)

- "Bridging Oligo Pools" plate (96-deepwell block)
- "Part Plate" (96-well PCR plate)
- Adhesive PCR plate seal

3.4.2 Buffers and Reagents

- DMSO
- 4.5 *M* betaine in nuclease-free deionized H_2O
- Ampligase DNA Ligase with 10 × reaction buffer (Epicentre, A3210K)
- LCR Mastermix (per reaction: 2.5 μL 10 × reaction buffer, 2.5 μL 4.5 *M* betaine, 2 μL DMSO).
- High-efficiency competent *Escherichia coli* cells (>10^9 cfu/mL/μg), such as NEB 5-alpha (NEB, C2987I).
- Lysogeny Broth (LB) agar plates with appropriate antibiotics (typically 100 μg/mL ampicillin or 50 μg/mL kanamycin).

3.4.3 Procedure

1. Prepare fresh LCR Mastermix (7 μL per reaction, plus 40 μL excess as dead volume) and transfer into a well of a 96-deepwell plate as "Mastermix Trough."
2. Place LCR Reaction plate (Destination Plate), Part Plate, Bridging Oligo Pools plate, and the Mastermix Trough in position on deck.
3. Run the LCR worklist to combine all components of each LCR in the defined wells on the Destination Plate, with mixing by aspiration/dispense cycling at each stage (~3 min per reaction).
4. Manually seal LCR plate with adhesive PCR plate seal.
5. The automated method automatically transfers the completed LCR plate to the integrated thermocycler (T-Robot). Alternatively, manually transfer the plate to an external thermocycler.
6. Run the LCR program (Table 1).

Table 1 LCR Cycling Conditions

Step	Temperature (°C)	Time (mm:ss)	Cycles
Initial denaturation	94	02:00	1×
Denaturation	94	00:10	50×
Annealing	55	00:30	
Ligation	65	01:00	
Hold	4	Hold	—

7. In a 96-well PCR plate, transform 0.5 μL of each LCR into separate 10 μL aliquots of high-efficiency competent cells by heat shock using a thermocycler. Recover in 100 μL SOC media for 1 h.
8. Spread all 110 μL of each transformation reaction onto separate selective agar plates.
9. Incubate at 37°C overnight.

3.5 Colony Screening and Sequence Verification

All LCR-assembled constructs need to be sequence verified to ensure correct assembly of parts and to check for any PCR-introduced mutations. The workflow below describes how we prescreen potential clones from transformation plates by plasmid purification and restriction analysis. Plasmids are designed to have a unique restriction site at the 5′ and 3′ termini (*Eco*RI and *Bam*HI, respectively) of the cloned insert (see Section 2.1), allowing the separation of vector backbone and recombinant pathway for analysis by electrophoresis (Fig. 4). Clones showing the correct restriction patterns are then dispatched for DNA sequencing.

3.5.1 Equipment

- Hamilton Microlab STAR robotics platform or equivalent
- Standard 96-well format thermocycler
- Fragment Analyzer (Advanced Analytical) capillary electrophoresis system or equivalent
- Multitron Pro plate shaker (Infors HT) or other shaking plate incubator
- 96-Well format deepwell plates ("Colony-Picked Cultures")
- 96-Well PCR plates ("Plasmid," "Restriction Digest," "Capillary Electrophoresis")
- 96-Well microtiter "Glycerol Stock" plate

3.5.2 Buffers and Reagents

- Restriction enzymes and 10 × reaction buffer
- Nuclease-free, deionized water
- Restriction digest Mastermix (5 U of each enzyme, 1 μL 10 × reaction buffer, nuclease-free deionized water for final 3 μL per reaction)
- QIAprep 96 Turbo Miniprep Kit (Qiagen, #27193) or equivalent
- Appropriate "Qualitative/Sizing Analysis Kit" for Fragment Analyzer (Advanced Analytical). Typically the "dsDNA 920" or "dsDNA 930" reagent kits are suitable
- LB with plasmid-selective antibiotics
- LB agar plates with selective antibiotics

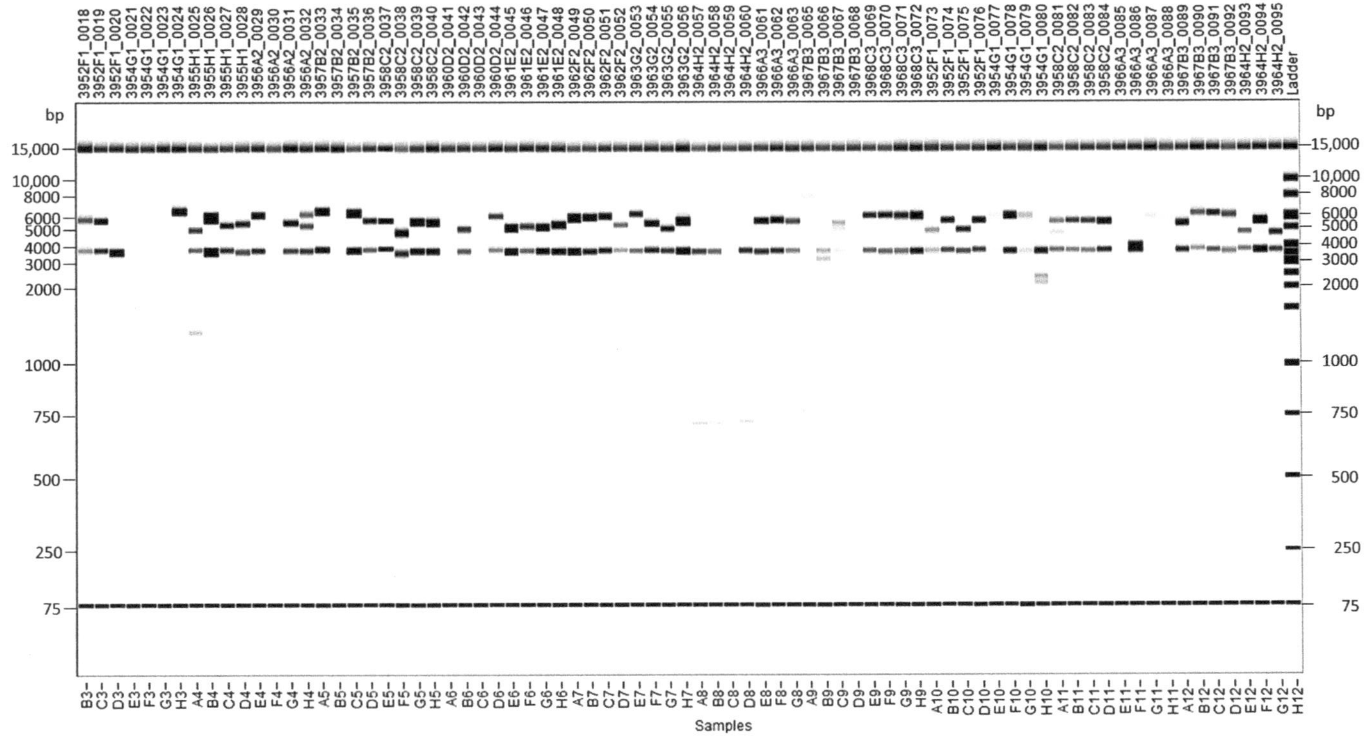

Fig. 4 Capillary electrophoretogram of restriction digestion analysis of candidate plasmid pathways from a multiplexed assembly. Plasmid sequences are designed to include unique cleavage sites (for the restriction enzymes *Eco*RI and *Bam*HI) that flank the assembled pathway, 96-well, purified plasmids are digested and analyzed using capillary electrophoresis to check for plasmid backbone (*lower band*) and pathways (*upper band*). Those exhibiting the correct banding pattern are subsequently sequence verified.

3.5.3 Procedure

1. Using automated colony picking, inoculate four colonies from each LCR transformation into 1 mL LB (with selective antibiotics), in defined wells in the "Colony-Picked Cultures" 96-deepwell plate.
2. Grow overnight at 37°C with shaking at 1000 rpm.
3. Transfer 100 µL of each well to a sterile 96-well "Glycerol Stock" microtiter plate containing 100 µL of 50% glycerol in each well. Freeze and store this stock plate at −80°C.
4. Perform a 96-well plate format plasmid preparation on the remaining culture, following kit instructions, eluting in 100 µL in the "Plasmids" plate.
5. Perform automated setup of a restriction digest analysis as follows:
   a. Prepare the restriction digest mastermix (3 µL per reaction, plus 40 µL excess as dead volume) and transfer 80 µL to each well of column 1 of a 96-deepwell plate (Digest Mastermix Trough).
   b. Place the plasmid miniprep plate, Digest Mastermix Trough, and Restriction Digest Plate in position on the robot deck.
   c. Automated transfer of 3 µL of restriction digest mastermix to each well of the Digest Plate.
   d. Automated transfer of 7 µL of each plasmid minipreps to the corresponding target wells in the Digest Plate, with mixing by aspiration/dispense cycling using the 96-channel head.
6. Seal the plate and incubate for 1 h at 37°C in a thermocycler.
7. Return the Digest Plate, an empty 96-well PCR plate (Capillary Electrophoresis Plate) and a 96-deepwell Mastermix Trough (containing 300 µL of 1 × TE dilution buffer in column 1) in position on the robot deck.
8. Perform automated setup of the Capillary Electrophoresis plate as follows:
   a. Transfer 22 µL of TE dilution buffer to each well of the Capillary Electrophoresis plate.
   b. Transfer 2 µL each plasmid from the Restriction Digest to the corresponding wells of the Capillary Electrophoresis plate. Note: well H12 is reserved for the sizing ladder.
9. Add 2 µL of sizing ladder to well H12 and mix.
10. Run the plate on the Fragment Analyzer, following the instructions for your specific reagent kit.
11. Select any samples which demonstrate the correct restriction digest pattern (correct number and size of fragments), indicating successful assembly (e.g., Fig. 4).

12. Repeat this procedure from step 1 for any designs for which no clones with the correct digest pattern are identified.
13. Sequence verify correctly sized clones for each pathway design (recommended to sequence at least two for each pathway).

3.5.4 Notes

Hamilton CO-RE grips can be modified to handle curved Petri dishes but for standard CO-RE grips or other robotic transfer arms (i.e., TECAN RoMA) 8-well square Nunc plates can be used.

4. SUMMARY AND CONCLUSION

DNA assembly is a core enabling technology of Synthetic Biology and one-step assembly of multiple DNA molecules allows the rapid production of new designed DNA constructs. Using this approach to prototype and optimize biochemical pathways allows the testing of very large design spaces. To do this effectively requires automation to facilitate HTP construction and verification of many pathways in parallel and replace laborious and error-prone manual operations for reproducibility and traceability. Here we have presented a suggested and optimized workflow for the simultaneous and efficient production of many plasmid-based pathways on automated robotics platforms.

The efficiency of this workflow varies with each pathway. Typically our assemblies have consisted of <10 DNA parts varying from 0.4 to 2.5 kb in size (mixed) and a final construct size of <12 kb. With these parameters this workflow has been successful in producing at least one colony with the correct, sequence-perfect pathway for each design (~40% of total colonies when building libraries of pathways) and many designs have shown 100% of colonies checked to be correct.

While we find a high success rate of assembly, there are some limitations of LCR such as the drop-off in efficiency as constructs increase in size (>12 kb) which could be limiting for large pathways such as antibiotic biosynthesis gene clusters or in the production of synthetic chromosomes. Typically these very large constructs are assembled in an iterative manner by assembly of increasing sized modules which could be produced by LCR and later assembled using yeast homologous recombination methods.

There are also bottlenecks in this workflow such as the part-specific requirement of custom primers for PCR amplification. When many parts are to be produced, this creates difficulty in primer design, selection, and tracking. Sequencing of many large plasmids also holds similar problems and we envisage the addition of a next-generation sequencing workflow for sequence verification would increase the ability to screen more clones and improve the efficiency of screening. Both of these issues highlight the importance to have a robust sample and data tracking system in place, standard manual labelling, and recording of information rapidly becomes impractical but development of these systems must be laboratory specific and is out of the scope of this chapter.

To automate any laboratory procedure is not a simple task and requires careful and iterative testing at every stage of the process. We hope that the guidelines in this chapter will aid others in rapidly producing similar automated workflows.

ACKNOWLEDGMENTS

The authors acknowledge the funding from the Biotechnology and Biological Sciences Research Council (BBSRC) under grants BB/M017702/1, "Centre for Synthetic Biology of Fine and Speciality Chemicals (SYNBIOCHEM)." This is a contribution from the Manchester Centre for Synthetic Biology of Fine and Speciality Chemicals (SYNBIOCHEM).

REFERENCES

Anderson, J., Dueber, J. E., Leguia, M., Wu, G. C., Goler, J. A., Arkin, A. P., et al. (2010). BglBricks: A flexible standard for biological part assembly. *Journal of Biological Engineering*, *4*(1), 1.

Carbonell, P., Currin, A., Dunstan, M., Fellows, D., Jervis, A., Rattray, N. J., et al. (2016). SYNBIOCHEM—A SynBio foundry for the biosynthesis and sustainable production of fine and specialty chemicals. *Biochemical Society Transactions*, *44*(3), 675–677.

Carbonell, P., Parutto, P., Baudier, C., Junot, C., & Faulon, J. L. (2013). RetroPath: Automated pipeline for embedded metabolic circuits. *ACS Synthetic Biology*, *3*(8), 565–577.

Casini, A., Storch, M., Baldwin, G. S., & Ellis, T. (2015). Bricks and blueprints: Methods and standards for DNA assembly. *Nature Reviews Molecular Cell Biology*, *16*(9), 568–576.

Chandran, S. (2017). Rapid assembly of DNA via ligase cycling reaction (LCR). *Methods in Molecular Biology*, *1472*, 105–110.

Currin, A., Swainston, N., Day, P. J., & Kell, D. B. (2015). Synthetic biology for the directed evolution of protein biocatalysts: Navigating sequence space intelligently. *Chemical Society Reviews*, *44*(5), 1172–1239.

de Kok, S., Stanton, L. H., Slaby, T., Durot, M., Holmes, V. F., Patel, K. G., et al. (2014). Rapid and reliable DNA assembly via ligase cycling reaction. *ACS Synthetic Biology*, *3*(2), 97–106.

Delépine, B., Duigou, T., Carbonell, P., & Faulon, J. L. (2018). RetroPath2.0: A retrosynthesis workflow for metabolic engineers. *Metabolic Engineering*, *45*, 158–170.

Engler, C., Kandzia, R., & Marillonnet, S. (2008). A one pot, one step, precision cloning method with high throughput capability. *PLoS One*, *3*(11), e3647.

Galanie, S., Thodey, K., Trenchard, I. J., Interrante, M. F., & Smolke, C. D. (2015). Complete biosynthesis of opioids in yeast. *Science*, *349*(6252), 1095–1100.

Gibson, D. G., Young, L., Chuang, R. Y., Venter, J. C., Hutchison, C. A., & Smith, H. O. (2009). Enzymatic assembly of DNA molecules up to several hundred kilobases. *Nature Methods*, *6*(5), 343–345.

Ham, T. S., Dmytriv, Z., Plahar, H., Chen, J., Hillson, N. J., & Keasling, J. D. (2012). Design, implementation and practice of JBEI-ICE: An open source biological part registry platform and tools. *Nucleic Acids Research*, *40*(18), e141.

Jeschek, M., Gerngross, D., & Panke, S. (2016). Rationally reduced libraries for combinatorial pathway optimization minimizing experimental effort. *Nature Communications*, 7, 11163.

Lee, T. S., Krupa, R. A., Zhang, F., Hajimorad, M., Holtz, W. J., Prasad, N., et al. (2011). BglBrick vectors and datasheets: A synthetic biology platform for gene expression. *Journal of Biological Engineering*, *5*(1), 12.

Le Feuvre, R. A., Carbonell, P., Currin, A., Dunstan, M., Fellows, D., Jervis, A. J., et al. (2016). SYNBIOCHEM Synthetic Biology Research Centre, Manchester—A UK foundry for fine and specialty chemicals production. *Synthetic and Systems Biotechnology*, *1*(4), 271–275.

Leonard, E., Chemler, J., Lim, K. H., & Koffas, M. A. (2016). Expression of a soluble flavone synthase allows the biosynthesis of phytoestrogen derivatives in Escherichia coli. *Applied Microbiology and Biotechnology*, *70*(1), 85–91.

Martin, V. J., Pitera, D. J., Withers, S. T., Newman, J. D., & Keasling, J. D. (2003). Engineering a mevalonate pathway in *Escherichia coli* for production of terpenoids. *Nature Biotechnology*, *21*(7), 796–802.

McLean, K. J., Hans, M., Meijrink, B., van Scheppingen, W. B., Vollebregt, A., Tee, K. L., et al. (2015). Single-step fermentative production of the cholesterol-lowering drug pravastatin via reprogramming of Penicillium chrysogenum. *Proceedings of the National Academy of Sciences of the United States of America*, *112*(9), 2847–2852.

Menon, N., Pásztor, A., Menon, B. R., Kallio, P., Fisher, K., Akhtar, M. K., et al. (2015). A microbial platform for renewable propane synthesis based on a fermentative butanol pathway. *Biotechnology for Biofuels*, *8*(1), 61.

Salis, H. M., Mirsky, E. A., & Voigt, C. A. (2009). Automated design of synthetic ribosome binding sites to control protein expression. *Nature Biotechnology*, *27*(10), 946–950.

Smanski, M. J., Bhatia, S., Zhao, D., Park, Y., Woodruff, L. B., Giannoukos, G., et al. (2014). Functional optimization of gene clusters by combinatorial design and assembly. *Nature Biotechnology*, *32*(12), 1241–1249.

Swainston, N., Dunstan, M., Jervis, A. J., Robinson, C. J., Carbonell, P., Williams, A. R., et al. (2018). PartsGenie: An integrated tool for optimising and sharing synthetic biology parts. *Bioinformatics*. bty105.

Yim, H., Haselbeck, R., Niu, W., Pujol-Baxley, C., Burgard, A., Boldt, J., et al. (2011). Metabolic engineering of Escherichia coli for direct production of 1,4-butanediol. *Nature Chemical Biology*, 7(7), 445–452.

CHAPTER FOURTEEN

Commodity Chemicals From Engineered Modular Type I Polyketide Synthases

Satoshi Yuzawa[*,†], **Amin Zargar**[*,‡], **Bo Pang**[†], **Leonard Katz**[†,‡], **Jay D. Keasling**[*,†,‡,§,¶,||,#,1]

[*]Biological Systems and Engineering Division, Lawrence Berkeley National Laboratory, Berkeley, CA, United States
[†]Joint BioEnergy Institute, Emeryville, CA, United States
[‡]QB3 Institute, University of California, Berkeley, CA, United States
[§]Department of Bioengineering, University of California, Berkeley, CA, United States
[¶]Department of Chemical and Biomolecular Engineering, University of California, Berkeley, CA, United States
[||]Novo Nordisk Foundation Center for Biosustainability, Technical University of Denmark, Hørsholm, Denmark
[#]Center for Synthetic Biochemistry, Shenzhen Institutes for Advanced Technologies, Shenzhen, China
[1]Corresponding author: e-mail address: keasling@berkeley.edu

Contents

Abstract

Reduced polyketides are a subclass of natural products that have a variety of medical, veterinary, and agricultural applications and are well known for their structural diversity. Although these compounds do not resemble each other, they are all made by a class of enzymes known as modular polyketide synthases (PKSs). The commonality of PKS domains/modules that compose PKSs and the understanding of the relationship between the sequence of the PKS and the structure of the compound it produces render modular PKSs as excellent targets for engineering to produce novel compounds with predicted structures. Here, we describe experimental protocols and considerations for modular PKS engineering and two case studies to produce commodity chemicals by engineered PKSs.

Methods in Enzymology, Volume 608
ISSN 0076-6879
https://doi.org/10.1016/bs.mie.2018.04.027

1. INTRODUCTION

Natural products and their derivatives have been a rich source of medicinal compounds for many diseases and continue to play an important role in drug discovery (Harvey, Edrada-Ebel, & Quinn, 2015). From 1981 to 2014, approximately half of the new chemicals approved by the US Food and Drug Administration (FDA) were natural products, their derivatives, or natural product analogs (Newman & Cragg, 2016). Reduced polyketides are a subclass of natural products that have a variety of medical, veterinary, and agricultural applications. Examples, shown in Fig. 1, include the FDA-approved compounds erythromycin (antibacterial), amphotericin (antifungal), and monensin (animal growth promoter), and the US Environmental Protection Agency-approved agents avermectin (anthelmintic) and spinosad (insecticide). Epothilone and salinosporamide have anticancer properties and are currently in clinical trials. All of these compounds are produced by Actinobacteria (a phylum of Gram-positive, high GC bacteria), and the most common genus that produces reduced polyketides is *Streptomyces*. The backbones of these chemically diverse compounds are made by modular polyketide synthases (PKSs) from simple metabolites such as malonyl-coenzyme A (CoA) and methylmalonyl-CoA. The most cited example of a modular PKS is 6-deoxyerythronolide B synthase (DEBS or the erythromycin PKS), which produces 6-deoxyerythronolide B (6-dEB), the polyketide backbone of erythromycin. As shown in Fig. 2, 6-dEB is produced by the stepwise condensation of one molecule of propionyl-CoA and six molecules of methylmalonyl-CoA in a manner similar to the biosynthesis of fatty acids. However, unlike fatty acid biosynthesis (Beld, Lee, & Burkart, 2015), each polyketide chain elongation step is carried out by a discrete set of catalytic domains defined as an extension module. Therefore, in general, the number of extension modules in a given PKS determines the size of the resultant polyketide produced. DEBS has six extension modules that utilize methylmalonyl-CoA in addition to an N-terminal loading module that catalyzes polyketide chain initiation and a C-terminal offloading domain that catalyzes polyketide release, which accounts for the 14-membered macrolactone ring and a methyl group at every other position in the ring. The largest known PKS contains 30 modules, which produces a linear polyketide containing 64 carbons in its backbone (Zhang et al., 2017). Since Donadio et al. provided the first evidence that modular PKSs function in the linear order indicated by their amino acid sequences (Donadio, Staver, McAlpine, Swanson, & Katz, 1991),

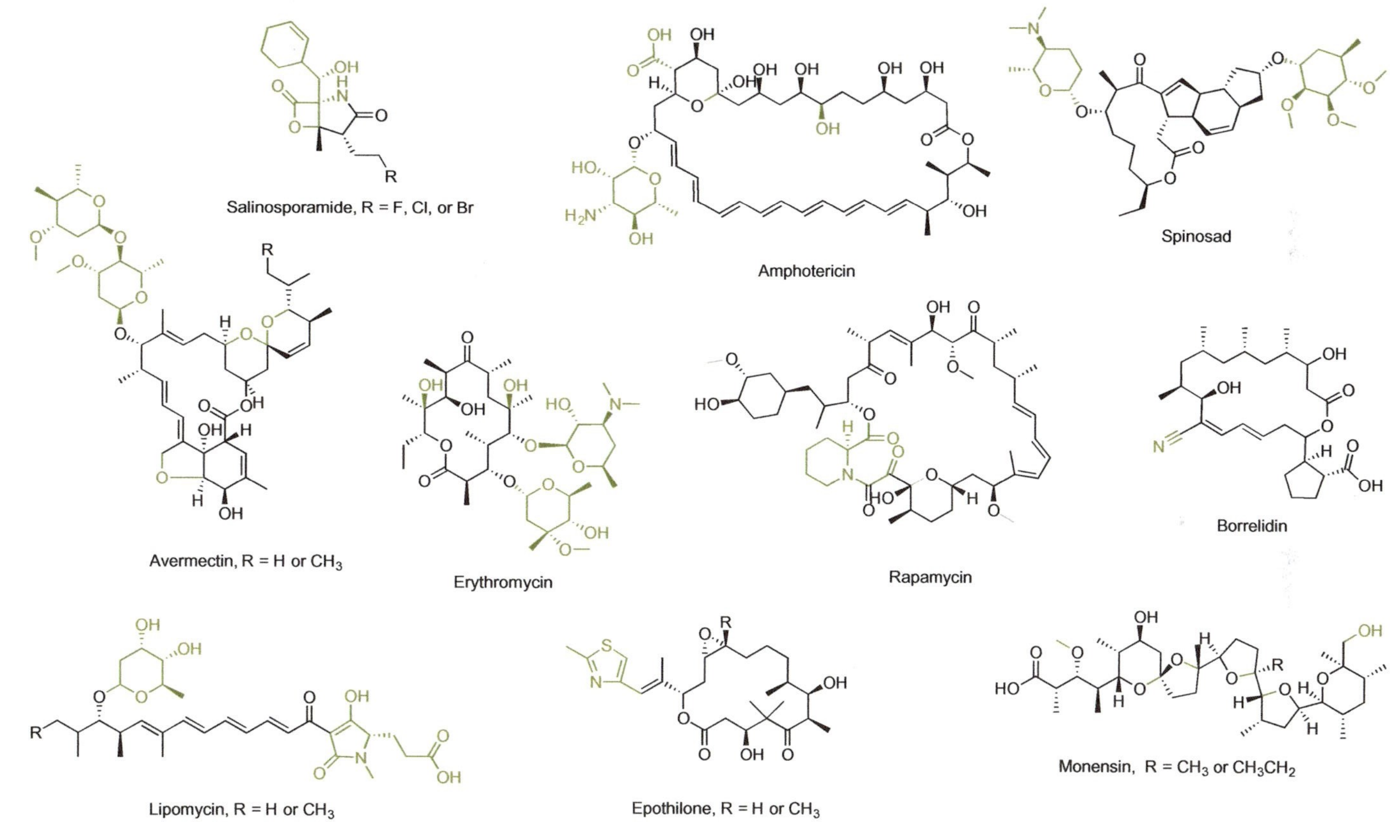

Fig. 1 Structures of natural polyketides. The parts of the compounds determined by the corresponding polyketide synthase are shown in *black*.

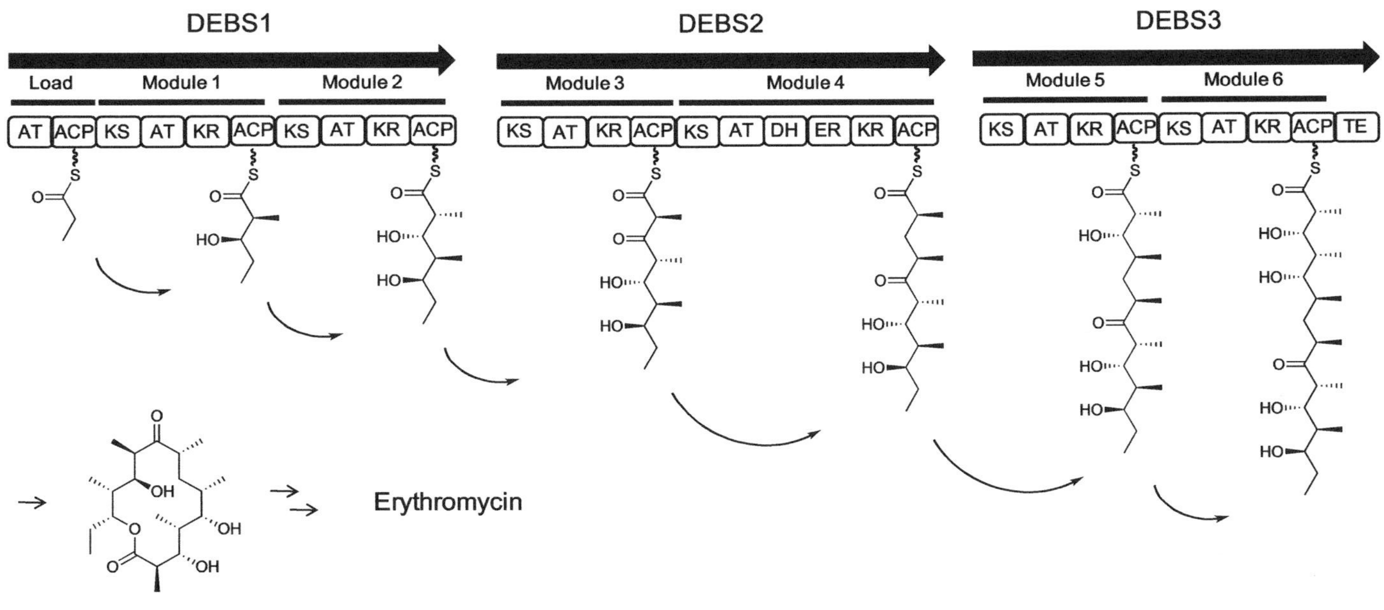

6-dEB

Fig. 2 Organization of the 6-deoxyerythronolide B synthase (DEBS) and structure of its product, 6-deoxyerythronolide B (6dEB). Stepwise growth of the polyketide chain is shown attached to the ACP domain of each module after full elaboration, including stereochemistry, by modular functions. Modules are labeled Load through Module 6 and are delineated by *solid black lines*. The proteins comprising DEBS are shown by *arrows*. The atoms labeled 1 and 13 in the acyl chain attached to module 6 undergo a lactonization reaction conducted by the TE domain to result in the production of the macrolactone 6-dEB, which undergoes further biochemical processing (not shown) to generate erythromycin (Fig. 1). Abbreviations: *ACP*, acyl carrier protein; *AT*, acyltransferase; *DH*, dehydratase; *ER*, enoyltransferase; *KR*, β-ketoreductase; *KS*, β-acyl ACP synthase; *TE*, thioesterase.

rational engineering of the PKSs has been attempted by numerous academic and industrial groups to produce structural analogs of polyketides with improved properties or novel bioactivities. Despite the large investment of time and money, however, these efforts have met with very limited success, undoubtedly due to our limited understanding of the enzymology of these multifunctional proteins. Cryo-electron microscopy-, small angle X-ray scattering-, and crystal structure-based visualization of PKS modules have enabled a more detailed understanding of the enzymology (Dutta et al., 2014; Edwards, Matsui, Weiss, & Khosla, 2014; Herbst, Jakob, Zahringer, & Maier, 2016; Whicher et al., 2014). We strongly believe that both structural and fundamental biochemical understanding of these enzymes are required for efficient production of novel compounds by engineered PKSs. Previously, we and others have highlighted the potential of PKSs to generate biofuels and commodity chemicals (Barajas, Blake-Hedges, Bailey, Curran, & Keasling, 2017; Cai & Zhang, 2017; Zargar et al., 2017). Here, we describe experimental protocols and considerations for modular PKS engineering and two case studies to produce commodity chemicals by engineered PKSs.

2. ENZYMOLOGY OF MODULAR POLYKETIDE SYNTHASE

Each PKS module minimally consists of a ketosynthase (KS) domain, an acyltransferase (AT) domain, and an acyl carrier protein (ACP) domain where the AT domain determines the specific acyl-CoA incorporated into the growing polyketide chain, and then transfers it to the ACP in the same module. The KS domain then catalyzes a decarboxylative Claisen-like condensation reaction between the growing polyketide chain and the extender unit on the ACP. Modular PKSs are categorized into *cis*- and *trans*-AT (or AT-less) systems but we will focus on *cis*-AT modular PKS engineering in this article since *trans*-AT PKSs are relatively new targets, and less engineering efforts have been conducted using them. After each chain elongation reaction, the newly generated β-keto group is reduced by a ketoreductase (KR) domain if present in the module. For example, DEBS contains a KR domain in modules 1,2,4,5, and 6 (Fig. 2). It also contains a reductase-incompetent but an epimerase-competent KR domain in module 3. Importantly, the stereochemistries of both β-hydroxyl- and α-alkyl side chains in the nascent polyketide generated are controlled primarily by the individual KR domains of each PKS module (Keatinge-Clay, 2016). KR domains are characterized based on their stereochemical outcomes. A-type and B-type KRs generate the L-β-hydroxyl and D-β-hydroxyl groups, respectively. Each KR type can also control stereochemical orientations of α-substituents. A1- or B1-type KRs produce the D-α-substituent, and A2- or B2-type KRs produce L-α-substituent. Finally, C1- or C2-type KRs are reductase-incompetent but retain epimerase activity; C1 produces the D-α-substituent, and C2 produces the L-α-substituent. The D/L naming system is not perfect but preferred over the *R*/*S* system because the assignment of *R* or *S* depends on the substituents present. The *trans*- or *cis*-double bonds within polyketides are generated by a dehydratase (DH) domain after KR-catalyzed reductions. DEBS has a DH domain in module 4 (Fig. 2). If a module further contains an enoyl reductase (ER) domain as seen in DEBS module 4, a reduced β-methylene center is generated. After all of these reactions, the elongated and reduced—if reductive loops (KR, KR–DH, or KR–DH–ER) are present—polyketide intermediate is translocated to the KS domain of the next module and similar reactions are repeated until the chain is fully elongated. Finally, a thioesterase (TE) domain releases a product by cyclization or hydrolysis. TE domains are usually embedded in the last module of the PKS, but polyether-producing PKSs such as the monensin PKSs employ a free-standing TE to cleave the final intermediates (Oliynyk et al., 2003). Several other less common termination mechanisms are also known (Du & Lou, 2010).

The polyketide chain initiation event is catalyzed by a set of domains preceding the first extension module, which is commonly called a loading module. The most common domain organization consists of a condensation-incompetent KS domain (e.g., KS^Q, where the Cys residue in the active site is mutated to Glu), an AT domain, and an ACP domain (Moore & Hertweck, 2002). The AT domain is strictly specific for malonyl-CoA or methylmalonyl-CoA and generates acetyl or propionyl starter units, respectively, after decarboxylation by the KS^Q. On the other hand, a loading module composed of an AT–ACP didomain tends to show broader substrate specificity. For example, as shown in Fig. 1, the avermectin PKS can start polyketide biosynthesis from either isobutyryl-CoA or 2-methylbutyryl-CoA (Ikeda, Nonomiya, Usami, Ohta, & Omura, 1999). Similarly, the lipomycin PKS and borrelidin PKS loading didomains have also been demonstrated to have broad substrate specificity in the chain initiation reaction (Hagen et al., 2014; Yuzawa, Eng, Katz, & Keasling, 2013, 2014).

3. POLYKETIDE SYNTHASE ENGINEERING

The modularity of PKSs and the commonality of PKS domains/modules that compose them render them as excellent targets for engineering to produce novel compounds with predicted structures. Theoretically, it should be possible to design PKS domains/modules required to produce a desired structure, build the corresponding DNA, introduce it into a desired host, and express it for target chemical production. For example, the online ClusterCAD database facilitates the selection of natural *cis*-AT PKS gene parts to design a novel hybrid PKS that produces a desired molecule (see Section 3.3 for details) (Eng et al., 2018). It should be noted that most of the reported efforts have been focused on producing novel antibiotics or other polyketides of medical interest. In these cases, the emphasis is on creating diversity and low-level production of target compounds is usually sufficient for assessment of its potential value. On the other hand, our laboratory has been engaged for repurposing modular PKSs to produce simpler molecules such as biofuels and commodity chemicals as drop-in replacements for currently used petroleum-based compounds. In these scenarios, the emphasis is on production levels. In general, the dollar values of these targeted molecules are low compared to pharmaceuticals. To compete with the same compounds resourced from petroleum, considerations of product titers, rates, and yields (TRY) are very important. In the following sections, we will describe successful and promising PKS engineering strategies to support high-level chemical production.

3.1 Active-Site Mutagenesis

Active-site mutagenesis is thought be a minimally invasive method to change the function of a given PKS module. Although the most successful examples are the inactivation of reductive loop (KR, DH, or ER) domains, a recent example (see later) describes an AT domain engineered mutation where the substrate specificity is almost exclusively converted to a targeted unnatural substrate.

3.1.1 Acyltransferase Domain

Comparative sequence analysis revealed well-conserved GHSIG and HAFH motifs for ATs that are specific for malonyl-CoA, and GHSQG and YASH motifs for ATs specific for (2*S*)-methylmalonyl-CoA (Fig. 3) (Yadav, Gokhale, & Mohanty, 2003). However, attempts to convert the substrate specificity of the AT domain of module 4 of DEBS from methylmalonyl-CoA to malonyl-CoA by site-directed mutagenesis (YASH to **H**A**F**H) resulted only in an incomplete shift (or relaxed substrate specificity) in extender unit choice in vivo (Reeves et al., 2001). Recently, Koryakina et al., using active-site mutated libraries, identified a single mutation (YASH to **R**ASH) that almost completely changed the substrate preference of AT6 of DEBS3, employing a standalone

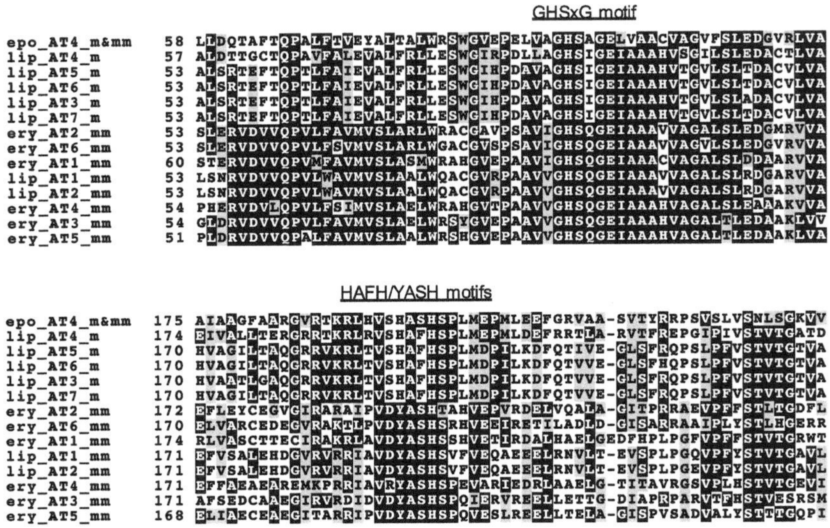

Fig. 3 Multiple sequence alignment of AT domains. Six AT domain sequences from the erythromycin (ery) PKS were aligned with homologs from the lipomycin (lip) PKS and the epothilone (epo) PKS. Methylmalonyl-CoA-specific ATs were indicated by "mm" and malonyl-CoA-specific ATs were indicated by "m."

dimodular PKS, from its natural substrate methylmalonyl-CoA to the unnatural propargylmalonyl-CoA with only modest reductions in product titers in vitro (Koryakina et al., 2017). It also should be noted that the epothilone PKS AT4, which naturally incorporates both malonyl-CoA and methylmalonyl-CoA, has GHSAG and HASH motifs in the corresponding region (Fig. 3) (Tang, 2000). Some AT domains accept other carboxyacyl-CoAs such as ethylmalonyl-CoA. These AT domains have different fingerprint residues (not conserved) at the corresponding regions.

3.1.2 KR Domain

The YAAAN motif is often seen in KR domain active sites (Fig. 4). The double mutants (YAAxN to **FEA**xN) of each of the KR16 and KR17 domains in the nystatin PKS resulted in production of the corresponding keto-nystatin derivatives at about 50% of the titers produced by the wild-type nystatin PKS in vivo (Brautaset et al., 2008). Since NADPH binds to a GxGxxG motif, this site is also a potential target site to inactivate the function.

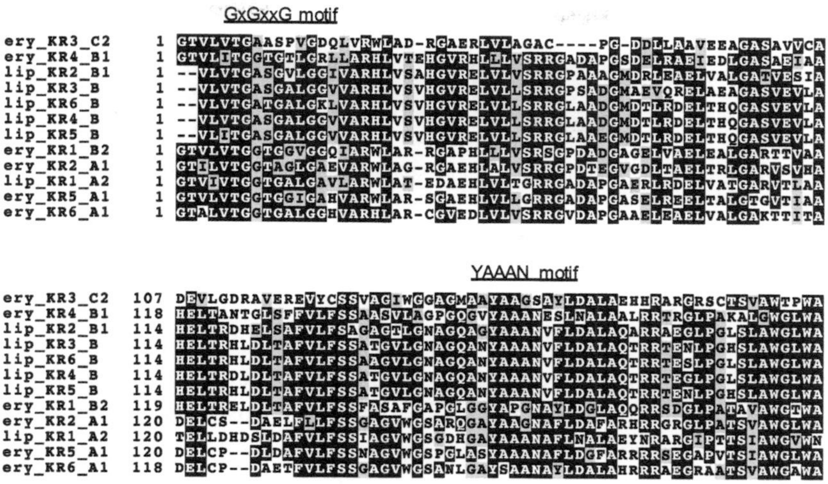

Fig. 4 Multiple sequence alignment of KR domains. Five KR domain sequences from the erythromycin (ery) PKS were aligned with homologs from the lipomycin (lip) PKS. These KR domains include A1, A2, B, B1, B2, and C2 types.

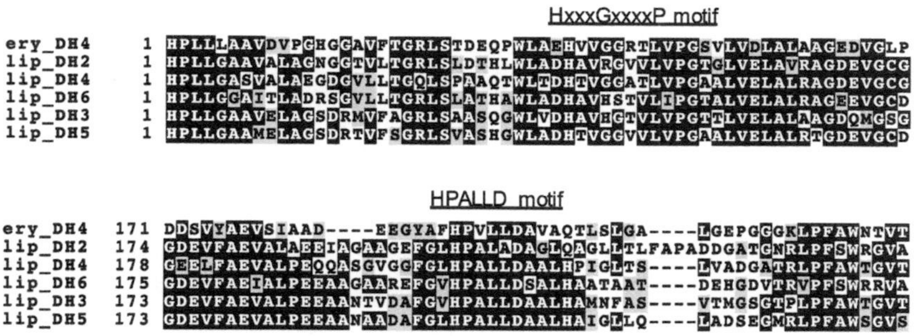

Fig. 5 Multiple sequence alignment of DH domains. The DH domain sequence from the erythromycin (ery) PKS was aligned with homologs from the lipomycin (lip) PKS.

3.1.3 Dehydratase Domain

DH domains have catalytic histidine and aspartic acid residues in <u>**H**</u>xxxGxxxxP and HPALL<u>**D**</u> motifs (Fig. 5). Histidine to phenylalanine or tyrosine mutations were successfully employed to inactivate DH domains in the nystatin and FR-008 PKSs, respectively, to produce the corresponding hydroxy analogs at about half or comparable levels produced by the wild-type counterparts in vivo (Brautaset et al., 2008; Zhou et al., 2008).

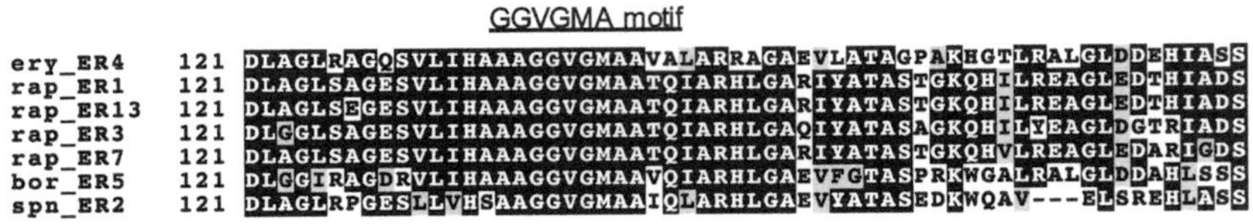

Fig. 6 Multiple sequence alignment of ER domains. The ER domain sequence from the erythromycin (ery) PKS was aligned with homologs from several other modular PKSs.

3.1.4 Enoyl Reductase Domain

NADPH binds to the conserved GGVGMA motif (Fig. 6). The double mutant (GGVGMA to **SP**VGMA) of the DEBS ER4 domain resulted in production of the corresponding anhydro-erythromycin analog although the titer was estimated to be one-fifth that of erythromycin produced by the wild-type PKS in vivo (Donadio, McAlpine, Sheldon, Jackson, & Katz, 1993).

In Sections 3.1.2–3.1.4, we only selected successful examples. More recently, KR, DH, and ER inactivations have been used to produce a library of 22 monensin analogs, but the highest titers observed were approximately 10-fold lower than those produced by the unmutated monensin PKS (Kushnir et al., 2012). Since site-directed mutagenesis is unlikely to affect the overall domain and module structure, protein–protein interactions within a module, or protein production levels in chassis, the reduced titers observed in these examples could be explained by substrate specificity of downstream catalytic domains that are involved with rate-determining reactions.

3.2 Domain Swapping

Domain swapping within a PKS module is a frequently used method to change module functions, which enables researchers to change product structures at specific sites. Several different engineering approaches have been developed to exchange catalytic domains in individual modules while minimizing a negative impact on other domains within the same module, which is summarized below.

3.2.1 Acyltransferase Domain

At least, 20 different malonyl-CoA analogs have been found to be incorporated into naturally occurring polyketides by *cis*-AT PKSs (Wilson & Moore, 2012). These analogs include short to medium-chain alkylmalonyl-CoAs, halogenated malonyl-CoAs, and benzylmalonyl-CoA. AT domain swapping may be the most reliable way to change substrate specificity of a given module because predicted structural analogs were actually produced in many cases (McDaniel et al., 1999). However, the resulting hybrid PKS systems often showed activity reduced by a few orders of magnitude compared with their wild-type counterparts (Dunn & Khosla, 2013). To address this issue, our laboratory evaluated fusion points for AT domain swapping using DEBS module 6 (M6) + TE using as an acceptor PKS in vitro since previous reports did not rigorously explore the role of interdomain linkers (Yuzawa et al., 2017). We selected a set of fusion points as described in Fig. 7 (see also Section 4).

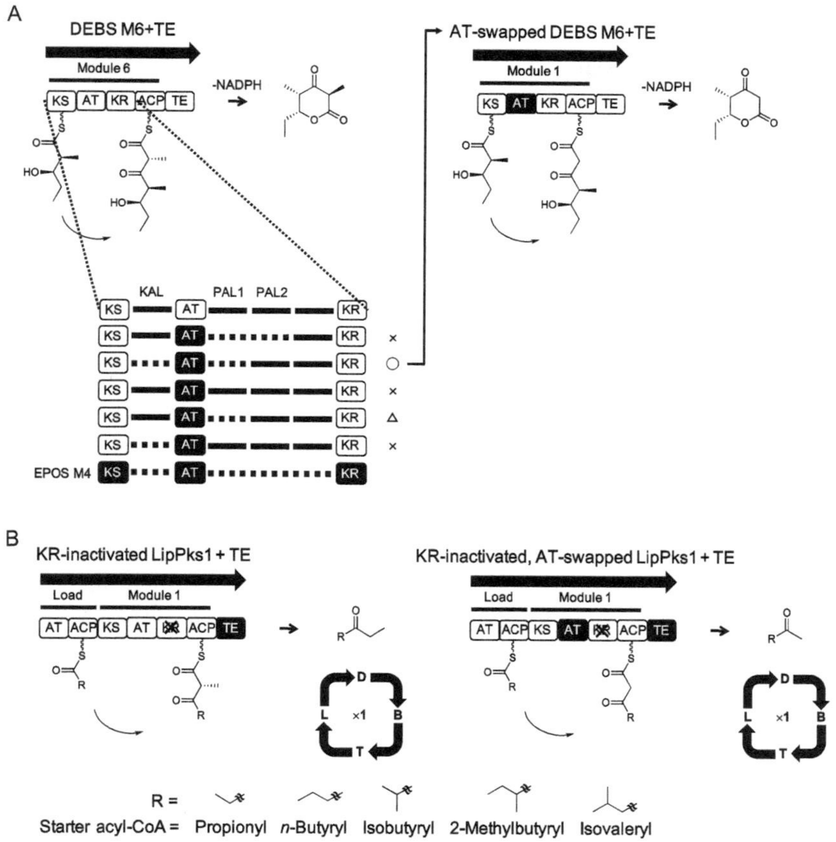

Fig. 7 Precise AT domain swapping of modular polyketide synthases (PKSs). (A) DEBS module 6 (M6) + TE was used as a model PKS system to optimize domain boundaries for AT domain swapping. The activities of the AT-swapped PKSs are shown in *circle* (the fastest k_{cat}), *triangle* (the second fastest k_{cat}), and *cross* (inactive). (B) KR-inactivated LipPks1 + TE produces ethyl ketones from various acyl starter CoAs and methylmalonyl-CoA (*left*). KR-inactivated, AT-swapped LipPks1 + TE produces methyl ketones from various starter acyl CoAs and malonyl-CoA (*right*). Heterologous catalytic domains are filled with black (*white letter*). Heterologous interdomain linkers are shown as *dashed lines*. Abbreviations: *EPOS*, epothilone PKS; *KAL*, KS to AT linker; *PAL*, post AT linker; all other abbreviations as in Fig. 2.

Using these fusion points, we replaced DEBS AT6 with AT4 of the epothilone PKS, which naturally incorporates both malonyl-CoA and methylmalonyl-CoA. The best mutant retained the wild type activity (85% in k_{cat}) when methylmalonyl-CoA was used as a substrate. The protein stability, which was inferred from the protein purification yield (72% compared to the wild-type PKS), was also maintained. These data indicated that including both KS to AT linker (KAL) and the first half of the post AT linker (PAL1) segment with the AT domain is essential to maintain both enzyme activity and protein stability. This KAL–AT–PAL1 swapping was also validated in another monomodular PKS system, LipPks1 + TE. The last half of PAL (PAL2), LPTY(A/P)FQ (H/R)xRYWL, is highly conserved in *cis*-AT PKSs and is known to interact with a KS domain along with a KAL within a module in DEBS (Tang, Chen, Kim, Cane, & Khosla, 2007; Tang, Kim, Mathews, Cane, & Khosla, 2006; Yuzawa, Kapur, Cane, & Khosla, 2012). Although well conserved, it appears that retaining a native PAL2 in an acceptor PKS is essential to keep structural integrity and activity of a module. For a domain boundary between the KS and the KAL, a conserved GTNAH motif was employed. In larger PKS systems, it is possible that the downstream KS domains might not accept unnatural polyketide intermediates generated by the AT-swapped upstream modules, which may explain significantly decreased product titers in the many previous efforts reported.

3.2.2 Reductive Loop

Reductive loop swapping was also successfully performed in vivo using DEBS1 + TE, a dimodular PKS system (Kellenberger et al., 2008). Kellenberger et al. exchanged the A1-type KR domain of the module 2 with various KR domains. The most successful KR domain swapping was A1 to B1 which generated the expected D-β-hydroxyl group with various product titers (17%–150%). However, neither A1 to A2 nor A1 to B2 swapping, which are supposed to alter the α-methyl group stereochemistry, produced the predicted polyketide compounds. More recently, a broader panel of A2-type KR domains was tested in the same acceptor PKS and a few KR domains were reported to give expected products at low yields (<15%) (Annaval, Paris, Leadlay, Jacob, & Weissman, 2015). On the other hand, our laboratory employed LipPks1 + TE, which contains an A2-type KR, to investigate A2 to A1 swapping in vitro (Eng et al., 2016). In contrast to the above report, we observed that the system retained 60%–80% of the activity after domain swapping and generated the opposite α-methyl group stereochemistry. We also demonstrated that retaining a native dimerization element that lies between the AT and KR domain in the acceptor PKS was critical to maintaining PKS activity (Eng et al., 2016;

Zheng, Fage, Demeler, Hoffman, & Keatinge-Clay, 2013). Using the DEBS1 + TE system, Kellenberger et al. also examined if the KR domain of the module 2 could be replaced with a KR–DH or KR–DH–ER reductive loop (Kellenberger et al., 2008). When KR–DH from the module 10 of the rapamycin PKS was used, the PKS produced the corresponding anhydro product with only modest reduction in product titers. DEBS1 + TE variants that contain a KR–DH–ER reduction loop in place of the KR2 domain also successfully produced expected products at reasonable titers (30%–40%). In Section 3.2.1, we described the importance of retaining a conserved LPTY(A/P)FQ(H/R)xRYWL sequence in an acceptor PKS for AT domain swapping. Our laboratory also tested the same hypothesis for reductive loop swapping and demonstrated the importance in the related PKS engineering (Fig. 8; see also Section 6) (Hagen et al., 2016).

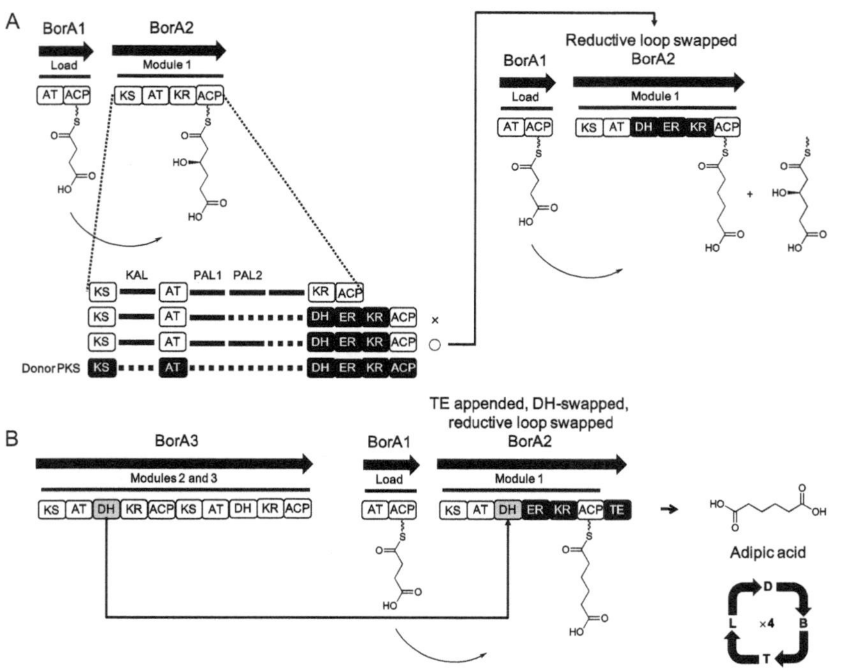

Fig. 8 Precise reduction domain swapping of modular polyketide synthases (PKSs). (A) BorA1 accepts succinyl-CoA in the polyketide chain initiation reaction, which is extended by BorA2. This PKS system was used to optimize domain boundaries for reduction domain swapping. The activities of the reductive loop-swapped PKSs are shown in *circle* (active) and *cross* (inactive). (B) The DH domain in the reduction reductive loop was replaced with DH from BorA3 (shown in *gray*) and the TE from DEBS was attached to the C-terminus of BorA2 to release the product. The resulting KR-swapped PKS produces adipic acid. Heterologous catalytic domains are filled with *black* (*white letter*). Heterologous interdomain linkers are shown as *dashed lines*. Abbreviations as in Fig. 2.

3.3 Computer-Aided Design

PKS engineering could be greatly aided by computational tools. Huge amounts of information have accumulated since the discovery of modular PKSs, and several PKS-related databases were developed to store and utilize this information for specific applications (Anand et al., 2010; Conway & Boddy, 2013; Eng et al., 2018). However, most of these computational tools were developed for genome mining to find novel natural products or PKS engineering to produce natural product analogs, directed toward increasing the size and diversity of natural product-based compound library.

For commodity chemical production, the scenario is different because the target structures are defined. ClusterCAD, developed by our laboratory, is a computational tool dedicated to PKS engineering (https://clustercad.jbei.org/). ClustarCAD will identify a PKS module with a chemical intermediate most similar to a desired target and can then identify potential donor parts for subsequent domain swapping. Currently, ClusterCAD contains 72 PKS clusters. A typical workflow of ClusterCAD is described in Fig. 9. The outputs of the above steps are the starting point of PKS engineering. After careful consideration of the parameters involved (sequence similarity, substrate specificity, domain boundaries etc.), hybrid PKSs may be Designed, Built, and Tested. After the data analysis (Learn), this D–B–T–L cycle should be repeated until the desired product TRY is achieved.

4. CASE STUDY 1: SHORT-CHAIN KETONE PRODUCTION

Short-chain ketones such as acetone, butanone (also called methyl ethyl ketone or MEK), and 4-methyl-2-pentanone (methyl isobutyl ketone or MIBK) are industrially important solvents that are used in a variety of applications, including paints, coatings, adhesives, magnetic tapes, inks, as well as for cleaning and extraction. According to ICIS (https://www.icis.com/), annual global production of these short-chain ketones are approximately 5 million tons (acetone) and 0.5 million tons (butanone and 4-methyl-2-pentanone), respectively (Srirangan et al., 2016). Although some microbes (mainly *Clostridium* strains) naturally produce acetone, no microbe or natural metabolic pathway, to our knowledge, is known to produce detectable levels of the other short-chain ketones, all of which, including acetone, are currently derived from petroleum sources.

3-Keto acids are chemically unstable and known to decompose to ketones spontaneously under mild conditions (Eng et al., 2016; Goh et al., 2014;

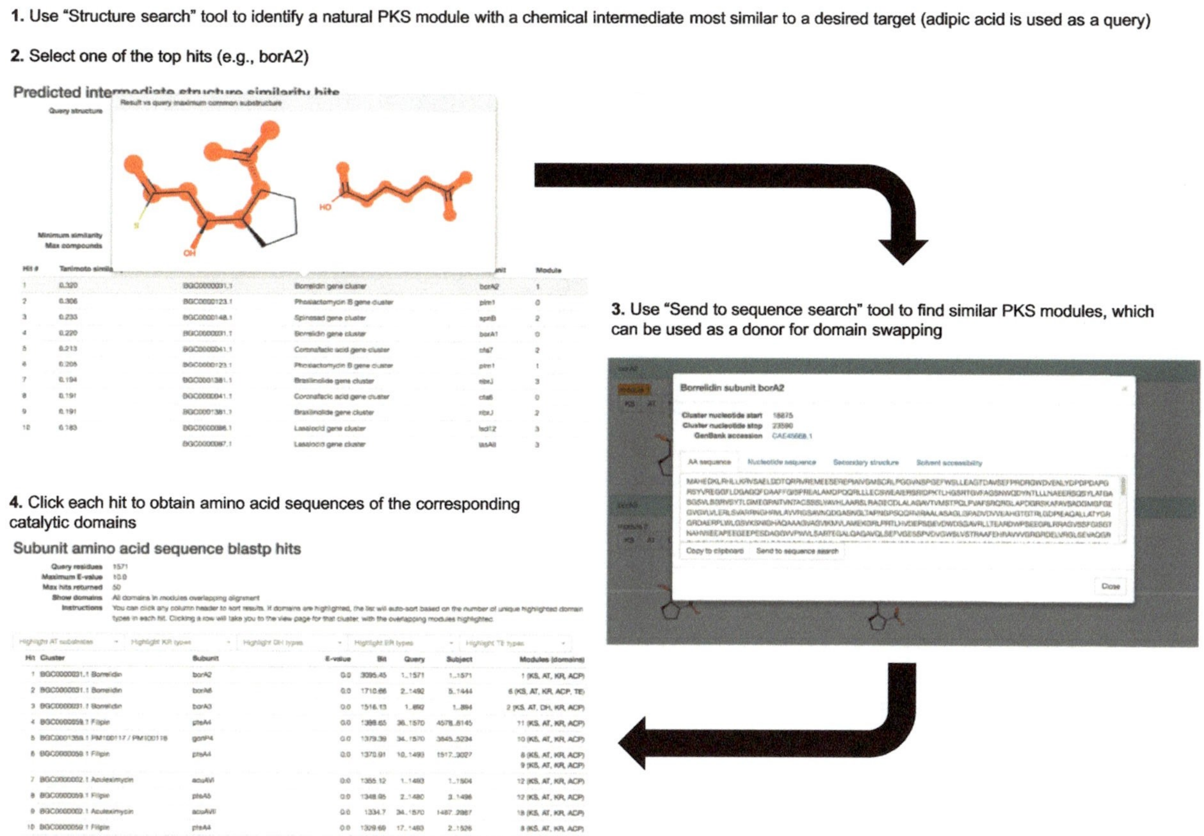

Fig. 9 A typical workflow in ClusterCAD.

Goh, Baidoo, Keasling, & Beller, 2012; Yuzawa, Katz, & Keasling, 2015), which suggests that MEK and MIBK could be produced from the corresponding 3-keto acids, 3-oxopentanoic acid, 5-methy-3-oxohexanoic acid, respectively. Recently, we created a hybrid PKS that produces various 3-keto acids from the corresponding starter acyl-CoAs (propionyl-CoA, *n*-butyryl-CoA, isobutyryl-CoA, 2-methylbutyryl-CoA, and isovaleryl-CoA) and malonyl-CoA, which were converted into five methyl ketones including MEK and MIBK in vitro (Yuzawa et al., 2017). Microbial production of MEK has also been achieved in *Escherichia coli* (4.1 mg/L). Here we describe the experimental protocols to design the MEK-producing PKS. Plasmids used in this case study are summarized in Table 1, which can be downloaded from the public version of JBEI registry (http://public-registry.jbei.org). The codons were optimized for *E. coli.*

1. *LipPks1 + TE.* We selected LipPks1 from the lipomycin synthase as an acceptor PKS. LipPks1 is composed of a loading module (AT–ACP), an extension module (KS–AT–KR–ACP), and a C-terminal docking domain. The C-terminal docking domain starting from YGPDP was replaced with the DEBS TE domain and its associated linker starting from QLDSGTPA to create LipPks1 + TE. We previously demonstrated in vitro that N-terminally histidine-tagged LipPks1 + TE is capable of producing various 3-hydroxy acids including 2-methyl-3-hydroxypentanoic acid, which is produced from propionyl-CoA and methylmalonyl-CoA in the presence of NADPH (Yuzawa et al., 2013).
2. *AT-swapped LipPks1 + TE.* We replaced the original AT and the associated linkers (KAL–AT–PAL1) with the corresponding region in module 1 of the borrelidin PKS (BorA2), which naturally incorporates malonyl-CoA. The conserved GTNAH motif in the C-terminus of the KS domain and the LPTY motif in the PAL2 linker in LipPks1 and BorA2 were used as domain boundaries for AT domain swapping (Fig. 4). The AT-swapped PKS was demonstrated to produce 3-hydroxypentanoic

Table 1 Plasmids Used in Case Study 1

Plasmid	Summary	Source of Reference
pSY074 (JPUB_006013)	LipPks1 + TE	Yuzawa et al. (2013)
pSY075 (JPUB_006015)	AT-swapped LipPks1 + TE	Yuzawa et al. (2017)
pSY088 (JPUB_007580)	KR-inactivated, AT-swapped LipPks1 + TE	Yuzawa et al. (2017)

acid from propionyl-CoA and malonyl-CoA in the presence of NADPH without losing biochemical activity in vitro (Yuzawa et al., 2017).

3. *KR-inactivated, AT-swapped LipPks1 + TE.* The KR domain was inactivated by mutating the active site Ser to Ala (LFSSIAG to LFS**A**IAG) to create a PKS that produces 3-ketopentanoic acid. The resulting PKS produced MEK in vitro from propionyl-CoA and malonyl-CoA (Yuzawa et al., 2017).

5. CASE STUDY 2: ADIPIC ACID PRODUCTION

The largest global use of adipic acid is to produce nylon 6,6, which accounts for approximately 70% of global output. Annual global production of adipic acid is approximately 2.5 million tons (https://www.icis.com/). A biological route to the production of adipic acid could provide an important alternative pathway, thereby lowering anthropogenic emissions of nitrous oxide, a potent greenhouse gas that is released during adipic acid production (Alini, Basile, Blasioli, Rinaldi, & Vaccari, 2007). While a number of biological routes to adipic acid have been developed, typically dependent on reversal of β-oxidation of dicarboxylic acids (Yu, Xia, Zhong, & Qian, 2014), a PKS system sets the stage for the specific production of higher diacids (e.g., suberic acid), which would not require downstream separation from mixtures of other medium chain diacids. Additionally, the modularity of PKSs may allow for the production of branched and/or oxidized diacids that are not readily accessible through conventional synthetic chemistry or the above biosynthetic routes.

Within the constraints of modular PKS-based biosynthesis, we proposed that adipic acid synthesis would most conveniently start from succinyl-CoA and undergo one round of extension with full reduction using malonyl-CoA to produce the adipoyl-ACP intermediate. Adipic acid would then be released by a TE domain. Due to its important role in the TCA cycle, succinyl-CoA is readily available in organisms capable of aerobic respiration, as is malonyl-CoA, which is used in fatty acid biosynthesis. Here we describe the experimental protocols to design a PKS to produce adipic acid. Plasmids used in this case study are summarized in Table 2, which can be downloaded from the public version of JBEI registry (http://public-registry.jbei.org). The codons were optimized for *E. coli*.

1. *Reductive loop swapped BorA2.* We selected BorA2 from the borrelidin PKS as an acceptor PKS (Fig. 9). BorA2 is composed of an N-terminal docking domain, an extension module (KS–AT–KR–ACP), and a C-terminal docking domain. The native KR domain was replaced with a KR–DH–ER reductive loop of module 2 of the spinosad PKS. The

Table 2 Plasmids Used in Case Study 2

Plasmid	Summary	Source of Reference
pARH162 (JPUB_006465)	Reductive loop swapped BorA2	Hagen et al. (2016)
pARH164 (JPUB_006469)	DH-swapped, reductive loop swapped BorA2	Hagen et al. (2016)
pARH176 (JPUB_006471)	TE appended, DH-swapped, reductive loop swapped BorA2	Hagen et al. (2016)

upstream site for replacement follows the conserved PAL2 linker of BorA2 (LPTYPFQHQHYW). The downstream site for replacement is after the GLASK motif (the ACP domain starts right after the sequence). The corresponding protein with an N-terminal maltose binding protein was catalytically active with regard to condensation, but the major acyl-ACP product was 3-hydroxy-adipoyl-ACP even after reductive loop swapping (Hagen et al., 2016).

2. *DH-swapped, reductive loop swapped BorA2.* We hypothesized that the DH domain from the spinosad PKS was not competent to dehydrate 3-hydroxy-adipoyl-ACP due to the presence of the 6-carboxyl group. We therefore sought to test the activity of a DH domain, which uses a substrate carrying a carboxyl group in its natural context. The DH domain of the reductive loop swapped BorA2 was replaced with BorA3 DH, which naturally accepts a substrate carrying a carboxyl group, using the junctions HPLLG (N-terminal) and TAESL (C-terminal). The resulting protein actually produced adipoyl-ACP as the major product (Hagen et al., 2016).
3. *TE appended, DH-swapped, reductive loop swapped BorA2.* To release adipic acid from the ACP, DEBS TE and the associated linker starting from SGTPA were appended to the construct. As expected, adipic acid was detected from the PKS in vitro (Hagen et al., 2016).

6. SUMMARY

In this chapter, we focused on modular PKSs and their engineering protocols. Although modular PKS engineering has made encouraging progress in the last few years, we believe that there are still significant barriers to overcome to achieve efficient production of complex compounds such as natural product analogs from engineered PKSs, because linking several modules in sequence is still extremely challenging. For example, Menzella et al. attempted in vivo combinatorial biosynthesis using various di- and trimodular PKSs connected by docking domains to produce new compounds (Menzella et al., 2005; Menzella, Carney, & Santi, 2007). Although approximately half of the combinations successfully produced the expected polyketides, more than 90% of the PKSs showed <10% activities compared to the wild-type counterparts, suggesting the difficulty in establishing productive interactions between heterologous modules. Controlling noncovalent intermodular interactions gives the greatest opportunity for generating large numbers of novel compounds, which would be a major achievement in the future PKS engineering.

On the other hand, commodity chemicals have relatively simple structures and usually require one or two modules, which require less debugging of the systems. In addition, unlike iterative PKSs, which are also attractive targets to produce commodity chemicals (Liu et al., 2015; Menendez-Bravo, Comba, Sabatini, Arabolaza, & Gramajo, 2014), successful examples of active-site mutagenesis and precise domain swapping in simple PKS systems demonstrated our ability to reprogram each polyketide chain elongation and backbone modification reaction in a predicted manner without sacrificing enzyme activity, as described earlier. The next challenge is to produce large amounts of target commodity chemicals biologically. Currently, *E. coli* and *Streptomyces* strains are often used as hosts for heterologous polyketide production by modular PKSs (Baltz, 2010; Yuzawa, Kim, Katz, & Keasling, 2012). The product titers are usually 0.5–50 mg/L in unoptimized shake flask conditions. In the long term, production levels need to be increased to 10–100 g/L to compete with the relatively inexpensive means of production of these chemicals from petroleum. Recently, *Streptomyces* genetic engineering tools such as strong promoters and ribosomal binding sites, a number of actinophage integrase-based site-specific gene integration systems, and CRISPR/Cas9 technologies have been developed (Bai et al., 2015; Baltz, 2012; Liu, Liang, Ang, & Zhao, 2017; Phelan et al., 2017); these may enable us to achieve commodity chemical production at economically viable costs via engineered PKSs.

ACKNOWLEDGMENTS

This work was supported by the Joint BioEnergy Institute, which is funded by the US Department of Energy (DOE), the Office of Science, Office of Biological and Environmental Research under Contract No. DE-AC02-05CH11231 between DOE and

Lawrence Berkeley National Laboratory. The publisher, by accepting the article for publication, acknowledges that the US government retains a nonexclusive, paid-up, irrevocable, worldwide license to publish or reproduce the published form of the manuscript, or allow others to do so, for US government purpose. J.D.K. has a financial interest in Amyris, Lygos, Demetrix, Constructive Biology, Maple Bio, and Napigen. L.K. has a financial interest in Lygos.

REFERENCES

Alini, S., Basile, F., Blasioli, S., Rinaldi, C., & Vaccari, A. (2007). Development of new catalysts for N_2O-decomposition from adipic acid plant. *Applied Catalysis B: Environmental, 70*, 323–329.

Anand, S., Prasad, M. V., Yadav, G., Kumar, N., Shehara, J., Ansari, M. Z., et al. (2010). SBSPKS: Structure based sequence analysis of polyketide synthases. *Nucleic Acids Research, 38*, W487–W496.

Annaval, T., Paris, C., Leadlay, P. F., Jacob, C., & Weissman, K. J. (2015). Evaluating ketoreductase exchanges as a means of rationally altering polyketide stereochemistry. *Chembiochem, 16*, 1357–1364.

Bai, C., Zhang, Y., Zhao, X., Hu, Y., Xiang, S., Miao, J., et al. (2015). Exploiting a precise design of universal synthetic modular regulatory elements to unlock the microbial natural products in Streptomyces. *Proceedings of the National Academy of Sciences of the United States of America, 112*, 12181–12186.

Baltz, R. H. (2010). Streptomyces and Saccharopolyspora hosts for heterologous expression of secondary metabolite gene clusters. *Journal of Industrial Microbiology & Biotechnology, 37*, 759–772.

Baltz, R. H. (2012). Streptomyces temperate bacteriophage integration systems for stable genetic engineering of actinomycetes (and other organisms). *Journal of Industrial Microbiology & Biotechnology, 39*, 661–672.

Barajas, J. F., Blake-Hedges, J. M., Bailey, C. B., Curran, S., & Keasling, J. D. (2017). Engineered polyketides: Synergy between protein and host level engineering. *Synthetic and Systems Biotechnology, 2*, 147–166.

Beld, J., Lee, D. J., & Burkart, M. D. (2015). Fatty acid biosynthesis revisited: Structure elucidation and metabolic engineering. *Molecular BioSystems, 11*, 38–59.

Brautaset, T., Sletta, H., Nedal, A., Borgos, S. E., Degnes, K. F., Bakke, I., et al. (2008). Improved antifungal polyene macrolides via engineering of the nystatin biosynthetic genes in Streptomyces noursei. *Chemistry & Biology, 15*, 1198–1206.

Cai, W., & Zhang, W. (2017). Engineering modular polyketide synthases for production of biofuels and industrial chemicals. *Current Opinion in Biotechnology, 50*, 32–38.

Conway, K. R., & Boddy, C. N. (2013). ClusterMine360: A database of microbial PKS/NRPS biosynthesis. *Nucleic Acids Research, 41*, D402–D407.

Donadio, S., McAlpine, J. B., Sheldon, P. J., Jackson, M., & Katz, L. (1993). An erythromycin analog produced by reprogramming of polyketide synthesis. *Proceedings of the National Academy of Sciences of the United States of America, 90*, 7119–7123.

Donadio, S., Staver, M. J., McAlpine, J. B., Swanson, S. J., & Katz, L. (1991). Modular organization of genes required for complex polyketide biosynthesis. *Science, 252*, 675–679.

Du, L., & Lou, L. (2010). PKS and NRPS release mechanisms. *Natural Product Reports, 27*, 255–278.

Dunn, B. J., & Khosla, C. (2013). Engineering the acyltransferase substrate specificity of assembly line polyketide synthases. *Journal of the Royal Society, Interface, 10*. 20130297.

Dutta, S., Whicher, J. R., Hansen, D. A., Hale, W. A., Chemler, J. A., Congdon, G. R., et al. (2014). Structure of a modular polyketide synthase. *Nature, 510*, 512–517.

Edwards, A. L., Matsui, T., Weiss, T. M., & Khosla, C. (2014). Architectures of whole-module and bimodular proteins from the 6-deoxyerythronolide B synthase. *Journal of Molecular Biology, 426*, 2229–2245.

Eng, C. H., Backman, T. W. H., Bailey, C. B., Magnan, C., Garcia Martin, H., Katz, L., et al. (2018). ClusterCAD: A computational platform for type I modular polyketide synthase design. *Nucleic Acids Research, 46*, D509–D515.

Eng, C. H., Yuzawa, S., Wang, G., Baidoo, E. E., Katz, L., & Keasling, J. D. (2016). Alteration of polyketide stereochemistry from anti to syn by a ketoreductase domain exchange in a type I modular polyketide synthase subunit. *Biochemistry, 55*, 1677–1680.

Goh, E. B., Baidoo, E. E., Burd, H., Lee, T. S., Keasling, J. D., & Beller, H. R. (2014). Substantial improvements in methyl ketone production in E. coli and insights on the pathway from in vitro studies. *Metabolic Engineering, 26*, 67–76.

Goh, E. B., Baidoo, E. E., Keasling, J. D., & Beller, H. R. (2012). Engineering of bacterial methyl ketone synthesis for biofuels. *Applied and Environmental Microbiology, 78*, 70–80.

Hagen, A., Poust, S., de Rond, T., Yuzawa, S., Katz, L., Adams, P. D., et al. (2014). In vitro analysis of carboxyacyl substrate tolerance in the loading and first extension modules of borrelidin polyketide synthase. *Biochemistry, 53*, 5975–5977.

Hagen, A., Poust, S., Rond, T., Fortman, J. L., Katz, L., Petzold, C. J., et al. (2016). Engineering a polyketide synthase for in vitro production of adipic acid. *ACS Synthetic Biology, 5*, 21–27.

Harvey, A. L., Edrada-Ebel, R., & Quinn, R. J. (2015). The re-emergence of natural products for drug discovery in the genomics era. *Nature Reviews. Drug Discovery, 14*, 111–129.

Herbst, D. A., Jakob, R. P., Zahringer, F., & Maier, T. (2016). Mycocerosic acid synthase exemplifies the architecture of reducing polyketide synthases. *Nature, 531*, 533–537.

Ikeda, H., Nonomiya, T., Usami, M., Ohta, T., & Omura, S. (1999). Organization of the biosynthetic gene cluster for the polyketide anthelmintic macrolide avermectin in Streptomyces avermitilis. *Proceedings of the National Academy of Sciences of the United States of America, 96*, 9509–9514.

Keatinge-Clay, A. T. (2016). Stereocontrol within polyketide assembly lines. *Natural Product Reports, 33*, 141–149.

Kellenberger, L., Galloway, I. S., Sauter, G., Bohm, G., Hanefeld, U., Cortes, J., et al. (2008). A polylinker approach to reductive loop swaps in modular polyketide synthases. *Chembiochem, 9*, 2740–2749.

Koryakina, I., Kasey, C., McArthur, J. B., Lowell, A. N., Chemler, J. A., Li, S., et al. (2017). Inversion of extender unit selectivity in the erythromycin polyketide synthase by acyltransferase domain engineering. *ACS Chemical Biology, 12*, 114–123.

Kushnir, S., Sundermann, U., Yahiaoui, S., Brockmeyer, A., Janning, P., & Schulz, F. (2012). Minimally invasive mutagenesis gives rise to a biosynthetic polyketide library. *Angewandte Chemie (International Ed. in English), 51*, 10664–10669.

Liu, Z., Liang, Y., Ang, E. L., & Zhao, H. (2017). A new era of genome integration-simply cut and paste! *ACS Synthetic Biology, 6*, 601–609.

Liu, Q., Wu, K., Cheng, Y., Lu, L., Xiao, E., Zhang, Y., et al. (2015). Engineering an iterative polyketide pathway in Escherichia coli results in single-form alkene and alkane overproduction. *Metabolic Engineering, 28*, 82–90.

McDaniel, R., Thamchaipenet, A., Gustafsson, C., Fu, H., Betlach, M., Betlach, M., et al. (1999). Multiple genetic modifications of the erythromycin polyketide synthase to produce a library of novel "unnatural" natural products. *Proceedings of the National Academy of Sciences of the United States of America, 96*, 1846–1851.

Menendez-Bravo, S., Comba, S., Sabatini, M., Arabolaza, A., & Gramajo, H. (2014). Expanding the chemical diversity of natural esters by engineering a polyketide-derived pathway into Escherichia coli. *Metabolic Engineering, 24*, 97–106.

Menzella, H. G., Carney, J. R., & Santi, D. V. (2007). Rational design and assembly of synthetic trimodular polyketide synthases. *Chemistry & Biology*, *14*, 143–151.
Menzella, H. G., Reid, R., Carney, J. R., Chandran, S. S., Reisinger, S. J., Patel, K. G., et al. (2005). Combinatorial polyketide biosynthesis by de novo design and rearrangement of modular polyketide synthase genes. *Nature Biotechnology*, *23*, 1171–1176.
Moore, B. S., & Hertweck, C. (2002). Biosynthesis and attachment of novel bacterial polyketide synthase starter units. *Natural Product Reports*, *19*, 70–99.
Newman, D. J., & Cragg, G. M. (2016). Natural products as sources of new drugs from 1981 to 2014. *Journal of Natural Products*, *79*, 629–661.
Oliynyk, M., Stark, C. B. W., Bhatt, A., Jones, M. A., Hughes-Thomas, Z. A., Wilkinson, C., et al. (2003). Analysis of the biosynthetic gene cluster for the polyether antibiotic monensin in Streptomyces cinnamonensis and evidence for the role of monB and monC genes in oxidative cyclization. *Molecular Microbiology*, *49*, 1179–1190.
Phelan, R. M., Sachs, D., Petkiewicz, S. J., Barajas, J. F., Blake-Hedges, J. M., Thompson, M. G., et al. (2017). Development of next generation synthetic biology tools for use in Streptomyces venezuelae. *ACS Synthetic Biology*, *6*, 159–166.
Reeves, C. D., Murli, S., Ashley, G. W., Piagentini, M., Hutchinson, C. R., & McDaniel, R. (2001). Alteration of the substrate specificity of a modular polyketide synthase acyltransferase domain through site-specific mutations. *Biochemistry*, *40*, 15464–15470.
Srirangan, K., Liu, X., Akawi, L., Bruder, M., Moo-Young, M., & Chou, C. P. (2016). Engineering Escherichia coli for microbial production of butanone. *Applied and Environmental Microbiology*, *82*, 2574–2584.
Tang, L. (2000). Cloning and heterologous expression of the epothilone gene cluster. *Science*, *287*, 640–642.
Tang, Y., Chen, A. Y., Kim, C. Y., Cane, D. E., & Khosla, C. (2007). Structural and mechanistic analysis of protein interactions in module 3 of the 6-deoxyerythronolide B synthase. *Chemistry & Biology*, *14*, 931–943.
Tang, Y., Kim, C. Y., Mathews, I. I., Cane, D. E., & Khosla, C. (2006). The 2.7-Angstrom crystal structure of a 194-kDa homodimeric fragment of the 6-deoxyerythronolide B synthase. *Proceedings of the National Academy of Sciences of the United States of America*, *103*, 11124–11129.
Whicher, J. R., Dutta, S., Hansen, D. A., Hale, W. A., Chemler, J. A., Dosey, A. M., et al. (2014). Structural rearrangements of a polyketide synthase module during its catalytic cycle. *Nature*, *510*, 560–564.
Wilson, M. C., & Moore, B. S. (2012). Beyond ethylmalonyl-CoA: The functional role of crotonyl-CoA carboxylase/reductase homologs in expanding polyketide diversity. *Natural Product Reports*, *29*, 72–86.
Yadav, G., Gokhale, R. S., & Mohanty, D. (2003). Computational approach for prediction of domain organization and substrate specificity of modular polyketide synthases. *Journal of Molecular Biology*, *328*, 335–363.
Yu, J. L., Xia, X. X., Zhong, J. J., & Qian, Z. G. (2014). Direct biosynthesis of adipic acid from a synthetic pathway in recombinant Escherichia coli. *Biotechnology and Bioengineering*, *111*, 2580–2586.
Yuzawa, S., Deng, K., Wang, G., Baidoo, E. E., Northen, T. R., Adams, P. D., et al. (2017). Comprehensive in vitro analysis of acyltransferase domain exchanges in modular polyketide synthases and its application for short-chain ketone production. *ACS Synthetic Biology*, *6*, 139–147.
Yuzawa, S., Eng, C. H., Katz, L., & Keasling, J. D. (2013). Broad substrate specificity of the loading didomain of the lipomycin polyketide synthase. *Biochemistry*, *52*, 3791–3793.

Yuzawa, S., Eng, C. H., Katz, L., & Keasling, J. D. (2014). Enzyme analysis of the polyketide synthase leads to the discovery of a novel analog of the antibiotic alpha-lipomycin. *Journal of Antibiotics (Tokyo)*, *67*, 199–201.

Yuzawa, S., Kapur, S., Cane, D. E., & Khosla, C. (2012). Role of a conserved arginine residue in linkers between the ketosynthase and acyltransferase domains of multimodular polyketide synthases. *Biochemistry*, *51*, 3708–3710.

Yuzawa, S., Katz, L., & Keasling, J. D. (2015). *Producing 3-hydroxycarboxylic acid and ketone using polyketide synthase*. US Patent Application, 20150307855.

Yuzawa, S., Kim, W., Katz, L., & Keasling, J. D. (2012). Heterologous production of polyketides by modular type I polyketide synthases in Escherichia coli. *Current Opinion in Biotechnology*, *23*, 727–735.

Zargar, A., Bailey, C. B., Haushalter, R. W., Eiben, C. B., Katz, L., & Keasling, J. D. (2017). Leveraging microbial biosynthetic pathways for the generation of 'drop-in' biofuels. *Current Opinion in Biotechnology*, *45*, 156–163.

Zhang, L., Hashimoto, T., Qin, B., Hashimoto, J., Kozone, I., Kawahara, T., et al. (2017). Characterization of giant modular PKSs provides insight into genetic mechanism for structural diversification of aminopolyol polyketides. *Angewandte Chemie (International Ed. in English)*, *56*, 1740–1745.

Zheng, J., Fage, C. D., Demeler, B., Hoffman, D. W., & Keatinge-Clay, A. T. (2013). The missing linker: A dimerization motif located within polyketide synthase modules. *ACS Chemical Biology*, *8*, 1263–1270.

Zhou, Y., Li, J., Zhu, J., Chen, S., Bai, L., Zhou, X., et al. (2008). Incomplete beta-ketone processing as a mechanism for polyene structural variation in the FR-008/candicidin complex. *Chemistry & Biology*, *15*, 629–638.

CHAPTER FIFTEEN

Modular Pathway Rewiring of Yeast for Amino Acid Production

Quanli Liu*,†, Tao Yu*,†, Kate Campbell*,†, Jens Nielsen*,†,‡,1, Yun Chen*,†

*Department of Biology and Biological Engineering, Chalmers University of Technology, Gothenburg, Sweden
†Novo Nordisk Foundation Center for Biosustainability, Chalmers University of Technology, Gothenburg, Sweden
‡Novo Nordisk Foundation Center for Biosustainability, Technical University of Denmark, Kongens Lyngby, Denmark
[1]Corresponding author: e-mail address: nielsenj@chalmers.se

Contents

Abstract

Amino acids find various applications in biotechnology in view of their importance in the food, feed, pharmaceutical, and personal care industries as nutrients, additives, and drugs, respectively. For the large-scale production of amino acids, microbial cell factories are widely used and the development of amino acid-producing strains has mainly focused on prokaryotes *Corynebacterium glutamicum* and *Escherichia coli*. However, the eukaryote *Saccharomyces cerevisiae* is becoming an even more appealing microbial host for production of amino acids and derivatives because of its superior molecular and physiological features, such as amenable to genetic engineering and high tolerance to harsh conditions. To transform *S. cerevisiae* into an industrial amino acid production platform, the highly coordinated and multiple layers regulation in its amino acid metabolism should be relieved and reconstituted to optimize the metabolic flux toward synthesis of target products. This chapter describes principles, strategies, and applications

Methods in Enzymology, Volume 608
ISSN 0076-6879
https://doi.org/10.1016/bs.mie.2018.06.009

of modular pathway rewiring in yeast using the engineering of L-ornithine metabolism as a paradigm. Additionally, detailed protocols for in vitro module construction and CRISPR/Cas-mediated pathway assembly are provided.

1. INTRODUCTION

1.1 Chemical Production in Cell Factories: Achieving Optimal Flux

Chemical production is an integral component of modern society and is typically mediated by traditional chemical production techniques. This includes processes such as chemical synthesis, plant extraction, and biotransformation, which can be cumbersome, nonsustainable and can lead to detrimental effects on the environment (Peralta-Yahya, Zhang, del Cardayre, & Keasling, 2012; Stephanopoulos, 2012). A renewable, sustainable, and scalable alternative to these methods is industrial bioprocessing, which converts via microbial fermentation abundant and renewable feedstock into high value or bulk chemicals, including biofuels, polymeric material, and cosmetic precursors or pharmaceuticals (Hossain et al., 2018; Nielsen, Larsson, van Maris, & Pronk, 2013). A major challenge, however, for pursuing this option is that there is currently no economically feasible method for rapidly generating cell platforms that are intrinsic to this process.

In the last decade, advances in metabolic engineering and synthetic biology have nonetheless provided promising new opportunities for using microbial cell factories as chemical production platforms. This has principally been driven by an improved understanding of basic cell metabolism as well as more proficient methods of genetic editing such as clustered regulatory interspaced short palindromic repeats (CRISPR/Cas) systems (Nielsen & Keasling, 2016). Genetic engineering methods can subsequently reconfigure the cell's metabolic network to convert substrates into the desired value-added product. However, as the metabolic network is highly interconnected, the manipulation of endogenous genes and the introduction of heterologous genes or pathways can easily provoke metabolic flux imbalances, which can lead to unfavorable outcomes, such as growth deficits, accumulation of intermediates, and formation of unwanted and potential toxic by-products (Lo et al., 2013; Solomon & Prather, 2011). Removal of these imbalances could therefore improve resulting strains with respect to cell growth, product titer, rate of production, and yield (TYR), all of which are of great importance when platform strains are used in industrial applications. Optimized pathway flux is subsequently a critical factor for strain engineering (Woolston, Edgar, & Stephanopoulos, 2013).

Optimal pathway flux can be achieved via implementation of combinatorial and global genome-wide strategies in strain construction as opposed to classic gene overexpressions or knockouts. For example, transcript levels of gene(s) can be modulated using promoter library (Alper, Fischer, Nevoigt, & Stephanopoulos, 2005; Shen, Hu, Li, & Liu, 2015) or transcription factors engineering CRISPR/Cas systems (Jensen et al., 2017; Lian, HamediRad, Hu, & Zhao, 2017). Direct genome-editing tools, such as multiplexed automated genome engineering (Wang et al., 2009), CRISPR/Cas (DiCarlo et al., 2013), and trackable multiplex recombineering (Warner, Reeder, Karimpour-Fard, Woodruff, & Gill, 2010), can all be implemented to generate significant phenotypic variations with short time frames. Such candidate strains are then subject to high-throughput screening (HTS) to rapidly select and identify strain(s) with the best performance. However, these methods only have a limited application in phenotypes available for HTS assays, such as growth-coupled phenotype (Shen et al., 2015) or colorimetric screens (Tanaka & Ohmiya, 2008).

Modular pathway rewiring (MPR), also known as multivariate modular metabolic engineering (Biggs, De Paepe, Santos, De Mey, & Kumaran Ajikumar, 2014), represents a promising and generalizable approach to ease pathway construction and balance expression in order to control flux in artificial biological systems. Using principles underlying the modular behavior of cell signaling pathways (Hartwell, Hopfield, Leibler, & Murray, 1999), MPR reassembles metabolic genes within a pathway, by transforming the predictable and quantitative regulatory elements at both the transcriptional and translational levels (Pirie, De Mey, Jones Prather, & Ajikumar, 2013), into individual modules based on pathway branching, compound chemistry, and enzyme turnover (Ajikumar et al., 2010; Yadav, De Mey, Lim, Ajikumar, & Stephanopoulos, 2012). Once these modules have been assembled, it is possible to systematically and efficiently rewire a specific metabolic feature with a minimal number of designs, thus alleviating the need for HTS. Since its first demonstration in boosting taxadiene synthesis by 15,000-fold improvement in production over the control strain in *Escherichia coli* (Ajikumar et al., 2010), much progress has been recorded and highlights the broad applicability of the MRP for the development and optimization of a variety of bioprocesses differing in module, constructions, target metabolites, and host microorganisms (Gao et al., 2018; Jensen, Eberhardt, & Wendisch, 2015; Jiang, Qiao, Bentley, Liu, & Zhang, 2017; Jo et al., 2017; Qin et al., 2015;

Wang, Wang, Zhang, Fan, & Tan, 2017; Wu et al., 2013; Xu et al., 2013; Zhou et al., 2012). Furthermore, depending on pathway design, MPR can permit the independent optimization of pathway modules by directly feeding and sampling each unit (Boock, Gupta, & Prather, 2015).

1.2 Cell Factories for the Production of Amino Acids

Amino acids find various applications in biotechnology in view of their importance in the food, pharmaceutical, feed, and personal care industries as nutrients, additives, and drugs, respectively. For example, L-glutamic acid and its salts, which cover nearly two-thirds of the amino acid market, are commonly used as flavor enhancers in the food industry in the form of monosodium glutamate (Ma et al., 2017; Wendisch, Jorge, Perez-Garcia, & Sgobba, 2016). The addition of amino acids can also improve the growth of animals and the quality of meat when added to animal feed (Hu et al., 2017). Moreover, aromatic amino acids derivatives, such as stilbenes, flavonoids, and alkaloids, exhibit various human health-promoting activities (Falcone Ferreyra, Rius, & Casati, 2012). L-Citrulline, an intermediate of arginine biosynthesis, plays an important role in human health and nutrition (Ikeda, Mitsuhashi, Tanaka, & Hayashi, 2009). The global demand is further expected to increase in the next decades, requiring researchers to focus more on developing advanced manufacturing techniques to meet the challenging demand of various amino acids and their derivatives.

For the large-scale production of amino acids, microbiological methods and processes are widely employed. Development and improvement of amino acid-producing strains in the past have mainly focused on the two major industrial production microorganisms, *Corynebacterium glutamicum* and *E. coli*, with the use of various approached such as the rational intuitive approaches and the systematic and rational–random approaches (Ma et al., 2017). The former category consists of classical metabolic engineering methods alongside application of more nascent synthetic biology processes. This approach targets the biosynthetic pathway of a certain amino acid to maximize their productivity by improving the uptake and utilization of carbon source, enhancing supply of precursors, eliminating of by-products that may act as energy sinks, supplying cofactors, and exploring utilization of other genes that may also improve production (Lee, Na, et al., 2012). The latter category mainly includes traditional random mutagenesis and screening programs, omics-based metabolic engineering techniques, as well as various adaptive evolutionary approaches for cases when there are no obvious target genes. These strategies either via large-scale analysis or HTS all share the same strategy of aiming to proceed quickly through cycles of design-build-test and -learn in order to enable optimal amino acid production (Nielsen & Keasling, 2016).

1.3 Building Yeast as a Cell Factory for Amino Acids

Over the past century, budding yeast *Saccharomyces cerevisiae* has proven to be an reliable workhorse of industrial biotechnology and has been applied to the production of a wide range of chemicals, biofuels, pharmaceutical proteins, and natural bioactive compounds (Nielsen et al., 2013). Compared to its prokaryotic counterparts, eukaryotic *S. cerevisiae* possesses attractive features for industrial-scale processes: principally its robustness and tolerance to harsh fermentation conditions and its approval to be used in the production of many food-grade products. In addition, it is generally recognized as safe status, amenable to genetic engineering with availability of diverse molecular biology tools (Gibson et al., 2008; Jensen et al., 2017; Lian et al., 2017; Mans et al., 2015; Shao, Zhao, & Zhao, 2009), and ability to efficiently express complex enzymes, such as covalent modification enzymes (Liu et al., 2018) and cytochrome P450-containing enzymes (Li, Schneider, Kristensen, Borodina, & Nielsen, 2016), make budding yeast become an even more appealing microbial host for production of various amino acid derivatives. For example, the pathways of aromatic amino acids including L-phenylalanine, L-tyrosine, and L-tryptophan in *S. cerevisiae* have been extensively engineered to produce a broad spectrum of value-added chemicals, including polymer-building blocks (Suastegui et al., 2017) and natural bioactive products such as flavonoids and stilbenes (Li et al., 2015; Liu et al., 2018). In addition, the catabolism of several amino acids, referred as the Ehrlich pathway, has also been exploited for the production of advanced biofuels branched-chain alcohols (Avalos, Fink, & Stephanopoulos, 2013) and 2-phenylethanol (Shen, Nishimura, Matsuda, Ishii, & Kondo, 2016), an aromatic alcohol commonly used in the cosmetic industry for its rose scent.

However, one bottleneck to transform *S. cerevisiae* into an industrial amino acid production platform is the highly coordinated and multiple layers regulation in its amino acid metabolism, which can often lead to low yield and productivity, preventing the overproduction of these aforementioned amino acid-derived compounds. In this study, we use the production of the L-arginine biosynthetic intermediate, L-ornithine (Qin et al., 2015), as a paradigm to demonstrate the effectiveness of the MPR strategy for rewiring and releasing the capacity of yeast amino acid metabolism. Moreover, we also discuss how new technologies, including the use of CRISPR/Cas system, can be integrated into this workflow, and how this may contribute to improving production of amino acids and other value-added chemicals using yeast.

2. CASE STUDY OF MPR OF AMINO ACID METABOLISM IN *S. CEREVISIAE*

2.1 Principles of Module Identification

The key for successful MPR is to identify the biochemical and genetic features of both catabolic and anabolic processes involved in metabolism of the target amino acid, i.e., map the metabolic pathway, identify potential rate-limiting steps, map genetic regulation of gene expression, and enzyme activity and interaction with other metabolic pathways. Understanding these aspects provides fundamental rationale in recasting the target metabolic pathway into different modules. As the end products of cellular primary metabolism, amino acids are synthesized by multigene metabolic processes derived from both carbon and nitrogen metabolism (Ljungdahl & Daignan-Fornier, 2012). Hence, the specificity of metabolic pathways presents one principle for module design, for example, the common glycolytic pathway and specific amino acid biosynthesis or degradation pathways should be assigned as different modules. Subcellular compartmentalization is another characteristic property of amino acid metabolism with the exchange of metabolites between cytoplasm and subcellular organelles meaning pathway partition should be taken into account (Qin et al., 2015). In addition, the balance between supply and usage of cofactors, including NADH, NADPH, and ATP, can be incorporated into an individual module (Wang, Chen, Fang, & Tan, 2017).

In one of our earlier studies, the complete pathway from uptake of the carbon source glucose to synthesis of L-ornithine could be refactored into three main modules as follows (Fig. 1). Module 1 addresses the competition of endogenous L-ornithine degradation or consumption pathways. Following export from the mitochondria to the cytoplasm, L-ornithine can then be channeled to the L-arginine biosynthesis or L-proline biosynthesis pathway by ornithine carbamoyltransferase (OTC; *ARG3*) and ornithine aminotransferase (*CAR2*), respectively. L-Ornithine production as a result of L-arginine catabolism by the arginase reaction (*CAR1*) is also included; a

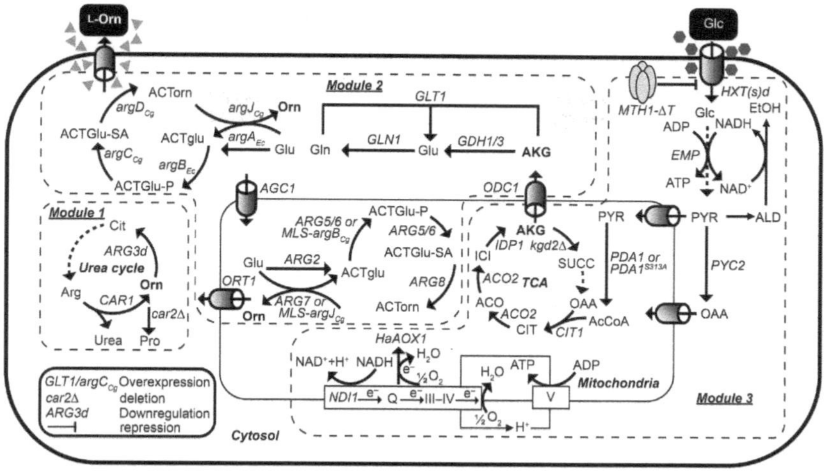

Fig. 1 Schematic overview of L-ornithine biosynthesis in *S. cerevisiae* and pathway modularization for the MPR. *Solid arrows* represent single reaction steps, while *dashed arrows* indicate multiple reaction steps. *Adapted from Qin, J. F., Zhou, Y. J. J., Krivoruchko, A., Huang, M. T., Liu, L. F., Khoomrung, S., & Nielsen, J. (2015). Modular pathway rewiring of* Saccharomyces cerevisiae *enables high-level production of L-ornithine.* Nature Communications, 6, *8224 and more details regarding gene function can be found thereof.*

second module (Module 2) describes more specifically the L-ornithine biosynthesis process, which consists of all biochemical reactions leading from the tricarboxylic acid (TCA) cycle intermediate α-ketoglutarate to L-ornithine. The L-ornithine biosynthesis pathway can also be termed as the acetylated derivative cycle, as *N*-acetyl-L-ornithine generated in the last step is recycled and used to activate the initial substrate L-glutamate; Module 3 addresses the need for increased availability of precursor α-ketoglutarate. This module has three main components: glucose uptake reactions and its regulation, glycolysis, and lastly the upstream part of the TCA cycle. In addition, the respiratory chain, which is closely related with the TCA cycle flux, is also included in this module.

2.2 General Molecular Manipulation Methods

Effective optimization of a given biosynthetic pathway generally involves perturbations of numerous genes with metabolic and/or regulatory functions, some that can be achieved through the implementation of the MPR strategy. To this end, methods for assembling individual genetic elements into a single strain rapidly and reliably are necessary. Compared to other microorganisms, a prominent feature of *S. cerevisiae* is its robust homologous recombination (HR) machinery, which finds extensive applications in various DNA manipulations, such as the assembly of exogenous multigene biosynthetic pathways (Shao et al., 2009). Here, we provide a general procedure for assembling of DNA expression cassettes and subsequent pathway assembly, combining the native capability yeast HR and emerging CRISPR/Cas systems (Fig. 2).

2.2.1 DNA Cassettes and Modules Assembly

In this section we briefly describe the procedure for generating and assembling genetic parts with minor modifications from Zhou et al. (2012).

(i) Design to pathways and modules of interest

Select the desirable transcriptional elements promoters and terminators and design the primers.

(ii) Amplification of genetic parts

PCR-amplify using the high-fidelity Phusion DNA polymerase and corresponding oligo primers each individual gene within each module, and an expression cassette including a promoter, a structural gene, and a terminator.

(iii) One-pot fusion of the DNA elements

Assemble individual expression cassettes using two-step overlapping extension-PCR (OE-PCR) to generate complete modules. The 5′-end of the first gene expression cassette should be designed to overlap with a targeted locus of a chromosome for integration or a vector, while the 3′-end should be designed to share homology with the second cassette. Each successive expression cassette harbors sequences homologous to the terminator of the front module or the promoter of the next module to enable in vivo HR in the following pathway assembly step. The size of homologous sequences is recommended to be 500–1000 bp.

(iv) Two-step OE-PCR procedure

a. First round PCR without oligo primers. To ensure specific amplification occur, purified DNA fragments are mixed with a molar ratio at 1:3:5:7:*X*:7:5:3:1, meaning the molarity of DNA fragment increases from termini to middle by a factor of 2, and the amount of terminal DNA amount is 50–100 ng/kb.

b. Use above DNA mixture as the template and prepare a PCR reaction mixture as follows:

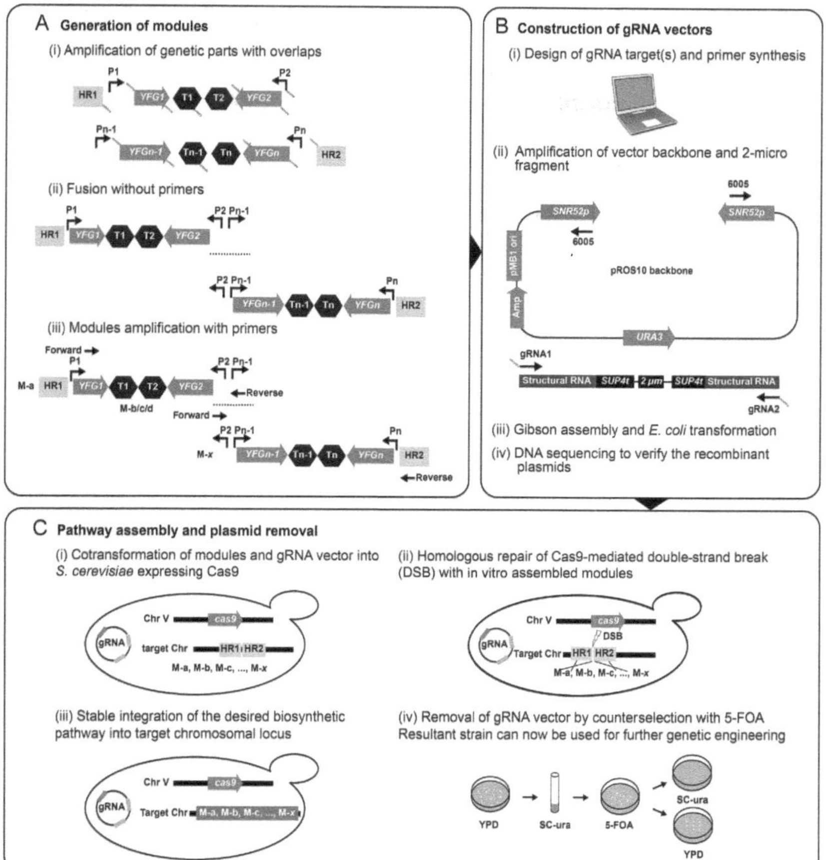

Fig. 2 Workflow for CRISPR/Cas9-mediated biosynthetic pathway assembly and genomic integration in *S. cerevisiae*. (A) Generation of designated modules. Individual genetic parts, including genome homologous sequences (HR1 and HR2), promoters, genes (Your Favorite Gene, *YFG*), and terminators, are PCR amplified with short overlaps. A two-step overlapping extension PCR (OE-PCR) procedure is employed to fuse DNA fragments in order to generate functional modules. (B) Construction of gRNA vector. gRNA sequences targeting genome can be automatically designed with webtools. Plasmid backbone with *URA3* marker and 2-microfragment is PCR amplified and subject to Gibson assembly. More detailed protocols can be found in Mans et al. (2015). (C) Pathway assembly and targeted integration via double-strand break (DSB) introduced by Cas9 cleavage. Assembled modules and gRNA vector are cotransformed into *S. cerevisiae*. Colony PCR is finally used to identify correct integration with the gRNA vector being removed by the counterselection with 5-fluoroorotic acid (5-FOA).

Component	Amount (μL)	Final Concentration
5 × HF buffer	5	
dNTPs (2.5 m*M* each)	3	300 μ*M*
PrimeSTAR HS polymerase	0.25	0.02 U/μL
DNA mixture	*X*	Varied
Distilled water	16.85 − *X*	
Total	25	

c. Use the following conditions for PCR

Step	Temperature (°C)	Time
1	95	3 min
2	98	10 s
3	56	15 s, 15 ×
4	72	1 min/kb
5	72	10 min
6	16	∞

d. Second round PCR. Take unpurified PCR products as starting template and prepare a PCR reaction mixture as follows:

Component	Amount (μL)	Final Concentration
5 × HF buffer	10	
dNTPs (2.5 m*M* each)	4	200 μ*M*
PrimeSTAR HS polymerase	0.5	0.02 U/μL
DNA mixture	2	Varied
Forward primer (20 μ*M*)	1.25	0.5 μ*M*
Reverse primer (20 μ*M*)	1.25	0.5 μ*M*
Distilled water	31	
Total	50	

e. Use the following conditions for PCR

Step	Temperature (°C)	Time
1	95	3 min
2	98	10 s
3	56	15 s, 35 ×
4	72	1 min/kb
5	72	10 min
6	16	∞

f. Run PCR products on 1% agarose gel and gel purify the respective bands.

Note: The Phusion DNA polymerase generally yields lower quantity of product in the one-pot fusion and amplification and is not recommended.

2.2.2 CRISPR/Cas-Mediated Pathway Assembly

The mitotic stability of heterologous gene and pathway expression is of great importance to allow platform yeast strains to be established in the chemical industry. To this end, genomic integration is preferred over expression of nonintegrative plasmids for cell factory construction using *S. cerevisiae*. Genomic integration removes the need for antibiotics or chemically defined medium to maintain selection pressure, as would be the cases with plasmids. Furthermore, it prevents recombinant strain from encountering potentially adverse effect caused by extrachromosomal DNA copy variability and structural instability (Flagfeldt, Siewers, Huang, & Nielsen, 2009). On the other hand, it should be kept in mind that the level of transcription for the same gene can vary when integrated in different yeast genomic regions. This suggests that, depending on the genomic region used, different gene expression regulation may occur. For instance, genes located in subtelomeric are known for epigenetics to affect expression levels by means of transcriptional silencing. Hence, recent work has aimed at scanning the *S. cerevisiae* genome to identify positions where stable gene expression is possible and therefore suitable for the integration of metabolic genes and pathways (Mikkelsen et al., 2012). The methods detailed here illustrate the use of CRISPR/Cas9-targeting of specific chromosomal sites to rapidly assemble biosynthetic pathways and create marker-free and scarless yeast cells with ease and within a short space of time.

(i) Design of gRNA sequence for targeting genome locus.

The CRISPR/Cas9 system that we use to assemble biosynthetic pathways was established by the Pronk lab and consists of budding yeast strain CEN.PK113-5D derivative IMX581 and a double gRNA plasmid pROS10 (*URA3* marker) (Mans et al., 2015). The IMX581 strain expresses a stable integrated copy of codon-optimized *Streptococcus pyogenes* species of Cas9 nuclease. In contrast to centromeric plasmid-borne *cas9* expression, this approach allows growth of strains on complex medium, which enables faster growth and efficient plasmid recycling (Mans et al., 2015). The design of the target gRNA sequences against selected chromosomal sites can be performed as recommended by the Pronk lab using the Yeastriction tool at http://yeastriction.tnw.tudelft.nl or other online software such as at http://chopchop.cbu.uib.no// and http://crispr.dbcls.jp/.

Use the following sequence as gRNA primer: TGCGCATGTTTCGGCGTTCGAAACTTCTCCGCAGTGAAAGATAAATGATC$\mathbf{N_{20}}$GTTTTAGAGCTAGAAATAGCAAGTTAAAATAAG.

(where $\mathbf{N_{20}}$ is the target sequence without the PAM sequence and flanking sequences overlap with pROS10 vector). Order the primer and standard desalted oligos (sufficient for the cloning procedure) and reconstitute to a working stock of 100 μ*M*.

(ii) Stepwise construction of recombinant gRNA vectors.

The stepwise construction of recombinant gRNA vectors, including construction of the 2 μm fragment, construction of the linearized backbone, assembly the 2 μm fragment with the backbone, and confirmation and amplification of constructed plasmid, is described in Mans et al. (2015).

(iii) Pathway assembly.

Cotransform equimolar amounts of purified individual modules (50–100 ng/kb) and pROS10-derived gRNA plasmids (300–500 ng) into *S. cerevisiae* according to the described protocol (Gietz & Woods, 2002).

(iv) Verification of the assembled pathways.

a. Pick colonies formed on the selection plates (SC-ura) and grow in 2 mL SC drop-out liquid medium at 30°C overnight with shaking.

b. Resulting culture is used for crude genome DNA isolation using methods by Looke, Kristjuhan, & Kristjuhan (2011).

c. Perform diagnostic PCR to confirm correct DNA integration events during transformation and prepare a PCR reaction mixture as follows:

Component	Amount (μL)	Final Concentration
2 × Sapphire mix	5	
Genomic DNA	0.5	
Forward primer (20 μ*M*)	0.25	0.5 μ*M*
Reverse primer (20 μ*M*)	0.25	0.5 μ*M*
Distilled water	4	
Total	10	

d. Use the following conditions for PCR

Step	Temperature (°C)	Time
1	98	1 min
2	98	10 s
3	56	10 s, 40 ×
4	72	10 s/kb
5	72	2 min
6	16	∞

e. Run PCR products on 1% agarose gel and check the size of the respective bands.

(v) Plasmid removal.

a. Inoculate the confirmed culture in 5 mL of nonselective YPD liquid medium.

b. Incubate the culture at 30°C until the exponential growth phase is finished.

c. Streak part of the culture on a 5-fluoroorotic acid (5-FOA) agar plate and incubate at 30°C until single colonies are clearly visible.

d. Restreak the obtained single colonies on nonselective plates (YPD) and selective plates (SC-ura) to confirm removal of the gRNA plasmid.

e. Transfer colonies that grow on nonselective, but not on selective medium agar plates to 5 mL nonselective liquid medium.

f. After sufficient cell growth, the culture can be stocked and stored at −80°C and used for another round of transformation.

2.3 Strategies for Implementing MPR

The main concept of metabolic engineering is to drive the metabolic flux into the production pathway. Various strategies have been developed to facilitate this step, including upregulating pathway enzymes, deleting or downregulating competing pathways, and adjusting the global metabolism by engineering transcriptional factors. While MPR shares the goal of conventional metabolic engineering, it accelerates the whole process in a highly integrated and standardized way to identify pathway bottlenecks and address the issue of flux imbalance. The following sections subsequently present and discuss several general strategies for implementing MRP to optimize amino acid production in *S. cerevisiae*.

2.3.1 Elimination of Competing Pathway

In yeast cells, amino acid metabolism consists of two processes: anabolic biosynthesis and catabolic utilization, which coordinate with one another to optimize cells' response to nutrient availability (Ljungdahl & Daignan-Fornier, 2012). To overproduce a specific amino acid, the abolishing or weakening of the catabolic pathway becomes an obvious target to engineer. In our study, Module 1 was designed to contribute to L-ornithine overproduction through two routes. The first involves blocking use of L-ornithine for L-arginine biosynthesis, blocking the utilization of, by abolishing the OTC activity, a method which proved a successful strategy for constructing a L-ornithine overproducer in *C. glutamicum*. Implementation of this modification in *S. cerevisiae* should lead to accumulation of L-ornithine as expected but also cause L-arginine auxotrophy. To ensure supply of L-arginine, the transcript level of Arg3p, which catalyzes the biosynthesis of L-arginine precursor L-citrulline from L-ornithine, can be downregulated by replacing the native promoter with weaker ones, thereby preventing auxotrophy. In our case, two promoters with different regulatory features were tested, the glucose-regulated *HXT1* promoter and the low-activity *KEX2* promoter. Accumulation and secretion of L-ornithine were observed for both engineered strains M1a (*HXT1* promoter) and M1b (*KEX2* promoter). While strain M1a produced 24 mg/L L-ornithine, strain M1b had a 76% higher titer of 42 mg/L (Fig. 3), both in batch flask cultivations. Meanwhile, the intracellular level of L-arginine decreased approx. 30% compared with that of the parental strain.

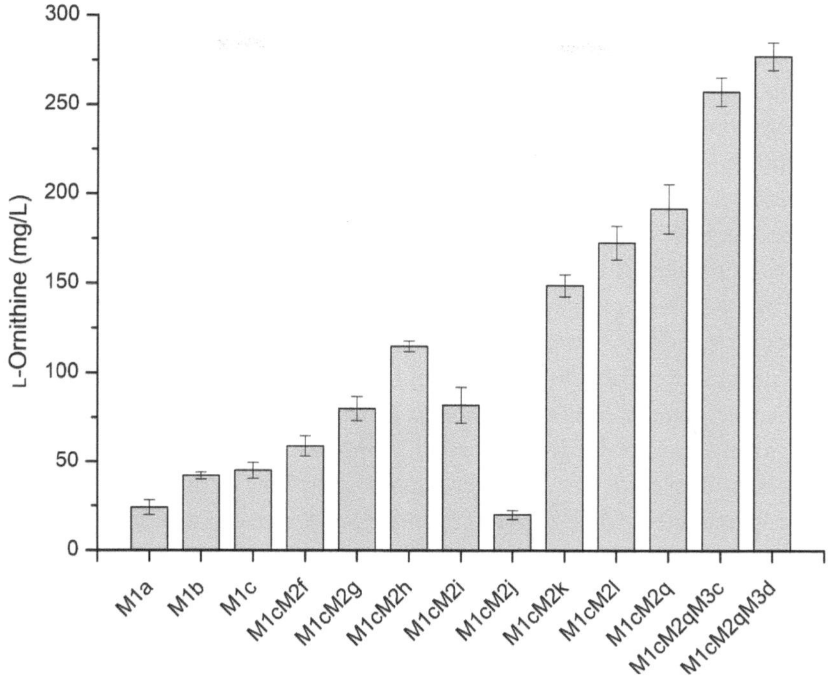

Fig. 3 L-Ornithine overproduction through implementing MPR strategies. Cells were grown in defined minimal medium with 20 g/L glucose as carbon source, and cultures were sampled after 72 h growth for L-ornithine measurement. Average values ± SD ($n \geq 3$) are shown. More genetic descriptions of recombinant strains can be found in the original publication by Qin et al. (2015).

In *S. cerevisiae*, the L-arginine catabolism begins in the cytosol with the hydrolysis of L-arginine by the arginase Car1p to form urea and L-ornithine (Fig. 1). L-Ornithine is then transaminated by the aminotransferase Car2p into L-glutamate gamma-semialdehyde, which can be further converted into L-proline or L-glutamate depending on whether molecular oxygen is present. Removal of *CAR2* was performed to improve L-ornithine accumulation by blocking this futile cycle. However, deletion of *CAR2* in M1b (strain M1c) resulted in only a moderate increase in the L-ornithine titer to 45 mg/L (Fig. 3), which can be ascribed to the lack of a functional transaminase activity because of low concentration of L-ornithine. Similar strategy has been employed for overproducing aromatic amino acid-derivative *p*-coumaric acid via deletion of *ARO10* and *PDC5* genes, embedded in the Ehrlich pathway responsible for amino acid degradation (Rodriguez, Kildegaard, Li, Borodina, & Nielsen, 2015).

2.3.2 Augmentation of Biosynthetic Pathway and Precursor

Multiple layers of regulation have been evolved in yeast to precisely regulate the flux through amino acid pathways and adapt to the changing growth environment. These mechanisms include but are not limited to, controlling the uptake of nitrogen sources, the general amino acid control pathway response to amino acid starvation as well as transcriptional regulation and end product feedback for specific amino acid biosynthesis (Ljungdahl & Daignan-Fornier, 2012). To boost the metabolic flux of target amino acid biosynthesis, these regulatory events must therefore be considered and if necessary abolished or attenuated.

A general approach for augmenting biosynthetic pathway activity is to increase the genetic copy of metabolic genes under the control of constitutively strong promoters. In the case of L-ornithine production, for example, genes responsible for the mitochondrial L-ornithine biosynthetic pathway from L-glutamate, *ARG56*, *ARG7*, and *ARG8*, were integrated to specific chromosomal sites and overexpressed by using strong promoters including *TEF1*p, *tHXT7*p, *TDH3*p, and *PGK1*p. As anticipated, resulting strain M1cM2f possessed improved L-ornithine production of 59 mg/L, representing a 31% increase compared with that of the parental strain M1c (Fig. 3). Moreover, when another gene *ARG2*, which encodes *N*-acetylglutamate synthase and complexes with *N*-acetylglutamate kinase, is also overexpressed, biosynthesis of L-ornithine is further enhanced to 80 mg/L (strain M1cM2g), which makes an additional 36% increase compared with the control strain M1cM2f (Fig. 3). Correspondingly, when this approach is applied to *p*-coumaric acid production, the alleviation of feedback inhibition, in this case for two metabolic enzymes in the aromatic amino acid pathway, Aro4p and Aro7p, also significantly enhanced the carbon flux through this pathway and improved production of *p*-coumaric acid (Rodriguez et al., 2015).

In terms of precursor for amino acid production, yeast cells provided with an appropriate source of carbon and ammonium can synthesize all L-amino acids used in protein synthesis. Here, L-glutamate and L-glutamine exclusively act as the donor of amino group and as a hub (Ljungdahl & Daignan-Fornier, 2012), increasing their synthesis would therefore be a promising lead for the amplification of flux for other amino acids. Hence, we set out to engineer L-glutamate biosynthesis in the cytosol to provide sufficient precursor for L-ornithine overproduction. Among three L-glutamate biosynthesis routes in *S. cerevisiae*, overexpression of the NADPH-dependent glutamate dehydrogenase isoform *GDH1* (strain M1cM2l) resulted in significant improvement in L-ornithine production, that is, a 16% rise in the final titer (173 mg/L) compared with the parental strain M1cM2k (Fig. 3). In addition, through implementing a multilevel approach to divert carbon flux from glycolysis toward the biosynthesis of the precursor erythrose 4-phosphate, high titers of aromatic amino acid-derived shikimic acid (2.5 g/L) and muconic acid (2.5 g/L) were achieved (Suastegui et al., 2017).

2.3.3 Subcellular Relocation

A main challenge for optimizing chemical production using microbial hosts is to limit cross talk between high-flux engineered metabolic pathways and the endogenous metabolism of the production host (Lee, DeLoache, & Dueber, 2012). Eukaryotes, including *S. cerevisiae*, address the problem of metabolic cross talk by compartmentalizing related proteins and metabolites into membrane-bound organelles, such as mitochondria, vacuole, and peroxisome, to (i) direct the activity of enzymes toward specific substrates, (ii) sequester toxic compounds, (iii) increase the availability of intermediates, and finally (iv) establish distinct physiological environments to allow, for example, site-specific redox states (DeLoache, Russ, & Dueber, 2016). However, the subcellular fractionation of biosynthetic pathways therefore necessitates rapid and efficient metabolite exchange between different subcellular compartments.

L-Ornithine biosynthesis in *S. cerevisiae* takes place in the mitochondria; however, it uses L-glutamate as substrate, which is imported from the cytosol. After being transported to the cytosol, L-ornithine is further converted to L-arginine. Therefore, biochemical activities responsible for internal trafficking of L-ornithine and L-glutamate need to be balanced to operate in a coordinated manner to reduce the loss of intermediates in the process of transorganelle transport. Indeed, we found that by increasing the expression level of *ORT1*, which encodes the transporter responsible for exporting L-ornithine from the mitochondria to the cytosol, led to the accumulation of L-ornithine to 115 mg/L (strain M1cM2h), representing a 44% increase relative to the parental strain M1cM2g (Fig. 3). Furthermore, the glutamate uniporter/aspartate–glutamate exchanger coding gene *AGC1* was also overexpressed to ensure sufficient supply of L-glutamate for L-ornithine biosynthesis. The resultant strain M1cM2k produced 149 mg/L of L-ornithine (Fig. 3), representing a further 30% increase.

In an alternative strategy, it is also possible to remove metabolic barriers for metabolite transport by relocating biosynthetic pathways (Avalos et al., 2013; DeLoache et al., 2016). Hence, to reduce the requirement of L-glutamate transport for L-ornithine synthesis (and possible metabolite and/or energy loss during this process), we relocated glutamate dehydrogenases NADPH-dependent Gdh1p (strain M1cM2i) or NADH-dependent Gdh2p (M1cM2j) into the mitochondria. In contrast to expectation, both modifications resulted in a significant decrease in L-ornithine titer (Fig. 3), a result which could be attributed to the inefficient import of metabolic enzymes into mitochondria or unfavorable physiological conditions for functional glutamate dehydrogenases in this compartment. On the other hand, rebuilding the complete L-ornithine biosynthetic pathway in the cytosol using prokaryotic metabolic genes (strain M1cM2q), where the precursor L-glutamate is synthesized, was demonstrated to improve the production of L-ornithine (Fig. 3). Similar efforts have been made to overproduce biofuel chemical derived from the aromatic amino acid degradation process. Compartmentalization of the Ehrlich pathway, for example, into the mitochondria increased isobutanol production by 2.6-fold, whereas overexpression of the same pathway in the cytoplasm only improved the yields by 10% and detailed investigation showed that higher local enzyme concentrations were achieved in the engineered strains (Avalos et al., 2013).

2.3.4 Coupling Carbon Metabolism and Cofactor Engineering

Cofactors, including NAD(H), NADP(H), and ATP, are involved in a large number of intracellular reactions and critically influence redox balance and cellular metabolism (Wang, Chen, et al., 2017). As the thermodynamic driving force for efficient carbon metabolism, cofactors play a vital role in redirecting metabolic flux to target products to improve their productivities and yields and thus represent potential targets for pathway optimization. Different cofactor regulation strategies have been developed to maintain redox balance as follows: regulating endogenous cofactor systems, supplementing the host with heterologous cofactor regeneration systems, modifying cofactor preferences, or creating synthetic cofactor systems (Wang, Chen, et al., 2017).

S. cerevisiae has evolved to perform alcoholic fermentation, rapidly converting excess sugar substrates to ethanol, even under fully aerobic conditions, which is termed as the Crabtree effect (Vemuri, Eiteman, McEwen, Olsson, & Nielsen, 2007). Previous studies showed that the Crabtree effect causes a low capacity of the TCA cycle, which is highly connected with amino acid biosynthesis (Heyland, Fu, & Blank, 2009). Hence, attenuation of the Crabtree effect should release the potential of α-ketoglutarate production by the TCA cycle and should subsequently improve the carbon flux to L-ornithine. As NADH generation is coupled to α-ketoglutarate biosynthesis, accelerated utilization and regeneration of NADH present a promising approach to alleviate the negative influence of the Crabtree effect. Overexpression of heterogeneous NADH mitochondrial alternative oxidase (*AOX*) from *Hansenula anomala* (*HaAOX1*) resulted in an L-ornithine titer of 258 mg/L (strain M1cM2qM3c) (Fig. 3), representing a 35% increase as compared with that of the parental strain M1cM2q. Additional overexpression of *NDI1*, which encodes the endogenous NADH dehydrogenase mediating the delivery of electrons to the respiratory chain, further increased L-ornithine production to 278 mg/L (strain M1cM2qM3d) (Fig. 3). These findings indicated that cofactor engineering was an efficient strategy to boost the TCA cycle flux for overproducing TCA cycle-derived amino acids.

3. CONCLUDING REMARKS

In this chapter, we introduced the concept of MPR in engineering of *S. cerevisiae* amino acid metabolism to boost the biosynthesis of specific amino acids or derivatives thereof. By connecting the central metabolism of yeast to the downstream steps of L-ornithine biosynthesis, a yeast cell factory with significantly enhanced L-ornithine production capacity was constructed. As a generalizable strain optimization technique that can be applicable for multiple hosts and various metabolic pathways (Biggs et al., 2014), the MPR holds promise for increasing the application of yeast platform strains for modulating the amino acid metabolism. Moreover, this process can be greatly facilitated by the emerging tools for controlling gene expression and assembling metabolic pathways. A major concern for strain and pathway optimization, however, relates to genetic and phenotypic stability, which could be improved by taking advantage of techniques directly work on the chromosome, such as the CRISPR/Cas system (Jensen et al., 2017; Lian et al., 2017; Mans et al., 2015). By coupling the MPR with other rational strain engineering approaches (Alper et al., 2005) and computational tools for pathway analysis and design (Wang, Dash, Ng, & Maranas, 2017), it is possible to systematize and streamline the metabolic engineering and strain optimization for amino acid overproduction in *S. cerevisiae* and therefore make this process more economically viable for the chemical production industry.

ACKNOWLEDGMENTS

We would like to thank Dr. Yi Liu and Dr. Xiaowei Li from the SysBio Lab, Chalmers University of Technology, for helpful comments. This work was financially supported by the Novo Nordisk Foundation, the Knut and Alice Wallenbergs Foundation, Vetenskapsrådet, FORMAS, Carl Tryggers Stiftelse, and Ångpanneföreningens Forskningsstiftelse.

REFERENCES

Ajikumar, P. K., Xiao, W. H., Tyo, K. E., Wang, Y., Simeon, F., Leonard, E., et al. (2010). Isoprenoid pathway optimization for Taxol precursor overproduction in *Escherichia coli*. *Science*, *330*(6000), 70–74.

Alper, H., Fischer, C., Nevoigt, E., & Stephanopoulos, G. (2005). Tuning genetic control through promoter engineering. *Proceedings of the National Academy of Sciences of the United States of America*, *102*(36), 12678–12683.

Avalos, J. L., Fink, G. R., & Stephanopoulos, G. (2013). Compartmentalization of metabolic pathways in yeast mitochondria improves the production of branched-chain alcohols. *Nature Biotechnology*, *31*(4), 335.

Biggs, B. W., De Paepe, B., Santos, C. N., De Mey, M., & Kumaran Ajikumar, P. (2014). Multivariate modular metabolic engineering for pathway and strain optimization. *Current Opinion in Biotechnology*, *29*, 156–162.

Boock, J. T., Gupta, A., & Prather, K. (2015). Screening and modular design for metabolic pathway optimization. *Current Opinion in Biotechnology*, *36*, 189–198.

DeLoache, W. C., Russ, Z. N., & Dueber, J. E. (2016). Towards repurposing the yeast peroxisome for compartmentalizing heterologous metabolic pathways. *Nature Communications*, 7, 11152.

DiCarlo, J. E., Norville, J. E., Mali, P., Rios, X., Aach, J., & Church, G. M. (2013). Genome engineering in *Saccharomyces cerevisiae* using CRISPR-*Cas* systems. *Nucleic Acids Research*, *41*(7), 4336–4343.

Falcone Ferreyra, M. L., Rius, S. P., & Casati, P. (2012). Flavonoids: Biosynthesis, biological functions, and biotechnological applications. *Frontiers in Plant Science*, *3*, 222.

Flagfeldt, D. B., Siewers, V., Huang, L., & Nielsen, J. (2009). Characterization of chromosomal integration sites for heterologous gene expression in *Saccharomyces cerevisiae*. *Yeast*, *26*(10), 545–551.

Gao, C., Wang, S., Hu, G., Guo, L., Chen, X., Xu, P., et al. (2018). Engineering *Escherichia coli* for malate production by integrating modular pathway characterization with CRISPRi-guided multiplexed metabolic tuning. *Biotechnology and Bioengineering*, *115*(3), 661–672.

Gibson, D. G., Benders, G. A., Axelrod, K. C., Zaveri, J., Algire, M. A., Moodie, M., et al. (2008). One-step assembly in yeast of 25 overlapping DNA fragments to form a complete synthetic *Mycoplasma genitalium* genome. *Proceedings of the National Academy of Sciences of the United States of America*, *105*(51), 20404–20409.

Gietz, R. D., & Woods, R. A. (2002). Transformation of yeast by lithium acetate/single-stranded carrier DNA/polyethylene glycol method. *Methods in Enzymology*, *350*, 87–96.

Hartwell, L. H., Hopfield, J. J., Leibler, S., & Murray, A. W. (1999). From molecular to modular cell biology. *Nature*, *402*(6761 Suppl), C47–C52.

Heyland, J., Fu, J., & Blank, L. M. (2009). Correlation between TCA cycle flux and glucose uptake rate during respiro-fermentative growth of *Saccharomyces cerevisiae*. *Microbiology*, *155*(Pt. 12), 3827–3837.

Hossain, G. S., Nadarajan, S. P., Zhang, L., Ng, T. K., Foo, J. L., Ling, H., et al. (2018). Rewriting the metabolic blueprint: Advances in pathway diversification in microorganisms. *Frontiers in Microbiology*, *9*, 155.

Hu, C. J., Jiang, Q. Y., Zhang, T., Yin, Y. L., Li, F. N., Deng, J. P., et al. (2017). Dietary supplementation with arginine and glutamic acid modifies growth performance, carcass traits, and meat quality in growing-finishing pigs. *Journal of Animal Science*, *95*(6), 2680–2689.

Ikeda, M., Mitsuhashi, S., Tanaka, K., & Hayashi, M. (2009). Reengineering of a *Corynebacterium glutamicum* L-arginine and L-citrulline producer. *Applied and Environmental Microbiology*, *75*(6), 1635–1641.

Jensen, J. V., Eberhardt, D., & Wendisch, V. F. (2015). Modular pathway engineering of *Corynebacterium glutamicum* for production of the glutamate-derived compounds ornithine, proline, putrescine, citrulline, and arginine. *Journal of Biotechnology*, *214*, 85–94.

Jensen, E. D., Ferreira, R., Jakociunas, T., Arsovska, D., Zhang, J., Ding, L., et al. (2017). Transcriptional reprogramming in yeast using *dCas9* and combinatorial gRNA strategies. *Microbial Cell Factories, 16*(1), 46.

Jiang, W., Qiao, J. B., Bentley, G. J., Liu, D., & Zhang, F. (2017). Modular pathway engineering for the microbial production of branched-chain fatty alcohols. *Biotechnology for Biofuels, 10*, 244.

Jo, S., Yoon, J., Lee, S. M., Um, Y., Han, S. O., & Woo, H. M. (2017). Modular pathway engineering of *Corynebacterium glutamicum* to improve xylose utilization and succinate production. *Journal of Biotechnology, 258*, 69–78.

Lee, H., DeLoache, W. C., & Dueber, J. E. (2012). Spatial organization of enzymes for metabolic engineering. *Metabolic Engineering, 14*(3), 242–251.

Lee, J. W., Na, D., Park, J. M., Lee, J., Choi, S., & Lee, S. Y. (2012). Systems metabolic engineering of microorganisms for natural and non-natural chemicals. *Nature Chemical Biology, 8*(6), 536–546.

Li, M., Kildegaard, K. R., Chen, Y., Rodriguez, A., Borodina, I., & Nielsen, J. (2015). *De novo* production of resveratrol from glucose or ethanol by engineered *Saccharomyces cerevisiae*. *Metabolic Engineering, 32*, 1–11.

Li, M., Schneider, K., Kristensen, M., Borodina, I., & Nielsen, J. (2016). Engineering yeast for high-level production of stilbenoid antioxidants. *Scientific Reports, 6*, 36827.

Lian, J., HamediRad, M., Hu, S., & Zhao, H. (2017). Combinatorial metabolic engineering using an orthogonal tri-functional CRISPR system. *Nature Communications, 8*(1), 1688.

Liu, X., Cheng, J., Zhang, G., Ding, W., Duan, L., Yang, J., et al. (2018). Engineering yeast for the production of breviscapine by genomic analysis and synthetic biology approaches. *Nature Communications, 9*(1), 448.

Ljungdahl, P. O., & Daignan-Fornier, B. (2012). Regulation of amino acid, nucleotide, and phosphate metabolism in *Saccharomyces cerevisiae*. *Genetics, 190*(3), 885–929.

Lo, T. M., Teo, W. S., Ling, H., Chen, B., Kang, A., & Chang, M. W. (2013). Microbial engineering strategies to improve cell viability for biochemical production. *Biotechnology Advances, 31*(6), 903–914.

Looke, M., Kristjuhan, K., & Kristjuhan, A. (2011). Extraction of genomic DNA from yeasts for PCR-based applications. *BioTechniques, 50*(5), 325–328.

Ma, Q., Zhang, Q., Xu, Q., Zhang, C., Li, Y., Fan, X., et al. (2017). Systems metabolic engineering strategies for the production of amino acids. *Synthetic and Systems Biotechnology, 2*(2), 87–96.

Mans, R., van Rossum, H. M., Wijsman, M., Backx, A., Kuijpers, N. G., van den Broek, M., et al. (2015). CRISPR/Cas9: A molecular Swiss army knife for simultaneous introduction of multiple genetic modifications in *Saccharomyces cerevisiae*. *FEMS Yeast Research, 15*(2), fov004.

Mikkelsen, M. D., Buron, L. D., Salomonsen, B., Olsen, C. E., Hansen, B. G., Mortensen, U. H., et al. (2012). Microbial production of indolylglucosinolate through engineering of a multi-gene pathway in a versatile yeast expression platform. *Metabolic Engineering, 14*(2), 104–111.

Nielsen, J., & Keasling, J. D. (2016). Engineering cellular metabolism. *Cell, 164*(6), 1185–1197.

Nielsen, J., Larsson, C., van Maris, A., & Pronk, J. (2013). Metabolic engineering of yeast for production of fuels and chemicals. *Current Opinion in Biotechnology, 24*(3), 398–404.

Peralta-Yahya, P. P., Zhang, F., del Cardayre, S. B., & Keasling, J. D. (2012). Microbial engineering for the production of advanced biofuels. *Nature, 488*(7411), 320–328.

Pirie, C. M., De Mey, M., Jones Prather, K. L., & Ajikumar, P. K. (2013). Integrating the protein and metabolic engineering toolkits for next-generation chemical biosynthesis. *ACS Chemical Biology, 8*(4), 662–672.

Qin, J. F., Zhou, Y. J. J., Krivoruchko, A., Huang, M. T., Liu, L. F., Khoomrung, S., et al. (2015). Modular pathway rewiring of *Saccharomyces cerevisiae* enables high-level production of L-ornithine. *Nature Communications*, *6*, 8224.

Rodriguez, A., Kildegaard, K. R., Li, M. J., Borodina, I., & Nielsen, J. (2015). Establishment of a yeast platform strain for production of *p*-coumaric acid through metabolic engineering of aromatic amino acid biosynthesis. *Metabolic Engineering*, *31*, 181–188.

Shao, Z., Zhao, H., & Zhao, H. (2009). DNA assembler, an *in vivo* genetic method for rapid construction of biochemical pathways. *Nucleic Acids Research*, *37*(2), e16.

Shen, H. J., Hu, J. J., Li, X. R., & Liu, J. Z. (2015). Engineering of *Escherichia coli* for lycopene production through promoter engineering. *Current Pharmaceutical Biotechnology*, *16*(12), 1094–1103.

Shen, L., Nishimura, Y., Matsuda, F., Ishii, J., & Kondo, A. (2016). Overexpressing enzymes of the Ehrlich pathway and deleting genes of the competing pathway in *Saccharomyces cerevisiae* for increasing 2-phenylethanol production from glucose. *Journal of Bioscience and Bioengineering*, *122*(1), 34–39.

Solomon, K. V., & Prather, K. L. (2011). The zero-sum game of pathway optimization: Emerging paradigms for tuning gene expression. *Biotechnology Journal*, *6*(9), 1064–1070.

Stephanopoulos, G. (2012). Synthetic biology and metabolic engineering. *ACS Synthetic Biology*, *1*(11), 514–525.

Suastegui, M., Ng, C. Y., Chowdhury, A., Sun, W., Cao, M. F., House, E., et al. (2017). Multilevel engineering of the upstream module of aromatic amino acid biosynthesis in *Saccharomyces cerevisiae* for high production of polymer and drug precursors. *Metabolic Engineering*, *42*, 134–144.

Tanaka, Y., & Ohmiya, A. (2008). Seeing is believing: Engineering anthocyanin and carotenoid biosynthetic pathways. *Current Opinion in Biotechnology*, *19*(2), 190–197.

Vemuri, G. N., Eiteman, M. A., McEwen, J. E., Olsson, L., & Nielsen, J. (2007). Increasing NADH oxidation reduces overflow metabolism in *Saccharomyces cerevisiae*. *Proceedings of the National Academy of Sciences of the United States of America*, *104*(7), 2402–2407.

Wang, M., Chen, B., Fang, Y., & Tan, T. (2017). Cofactor engineering for more efficient production of chemicals and biofuels. *Biotechnology Advances*, *35*(8), 1032–1039.

Wang, L., Dash, S., Ng, C. Y., & Maranas, C. D. (2017). A review of computational tools for design and reconstruction of metabolic pathways. *Synthetic and Systems Biotechnology*, *2*(4), 243–252.

Wang, H. H., Isaacs, F. J., Carr, P. A., Sun, Z. Z., Xu, G., Forest, C. R., et al. (2009). Programming cells by multiplex genome engineering and accelerated evolution. *Nature*, *460*(7257), 894–898.

Wang, M., Wang, G., Zhang, T., Fan, L., & Tan, T. (2017). Multi-modular engineering of 1,3-propanediol biosynthesis system in *Klebsiella pneumoniae* from co-substrate. *Applied Microbiology and Biotechnology*, *101*(2), 647–657.

Warner, J. R., Reeder, P. J., Karimpour-Fard, A., Woodruff, L. B., & Gill, R. T. (2010). Rapid profiling of a microbial genome using mixtures of barcoded oligonucleotides. *Nature Biotechnology*, *28*(8), 856–862.

Wendisch, V. F., Jorge, J. M. P., Perez-Garcia, F., & Sgobba, E. (2016). Updates on industrial production of amino acids using *Corynebacterium glutamicum*. *World Journal of Microbiology and Biotechnology*, *32*(6), 105.

Woolston, B. M., Edgar, S., & Stephanopoulos, G. (2013). Metabolic engineering: Past and future. *Annual Review of Chemical and Biomolecular Engineering*, *4*, 259–288.

Wu, J., Liu, P., Fan, Y., Bao, H., Du, G., Zhou, J., et al. (2013). Multivariate modular metabolic engineering of *Escherichia coli* to produce resveratrol from L-tyrosine. *Journal of Biotechnology*, *167*(4), 404–411.

Xu, P., Gu, Q., Wang, W., Wong, L., Bower, A. G., Collins, C. H., et al. (2013). Modular optimization of multi-gene pathways for fatty acids production in *E. coli*. *Nature Communications*, *4*, 1409.

Yadav, V. G., De Mey, M., Lim, C. G., Ajikumar, P. K., & Stephanopoulos, G. (2012). The future of metabolic engineering and synthetic biology: Towards a systematic practice. *Metabolic Engineering*, *14*(3), 233–241.
Zhou, Y. J. J., Gao, W., Rong, Q. X., Jin, G. J., Chu, H. Y., Liu, W. J., et al. (2012). Modular pathway engineering of diterpenoid synthases and the mevalonic acid pathway for miltiradiene production. *Journal of the American Chemical Society*, *134*(6), 3234–3241.

adav, V. G., De Mey, M., Lim, C. G., Ajikumar, P. K., & Stephanopoulos, G. (2012). The future of metabolic engineering and synthetic biology: Towards a systematic practice. *Metabolic Engineering*, *14*(3), 233–241.
hou, Y. J. J., Gao, W., Rong, Q. X., Jin, G. J., Chu, H. Y., Liu, W. J., et al. (2012). Modular pathway engineering of diterpenoid synthases and the mevalonic acid pathway for miltiradiene production. *Journal of the American Chemical Society*, *134*(6), 3234–3241.